Handbook of Military and Defense Operations Research

Series in Operations Research

Series Editors

Malgorzata Sterna and Marco Laumanns

About the Series

The CRC Press Series in Operations Research encompasses books that contribute to the methodology of Operations Research and applying advanced analytical methods to help make better decisions.

The scope of the series is wide, including innovative applications of Operations Research which describe novel ways to solve real-world problems, with examples drawn from industrial, computing, engineering, and business applications. The series explores the latest developments in Theory and Methodology, and presents original research results contributing to the methodology of Operations Research, and to its theoretical foundations.

Featuring a broad range of reference works, textbooks, and handbooks, the books in this Series will appeal not only to researchers, practitioners, and students in the mathematical community, but also to engineers, physicists, and computer scientists. The inclusion of real examples and applications is highly encouraged in all of our books.

Rational Queueing
Refael Hassin

Introduction to Theory of Optimization in Euclidean Space
Samia Challal

Handbook of the Shapley Value
Encarnación Algaba, Vito Fragnelli, and Joaquín Sánchez-Soriano

Advanced Studies in Multi-Criteria Decision Making
Sarah Ben Amor, João Luís de Miranda, Emel Aktas, and Adiel Teixeira de Almeida

Handbook of Military and Defense Operations Research
Natalie M. Scala and James P. Howard, II

For more information about this series please visit: https://www.crcpress.com/Chapman-HallCRC-Series-in-Operations-Research/book-series/CRCOPSRES

Handbook of Military and Defense Operations Research

Edited by
Natalie M. Scala
Towson University

James P. Howard, II
Johns Hopkins Applied Physics Laboratory

CRC Press
Taylor & Francis Group
Boca Raton London New York

CRC Press is an imprint of the
Taylor & Francis Group, an **informa** business

A CHAPMAN & HALL BOOK

CRC Press
Taylor & Francis Group
6000 Broken Sound Parkway NW, Suite 300
Boca Raton, FL 33487-2742

First issued in paperback 2021

© 2020 by Taylor & Francis Group, LLC
CRC Press is an imprint of Taylor & Francis Group, an Informa business

No claim to original U.S. Government works

ISBN-13: 978-1-138-60733-0 (hbk)
ISBN-13: 978-1-03-217403-7 (pbk)
DOI: 10.1201/9780429467219

Visit the Taylor & Francis Web site at
http://www.taylorandfrancis.com

and the CRC Press Web site at
http://www.crcpress.com

To my parents: Michele and Michael Scala
\- Natalie

For Trixie, Kermit, and Nina
\- James

Contents

Preface

Introduction and Objectives

Operations research (OR) is a core discipline in military and defense management. Coming to the forefront initially during World War II, OR provided critical contributions to logistics, supply chains, and strategic simulation, while enabling superior decision making for Allied forces. Since these early days, OR has grown to include analytics and many applications, including artificial intelligence, cybersecurity, and big data, and is the cornerstone of management science in manufacturing, marketing, telecommunications, and many other fields. The *Handbook of Military and Defense Operations Research* presents the voices leading OR and analytics to new heights in security through research, practical applications, case studies, and lessons learned in the field.

No longer constrained to printed tables and traditional optimization problems, OR and analytics support a dynamic problem space that has applications to national security, policing, cyberspace, foreign policy, terrorism, and homeland security. Because of this, there is a critical need for students, practitioners, military, and security analysts to understand the problem space, best practices, and lessons learned from real-life problems. Contextual milieu is key to understanding and developing cutting-edge analyses that transform national security and advance both traditional and emerging military operations research and analytics applications.

The objective of this book is to provide a set of approaches, case studies, essays, and positional papers to illustrate applications of operations research and analytics techniques within the military and national defense sectors. The content can then be used by educators to facilitate and enhance classroom discussion as well as by practitioners for guidance on best practices.

Organization of the Volume

To achieve these goals, we arrange the book into four sections: approaches, soft skills and client relations, applications, and perspectives. We present approaches that range from general to specific and provide detailed guidance on employing traditional OR and analytics techniques to the military and defense problem space. This includes using mainstream software tools, such as Microsoft Excel, to develop analyses, applying analytics and data science methods, leveraging utility theory and decision analysis, planning workforce and manpower needs, and performing assessments. We also present forward thinking and new approaches, such as threatcasting, to examine needs and potential adversarial actions in the long-term future. Additional approaches include modeling of stochastic systems, graph theoretic analyses of social networks, process optimization, simulation, and test planning for defense acquisition.

Applications of theory can only be successful when client needs and objectives are clearly understood and modeled in the problem. Broadly applying techniques while ignoring the present state can lead to client distrust in the results, models that perform poorly when implemented, and other failures. All client problems and needs cannot be solved overnight or with one model; continuous improvement and systematic steps are needed. Military and defense OR analysts in practice must build models and analyses that fit into the problem space and client objectives. To this end, we include two chapters on soft skills and client relations, including discussions on the how and why of model implementation and team dynamics. Practitioners should pay special attention to client culture and problem context to create solutions that enable strategic advantage.

Next, we present case studies that illustrate OR and analytics in action as well as examples that include applying decision analysis principles to cybersecurity metrics and best practices, information theory to critical information in the NATO alliance, and time series and Markov models to terrorist group activity. We also include a well-used systems engineering model for cost and risk analysis. We intend these examples to deepen students' understandings of the approaches presented in this book and to kick-start practitioners in implementing cutting-edge military and defense OR and analytics in their organizations.

Finally, the perspectives section identifies frameworks for data science, trends in OR education, needs in military OR research, and an understanding of where the field has been. We intend for these position papers to provide guidance for future directions and practices in military and defense operations research.

What This Handbook Does

The target audience for this handbook is undergraduate students in operations research and analytics, in order to facilitate understanding of the use of these tools and methods in practice. Students at military service academies may especially benefit from the collection of case studies and approaches in the military and defense context. Furthermore, practitioners in government military and defense roles, both civilians and contractors, as well as military service members assigned to operations research and systems analysis (ORSA) roles, will benefit from this handbook's discussion on approaches, client relations, applications, and perspectives.

Instructors will benefit from this handbook through approaches and case studies that can be utilized directly in the classroom. Students will benefit from access to hands-on and practical experiences from analysts and researchers in the field. Regardless of background, readers of this handbook will find solutions for problems that face the military and defense problem space every day.

The military and defense application area is evolving beyond traditional OR and wargaming into analytics applications such as big data, cybersecurity, and insider threats. This handbook addresses both traditional and emerging areas and is essential for understanding both tradition and new trends in the field. We apply the experiences of educators and practitioners working in the field. By doing so, we employ the latest technology developments into case studies and applications while identifying best practices unique to the military, security, and national defense problem space. Finally, we highlight similarities and dichotomies between analyses and trends that are unique to military, security, and defense problems.

We sincerely appreciate your interest in this handbook as a reader, student, practitioner, or instructor. We intend for you to find the content in each section to be valuable and essential in understanding analyses you need to build and techniques you need to employ. We welcome your comments and look for your inputs to support future volumes of this handbook.

Acknowledgments

We would like to thank Sarfraz Khan, Callum Fraser, Mansi Kabra, Rob Calver, and many others at CRC Press, along with Andrew Corrigan, Rachel Cook, and the Deanta Global team for seeing this book through to completion. We sincerely appreciate the assistance from Lorraine Black, Katerine Delgado Licona, and Yeabsira Mezgebe when building the index. We also want to thank our many contributors, authors, and reviewers for making this volume a reality. Finally, we thank our families and colleagues for supporting this project from inspiration to fruition.

– Natalie M. Scala and James P. Howard, II

Editors

Natalie M. Scala is an associate professor and director of the graduate programs in supply chain and project management in the College of Business and Economics at Towson University. She earned PhD and MS degrees in industrial engineering from the University of Pittsburgh. Her primary research is in decision analysis, with foci on military applications and cybersecurity. Dr Scala frequently consults to government clients and has extensive professional experience, including positions with Innovative Decisions, Inc., the United States Department of Defense, the RAND Corporation, and the FirstEnergy Corporation. She is currently serving as the president of the INFORMS Military and Security Society.

James P. Howard, II is a scientist at the Johns Hopkins Applied Physics Laboratory. Previously, he worked for the Board of Governors of the Federal Reserve System as an internal consultant on statistical computing. He has also been a consultant to numerous government agencies. Additionally, he has taught mathematics, statistics, and public affairs since 2010. He has a PhD in public policy from the University of Maryland, Baltimore County.

Contributors

Darryl K. Ahner
Office of the Secretary of Defense
Scientific Test and Analysis
Techniques Center of Excellence

Chris Arney
United States Military Academy
(Emeritus)

Lynette M.B. Arnhart
United States Army (Retired)

Nathaniel D. Bastian
United States Military Academy

David M. Bernacki
United States Air Force

Roger Chapman Burk
United States Military Academy

Richard F. Deckro
Air Force Institute of Technology

Walt DeGrange
CANA Advisors

Paul L. Goethals
United States Military Academy

Andrew O. Hall
United States Military Academy

Shane N. Hall
OptTek Systems, Inc.

Robert E. Hamm Jr.
United States Air Force (Retired)

Raymond R. Hill
Air Force Institute of Technology

James P. Howard, II
Johns Hopkins Applied Physics
Laboratory

Brian David Johnson
Arizona State University

Christine M. Schubert Kabban
Air Force Institute of Technology

Marvin L. King III
United States Army

Joseph M. Lindquist
United States Military Academy

Raymond Madachy
Naval Postgraduate School

David R. Mandel
Defence Research and Development
Canada

F. Freeman Marvin
Innovative Decisions, Inc.

Fairul Mohd-Zaid
Air Force Research Laboratory

Jonathan D. Nelson
University of Surrey

Greg H. Parlier
United States Army (Retired)

Wilson L. Price
Université Laval (Emeritus)

Vasanthan Raghavan
Qualcomm, Inc.

Natalie M. Scala
Towson University

Gina Sigler
Office of the Secretary of Defense
Scientific Test and Analysis
Techniques Center of Excellence

Charles A. Sulewski
United States Military Academy

Benjamin G. Thengvall
OptTek Systems, Inc.

Mark A.C. Timms
Department of National Defence, Canada

Ricardo Valerdi
University of Arizona

Natalie Vanatta
United States Military Academy

Brian M. Wade
United States Army Training and
    Doctrine Analysis Center – Monterey

Hung-da Wan
University of Texas at San Antonio

Krista Watts
United States Military Academy

Michael Yankovich
United States Military Academy

Section I

Approaches

Chapter 1

Modern Data Analytics for the Military Operational Researcher

Raymond R. Hill

1.1 Introduction

The Department of Defense (DoD) has been using analytics for a long time, probably as long as there has been human conflict (see Hill & Miller, 2017 for a similar discussion on military simulation). The profession of operations research provides the DoD its primary analytic workforce and arguably grew out of preparations for World War II (WWII) and subsequent efforts during the war (Miser, 2000). Tasha Inniss is quoted as stating, "Analytics can be thought of as an umbrella term to describe all quantitative decision-making" (Samuelson, 2018).

Big data, data science, and machine learning seem to dominate current discussions, even those focused on defense analytics. But what exactly do these terms mean and imply for defense analytics? That is a question that unfortunately seems to be of less interest. Another topic insufficiently discussed is whether these are terms for the same concepts, for similar concepts, or for very different concepts. This chapter addresses these questions and other concerns that arise in answering the question.

This chapter is not a tutorial on defense analytics, data science, or machine learning. This chapter is a broad definitional work using some generally defined examples and references from military applications.

The purpose of this chapter is to clearly define key statistical concepts, define what some of these newer terms mean, and place all this collectively into a modern defense analytics context. Thus, our target audience is the new analyst or manager/leader needing a better appreciation of data analysis and all that term involves. Ultimately, the reader will gain an understanding of applied statistical methods as well as applied machine learning methods and understand how these fit into the realm of modern defense analytics and data analytics in general. To this end, the discussion is narrative, avoiding mathematical details and fancy pictures.

1.2 Why the New Emphasis

If defense analytics has been around so long, and is seemingly a catch-all term for all things quantitative. Then why the current dialog concerning efforts to exploit the new techniques of big data and data analytics? Are we really discussing something new? Will these new methods provide some fundamental change we cannot achieve with current approaches? There is probably no universally accepted answer to these questions, even though it has been a while since Tukey (1962) first tied data analysis to applied statistical methods. The view taken in this chapter is that the current movement is due to a variety of reasons, some of which are defined and presented below.

Speed and Complexity. Everything happens quicker now. Information is collected continuously on just about everything. Computers process this information quickly to support our rapid pace of decision-making. However, there are challenges in just how much data can be processed in short periods of time and how much of that processing can be properly interpreted for use. Thus, resources are being expended to further expand data systems to accommodate the size and complexity of the data stream and couple this expansion with analytical methods tuned to provide the information needed to produce necessary insights in a timely manner. It is thus fully reasonable to see such computer-integrated solutions in emerging DoD areas such as:

- autonomous operations
- image processing
- target detection
- real-time system health monitoring
- cyber intrusion detection

Various computer algorithms afford us an ability to process the data from such systems in the timeframe required for decision-making. The challenge in the DoD is how to actually extend the use of these algorithms to gain the military advantages when leveraging the big data emanating from our various systems. Meeting the challenge will involve research and an understanding of feasible outcomes.

Big Data and Business Practices. The exploitation of streaming data can drive commercial industry decision-making. Point of sales data collection and processing can help ensure accurate inventory management and targeted sales campaigns. Online transaction processing systems adjust prices to quickly meet changing demands and improve profits. Data analytical results can be used to anticipate sales and pre-position inventory at warehouses to improve order response. Authors such as Hazen et al. (2014, 2016) discuss this tremendous potential.

The DoD has realized it also has data challenges. For instance, diagnostic data from health management systems aboard aircraft, satellite imagery streaming to ground collection stations, or the wealth of weather data available provide just as significant challenges as the problems found in industry.

These instances provide data storage challenges as well as data analytical challenges. There are also challenges associated with static datasets. Personnel data, inventory data, and maintenance data are large data sets often spread among disparate database systems. Challenges presented in these instances include how to consolidate the data as well as analyze the data to gain necessary insights.

Shiny Object Syndrome. The DoD is not immune to grasping onto new approaches or methods because of promises of greater competencies or capabilities. A properly constructed presentation coupled with a believable delivery has led to some forgettable analytical acquisition efforts (none of which will be named here). In the current setting, big data, machine learning, advanced data analytics, and various visualization tools, while absolutely legitimate tools and topics, are surrounded by a good amount of hype and promise. There is sometimes not enough focused discussion on what these topics really represent, what the technology can actually achieve, and what is truly required to achieve the presented capabilities. This is not to say everything presented is bad. The problem is perception – leaders sometimes perceive greater utility and greater novelty in the "new" methods than actually exists. The benefits of data analytics cannot be shortchanged due to a misunderstanding with respect to those benefits.

There are surely other reasons for the push on big data analytic methods and many are quite valid. The reasons presented above suffice for the current purpose – to clarify the discussion on the topic of data analytics, specifically point out how the discussion relates to the well-established body of statistical methods, and ultimately help ensure everyone understands the suite of technologies enough to increase the chance of successfully implementing these very useful, extremely powerful methods of data analytics and the more general term, data science.

1.3 Terminology and Fundamental Concepts of Data Science

Navigating the realm of data science requires understanding the key concepts and terminology associated with data science and data analytics. Data analysis ties back to the seminal article by Tukey (1962) where he characterized it as effectively applied statistics. Thus, the fundamentals of probability and statistics provide an appropriate foundational discussion for this chapter.

A *population* is the complete set of objects of interest. Sometimes the population can be fully examined, such as the population of employees of a company. More often, the population cannot be practically examined, making the population a more theoretical construct. An example of the latter is the population of small birds in the United States; we certainly cannot examine all of the population members but we may want knowledge regarding the population size and behaviors. Thus, various attributes of a population may still be of interest. For instance, consider an example of the former as the population of F-16 engines. An attribute of interest might be flight hours between major engine failures. The challenge is how to determine, or estimate, the population attribute of interest without having to examine all members of the population.

A *sample* is a representative subset of a population. Samples, properly acquired, can be used to gain insight into population attributes of interest, such as means and variances. For example, opinion polls are samples of the voter population and should be used to gain insight into the population dynamics. Attribute values of the sample objects should lead to estimates of the corresponding population attributes. Going back to the engine example, representative F-16 engines might be run on test stands to estimate the operating hours until failure. This information can be used to establish standards such as preventative maintenance policies. Such policies evolve from the statistical inferences developed based on the sample data.

The *probability* of an event is the likelihood of that event occurring (such as in an experiment). A probability function assigns the likelihood of each event in a population. When the events are discrete, each event has some probability value. When the events are continuous, the probabilities are associated with intervals between values. A probability distribution describes the likelihood for the population elements. There are a variety of well-established, theoretical probability distributions. These include the normal, t-, F-, and chi-square distributions (see Montgomery & Runger, 2011 for details). In the reliability modeling world, exponential and Weibull distributions are commonly employed. Given we assume our data follows some specific distribution, we use that data to estimate the distribution parameters and continue with the analyses. We will not delve into the specifics of these distributions in this chapter.

The population subset consists of *random samples* or *random variates* from the population of interest. Numerical values, such as mean and variance, generated based on the sample data are *statistics* and are random variables. Statistics derived from representative samples estimate the population attributes, also known as the parameters of the population distribution. Since samples are merely representative random samples, the statistics have uncertainty with respect to the population values. The distributional assumptions of the sample (based on the population) provide a way to model, or propagate this uncertainty. These mechanisms yield various statistical intervals and tests of hypotheses; confidence intervals, t-tests, and F-tests are among the most common.

The statistics used in conjunction with the distributional information are referred to as inferential statistics; we are making claims regarding unknown distributional parameters of the population.

> *A statistical inference is an inference justified by a probability model linking the data to the broader context*
>
> Ramsey, 2002

Descriptive statistics also provide a numerical summary of the sample data, but there is no formal inference made with respect to the statistics, to a population distribution, or the parameters of a population distribution. In practice, both descriptive and inferential statistics are employed. Which to use depends upon the analytical needs and the questions driving the study. Clearly the choice of statistics can influence the knowledge gained from the sample.

Significance, practical and statistical, needs to be understood. Practical significance pertains to how the statistical values relate to actual systems and practice. This requires knowledge of the system or process. Statistical significance is a determination that the parameters estimated based on the particular hypothesis constructed are different from the error estimated from that sample. Practical significance is a function of the system of interest. Statistical significance is a function of the data collected. These distinctions are important in practice but are too often viewed as equivalent. Something may have statistical significance but may be of no practical importance.

In fact, statistical significance is often incorrectly used. See Amrhein, Greenland & McShane (2019) for a nice summary of the issues associated with the use of statistical significance.

Hypothesis testing is used to judge the statistical significance of the statistics of interest. Such testing is also used to compare sets of statistics. Hypotheses are formed so that parameters are assumed negligible; they are merely estimates of noise or error. The data collected and subsequent parameters estimated hopefully support a rejection of that negligible assumption – the parameter is not likely to be just estimating error as assumed (the likelihood, or probability, of that parameter being noise as assumed is the p-value of the test). Hypothesis tests are constructed so they can be rejected in favor of what we are trying to prove.

The p-value is more carefully defined as "the probability that randomization alone leads to a test statistic as extreme or more extreme than the one observed" (Ramsey, 2002). Generally speaking, p-values less than 0.10 (90% significance) or 0.05 (95% significance) are judged as small enough to reject the negligible parameter assumption, but such levels are often arbitrarily selected. In practice, p-values deemed significant will vary by domain. Some domains, such as some social sciences, use larger p-values than are found in other domains, such as engineering. A simple interpretation of the p-value is that it represents the likelihood the null hypothesis is true, based on the random sample assessed. Such interpretations date back to the early 1900s. However, modern statistical thinking calls for a more measured use of p-values as just one component of the inferential process. Thus, modern data analysts must have a firm grasp of what they are testing and the devices employed to make inferences based on those tests.

Classical statistical-based methods are strongly tied to the inferential methods. Inferential methods are quite powerful under the distributional assumptions made and are particularly useful in the presence of the smaller samples of data we often must deal with in defense applications. These methods are especially powerful in the test and evaluation domain where sample sizes are never as large as desired and the focus is on learning about the population of objects to support decisions made about those objects, or systems.

However, as Breiman (2001) points out, there are two cultural views in data analysis/data science. The inferential statistics view, what he calls the data model view, interprets statistics as estimating the parameters of some underlying true model. The alternate view, what he calls the algorithmic view, aligns with the modern data analyst using a machine learning approach, the role predicted by Tukey (1962), and focuses on the predictive capability of the model with no real consideration to some underlying true model. This algorithmic view has been widely adopted among the computational analysis crowd and has introduced a variety of new terminology.

Among the new terminology is the concept of *learning*. In particular, *machine learning* and *statistical learning*. Learning is the iterative process of building knowledge. In the data analytics realm, it is the building of a specific function or model, based on the data provided, while proceeding from some pre-specified function or model form (not necessarily assumed to be some true model form). Machine learning and statistical learning are the set of techniques employed to estimate the function or model parameters.* The techniques used are algorithms that employ an optimization-type, or search function, to estimate the unknown parameters to arrive at the required function or

---

* There are definitions of machine learning not associated with algorithms, but these are not as widespread.

model. One differentiation among the two terms is that statistical learning algorithms seek parameters for some algorithm while the machine learning algorithms seek the best algorithm.

These learning approaches look to find the best set of algorithm parameters to produce the most accurate forecasts or predictions of the data. A common task framework concept is often used in machine learning applications to reinforce that learning has a focus on attaining best performance based on some metric of predictive capability. The underlying algorithms are the common statistical approaches for the most part.

Examples of machine learning approaches include:

- regression techniques
- clustering techniques
- classification methods
- pattern recognition methods

A machine learning algorithm intelligently searches for some best set of function or model parameters which then specify that particular function or model. Generally, the approach is to use an iterative improvement approach embedded within the overall algorithm. The algorithms used are not truly novel concepts; they have a long history of use in numerical analytics, applied operations research, and heuristic search (see Hill & Pohl, 2009).

Learners are *supervised* or are *unsupervised*. Supervision pertains to whether or not the dataset provided to the algorithm helps guide the parameter value search. The term does not pertain to any human involvement in the algorithm processing. If the dataset includes data elements that help guide the search process, it is a supervised learning algorithm; otherwise it is an unsupervised learning algorithm. Frumosu & Kulahci (2018) discuss semi-supervisory methods that employ both supervised and unsupervised methods in a manner to enhance the analytical approach.

Examples help to clarify the supervised versus unsupervised distinction. A regression model dataset includes responses of interest and data presumed to have some relationship to the response, which we call the independent dataset. For example, housing characteristics are collected as independent variables to predict housing prices (the dependent or response variable). The algorithm learns the best parameters in the regression equation to predict the housing prices such that the sum of the squared errors is minimized (the metric selected to guide the algorithm). As an example of this, Hill et al. (2017) use fragment speed, angle of impact, and fragment size characteristics to predict aircraft panel penetration by the fragment shot at the panel and learn a model suitable to represent explosion flash size due to fragment impact against an aircraft to support aircraft survivability analyses.

The regression model is specified but includes unknown regression coefficients. From a machine learning perspective, the model fitting approach finds those parameter values applied to the independent data that provide the best predictions of the response data. The response values are used to guide the learning process searching for the best values of the unspecified parameters in the regression model.

Another set of data includes classification values for objects in a set, along with attributes for the objects. A clustering algorithm finds some best way to group the objects based on the attribute values to best predict the classification values of the objects. Customer attributes collected among mortgage loan applicants can be clustered according to approval status to determine those attributes that cluster toward approval and those that cluster toward disapproval. Uhorchak (2018) used applicant attributes to derive clusters to predict candidate success in Special Forces operations training programs. Boyd & Vandenberghe (2018) list the various clustering applications which include:

- topic discovery based on document analyses
- clustering of patient by illness
- customer marketing segmentation
- weather zone definitions
- energy use patterns

Once again, the data provided helps guide the clustering learning process.

An unsupervised method does not have that set of data to help guide the learning. For instance, a third dataset contains attributes of some set of objects. The goal is to group the objects into n disjoint sets based on the attributes of the objects. The groupings thus obtained are used by the analysts to gain insight into how the objects are distinguished based on their attributes. Various clustering approaches proceed in such an unsupervised fashion.

The final term discussed here is *training* an algorithm and is best understood by considering two subsets of data drawn from some overall set, the training set and the validation set. Training is the use of the training set data to allow an algorithm to "learn" the best parameter settings to predict that training data. However, the real metric is how well the training algorithm predicts that other set of data from the target population of interest, the validation set. Training an algorithm involves an iterative improvement process in which some measure or metric is systematically improved by changing the parameters of interest in the algorithm.

The training set/validation set preference seems to help distinguish inferential methods versus learning methods. Inferential methods tend to use the entire dataset to better estimate the true model parameters. The learning methods really focus on using the training sets to predict the validation sets in some optimal fashion based on a metric of prediction accuracy.

1.4   Data, Data Everywhere

Scan a recent technology periodical or a quantitatively focused journal and you will likely see something written about data. Readers of this chapter are likely very familiar with what data represents, so it is assumed no definition of data is needed. Most readers would surely agree that poor-quality data is not very useful and gathering more of that same poor-quality data is not going to improve things. Quality analytics requires quality data. Ensuring quality data is particularly challenging since there is now so much data available and it is all too often collected in myriad formats. There are a good number of characterizations of data and data science, far too many to fully review here. The following presents a sampling of these works meant to provide reasonable coverage of the body of the conceptual work.

Hazen et al. (2014) summarize four dimensions of data quality as:

- accuracy
- timeliness

- consistency
- completeness

The accuracy dimension refers to how well the data compare to reality; does the data truly represent the system or process of interest. The timeliness dimension refers to whether the data are up-to-date when actually employed. The consistency dimension relates to how data are represented and whether representations of the same data differ. Finally, completeness refers to whether the dataset is fully populated or does the dataset suffer from missing data. Naturally, these dimensions will not always be of equal importance in a particular context or for a particular domain of interest. They will all, however, require consideration.

Analytic endeavors generally require data. Understanding the dimensions of data quality promotes actions to ensure quality data are inputted to the analytical endeavor. Data engineering efforts include data quality concerns.

Data engineering is the function that builds the data and analytical infrastructure (i.e., the databases), loads the collected data into that infrastructure, and applies metrics associated with the dimensions of data quality to ensure the data are correct in content and form. Data engineers are the 21st century database developers and experts. Clearly, the data engineering needs of today are much more complex, while the data engineering role in analytics is steadily growing more important. Data engineering is as important to analytics as software engineers are to application development.

Data quality must be assured. However, the process of obtaining quality data is often quite labor-intensive and time-consuming. Fortunately, there are a variety of tools and programming languages to assist the "data wrangling" process. Data cleaning involves not only reconciling erroneous or incomplete records, but it also involves verifying data formats and data content.

What is *big data*? Is big data merely databases too large for our spreadsheets? Is it data obtained from some continuous data feed? Big data is a ubiquitous term arguably best defined using the five Vs of big data (Wilson, 2018):*

- volume
- variety
- veracity
- velocity
- value

Volume refers to the scale of the data and just how much data are being collected. Variety refers to the many forms by which the data are obtained. Veracity refers to how certain we are regarding the data; what level of error is inherent in the data. Velocity refers to the speed by which the data arrives. Finally, a fifth V often discussed refers to the utility of the data, meaning how much value can be generated by using the data in the analytics to support the decision-making processes.

Relating these recent ideas to defense applications helps differentiate big data from simply large datasets. Personnel data may involve hundreds of fields per record and

---

* There are a variety of works offering a number of these Vs, the largest number being 42 Vs (Shafer, 2017).

hundreds of thousands of records.* However, the velocity of data change may be quite slow, possibly even static. This is more of a large dataset. Sensor data from an airborne platform may consist of constantly streaming hyper-spectral imaging data on a variety of ground-based objects. This is more a big data type of dataset. Maintenance functional data related to repair processes and associated manpower efforts represent large datasets. The equipment in operations may continuously transmit live maintenance and status information, which becomes a big data issue. Zhao, MacKinnon & Gallup (2015) offer a nice article on big data and the DoD.

The *data scientist* is a highly sought-after individual. Unfortunately, such a person is unlikely to actually exist simply because the world of the data scientist involves too diverse a range of skills to realistically exist as a single person (Waller, 2013). Instead, one should seek the team of data scientists whose collective skills provide the competencies to build, populate, analyze, and exploit large data and big data applications. Data analytics needs to be considered a team endeavor. Common skills among these team members include (Woods, 2012):

- mathematical and statistical skills
- technical skills
- communication skills
- teamwork and collaboration skills
- expertise in the particular areas of interest

The language of data analysis is statistics. This dates back to the seminal paper by Tukey (1962) and his later paper (Tukey, 1997). Rappa, as interviewed in Woods (2012), reinforces this and notes that data science is not really new, as it is grounded in statistics.

The advanced algorithms being used have strong mathematical foundations. Thus, mathematical and statistical skills are an absolute necessity to the data science team. The technical skills aspect builds on the mathematical and statistical basis in that knowledge of the numerical analytics or computer algorithms helps in the application of these advanced algorithms. In addition, expertise in database development, cloud programming, data manipulation, and web programming are beneficial. This reinforces the computer science skill set. As alluded to above, this required skill set really represents the collective skill set of a data science team. This team aspect is particularly crucial when the work involves the particular expertise associated with the domain. For instance, a big data sensor application may require the technical skills to collect, store, prepare, and process the data, but an experienced intelligence analyst must do that final check to ensure everything is done correctly and provides meaningful results and valid insights.

So this leads to defining data analytics. Is this really just statistical analytics? Is it simply applied statistics? Does this differ from predictive analytics, business analytics, or even operational research? There is no single view; thus, I will establish my view to continue the discussion.

Data analytics is that subset of the data science realm focused on gaining insight from data. Thus, the perspective here is that data analytics and statistical analytics are effectively the same. If that insight involves statements of fact based on the data, let's call that *descriptive data analytics*. If the insights are used for any sort

---

* I will often refer to each of these records as a data object.

of forecasting purpose, let's define that as *predictive data analytics*. This distinction aligns with Breiman's two culture concept (Breiman, 2001).

Analytics also requires context; where is it being applied (Nelson, 2018)? Is it business or government, supply chain management, production planning, or distribution execution? The list gets big very fast. Adding context yields specific areas such as:

- business analytics
- supply chain analytics
- retail analytics
- sports analytics
- defense analytics

Extending this further, substitute in descriptive analytics or predictive analytics and our analytic naming framework grows to accommodate a plethora of applications. This is a nice framework for understanding how the data science team is using the data along with their analytical skills to help support informed decisions over a wide range of applications each with a particular title encompassing a common set of goals and tools. The final topic to consider is some of the methods employed within the analytics.

1.5 Some Objectives of Data Analytics

The methods of multivariate analysis, or machine learning, offer a variety of analytical approaches. The choice and subsequent use of the method(s) are a function of the analytical objectives of the study. Consider the following three broad objectives of a study now as three categories of analytical tools:

- prediction
- classification
- dimension reduction

Each category uses various metrics to assess algorithm performance. Prediction approaches use some measure of prediction error. Classification algorithms often use measures of misclassification rates. Finally, dimension reduction methods focus on the variance explained per dimension modeled; the user selects that specific lower dimension explaining enough of the variability in the data.

Prediction objectives focus on estimating new response variables based on new input values. Analytically we gather a variety of input values, and their associated response values, and develop a model that best predicts those response values as a function of the input values. Response variables come in a range of data types. The data may be continuous, integer, even binary. In some cases, we may have response variables that are categorical in nature. Sometimes the data have a time component as well. All types are accommodated within predictive model development and subsequent use.

There is an interesting dilemma in the current machine learning-driven analytic environment in prediction – are you predicting from a sample or do you have the population? Inferential statistics assume you have a sample, with some error due to it being a sample, and thus you use some measures of risk when making inferences about the population. Newer methods effectively ignore any inference to some true model form, presume the sample is large enough that it is the population thereby removing the need for assuming some sample error, and focus on predicting that population data. These approaches, inferential versus predictive, are quite different in terms of basic philosophy, and the analytical team, as well as leadership, should reconcile which approach fits the current situation.

Classification objectives focus on how input values help define membership characteristics among the objects. Analytically, we take the set of input values, using as the response some membership characteristic, and our subsequent model "classifies" each object based on their specific input values. The classification rules can then be used to derive theory regarding membership or to predict membership of new objects based on their input values.

Data dimension reduction objectives focus on replacing some large dataset with some smaller, equally descriptive, and ideally more informative set of data. Analytically, we are examining some dataset viewed in some n-dimensional space and we look for some $(n - p)(p > 0)$ dimensional space that explains that data as completely as possible. A primary purpose of the reduced dimensional space is to provide a clearer interpretation of the data, what it represents, and ultimately generate more useful insights.

1.5.1   Sample Predictive Tools

Regression analysis builds a model relating predictor variables to the response variables. The goal is to find those unknown parameters of the model that when used to create a linear combination of the predictor variables provide some best predictive capability for the response variables. Since all data are provided, regression is a supervised machine learning method.

Given some data in $n \times m$ matrix X, and the $n \times 1$ vector of responses Y, the linear regression model:

$$\hat{Y}_{ij} = b_0 + \sum_{i=1}^{m} b_i x_i + e_{ij} \qquad \forall j = 1,\ldots,n$$

relates the independent data in X to the responses Y via the predicted values $\hat{Y}$ such that the sum of the errors squared is minimized. The most common algorithm used to find the estimated coefficients is called least squares estimation. The machine learning interpretation is that the algorithm searches for those optimal values of the coefficients. Hill et al. (2017) used regression methods to model the flash events due to missile fragment impacts against aircraft material.

Logistic regression is used when the response variable is binary, or can be mapped to a binary outcome. The model parameters provide a linear combination of the predictor variables that provide the likelihood of the response being in either of the two response states. Schofield et al. (2018) used this method to predict Air Force officer career retention as a function of multiple personal characteristics within each service member data object, while Wolf, Hill & Pignatiello (2014) used the method to predict small unmanned aerial vehicle failure rates.

Time series models are used to predict patterns in time series response data. The data employed are collected over time and are related by some time interval (called autocorrelation). This time dependence among the data is exploited to build a model that best predicts the available data. This best-fitting model is then extrapolated into future time periods to provide a prediction of future response values. There are a variety of time series model forms, too many to recount here. Models predict the time series data, Y_t, as functions of past responses, past model errors, patterns due to seasonal variations, even other independent variables. McDonald, White & Hill (2018) used forecasting methods to predict Army recruiting targets or goals.

Predictive tools fall into the supervised machine learning set of techniques. Some issues with such models include making predictions (which is always hard), ensuring the validity of the data (its accuracy and veracity), and determining the appropriate form of the final model employed.

1.5.2　Sample Classification Tools

Classification is important since analysts always seek some order to their data. For instance, given some performance indications, what type of failure might we expect? Given various sensor input readings, what kind of target are we sensing? Given some set of candidates for an important position, each with their own strengths and weaknesses, which of the candidates are likely to succeed?

Every object in the dataset employed is a point in a multidimensional space. Groupings are clustered together in this space. Nearest-neighbor methods classify objects into groups by using the distance from objects to grouping locations and associating the objects with the closest location. This is a relatively simple approach solved by iteratively defining group locations until those locations are deemed to provide the most accurate classification approach. New objects are predicted to fall in that group to which they are closest.

Discriminant analysis develops a function that separates objects into their predicted groupings (or categories) based on the object characteristics provided. Characteristics associated with financial health can discriminate loan customer risk. Personality profile scores can discriminate among candidates for high stress positions. Early career indicators can discriminate among future defense senior leaders. Object attributes, expressed as a linear combination that is the discriminant function, are provided a value for the data object which is then used in some rule to place that object into its predicted category. Tetrault (2016) used discriminant analysis in a study of various healthcare factors associated with respiratory problems among pilots flying high-performance military aircraft.

Support vector machine (SVM) algorithms are quite new, quite novel, and quite a powerful classification method. While discriminant analysis employs a linear function to separate groups, SVM employs a non-linear function to provide greater fidelity in separating the groups within the dataset. This function is obtained by projecting the n-dimensional data into some higher space, finding the linear separator in that higher space, and bringing that now non-linear function back into the original space. The result is a more accurate classifier on the specific data employed.

Classification methods are supervised methods seeking some way to accurately classify objects into their groupings. The larger the dataset, ideally, the better the classification model.

1.5.3　Sample Dimension Reduction Tools

Analysts typically have a more difficult time comprehending high dimensional data. We like lower dimensions, especially if those lower dimensions promote some graphical display. We also like to summarize data. Thus, the final set of data analytic methods are unsupervised methods that seek ways to reduce the dimensions of the dataset employed, while helping to provide useful insights.

Principal components analysis (PCA) takes some set of data in n-dimensions, and determines weights such that the linear combinations of the variables explain as much of the variance in the data as possible. The first principal component extracted from the data set explains the most variance. Of the remaining variance, the second principal component explains the most variance. This continues for all n principal components. In practice only a few of the principal components are really needed. The analyst can then use these smaller numbers of principal components to explain the full set of data. In practice, the analyst will seek some logical interpretation of each principal component since it is a combination of the actual collected variables, even though this clean interpretation is not achievable. PCA analysis is based on the eigenstructure of the variance structure of the data and thus requires some mathematical expertise.

The Factor Analysis (FA) method is similar in some ways to PCA. However, in FA we presume there are underlying factors that explain the data; we just do not know what those factors are or how they explain the data. As with PCA, an eigenvector analysis of the data structure (using the correlation structure) is employed to uncover the factors that do the best job of explaining the correlation among the data. Unlike PCA, attributing a relationship between the data and the factors is facilitated by information on how well each variable associates with each factor. Thus, the original data are summarized using the fewer, aggregate factors.

Multidimensional scaling (MDS) is the final data reduction method presented. Seemingly not as popular as PCA or FA, when appropriate, MDS can provide extremely useful results. The premise of MDS is that objects in their high dimensional space have some proximity to each other. If we project those points down into a two-dimensional space, but retain the relative proximity among the points found in the higher dimensions, then we can visualize the clustering of the data and from those clusters, use the common attribute values of the cluster to explain the underlying cause of the clustering, thereby reducing the dimensions required to explain the data.

1.5.4   Caveat on Tools

This was a very brief survey of some of the more useful tools in the data analytics tool kit. There are actually many methods available. This survey did not cover some of the other learning methods such as neural networks, popular in pattern recognition, or wavelets, a technique rapidly gaining interest in a wide variety of applications. Those are found in some of the new machine learning or statistical learning books that are available with method implementations found in a variety of software tools, both commercially available and in the public domain.

1.6   Warnings and Final Thoughts

Two apologies are in order. First, the section title deserves comment. Managers do not need warnings, but they should be aware of potential pitfalls in data analytics and some promises made regarding the use of analytical tools. Second, some of the following examples may seem too obvious, but sometimes it is the obvious that gets overlooked and causes problems.

Beware of the non-technical salesperson. Presentations on big data, data science, Internet of Things, etc., can be quite entertaining. Unfortunately, presenters can often get by with just a cursory knowledge of the topics they are pitching. If the presentation seems overly focused on how novel the approach is, beware. If they focus on their state-of-the-art machine learners, again beware. If their approach will solve all your problems, definitely beware. Ask the presenters pointed detail-oriented questions to avoid future problems.

Beware of the ultimate tool. Algorithms accept data and return results that need interpretation. No algorithm can accept your data and return your answers. Any pitch that claims this should be dismissed. Data must always be cleaned of errors, focused on the particular needs of the application, and formatted for use in the algorithm. This can be a very time-consuming process, even for algorithms or tools that just present data and information. Algorithms will never replace human knowledge and experience. Successful analytical endeavors will always involve the team of analysts, subject matter experts, and the software programmers. The final solution to important problems will evolve from algorithm results iteratively improved based on human team learning and actions. There will always be the need to interpret tool output; there will never be a perfect black box for analytics.

There is no "easy" button. Easy problems are easy, hard problems are hard. Obvious, but often overlooked. The DoD is facing really hard problems and is hopeful that investments in big data approaches will solve those problems. Rarely are easy solutions applicable to hard problems. As the DoD moves forward, leadership needs to invest not only in the intelligent use of modern computing algorithms, but also the requisite human capital needed to develop the final solution and the time required to arrive at that solution. Nobody has that black box whose start button functions like the mythical "easy" button found at popular office supply outlets. There will be times when particularly novel tools are not required; sometimes simple data presentation methods may suffice.

Admit to what is unclear. As this chapter has tried to emphasize, an important aspect of the big data allure is the terminology used. Unfortunately, some will use the terminology to imply delivery of more capability than is realistically achievable. Always ensure complete understanding of what is really being presented, what will really be delivered, what will really occur over the time period promised. There will be instances in which the incorrect technique is applied to a particular problem with the belief that the results are correct or because it simply happens to be the tool of choice. In either case, ensure personnel are available to provide a sanity check on the methods and results. DoD leadership must break through the "puffery" to get to the "no kidding" details of what is being provided.

Big data analytics has brought to the forefront the myriad useful algorithms with which to tackle pressing challenges in DoD analytics. Database developers, operational analysts, computer programmers, and statisticians are the data scientific team needed by the DoD. This chapter, while critical in parts, has sought to clarify the discussion on big data analytics to ensure DoD leaders and data scientists speak from the same basis of understanding.

There is a lot more to this topic than can possibly be covered in this short, high-level narrative. Great initial reads are those by Tukey (1962, 1997) and Breiman (2001). For more details see the excellent texts by James et al. (2013) and Schumacker (2016), both having a focus on the use of the R programming language, along with a reasonable discussion of the techniques. For more details on the primary multivariate techniques, the classic from Dillon & Goldstein (1984) is still hard to beat. For those who like survey papers or tutorials, there are actually not many of a general nature; exceptions are Chen et al. (2016) who address big data and data science and Cao (2017) who focuses on data science. A timely piece that provides a very objective view is the report by Davis (2019). Finally, a Google search on terms such as big data, data science, statistical learning, machine learning, or data analytics will yield myriad links to various magazine and periodical pieces that provide individual opinions and perspectives.

Acknowledgments

This work was supported by the Office of the Secretary of Defense, Directorate of Operational Test and Evaluation, and the Test Resource Management Center under the Science of Test research program.

Disclaimer: The views expressed in this chapter are those of the author and do not reflect the official policy or position of the United States Air Force, Department of Defense, or the US Government.

References

Amrhein, V., Greenland, S., & McShane, B. (2019). Retire statistical significance. *Nature*, 567, 305–307.

Boyd, S., & Vandenberghe, L. (2018). *Introduction to applied linear algebra: Vectors, matrices, and least squares*. UK: Cambridge University Press.

Breiman, L. (2001). Statistical modeling: The two cultures. *Statistical Science*, 16(3), 199–231.

Cao, L. (2017). Data science: A comprehensive survey. *ACM Computing Survey*, 50(3), 43:1–43:42.

Chen, Y., Chen, H., Gorkhali, A., Lu, Y., Ma, Y., & Li, L. (2016). Big data analytics and bid data science: A survey. *Journal of Management Analytics*, 3(1), 1–42. doi:10.1080/23270012.21 06.1141332

Davis, Z. S. (2019, March). Artificial intelligence on the battlefield: An initial survey of potential implications for deterrence, stability, and strategic surprise (Tech. Rep.). Center for Global Security Research, Lawrence Livermore National Laboratory.

Dillon, W. R., & Goldstein, M. (1984). *Multivariate analysis: Methods and application*. New York: Wiley.

Frumosu, F. D., & Kulahci, M. (2018). Big data analytics using semi-supervised learning methods. *Quality and Reliability Engineering International*, 34, 1413–1423.

Hazen, B. T., Boone, C. A., Ezell, J. D., & Jones-Farmer, L. A. (2014). Data quality for data science, predictive analytics, and big data in supply chain management: An introduction to the problem and suggestions for research and application. *International Journal of Production Economics*, 154, 72–80.

Hazen, B. T., Skipper, J. B., Ezell, J. D., & Boone, C. A. (2016). Big data and predictive analytics for supply chain sustainability: A theory-driven research agenda. *Computers and Industrial Engineering*, 101, 592–598.

Hill, R. R., Ahner, D. K., Morrill, D. F., Talafuse, T. P., & Bestard, J. J. (2017). Applying statistical engineering to the development of a ballistic impact flash model. *Quality Engineering*, 29(2), 181–189.

Hill, R. R., & Miller, J. O. (2017). A history of United States military simulation. In *Proceedings of the 2017 winter simulation conference* (p. 346–364). Piscataway, New Jersey.

Hill, R. R., & Pohl, E. A. (2009). An overview of meta-heuristics and their use in military modeling. In M. U. Thomas & A. B. Badiru (Eds.), *Handbook of military industrial engineering* (p. Chapter 25).

James, G., Witten, D., Hastie, T., & Tibshirani, R. (2013). *An introduction to statistical learning with applications in r.* New York, NY: Springer.

McDonald, J., White, E., & Hill, R. R. (2018). Forecasting U.S. army enlistment contract production in complex geographical marketing areas. *Journal of Defense Analytics and Logistics*, 1(1), 69–87.

Montgomery, D. C., & Runger, G. C. (2011). *Applied probability and statistics for engineers, fifth edition.* Hoboken, NJ: John Wiley & Sons.

Schofield, J. A., Zens, C. L., Hill, R. R., & Robbins, M. J. (2018). Utilizing reliability modeling to analyze United States air force officer retention. *Computers & Industrial Engineering*, *117*, 171–180.

Schumacker, R. E. (2016). *Using r with multivariate statistics.* Los Angeles, CA: Sage.

Tetrault, A. (2016). *The effects of high performance aircraft respiratory systems on pilots* (Unpublished master's thesis). Air Force Institute of Technology, Wright-Patterson AFB, OH.

Tukey, J. W. (1962). The future of data analytics. *The Annals of Mathematical Statistics*, 33(1), 1–67. doi:10.1080/10618600.2017.1384734

Tukey, J. W. (1997). More honest foundations for data analysis. *Journal of Statistical Planning and Inference*, 57(1), 21–28.

Uhorchak, N. M. (2018). *Comparative analysis of incomplete and complete SOCOM selection data* (Unpublished master's thesis). Air Force Institute of Technology, Wright-Patterson AFB, OH.

Wolf, S. E., Hill, R. R., & Pignatiello, J. J. (2014). Using neural networks and logistic regression to model small unmanned aerial system accident results for risk mitigation. *Military Operations Research*, *19*, 27–39.

Woods, D. (2012). What is a data scientist?: Michael Rappa, institute for advanced analytics.

Zhao, Y., MacKinnon, D. J., & Gallup. (2015). Big data and deep learning for understanding DoD data. *CrossTalk: The Journal of Defense Software Engineering*, Vol. 28, No. 4, pp. 4–11.

Chapter 2

Microsoft Excel: The Universal Tool of Analysis

Joseph M. Lindquist and Charles A. Sulewski

2.1   Introduction and Problem Definition

In a 2017 survey of Military Operations Research Society (MORS) members, more than 55% of respondents revealed that out of 15 analytical techniques presented, Data Analytics, Statistics, Simulation, and Optimization formed the principal techniques used in the execution of their analytical duties (Goerger, 2017). To enable quality analysis, analysts nearly always rely upon software to uncover insights that might not otherwise be accessible by manual methods. Underscoring the importance of software in the practice of operations research, consider that the Institute for Operations Research and the Management Sciences (INFORMS) publishes a bimonthly member newsletter that dedicates a portion of each issue to trends in specialized tools (software) to facilitate these analytical techniques. A simple internet search finds a myriad of books, mini-courses, and other resources available to practitioners to enable use of these tools.

While selecting the "right" software to accomplish the desired analysis should always be the goal, there are situations where an operations researcher may have to respond to a query without their tool of choice. For situations such as this, an operations researcher could:

1. Politely let the stakeholder know that they will "get back with them" after reconnecting with their specialized software.
2. Adapt to find a timely response using an alternate tool.

In practice, examples abound where an analyst was asked to provide insights on a topic where a specialized piece of software would be preferred, but for one reason or another it wasn't available. To underscore this, consider that the Center for Army Analysis published the Deploying Analyst Handbook in 2016, "developed from the experiences of ORSA analysts who deployed to operational theaters [over] ... the past 14-plus years" (Bahm et al., 2016, p. i). This handbook contains a section on software utilization where a subsection is devoted to MS Excel built-in functions, add-in software, and macros. Additionally, throughout the document are actual graphics used by analysts that were created using MS Excel. Contrast this treatment with another subsection that lumped many specialized software packages together to provide "examples of available software packages." The message is clear – while specialized software provides a preferred path for analysts, there will likely be times when an analyst must be prepared to use MS Excel.

The author is reminded of a time in 2011 when he faced this precise situation while assigned to conduct analytical support to operations in the Southern Philippines. Following some quantitative public perception survey work conducted with the US Embassy, the Joint Task Force, and Armed Forces of the Philippines (AFP) Partners, results were slated to be briefed to stakeholders on a small basecamp accessible only by boat. Upon arrival, the author was alerted to a specific topic that the AFP Commander was seeking to understand. All prepared analysis had been conducted using SPSS at the Joint Task Force Headquarters, and unfortunately, the only tool available on the basecamp computers was Microsoft Excel. In this case, using the optimal tool would have delayed the desired analysis, while using the "universal" tool of analysis could provide those insights on demand. In some cases, delayed analysis may be perfectly acceptable to the stakeholder, but if not, every operations researcher should be proficient in skills to provide an *adequate* answer until a *superb* answer can be delivered.

By nature, military computer networks are designed to be secure. Also by nature, network managers are skeptical of software or systems that are not typical for a common Department of Defense (DoD) user. Unfortunately for the operations researcher, much of the specialized software to conduct analysis might appear as a risk to the network – simply because it is not common to all users. To this end, the DoD and each service closely control IT systems to minimize security risks of software that has not been formally certified as "networthy." While processes exist to certify any specialized software, and many already *are* certified as networthy, much friction still exists between analyst requirements for specialized tools and the network connectivity to enable the analysis.

This chapter will address how to conduct several broad threads of analysis using MS Excel – something that we will dub the "universal" tool of analysis. This title is fitting for MS Excel, as it has an ability to do many analytical tasks adequately, but may not be the preferred tool for a *particular* task (Figure 2.1). To illustrate this analogy, if an analyst's "toolbox" contains specialized tools (such as the hammer, screwdriver, and wrench), then MS Excel is a universal tool (the multi-tool).

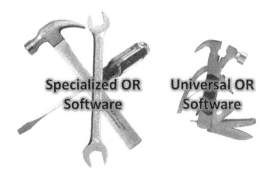

FIGURE 2.1: MS Excel: The multi-tool of OR Software.

The following is an outline of the rest of the chapter: in Section 2.2 we will show how MS Excel can be used to conduct data analysis (to include data exploration using pivot tables, summary statistics, regression, and visualization). In Section 2.3 we show how to solve a subset of linear and nonlinear optimization problems using MS Excel's built-in solver functionality. Finally, in Section 2.4, MS Excel is used to create a Monte Carlo Simulation, suitable for various queuing models encountered by an operations researcher. As a reference, Microsoft Excel 2013 (the current DoD standard at the time of publishing) is used for all screen shots and instructions. It is assumed that the reader has a basic working knowledge of MS Excel.

2.2 Data Analysis Using MS Excel

Consider a situation where you have been assigned to conduct analysis on survey data in support of the task force mission described in Section 2.1. You have been provided with data that contains responses from 3,000 Philippine respondents on a wide variety of topics and would like to explore the data to understand a few items of interest to stakeholders such as:

- respondent ages
- number of family members in the household
- monthly cost of electricity
- perception of electricity services
- working status of respondents
- household income
- educational attainment

2.2.1 Exploratory Data Analysis with Pivot Tables

For reference, consider a truncated data source shown in Figure 2.2 that contains only selected components of the survey data relevant to the items of interest. As a first step to understanding this survey data, we conduct exploratory data analysis (EDA). In practice, EDA is less an algorithmic procedure than an approach to examining the

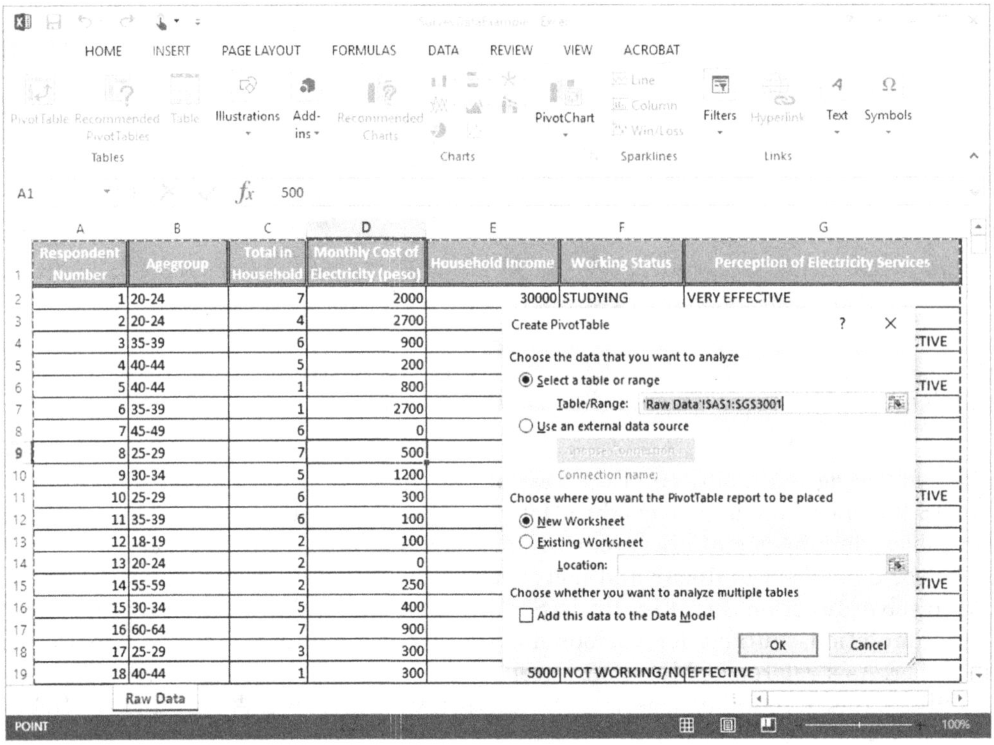

FIGURE 2.2: Respondent data to examine items of interest to a stakeholder.

data and framing the precise analytical questions that might be of interest. One of the most powerful MS Excel tools to conduct EDA is the Pivot Table. A Pivot Table has the ability to count, sum, average, sort, filter, and many other helpful actions *without* affecting raw data. We begin by selecting any cell in the data and then clicking **INSERT-Pivot Table** from the MS Excel menu bar. This brings up the dialog box as shown in Figure 2.2. Select the data appropriate to include in the analysis (in our case the entire table) and click OK.

The principle mechanism of analysis for a Pivot Table is the **VALUES** area of the tool. In the case of our survey data, selecting Respondent Number from the **PivotTable Fields**, returns the sum of values in the field Respondent Number (4,501,500) as shown in Figure 2.3. Perhaps it is more useful to know the number of respondents than the sum – in this case, the user would adjust **Sum of Respondent Number** to **Count of Respondent Number** to find that there were 3,000 respondents in the survey data.

This fact, of course, was already known – but it provided a useful start to understanding the first item of interest to the stakeholder, specifically respondent ages. For this, dragging **Agegroup** down into the **ROWS** area of the tool creates a distribution of age groups as shown in Figure 2.4.

One may find the tabular data provided by the pivot table sufficient; however, another powerful component of MS Excel's Pivot Table functionality is the Pivot Chart. A pivot chart is a dynamic visualization of the pivot table. To create a pivot chart, from the MS Excel menu, select ANALYZE-PivotChart. A dialog box that contains a wide variety of

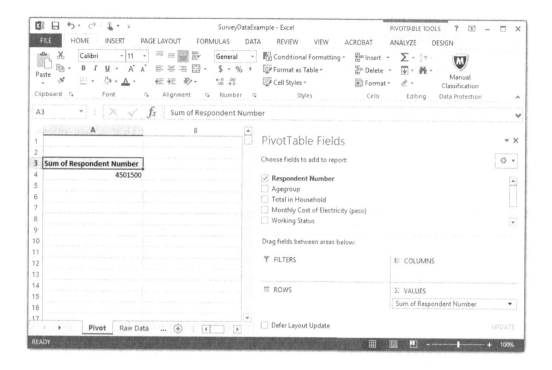

FIGURE 2.3: Computing the sum of respondent numbers in the survey data using a Pivot Table.

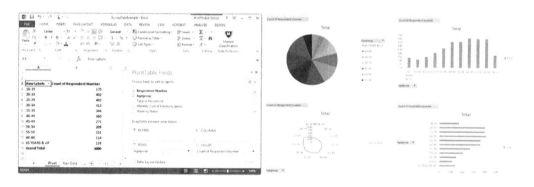

FIGURE 2.4: Visualizing the age distribution of respondents.

potential visualizations appears and the user may select one that conveys the essence of the data.[1] A variety of these visualizations is shown in Figure 2.4.

As previously described, a pivot chart is a *dynamic* visualization of the pivot table. To see this functionality, we introduce another capability of the pivot table – namely filtering and sorting. Suppose that the demographic of 18–49 was particularly important, and further that it should be displayed in descending order. To accomplish this,

[1] For a robust treatment on the visualization of data, see the seminal work by Tufte (Tufte, 1986).

note that on the pivot table near the header **Row Labels** resides a pull-down where a dialog box containing several options for sorting and filtering appears. As observed in Figure 2.5, the analyst can now choose from the many options to customize the data – while simultaneously updating the table of data *and* the visualization.

At this point, the creativity and curiosity of the analyst becomes important, as the conduct of EDA is highly dependent on the structure of the data being examined. In the case of our data, we note that much of the structure of ***Agegroup*** appears to be revealed without significant effort or adjustment, namely that there doesn't appear to be a high density of a particular age consolidated in a single "bin." If there were, an analyst might consider subdividing until the true structure of the data is revealed.

We turn our attention to the ***Monthly Cost of Electricity (peso)*** with the goal of understanding the portion of household income spent on electricity. We begin by unclicking ***Agegroup*** and clicking ***Monthly Cost of Electricity (peso)***. This results in several hundred "bins" of electricity costs – visualized in the pivot chart shown in Figure 2.6. One of the key features of the data revealed in this analysis is that there are several expenditures of electricity that have significantly more responses than others in close proximity, yielding an ostensibly multi-modal distribution. The pivot table entries show that these high-density bins correspond with values rounded to tens or hundreds of pesos. It would seem that when respondents were asked to provide their electricity costs, some may have provided a rounded estimate while others may have provided a more precise value. The take-away for the analyst should be that the data is not multi-modal – instead that more analysis is required.

To provide a more representative view of this electricity data, we wish to group our household expenditures on electricity into larger "bins." To do this, we right-click on any one of the expenditures in the pivot table (say, the 0 for no expenditure on electricity) to reveal another dialog box as shown in the right portion of Figure 2.5. By grouping from 0 to 2,500 by bins of 250, we gain a better sense for the expenditures of electricity by respondents, as shown in Figure 2.7. To conclude this analysis on household

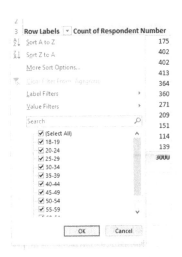
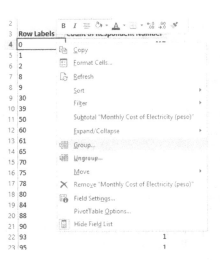

FIGURE 2.5: Sorting and filtering options in a Pivot Table (left) and grouping options in a Pivot Table (right).

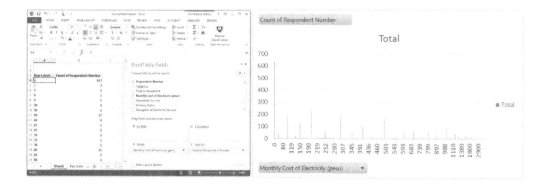

FIGURE 2.6: Visualizing the monthly cost of electricity.

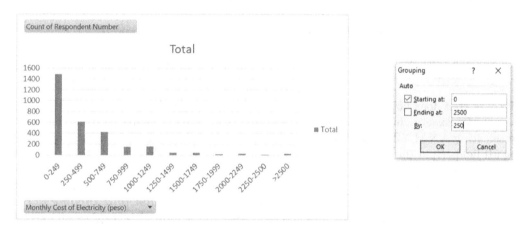

FIGURE 2.7: Representative view of household expenditures on electricity.

expenditure of electricity, we ask the question – what portion of household income is spent on electricity? It might not be surprising that a pivot table can provide insights on this question using built-in functionality to create cross tables and calculated fields.

Another significant benefit of the pivot table is in creating cross tables (also known as two-way or multi-dimensional tables). In the case of the survey data, we wish to see how expenditures of electricity compare for various household incomes. To begin examination of this question, we can create a cross table where data is divided into ***Monthly Cost of Electricity (peso)*** (by row) and by ***Household Income*** (by column). This is completed simply by dragging ***Household Income*** into the **COLUMNS** area of the pivot table.

The result is a count of the number of households in a particular income group that spend a particular monthly amount on electricity. This can be observed in Figure 2.8, visualized using a 3-D Surface plot (available from the MS Excel menu by selecting **ANALYZE – PivotChart – Surface**). While informational, perhaps a stakeholder wishes to see if lower-income households spend a greater proportion of their income on electricity than high-income households. In this case, we employ one final feature of pivot tables, namely the ability to compute values based on pivot table fields.

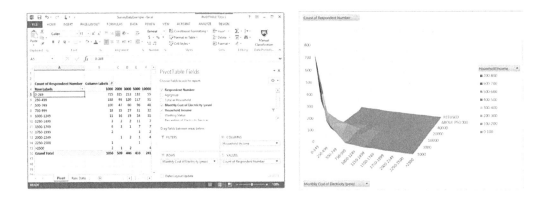

FIGURE 2.8: Use and visualization of cross tables.

To answer the stakeholder's question, we must find a way to compute:

$$\text{Portion of income spent on electricity} = \frac{\text{Monthly Cost of Electricity (peso)}}{\text{Household Income}}$$

in a way that does not alter the raw data while providing valuable insights.[2] For this, we use MS Excel's ability to create a calculated field. This is available from the menu by selecting **ANALYZE – Fields, Items & Sets – Calculated Field**. Begin by clearing all fields from the pivot table and navigating to the Calculated Field dialog box as described above. Next, we name our field and create the formula as shown in Figure 2.9. Once complete, the tool returns you to the Pivot Table where the user can drag the newly created field to the **VALUES** area of the Pivot Table, and *Household Income* to the **ROWS** area. Adjusting **Sum of electHHI** to instead compute **Average of electHHI**

FIGURE 2.9: Use and visualization of calculated fields.

[2] In this example, it would be quite simple to create a column in the original data that divides the electricity cost by household income. Once complete, the analyst could complete tasks as have been demonstrated in this chapter to arrive at the same conclusions. This, however, would increase the size of the data file unnecessarily, perform unnecessary computations, and potentially slow computation. The pivot table, in contrast, efficiently evaluates only the minimum of cells required, and without increasing the size of the data file.

now clearly reveals that on average, low-income households spend significantly more on electricity than do high-income households.

Many other features of the pivot table have been left for the reader to explore. For example, there are additional options for summarizing data in the **Value Field Settings**, working with **FILTERS**, as well as using the multi-dimensional table computed by adding secondary fields to rows or columns. Any of these topics could be self-taught using a host of freely available internet resources.

2.2.2 Point and Range Estimates

We continue the data analysis section of this chapter with a brief discussion of built-in functionality that may be used to compute point and range estimates of data. Suppose that we continue working with our survey data and wish to gain some insights on the households surveyed. For reference, consider a truncated data source shown in Figure 2.10 that contains only selected components of the survey data relevant to the items of interest. Included are fields for the household income, age of the head of household (HoH), HoH education level, and the total number of members in the household.

We begin by considering a stakeholder question concerning the HoH education level in the area where respondents were surveyed. Many references can be found to cover topics on inferential statistics and will not be specifically addressed here, except to describe the process to form a 95% confidence interval for the *true mean education level* of the population based on the sample provided. It can be shown that a confidence

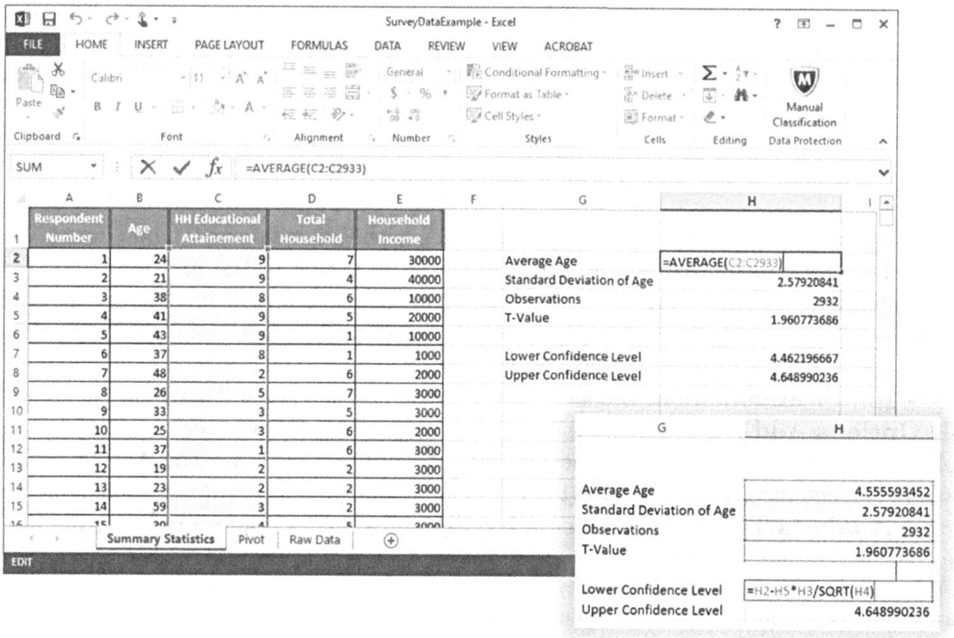

FIGURE 2.10: Potential predictors of household income.

interval for a population mean stemming from an unknown distribution and variance is computed using:

$$\bar{x} \pm t_{a/2} \frac{s}{\sqrt{n}},$$

where $\bar{x}$ is the sample mean, $t_{a/2}$ is the $t - value$ for a given significance level α (in our case, 0.05), s is the sample standard deviation, and n is the sample size.[3]

To compute the sample mean x, we use the MS Excel built-in command **AVERAGE()** as shown in the formula bar displayed in Figure 2.10. The sample standard deviation, size, and $t - value$ are computed similarly using the **STDEV()**, **COUNT()**, and **TINV()** built-in commands. Putting this all together to find the lower confidence bound is computed by entering a formula in cell H7 that references the locations of each component of the confidence interval. This is shown in the call-out portion of Figure 2.10. MS Excel has a built-in function **CONFIDENCE.T()** that can be used to compute $t_{a/2} \frac{s}{\sqrt{n}}$, yielding the exact same results as described above. As before, many other built-in functions for examining point estimates have been left for the reader to explore using MS Excel's robust help capability.

2.2.3   Regression Modeling

Our final strand of data analysis will demonstrate how to perform multiple linear regression (MLR) using MS Excel. Suppose our final stakeholder question deals with understanding the relationship of Household Income and the three predictors we have been examining. The stakeholder assumes that age, educational attainment, and total number of members in the household are *positively* correlated with the household income *and* that these three predictors can be used to predict the income of a randomly selected member of the population. In short, we are seeking to find the coefficient values, β_i, for the linear model that best fits our data, given by:

$$\text{Income} = \beta_0 + \beta_{\text{Age}} \cdot \text{Age} + \beta_{\text{Educ}} \cdot \text{Education} + \beta_{\text{Members}} \cdot \text{Members}.$$

The resulting model could then be tested to see if the βs are positive, how much variation in the data can be explained by the model, and which of the predictor variables are statistically different than 0. As with the confidence interval for the mean, many references can be found that cover the mathematical underpinnings of MLR in detail and will not be specifically addressed here.

To begin our examination, the user must load the **Analysis ToolPak** by clicking **FILE-Options-Add-Ins** from the menu bar and clicking Go adjacent to the Manage: Excel Add-ins option as shown on the left side of Figure 2.11. Once the Add-Ins dialog box opens, select the **Analysis ToolPak**. There are more than a dozen tools available with the **Analysis ToolPak** for common statistical inquiry. Before delving too deep,

---

[3] MS Excel computes summary statistics based on numeric values. In the original data, some respondents refused to provide their household income and this was coded as **REFUSED**. Special handling of non-numeric data is common to virtually all analysis, therefore the analyst must make certain assumptions prior to proceeding. For the sake of simplicity in presenting this material, we have removed (complete imputation) any non-numeric data, leaving a total of 2932 observations of our original data. A robust discussion about handling missing data can be found in Horton and Kleinman's manuscript (Horton & Kleinman, 2007).

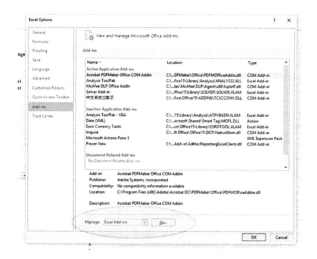 

FIGURE 2.11: Loading the MS Excel Analysis ToolPak.

note that while these tools perform well for their designed purpose, there are two principal downsides to the built-in functionality:

- Customization: here, the user is typically restricted to common parameter values chosen by the developer
- Sophistication: MS Excel provides analysis using only basic least squares modeling – not more advanced regression techniques that involve fixed effects, random effects, instrumental variable models, and the like

Among MS Excel's data analysis functionality is the Regression tool. To begin, click on DATA-Data Analysis on the MS Excel menu bar. This brings up a dialog box as shown on the left side of Figure 2.12. The Input Y Range is populated with data on household income and the Input X Range is populated with the data on age, educational attainment, and total number of members in the household. Several model diagnostic options are available, although for this example, we simply wish to view the default output. The formulation and resulting output are shown in Figure 2.12.

One of the first items that we observe from the output is that the r^2 and adjusted r^2 is quite low (≈ 0.18). In words, roughly 18% of the variability in household income can be explained by the predictor variables given in the model. Conversely, 82% of the variability in household income is still unexplained. Perhaps the occupation or region where the survey took place could help account for variability in the income of a household and increase the usefulness of the model to our stakeholder.

We next use the regression output to form our MLR model as:

$$\text{Income} = -1372.7 + 14.1 \cdot \text{Age} + 1093.0 \cdot \text{Education} + 96.0 \cdot \text{Members}$$

observing that each of the predictor coefficients are positive. At first glance, this provides reinforcement for the stakeholder belief that the three predictors are positively correlated with household income, although a bit more analysis is required.

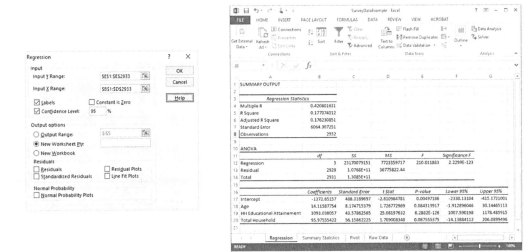

FIGURE 2.12: Loading the MS Excel Analysis ToolPak.

We turn our attention to the significance of each regression coefficient (provided in the *P*-value portion of the output.) We note that the p-values of this regression output measure significance of the hypothesis test:

$$H_0 : \beta_i = 0$$

$$H_a : \beta_i \neq 0.$$

"Small" *p*-values indicate strong evidence against the null hypothesis – that the true coefficient value β_i *is not* 0. "Large" *p*-values indicate weak evidence against the null hypothesis and would suggest that a plausible value for the true coefficient value *is* 0. While terms such as "small" or "large" are quite subjective, many sources consider *p*-values < 0.05 to be small and large otherwise. For our output, we note that the Intercept and Education coefficients are both small, while Age and Members are not (although they are close to the 0.05 cutoff.)

To bring this analysis to a close, we return to the stakeholder's questions to report:

- strong evidence exists to suggest that the head of household's educational attainment level is positively correlated with household income
- evidence that age and total number of members in the household is positively correlated with household income is weak
- there are likely other predictors that could account for variability in household income

2.3 Optimization

Another powerful component of MS Excel is its ability to find solutions to a wide variety of optimization problems through its Solver Add-in. At the time of publishing this book, Microsoft described this tool as follows:

Solver is a Microsoft Excel add-in program you can use for what-if analysis.
Use Solver to find an optimal (maximum or minimum) value for a formula in
one cell – called the objective cell – subject to constraints, or limits, on the val-
ues of other formula cells on a worksheet. Solver works with a group of cells,
called decision variables or simply variable cells that are used in computing the
formulas in the objective and constraint cells. Solver adjusts the values in the
decision variable cells to satisfy the limits on constraint cells and produce the
result you want for the objective cell.

Put simply, you can use Solver to determine the maximum or minimum value
of one cell by changing other cells

(Microsoft Corporation, 2018b).

2.3.1 Constrained Optimization

To begin demonstration of this capability, consider the problem of finding the shortest distance between the curve $y = 5x^2 + 2$ to the origin, as shown in Figure 2.13.
Employing basic calculus or even graphical inspection, this distance should be 2 units
and located at the coordinates (0, 2). In this "toy" example, we show how Solver can
quickly and accurately get this same result.

First, we consider the mathematical formulation of the problem – specifically to find
x_2 such that d is minimized. We also recognize that y_2 must lie on the curve $5x^2 + 2$. Our
formulation is as follows:

Variables:

$$(x_2, y_2) \equiv \text{ Point that lies on } 5x^2 + 2$$

$$d(x_1, y_1, x_2, y_2) = \sqrt{(x_2 - x_1)^2 + (y_2 - y_1)^2} \equiv \text{ Distance between } (x_1, y_1) \text{ and } (x_2, y_2)$$

Objective:

$$\text{MIN } d(0, 0, x_2, y_2)$$

Constraints:

$$y_2 = 5x^2 + 2$$

We now consider how to solve this problem using MS Excel. We begin by setting up
a "guess" for x_2 and y_2. These values need not satisfy any constraints, but should be a
reasonable estimate. In our case, we suspect that an estimate for x_2 is −0.5 and y_2 is 3.
We also set up the distance function as articulated in the problem formulation. A screen
shot of this setup is shown in Figure 2.14. At this point, the user must load the Solver
Add-in in a similar fashion, as described for the Analysis ToolPak in Section 2.2.3 and
as shown in Figure 2.11.

To begin analysis, click on DATA-Solver on the MS Excel menu bar. This brings up a
dialog box as shown in Figure 2.14. Populate the Set Objective with the cell containing
the computation of d, select Min for the desired optimization, and specify cells B1 and B2
as variable. Finally, we require cell B2 (y_2) to lie on the curve $5x + 2$. When we click on
Solve we are greeted with the solution that we expected ($x_2 = 0$ and $y_2 = 2$) and a message

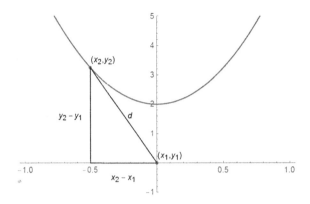

FIGURE 2.13: Minimizing the distance from curve to origin.

FIGURE 2.14: Setup of spreadsheet.

that indicates Solver found a solution. All Constraints and optimality conditions are satisfied. We will explore the Reports portion of this output in the section that follows.

2.3.2   Optimal Base Positioning

Now, let us turn to a simplified military example. Suppose that you are part of a task force in Afghanistan and must find an optimal location (lat – long) to place a new base that will support aerial medical evacuation (MEDEVAC) to three existing combat outposts (COPs). The task force commander has provided the following guidance:

- The new base placement must exist on a key ground supply route adjacent to the Panjshir river. This route is roughly modeled by the equation:

$$\text{lat(long)} = 0.6565 \text{ long} - 10.3201$$

for longitude values in the vicinity of the COPs. This route is highlighted in Figure 2.15.

- The new base must enable the "golden hour" standard for medical care (Welch, 2015). This standard specifies that any MEDEVAC departing from the new base must arrive at any of the COPs and get to advanced medical treatment facility (MTF) at Bagram airfield within one hour. MEDEVAC planners indicate that this corresponds to a combined flight path maximum of 150 km.

We begin by formulating the optimization problem under the assumption that we wish to minimize the sum of all flight distances from the new base to the requesting COP to the MTF. Of course, there are alternative formulations that could be explored, such as minimizing the maximum flying distance, placing a higher priority on one particular COP, minimizing the distance from the new base to the MTF, or something completely different. Our formulation is as follows:

Data:

Latitude and Longitude of COPs 1-3, MTF

Variables:

$long \equiv$ Longitude location for the new base.
$lat \equiv$ Latitude location for the new base.
d_i *(lat, long)* $\equiv$ Flying distance from new base location to MTF via COP i

Objective:

$$\text{MIN} \sum_{i=1}^{3} d_i(\text{lat}, \text{long})$$

Constraints:

$$\text{lat(long)} = 0.6565 \text{ long} - 10.3201, \quad \text{long} \in (69.5, 70.5)$$

$$0 \leq d_i \leq 150 \, \forall i$$

Computation of distances between locations on the surface of the Earth presents a mild challenge as the Earth's curvature should be considered. While there are several approximating functions discussed in the literature, we will adopt the approach taken by the National Oceanic and Atmospheric Administration (NOAA) in their computation of hurricane modeling (National Oceanic and Atmospheric Administration, 2018). While the details are straightforward, we will not discuss them here, as the main purpose of this discussion is to demonstrate the utility of MS Excel to solve this nonlinear optimization problem. All distances described below are computed in MS Excel using a cell formula that follows the NOAA model.

We set up a spreadsheet as shown in Figure 2.16 that contains the following:

- location data for each COP and the MTF (Columns B and C)
- placeholder for LAT/LONG of new base (Cells H3 and I3)

- computed distance from the new base to each COP (Column D)
- computed distance from each COP to the MTF (Column E)
- distance from the new base to the MTF via each COP (Column F)
- combined distance from each base to the MTF via each COP (Cell F6)

At this point, an analyst could simply try various locations along the ground supply route by manually changing the values in Cells H3 and I3 to compare dynamically computed distances that result in a "small" value for Cell F6. Of course, to find a feasible solution the analyst would also have to ensure that the values in Column F were always less than 150 km. This, however, is tedious and may not yield an optimal solution. Fortunately, MS Excel's Solver provides a robust tool to examine this problem in a methodical and mathematically justifiable manner. As discussed previously, invoke Solver through the menu by selecting **DATA – Solver** to produce the dialog box shown on the left side of Figure 2.17.

The four constraints are entered by clicking on the Add button. Note that the user can select from one of three solving methods – each is appropriate for a different family of optimization problems. Since our objective function (ultimately defined by the equations

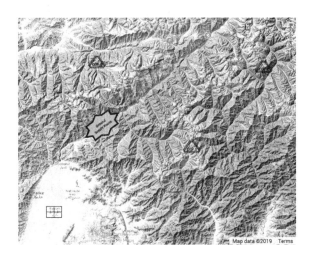

FIGURE 2.15: Situation sketch for optimal base placement (Google Maps, 2018).

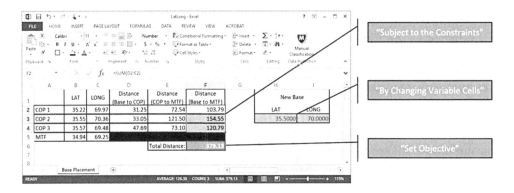

FIGURE 2.16: Setup for optimal base location problem.

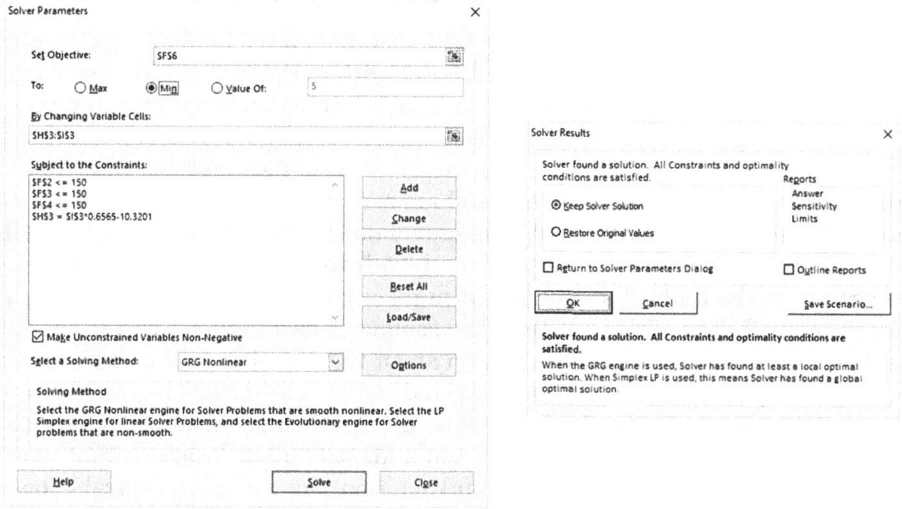

FIGURE 2.17: Setup of the Solver dialog box.

generated through the NOAA model described above) is nonlinear and smooth, we choose GRG Nonlinear and click Solve to find that Solver was able to find an optimal solution that satisfied all constraints. Three potential reports can be generated by selecting Answer, Sensitivity, or Limits from the right-hand side of the Solver Results dialog box.

As one example, we select the Sensitivity Report to reveal the output shown in Figure 2.18. We note that final values for the base location are shown in cells D9 and D10. Additionally, the final values for each of the constraints are shown in cells D15–D18. We see that (within tolerance) each of the total flight distances from the proposed base location to the MTF are within 150 km.

	A	B	C	D	E
1	**Microsoft Excel 15.0 Sensitivity Report**				
2	**Worksheet: [LatLong.csv]Base Placement**				
3	**Report Created: 6/29/2018 5:06:23 PM**				
4					
5					
6	Variable Cells				
7				**Final**	**Reduced**
8		**Cell**	**Name**	**Value**	**Gradient**
9		H3	COP 2 LAT	35.70857883	0
10		I3	COP 2 LONG	70.11222975	0
11					
12	Constraints				
13				**Final**	**Lagrange**
14		**Cell**	**Name**	**Value**	**Multiplier**
15		F2	COP 1 Distance (Base to MTF)	128.372591	0
16		F3	COP 2 Distance (Base to MTF)	150.0000781	-6.641832942
17		F4	COP 3 Distance (Base to MTF)	132.2745739	0
18		H3	COP 2 LAT	35.70857883	661.7170879
19					

FIGURE 2.18: Sensitivity Report for optimal base location problem.

Additional information is provided in the Lagrange Multiplier section of the report. Note the interpretation of the value for COP 2 Distance (Base to MTF) is that giving a 1 km *increase* in the constraint (combined flight path maximum of 151 instead of 150) results in a total *decrease* of 6.64 km in the objective function (combined distance from each base to the MTF via each COP). This is important in that it allows decision-makers to prudently assume risk under uncertainty of some assumptions (like 150 km to meet the "golden hour" standard). While 150 km is a planning factor for flight distance, atmospheric conditions, aircraft condition, crew familiarity with the area may all play an important role in the time it takes for an aircraft to traverse 150 km. In this case, a small increase in the flight distance might make a nearby (but mathematically infeasible) location much more desirable than the computed location.

The **Answer** report provides information about the variables and constraints in the model, specifically, whether a constraint is binding, or has slack. This report is useful to determine the best operational solution among other optimal solutions. In the context of the optimal base placement, this report could identify other locations that may have the same total flight distance as the optimal solution, but are more desirable based on operational considerations.

The **Limits** report provides an upper and lower limit for variables to ensure a feasible solution. This may be useful if a Solver solution is mathematically feasible, but not practically feasible – in the context of the optimal base placement this might include emplacement on a sharp bend in the road, or on an undesirable terrain feature.

2.3.3   Solving Method Choice

It should be noted that in each of these examples we chose the GRG Nonlinear (generalized reduced gradient) solving method. A would-be Solver analyst naturally may wonder – which method should I choose for my given problem? Frontline Solvers, the developer for MS Excel's Solver tool describes the choice as follows:

> *Solver's basic purpose is to find a solution – that is, values for the decision variables in your model – that satisfies all of the constraints and maximizes or minimizes the objective cell value (if there is one). The kind of solution you can expect, and how much computing time may be needed to find a solution, depends primarily on three characteristics of your model:*
>
> * *Your model size (number of decision variables and constraints, total number of formulas)*
> * *The mathematical relationships (e.g. linear or nonlinear) between the objective and constraints and the decision variables*
> * *The use of integer constraints on variables in your model*
>
> *… Your model's total size and the use of integer constraints are both relatively easy to assess when you examine your model. The mathematical relationships, which are determined by the formulas in your model, may be harder to assess, but they often have a decisive impact on solution time*
>
> (Frontline Solvers, 2019).

To summarize, selecting the "best" solver for the task at hand involves two main tradeoffs – time and accuracy. Some succinct advice for the analyst follows:

- If your objective and constraints involve linear functions of the decision variables and computing time is not a concern, use Simplex LP. This will ensure a globally optimal solution regardless of your starting guess
- If your objective and constraints involve linear functions of the decision variables, you have a pretty good idea of where the optimal solution lies, but computing time is a concern, then use GRG Nonlinear. With GRG Nonlinear, you can be confident that you can find a locally optimal solution reasonably quickly
- If your objective and/or constraints involve nonlinear functions of the decision variables, you have a pretty good idea of where the optimal solution lies, and computing time is a concern, use GRG Nonlinear. As indicated above, you can be confident that you will find a locally optimal solution reasonably quickly
- If your objective and/or constraints involve nonlinear functions of the decision variables, you have uncertainty where the optimal solution lies, but computing time is not a concern, use Evolutionary. This method seeks to find the best solution over a wide set of decision variables. The principle benefit of this technique is the robustness of the method, but it is slow to converge
- If your objective, constraints, or cells containing formulas used in computing either contain logical or conditional functions (AND, OR, NOT, IF, LOOKUP, etc.), use Evolutionary. While it is quite doubtful that it will find the global optimal solution, it will likely find a "good" solution

2.3.4 Other Applications

Many other applications of MS Excel's Solver have been applied to military problems including:

- shortest path planning of convoys
- network interdiction (critical node and edge analysis)
- optimal assignment given preferences (bin packing problem)

This section provided two worked examples of how an analyst could take a nonlinear optimization formulation and use MS Excel to solve them. It should be noted that MS Excel's Solver limits the number of constraints to 100 and decision variables to 200. The authors acknowledge that for large-scale problems that approach the limits of the tool, performance is degraded and modifications may be required to reach a solution. Such modifications may include omitting constraints and/or variables, linearizing constraints, or making simplifications to the objective function. While making the problem mathematically easier to solve, these simplifications may result in lost insights found by solving the unsimplified problem using a more robust tool.

Despite these drawbacks, the authors contend that given another optimization formulation (complete with decision variables, constraints, and objective function), a user could easily adapt the foundation described in this chapter to provide an "adequate" solution to any given problem.

2.4 Simulation

When a lack of time, money, or resources prohibits the collection of data, or perhaps when there is a lack of data itself, then the Operations Research Analyst may turn to an application known as simulation. Analysts use simulations to generate data, useful for statistical inference when analytic insight is required to inform key leaders making critical decisions. A simulation is defined as "a technique that imitates (and replicates) the operation of a real-world system as it evolves over time" (Winston & Goldberg, 2004, p. 1145). There exist different types of simulations. Examples of simulations include human in the loop exercises, physics-based or agent-based computer algorithms, commercial off-the-shelf preprogrammed packages, and individual programming techniques capable of being performed within an array of software packages.

This section will focus on a technique referred to as a Monte Carlo Simulation. A Monte Carlo Simulation generates a random sample of size n with replacement, from a specified probability distribution using a sequence of independent and identically distributed random numbers, to model real-world occurrences at a particular point in time (Winston & Goldberg, 2004, pp. 1147–1160). To demonstrate the utility of this method, consider a simplified example whereby a customer parks a car on the street outside a diner using a parking meter. The parking meter only takes quarters and each quarter provides 15 minutes of time. The customer typically does not carry loose change but finds three-quarters in the car to provide 45 minutes of parking. The customer must decide if 45 minutes is enough time to eat a meal and dessert at the diner. In real life, the customer would most likely not need a simulation to inform a decision when parking the car, while this example is provided as a mechanism to understand Monte Carlo Simulation in a familiar context. For this example, assume the customer knows the following information from previous visits to the same establishment:

- time to walk from the parking meter and sit down at the diner counter is always 2 minutes
- sitting down at the counter guarantees immediate service from the staff
- the average time to order, wait, and eat a meal follows a Normal Distribution with a mean time of 20 minutes and a standard deviation of 6 minutes
- the average time to order, wait, and eat a dessert follows a Gamma Distribution with a mean time of 8 minutes and a standard deviation of 4 minutes
- excellent customer service guarantees a lag time between eating a meal and ordering dessert to be at most 1 minute
- it always takes exactly 4 minutes to leave the diner counter, pay at the register, and return to the car

From the above information, one may model the time it takes to walk from the parking meter into the establishment, order their food, and return to the cars as follows:

- let W be the time of the event to walk from the parking meter to the counter. $W = 2$
- let M be the time of the event to order, wait, and eat a meal
- Event M has a Normal Distribution with $\mu = 20$, $\sigma = 6$
- let L be the lag time of the event between events M and D. $L = 1$
- let D be the time of the event to order, wait, and eat a dessert
- Event D has a Gamma Distribution with $\alpha = 4$, $\beta = 2$. where $\mu = \alpha * \beta = 8$ and $\sigma^2 = \alpha * \beta^2 = 16$
- let R be the time of the event to return to the car after walking away from the counter and paying at the register. $R = 4$
- T = Total Time = W + M + L + D + R

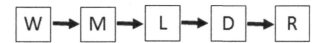

FIGURE 2.19: Linear additive model of all events.

The goal is to determine the probability that the customer will be able to return to the car in less than 45 minutes. To examine this goal, we wish to simulate the total time it takes the customer to enter the establishment, order a meal and a dessert, and return to the car once, and then replicate the simulation multiple times, while computing key metrics to help inform the customer's decision. The paragraphs that follow provide a cursory look at the process associated to this example while Sections 2.4.2 and 2.4.3 provide an in-depth explanation to a separate Monte Carlo Simulation modeling a different scenario.

2.4.1 Simulation, Replication, and Insights

The Normal Distribution and the Gamma Distribution described in this example are only two of the many commonly named probability distributions that users may apply in modeling real life scenarios. MS Excel provides a way to evaluate probabilities associated with many commonly used distributions.[4] Additional functionality is built-in to return the *value* associated with a specified distribution for a given *probability* (known as the inverse value of the distribution). The user may type "returns the inverse" into the Insert Function dialog box, or type the following examples in any cell box to examine further.

This functionality also enables the generation of data referred to as *random variates*. A random variate is, quite simply, an instance of a randomly generated number from a given probability distribution. As an example, to calculate a normally distributed random variate with a mean of 20 and standard deviation of 6, we type the following command into the formula bar:

$$= \text{NORM.INV}(\text{RAND}(), 20, 6).$$

To complete a single "meal" simulation as described above, we type the following command into the formula bar:

$$= \underbrace{2}_{W} + \underbrace{\text{NORM.INV}(\text{RAND}(), 20, 6)}_{M} + \underbrace{1}_{L} + \underbrace{(\text{GAMMA.INV}(\text{RAND}(), 4, 2)}_{D} + \underbrace{4}_{R}.$$

This single iteration provides an estimate of the total time to complete a meal on a particular visit. One might expect, however, that these results would change based on a number of situational factors at the diner. In order to provide true insights, the simulation must be replicated to capture these variations in the environmental conditions.

[4] Some MS Excel command functions for commonly named distributions include: =BINOM.DIST(), =POISSON.DIS, =NEGBINOM.DIST(), =HYPGEOM.DIST(), =NORM.DIST(), =NORM.S.DIST(), =GAMMA.DIST(), =LOGNORMAL.DIST(), and =CHISQ.DIST(). The reader is encouraged to refer to a probability textbook to ensure distribution assumptions and conditions are fully met in the context of modeling specific real-world events.

For the purpose of this example, we consider 100 iterations – the equivalent of 100 meals, but without the hassle (and 75 cents) for parking.

Each of the first ten iterations displayed in Figure 2.20 reveals a different total time in Column I based upon the Normal and Gamma-distributed random variables displayed in Columns B and C, respectively. Replicating the simulation with 100 iterations provides a distribution of values representing T (total time) in Column I of Figure 2.20 and visualized (using a Pivot Chart) in Figure 2.21.

	A	B	C	D	E	F	G	H	I	J
1	Normal Dist		Gamma Dist				Parking Meter Time = 15 Min per 1 Quarter			
2	Mu	SD	Mu	SD	Alpha	Beta				
3	20	6	8	4	4	2	P (T<	45	) =	0.880
5	Itteration	Random Number for Norm Dist	Random Number for Gamma Dist	Time to Walk from Parking Meter to Table	Meal Time (Order, Wait, & Eat)	Lag time Between Meal & Dessert	Dessert Time (Order, Wait, & Eat)	Time to Walk From Table, Pay Register, & Return to Car	Total Time	Is Time < 45 Min? Yes = 1 No = 0
6	1	=rand()	0.282	2	15.967	1	5.362	4	28.329	1
7	2	0.520	0.211	2	20.304	1	4.697	4	32.001	1
8	3	0.396	0.600	2	18.422	1	8.352	4	33.774	1
9	4	0.928	0.833	2	28.758	1	11.662	4	47.421	0
10	5	0.487	0.626	2	19.809	1	8.630	4	35.439	1
11	6	0.289	0.800	2	16.654	1	11.029	4	34.683	1
12	7	0.637	0.975	2	22.110	1	17.545	4	46.655	0
13	8	0.476	0.924	2	19.638	1	14.224	4	40.862	1
14	9	0.526	0.437	2	20.393	1	6.762	4	34.155	1
15	10	0.944	0.735	2	29.552	1	9.996	4	46.548	0

=NORM.INV(B6,A3,B3)   =GAMMA.INV(C6,E3,F3)   =SUM(D6:H6)

FIGURE 2.20: First 10 iterations of MC simulation output where n = 100.

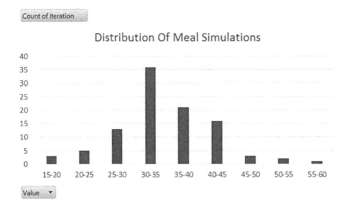

FIGURE 2.21: Histogram of simulated meal times.

Underpinning the decision analytically with key metrics derived from multiple iterations will provide a better-informed decision. For example, we might be interested in the average time to complete the meal over the course of 100 visits. From our data, we can quickly compute the average of the times in Column I to find that, over the course of 100 visits, our average total time was 35.363 minutes. Of course, from basic probability theory, we know that the expected value of the sum of random variables is equal to the sum of the expected values. In our case:

$$E[T] = E[W + M + L + D + R]$$
$$E[T] = E[W] + E[M] + E[L] + E[D] + [R]$$
$$E[T] = E[2] + \mu + E[1] + \alpha\beta + E[4]$$
$$E[T] = 2\,\text{min} + 20\,\text{min} + 1\,\text{min} + 8\,\text{min} + 4\,\text{min} = 35\text{min}.$$

Perhaps we are interested in the probability that in a single visit, the sequence of events will take less than 45 minutes. Observe in cell J3 of Figure 2.20, the probability $P(T < 45) = 0.88$. This is calculated by determining the proportion of the 100 iterations that are less than 45 minutes. Using the conditional "if()" command in column J assists with this calculation by then summing all values in column J and dividing by the total number of iterations. While column J displays seven iterations less than 45 minutes for the first ten iterations of the full Monte Carlo Simulation, for the full simulation of 100 iterations, there were 88 where the total time T was less than 45 minutes.

The user is now armed with two key metrics. The first is the empirical average of 35.363 minutes and the second metric is the 0.88 probability of returning to the car before the parking meter expires in 45 minutes. The patron is now able to make a decision to enter the establishment and eat based upon how risk-averse they are about receiving a ticket should the parking meter expire before their return.

This example provided a cursory look at how to build a Monte Carlo Simulation, and focused on two key metrics and visualization as an overview of what the method is capable of delivering. The next section of this chapter goes into more depth on the Monte Carlo Simulation, covers the theory behind the process, and provides a second example.

2.4.2 Generating Random Variables

When discussing random variate generation above, we presented MS Excel code without explanation of the underlying theory. In this section, we provide an overview on the mechanism to create inputs to the Monte Carlo Simulation. Most often, random variates are generated using a computer algorithm (making them truly only pseudo-random) (Winston & Goldberg, 2004, p. 1155). For brevity and clarity, we will refer to pseudorandom variates as random variates for the remainder of this section. As indicated in Table 2.1, MS Excel has an ability to generate random variates from several common distributions. The problem may arise, however, where the desired distribution is not one of the "common" precoded functions. For this, we rely on a technique known as the inverse transformation method to generate random variates.

As an example to conceptually understand the inverse transformation method, we begin with a cumulative distribution function (CDF) visualized in Figure 2.22 and generated from a Normal Distribution where $\mu = 20$ and $\sigma = 6$. Note that for $x = 20$, $N(20) = 0.5$, implying $P(X < 20) = 0.5$. Suppose further that $N(x)$ has an inverse, $N^{-1}(r)$,

TABLE 2.1: Inverse Distribution Commands for Common Distributions

Distribution Name	MS Excel Command
Beta	=BETA.INV()
Chi Squared	=CHISQ.INV()
F	=F.INV()
Gamma	=GAMMA.INV()
Lognormal	=LOGNORMAL.INV()
Normal	=NORM.INV()
Standard Normal	=NORMS.INV()
t	=T.INV()

where r is a probability. This implies that $N^{-1}(0.5) = 20$. In this example, r is the input, while $N^{-1}(r)$ serves as the method, and 20 becomes the output.

Continuing with this setup, suppose that R is a uniformly distributed random number on the interval [0,1]. It can be proven (Ross, 2003, p. 644) that a random variate X from the continuous distribution N (when N^{-1} is computable) is created by setting $X = N^{-1}(R)$. Repeating this process stochastically implies that for each iteration the analyst will draw a random number, represented graphically along the *y-axis*, and employ the inverse transform method to determine a corresponding random variate along the *x-axis*. For all iterations the analyst may record each random number as input data and each corresponding random variate as output data for statistical analysis. Replicating this process for a desired sample size n generates the key components of the Monte Carlo Simulation.

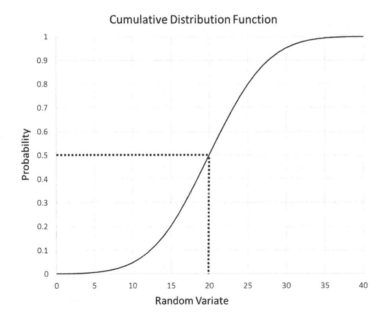

FIGURE 2.22: Cumulative distribution function for $N(x)$.

2.4.3 Optimal Personnel Allocation

Suppose that you are an Operations Research Analyst assigned to a division-level unit with multiple subordinate brigades. Over the course of a week, many serious incident reports (SIRs) occur, requiring a systematic administrative response from the brigade staff. SIRs are segregated into four categories of severity – we will label these as SIR-A, SIR-B, SIR-C, and SIR-D. Within each brigade there are ten personnel trained to respond to SIRs, and on average they collectively spend about 520 hours per year responding to SIRs. For these personnel, this effort is in addition to primary assigned duties.

Currently two or three personnel are assigned to an SIR response team, each team focused strictly on a single SIR category. The current allocation of personnel is based on the frequency of SIR types (SIR categories B and C seem to occur more frequently than A and D) and is shown in Table 2.2.

Recognizing you do not have the authority to alter the division command unit assets, you propose changing the number of personnel assigned to each subordinate team to better distribute the workload and improve the quality of response for each category of SIR. The overarching objective of this analysis is to minimize the total time that a *team* spends on responding to all SIR events. Adjusting the number of personnel assigned to each SIR response team could balance the workload between teams. Here, we assume that the amount of time a team spends on a task is proportional to the time an individual spends on a task. If there are x individuals assigned to an SIR that normally requires a seven man-hour response effort, it should take $\dfrac{7}{x}$ hours for the team to respond. We also assume that the time to complete a task is independent of the person assigned to the task.

We recognize that the number of SIRs and the time that a team must spend on the response are stochastic in nature, but may be informed by data. To determine the distribution of each, we assume the organization has a database of SIRs for which you can query a recent time period reflecting 200 SIRs binned into four unique categories labeled SIR-A, SIR-B, SIR-C, and SIR-D. Figure 2.23 displays eight of 200 notional SIRs cataloged between January 2018 and December 2018 that we use to inform our analysis.

To employ the Monte Carlo Simulation we organize the data (potentially using a Pivot Table), obtaining summary statistics that determine associated parameter values. While this section will highlight each of these techniques, a reference to previous chapter sections will provide additional instructions for each technique. The full problem formulation is shown in Figure 2.24.

TABLE 2.2: Personnel Allocation for SIR Categories

SIR Type	# Personnel Assigned
A	2
B	3
C	3
D	2

SIR #	SIR Category	Brigade	Battalion	Company	Date of Incident
n/a	n/a	n/a	n/a	n/a	1-Jan-18
18-001	SIR-B	6TH BDE	3rd BN	A-Co	2-Jan-18
18-002	SIR-B	2ND BDE	2nd BN	B-Co	2-Jan-18
18-003	SIR-A	5TH BDE	1st BN	C-Co	3-Jan-18
18-004	SIR-B	5TH BDE	2nd BN	B-Co	4-Jan-18
n/a	n/a	n/a	n/a	n/a	5-Jan-18
n/a	n/a	n/a	n/a	n/a	6-Jan-18
18-005	SIR-A	5TH BDE	1st BN	A-Co	7-Jan-18
18-006	SIR-B	5TH BDE	1st BN	B-Co	7-Jan-18
18-007	SIR-C	5TH BDE	1st BN	A-Co	7-Jan-18
18-008	SIR-D	3RD BDE	3rd BN	A-Co	8-Jan-18

FIGURE 2.23: Example data contained in the SIR database.

Data:

200 SIRs with data on type, unit, and date

Decision Variables:

$TeamCount_n \equiv$ Total number of personnel assigned to team n

Stochastic Variables:

$SIRCount_n \equiv$ Number of SIRs in category n.

$TeamTime_n \equiv$ Total time spent by a team on a category n SIR.

$IndTime_n \equiv$ Time spent by individual team member on a category n SIR.

Objective:

$$\text{MIN } TotalTime = \sum_{n=A}^{D} SIRCount_n * TeamTime_n$$

Constraints:

$$TeamTime_n = \frac{1}{TeamCount_n} * IndTime_n$$

FIGURE 2.24: Complete SIR allocation problem formulation.

A recommended first step is to determine the distribution of the SIR arrivals and a second distribution reflecting the amount of time required to respond administratively to each SIR. While we ultimately apply the Gamma Distribution to model the number of SIR events and the Triangular Distribution to model how many hours personnel respond to each event with administrative task requirements, justification for these choices is provided.

While several distributions could be chosen to model the number of SIRs that arrive to the headquarters, we choose the Gamma Distribution (2.1) because it is a flexible when modeling multiple arrivals within a specific amount of time.

$$g(x) = \begin{cases} \dfrac{1}{\beta^a \Gamma(a)} x^{(a-1)} e^{-x/\beta} & \text{if } x > 0 \\ 0 & \text{otherwise} \end{cases} \tag{2.1}$$

where

$$\Gamma(a) = \int_0^\infty x^{a-1} e^{-x} dx \quad \text{for } a > 0$$

The Gamma Distribution provides a two-parameter model (α, β) with a positive support ($0 \leq X \leq \infty$), tolerates for skew, and allows for a system of equations to determine each parameter. As briefly described in the diner example, the mean and variance are computed as follows

$$\mu = a * \beta$$

$$\sigma^2 = a * \beta^2.$$

With some simple algebra and summary statistics (for the number of SIRs each day of the year) on the sample mean ($\bar{x}$) and sample variance (s^2), we can approximate a and β as:

$$a = -\frac{\bar{x}^2}{s^2}$$

$$\beta = \frac{s^2}{\bar{x}}.$$

(2.2)

Once an event occurs, personnel must respond to the event with administrative task requirements. In this scenario, there is limited data documenting how many hours are required to accomplish each of the subsequent tasks. As an alternative, since there are only ten trained personnel in the entire brigade, a quick survey may gain their opinion, as subject matter experts, with respect to the number of hours spent responding to SIRs with administrative task requirements. If you can collect their opinion with respect to the minimal amount of time required, the maximum amount of time required, and the most likely time required to accomplish required tasks, then a Triangular Distribution is suitable for estimating the amount of time spent conducting all administrative task requirements.

The Triangular Distribution is useful when attempting to estimate values given limited information such as outlined in this scenario. Let parameters a, b, and c represent the least amount of time, most likely amount of time, and maximum amount of time required to accomplish each task. Refer to Equation (2.3) and Figure 2.25 to observer the Triangular density function, where the height denoted as $f(b)$ must always equal $2/(c - a)$ since the area under $f(x)$, over the support, must equal 1.

$$f(x) = \begin{cases} \dfrac{2(x-a)}{(b-a)(c-a)} & \text{if } a \leq x \leq b \\ \dfrac{2(c-x)}{(c-b)(c-a)} & \text{if } b < x \leq c \\ 0 & \text{otherwise} \end{cases}$$

(2.3)

In Figure 2.26, we set up the Triangular Distribution Parameter Values with survey response times on the left, the number of personnel currently assigned to each

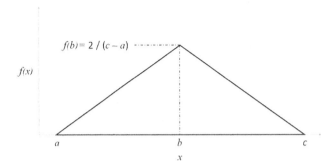

FIGURE 2.25: Triangular PDF.

	Survey Response Times for each SIR Category Administrative Requirement Task				Triangular Distribution Paramater Values		
	Min # of hours to accomplish task(s) due to event (per person)	Most likely # of hours to accomplish task(s) due to event (per person)	Max # of hours to accomplish task(s) due to event (per person)	Current # of Personnel Assigned to Each Team	*a* Min # of hours to accomplish task(s) due to event (per team)	*b* Most likely # of hours to accomplish task(s) due to event (per team)	*c* Max # of hours to accomplish task(s) due to event (per team)
SIR-A	1	3	6	2	0.500	1.500	3.000
SIR-B	4	4	6	3	1.333	1.333	2.000
SIR-C	5	6	9	3	1.667	2.000	3.000
SIR-D	8	12	16	2	4.000	6.000	8.000
Total # of Personnel Trained for SIR Administrative Response Tasks				10			

FIGURE 2.26: Initial triangular distribution parameter input values.

SIR category response team centered in gray, and the corresponding triangular distribution parameter values for each SIR category on the right. For each SIR category, we assumed that the number of hours per team to complete a task is proportional to the number of personnel assigned to each team. As an example, the minimum time for a two-person team to accomplish a task responding to SIR-A is modeled by the Triangular Distribution such that parameter a (0.50 hours per team) is equal to half of one hour per person. Similarly, the maximum time to respond to SIR-B is represented by parameter c (2 hours per team) which is one-third of 6 hours to accomplish a task per person.

Once the appropriate distributions have been chosen (Gamma and Triangular), and the appropriate parameter values for each distribution computed, the next step is to generate the random variates as outlined in Section 2.4.2. For the Gamma, instead of computing the inverse CDF (which is extremely cumbersome to compute manually) we use the built-in functionality for the Gamma Distribution described in Table 2.1. Putting this all together we can create a Gamma-distributed random variable with parameter values determined using the process described in Equations (2.2). Given the initial SIR-A data indicated an expected number of 0.112 SIRs with a standard deviation of 0.316 SIRs each day, and using a system of equations to determine the parameter values $\alpha = 0.126$ and $\beta = 0.890$, we can generate the simulated daily SIR-A requirements using the function: "=GAMMA.INV(rand(),0.126,0.890)".

Unfortunately, MS Excel lacks this built-in capability for the Triangular distribution since there is a unique piecewise function for different a, b, and c parameters. For the Triangular Distribution we must develop the CDF, the inverse CDF, and then manually code the inverse transform function into MS Excel. First refer to Equation (2.4) to observe the associated CDF to the Triangular PDF. Then refer to Equation (2.5) to observe the CDF with parameter values $a = 0.50$, $b = 1.50$, and $c = 3.00$ corresponding to the minimum, mode, and maximum times required for assigned teams to react to SIR-A events. Since the Triangular CDF (2.4) is a piecewise function, then its inverse transform function will also be a piecewise function. Equation (2.6) displays the inverse transform function of the Triangular CDF with respect to p, where p is a random number (probability) drawn from U (0,1) and $F^{-1}(p)$ is the random variate evaluated for each value of p.

$$F(x) = \int_a^x f(x)dx = \begin{cases} 0 & \text{if } x \leq a \\ \dfrac{(x-a)^2}{(b-a)(c-a)} & \text{if } a < x \leq b \\ 1 - \dfrac{(c-x)^2}{(c-b)(c-a)} & \text{if } b < x \leq c \\ 1 & \text{if } x > c \end{cases} \tag{2.4}$$

$$F(x) = \begin{cases} 0 & \text{if } x \leq 0.50 \\ \dfrac{(x-0.50)^2}{(1.50-0.50)(3.00-0.50)} & \text{if } 0.50 < x \leq 1.50 \\ 1 - \dfrac{(3.00-x)^2}{(3.00-1.50)(3.00-0.50)} & \text{if } 1.50 < x \leq 3.00 \\ 1 & \text{if } x > 3.00 \end{cases} \tag{2.5}$$

$$F^{-1}(p) = \begin{cases} a + \sqrt{p * (b-a) * (c-a)} & \text{if } p < \dfrac{b-a}{c-a} \\ c - \sqrt{(1-p) * (c-a) * (c-b)} & \text{if } p \geq \dfrac{b-a}{c-a} \end{cases} \tag{2.6}$$

Note the careful selection of the roots when solving for $F^{-1}(p)$ displayed in Equation (2.6) such that all random variates fall within the support $a \leq x \leq c$. Refer to Figure 2.27 to see how the "=if()" command in MS Excel replicates this piecewise inverse transform function and helps to determine the appropriate output (random variate) within the support $a \leq x \leq c$ from every corresponding input (random variable) drawn from U (0,1).

Figure 2.27 displays the first 18 of 1000 iterations for this Monte Carlo Simulation showing the total time, in hours, for each iteration shown in column "D" as calculated by our objective function. However, Figure 2.27 omits the additional columns of the same spreadsheet reflecting the simulated number of events per day, and the number of hours per day reacting to each event for SIR-B, SIR-C, and SIR-D. Column "E" reflects the number of SIR-A events per day for each iteration of the simulation using the inverse transform function to the Gamma CDF with its associated parameter values α and β previously calculated for SIR-A. Column "F" reflects the amount of hours needed to react to each SIR-A event using the inverse transform function to the Triangular CDF with its associated parameter values a, b, and c also previously calculated. A separate random variable denoted in Columns "G" and "H" is applied to

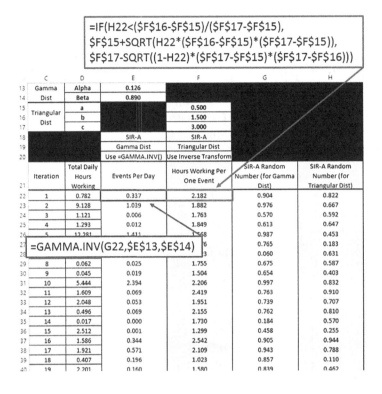

FIGURE 2.27: Monte Carlo Simulation screen shot

separate inverse transform functions to create independence between values displayed in Columns "E" and "F".

Though the command "=rand()" may be embedded within both inverse transform functions, using this command in that fashion adds to the overall runtime of the simulation due to repetitive calculations. To decrease the runtime the user may cell reference a sequence of random numbers listed in columns "G" and "H". This technique also helps troubleshoot potential mistakes. To do this, first generate a sequence of random numbers using the command "=rand()" in columns "G" and "H". Then "copy" the entire sequence of random numbers and "paste special" using "values" from the drop-down ribbon to paste these values on top of the same cells originally programmed with the "=rand()" command. This technique decreases computer runtime, and enables the user to specifically see which random number provides specific output elsewhere on the spreadsheet. The reader must also duplicate columns "E," "F," "G," and "H" to the other columns on the spreadsheet which model the other three SIR categories before calculating the total time in hours displayed in column "D." Column "D" labeled as "total hours" reflects the objective function.

The Monte Carlo Simulation is complete once row "22" is replicated to create 1,000 independent iterations (corresponding to 1,000 simulated years of SIR processing). Reflecting back to the primary reason for conducting a simulation, it is to generate necessary data for statistical inference and analytic insight when lacking real-world data. With a sample size $n = 1,000$, there exists enough data to develop relevant summary statistics with the confidence required for the model output to represent reality.

The average of total (daily) team hours calculated from 1,000 iterations when two personnel are assigned per team to react to SIRs A and D and three personnel are assigned per team to react to SIRs B and C is 1.43 hours. This would correspond to 521.47 hours per year. With 10 personnel in the unit, the model does replicate reality as personnel indicated that they spent around 520 hours per year responding to SIRs. We now have confidence that our model replicates reality and may turn our attention to conduct further analysis.

Recall the objective is to minimize the total number of hours spent reacting to all SIRs by changing the number of personnel assigned to each team responding to specified SIR categories. With this scenario modeled in MS Excel, you can quickly rerun the simulation to gain a separate average number of the total man-hours by changing the team sizes and see which combination yields the best results. As depicted in Figure 2.28, with ten personnel assigned to four separate teams with at least one person per team, this amounts to finding the number of ways to emplace three dividers across nine gaps or $\binom{9}{3} = 84$ options. The command "=COMBIN(9,3)" in MS Excel identifies 84 combinations for which you may not want to compute all results. However, manually changing a few combinations may provide utility in sharing insight with unit leadership.

Instead of manually changing all 84 combinations, we can instead minimize the current objective function subject to the following constraints: you have ten gainfully employed personnel; each team consists of whole members; and at least one person is assigned per team. Figure 2.29 displays the optimization model with all constraints, while Figure 2.30 applies the Solver Tool discussed in previous sections to minimize the objective function by changing team sizes. Changing the team sizes ultimately changes the Triangular Distribution parameter input values. A different average total time computed from 1,000 iterations is ready for your comparative analysis.

Of the two probability distributions modeling this scenario, you have no control over the average number of SIR events per day dictated by the data, therefore the α and β parameters determined for the Gamma Distribution remained fixed to replicate reality. However, changing the number of personnel assigned to each team directly changes the Triangular Distribution parameter value inputs a, b, and c denoted in Figure 2.26 which models the amount of time each team requires to accomplish tasks.

Applying the Solver Tool as displayed in Figure 2.30, you may optimize the objective function subject to the defined constraints. As a result, there is a decrease in the average number of hours from about 521 to roughly 423 man-hours. The decrease of 98 total

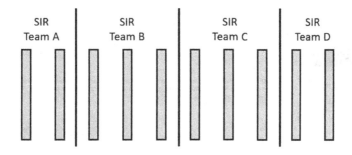

FIGURE 2.28: One of 84 potential ways to distribute SIR teams

Microsoft Excel: The Universal Tool of Analysis

Data:

200 SIRs with data on type, unit, and date

Decision Variables:

$TeamCount_n \equiv$ Total number of personnel assigned to team n

Stochastic Variables:

$SIRCount_n \equiv$ Number of SIRs in category n.

$TeamTime_n \equiv$ Total time spent by a team on a category n SIR.

$IndTime_n \equiv$ Time spent by individual team member on a category n SIR.

Objective:

$$\text{MIN } TotalTime = \sum_{n=A}^{D} SIRCount_n * TeamTime_n$$

Constraints:

$$TeamTime_n = \frac{1}{TeamCount_n} * IndTime_n$$

$$10 = \sum_{n=A}^{D} TeamCount_n$$

$$TeamCount_n = \text{integer } \forall n$$

$$1 \leq TeamCount_n \ \forall n$$

FIGURE 2.29: Complete SIR allocation problem formulation.

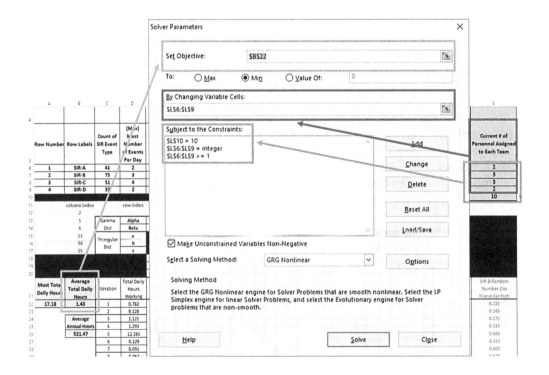

FIGURE 2.30: Monte Carlo Simulation with Solver GUI.

	Survey Response Times for each SIR Category Administrative Requirement Task				Triangular Distribution Paramater Values		
	Min # of hours to accomplish task(s) due to event (per person)	Most likely # of hours to accomplish task(s) due to event (per person)	Max # of hours to accomplish task(s) due to event (per person)	Current # of Personnel Assigned to Each Team	*a* Min # of hours to accomplish task(s) due to event (per team)	*b* Most likely # of hours to accomplish task(s) due to event (per team)	*c* Max # of hours to accomplish task(s) due to event (per team)
SIR-A	1	3	6	1	1.000	1.500	3.000
SIR-B	4	4	6	3	1.333	1.333	2.000
SIR-C	5	6	9	3	1.667	2.000	3.000
SIR-D	8	12	16	3	2.667	4.000	5.333
Total # of Personnel Trained for SIR Administrative Response Tasks				10			

FIGURE 2.31: Optimized triangular distribution parameter input values.

average man-hours is obtained by changing the a, b, and c Triangular Distribution parameter input values denoted in Figure 2.31. The time savings is ultimately accomplished by reassigning one person to the team responsible to reacting to the SIR-A category, three personnel to the SIR-B reaction team, three personnel to the SIR-C reaction team, and three personnel to the SIR-D reaction team.

The decrease of 98 total average man-hours across ten members is not a huge saving per person, but does represent a nearly 20% decrease in the team time required to respond to SIRs. Specific insight to this scenario suggests that the length of response time to each SIR category is more significant than the inter arrival time and the raw number of SIRs taking place in each category. Even though there was less than half of the number of SIRs in category D as compared to category B, the research supported by the simulation unveiled that the length of time responding to each SIR was a significant factor and perhaps previously overlooked by unit leadership when assigning personnel across teams.

Additionally, the Operations Research Analyst now has a tool that helps foster additional analytic insight by changing distribution parameters, or even the total number of personnel assigned to the unit. As an example, the analyst may alter the Gamma input parameters α and β to forecast required personnel changes due to an increase or decrease in future SIR activities. The user may desire changing the Triangular Distribution input parameters for an array of reasons. Perhaps one reason reflects unit training altering individual personnel times accomplishing administrative tasks. At this point, the analyst integrates quantitative analysis to develop analytic results necessary to inform unit leadership and stakeholders not only with sound recommendations for personnel changes, but other possible changes too. The Operations Research Analyst is able to provide additional comprehensive recommendations across several courses of action to stakeholders, reinforced by quantitative analysis and analytical insight.

2.5 Other Uses

Microsoft Excel has been (and continues to be) used in a host of other ways to facilitate analysis conducted in support of operational missions. While not an exhaustive list, what follows are some additional features and examples where the tool has been used to enable analysis. A quick internet search can yield a plethora of tutorials to aid the analyst in conducting the step-by-step process to execute the tasks outlined below.

2.5.1 Macros and Visual Basic for Applications

Microsoft Excel has a useful feature that allows users to automate mundane actions that might be required for preparing periodic reports. Suppose, for example, that each week an analyst receives a dataset designed to satisfy several organizational requirements. To provide an operational environment context, consider a data query that includes all significant activities (SIGACTS) for the operational area that contains dozens of structured fields that might include:

- nature of SIGACT (enemy initiated attack, indirect fire, improvised explosive device, found enemy cache, etc.)
- location of SIGACT (village, province, regional command, lat/long, etc.)
- Date Time Group of SIGACT (to include any updates)
- casualties (coalition, host nation, enemy, or civilian; killed or wounded, etc.)
- amplifying information (type of attack, type of explosive device, etc.)

Further, suppose the analyst is required to develop regionally focused reports that break the data down in several different manners – but in an identical fashion each week as illustrated in Figure 2.32. The analyst must invest a significant amount of time preparing the data and generating common graphics that ultimately takes away from the time to conduct true *analysis*. These specific, identical procedures that must be followed before interpreting the data for insights are often considered by analysts to be a "cost" of the analysis.

In this case, MS Excel's macro and VBA capabilities can be used to automate these repeated tasks. Microsoft describes the macro feature as follows:

> *If you have tasks in Microsoft Excel that you do repeatedly, you can record a macro to automate those tasks. A macro is an action or set of actions that you can run as many times as you want. When you create a macro, you are recording your mouse clicks and keystrokes. After you create a macro, you can edit it to make minor changes to the way it works*
>
> (Microsoft Corporation, 2018a).

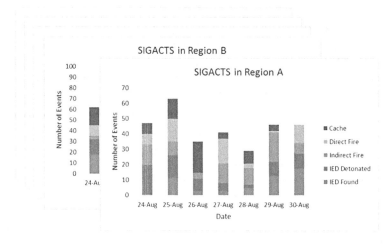

FIGURE 2.32: Example of regionally focused, recurring reports.

Once created, the macro code (recorded in Visual Basic for Applications – VBA) can be modified to suit the needs of the analyst. In truth, an analyst with *any* programming experience in *any* language can likely understand the VBA code generated by recording a macro with little effort. Examples abound where an analyst invested a small amount of time to automate a process using this method to enable having time to spend on *true* analysis and interpretation of the underlying data. While many online references are available to demonstrate this feature through an internet search of "VBA automate PowerPoint," one particularly useful site (including sample code and screen shots) can be found at chandoo.org (Kessler, 2011).

2.5.2 Linking Data

Another key feature of MS Excel is the ability to *link* data tables and charts to PowerPoint presentations. Continuing with the SIGACTS example, suppose that a standard task force product must be generated for key stakeholders each week. Further, suppose that the product relies, in part, on tables and graphics that were automated using macros and VBA described above. Often these presentations are updated by a staff section by copying and pasting the desired content into the presentation. An alternative mechanism to update these presentations is to *link* the PowerPoint slide to the chart, enabling dynamic updating of the slide.

In reality, this is a simple task that simply requires special handling when pasting the graphic for the first time. Specifically, when pasting, an analyst can choose to paste the graphic in a standard manner, or one whereby it links the slide to the graphic contained in MS Excel. This is done in many ways, but one of which is to use the Paste Options to keep source formatting and link the data as shown in Figure 2.33. The key advantage to this is that whenever the underlying MS Excel file is updated, the product is as well.

2.5.3 Additional Add-Ins

To conclude this section, we leave the reader with an encouragement to explore the many freely accessible, and for-fee MS Excel add-ins developed by Microsoft as well as

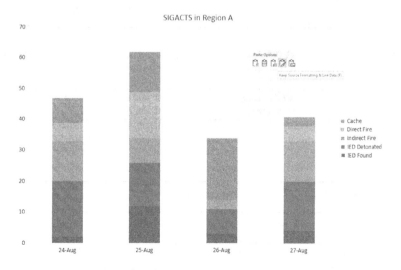

FIGURE 2.33: Linking graphics between MS Excel and MS PowerPoint.

"crowd-sourced" through the analytic community. These add-ins can aid with many tasks to include:

- creating basic geographic heat maps
- visually representing data in charts not organic to MS Excel
- creating network flow diagrams
- forecasting
- additional capability to conduct Monte Carlo Simulation
- additional capability to optimize using alternative algorithms

It is quite possible that if an analyst has an idea of a capability that MS Excel "should" have, that an add-in has likely been created to perform that task.

2.6   Parting Thoughts

This chapter began with a discussion about how MS Excel may not be the desired tool for conducting a particular strain of analysis, but in many cases, and for many reasons, it may be the *only* tool available. Through several examples inspired (at a minimum) by operational experiences by military operations research personnel, we have shown how MS Excel has earned the right to be called the "universal" tool of analysis.

References

Bahm, V., Collins, G., Kollhoff, R., Lindquist, J., Stollenwerk, M., Gaul, M., et al., (2016). *Deployed analyst handbook*. Center for Army Analysis, Fort Belvoir, VA.

Frontline Solvers. (2019). Excel Solver online help. https://www.solver.com/excel-solver-online-help. [Online; accessed 13-March-2019].

Goerger, S. (2017). MORS 2017 annual membership survey. *Phalanx*, *50*(4), 8–9.

Google Maps. (2018). Afghanistan area map. https://www.google.com/maps [Online; accessed 27-June-2018].

Horton, N. J., & Kleinman, K. P. (2007). Much ado about nothing: A comparison of missing data methods and software to fit incomplete data regression models. *The American Statistician*, *61*(1), 79–90.

Kessler, D. (2011). Create PowerPoint presentations automatically using vba. https://chandoo.org/wp/create-powerpoint-presentations-using-excel-vba/.

Microsoft Corporation. (2018a). Create a macro. https://support.office.com/en-us/article/quick-start-create-a-macro-741130ca-080d-49f5-9471-1e5fb3d581a8. [Online; accessed 24-August-2018].

Microsoft Corporation. (2018b). Define and solve a problem by using Solver. https://support.office.com/en-us/article/define-and-solve-a-problem-by-using-solver-5d1a388f-079d-43ac-a7eb-f63e45925040. [Online; accessed 27-June-2018].

National Oceanic and Atmospheric Administration. (2018). Latitude/longitude distance calculator. https://www.nhc.noaa.gov/gccalc.shtml [Online; accessed 27-June-2018].

Ross, S. M. (2003). *Introduction to probability models*. Cambridge: Academic Press.

Tufte, E. R. (1986). *The visual display of quantitative information*. Cheshire, CT: Graphics Press.

Welch, A. (2015). "Golden hour" policy saved hundreds of U.S. troops. *CBS News*. https://www.cbsnews.com/news/golden-hour-policy-decreased-combat-deaths-among-u-s-troops/

Winston, W., & Goldberg, J. (2004). *Operations research: Applications and algorithms*. Thomson Brooks/Cole.

Chapter 3

Multiattribute Decision Modeling in Defense Applications

Roger Chapman Burk

3.1 How Formal Decision Analysis Fits into Military Operations Research

Operations research (OR) generally centers on building a mathematical model of an operational system. In contrast, the branch of OR called decision analysis (DA) centers on building a mathematical model of the preference structure of a decision maker. The goal of DA is to reveal to him or her what decision should be made by eliciting details about what he or she wants to accomplish in a complex situation, perhaps one with many uncertainties and many competing priorities. The purpose of this chapter is to outline how DA fits into military operations research as a whole, and then to present some standard best practices that will cover multiattribute decisions, one of the most common types of military and defense problems that calls for DA.

There are many reasons why decisions can be perplexing enough to require formal analysis, such as poor problem definition, unclear alternatives, uncertain outcomes, and linkages over time. This chapter focuses on *multiattribute decision analysis* (MADA, sometimes called MODA (multi-objective decision analysis) or MCDM (multi-criterion decision making)). MADA addresses problems that perplex because of multiple competing objectives, so that not all can be fully satisfied and some tradeoff must be made. The origins of MADA are traced out in section 2-2 of Zeleny (1982), from its beginning with inquiries in economics and mathematics in the 1940s and 1950s through the development in the 1960s and 1970s of a recognized field of application. The classic text by Keeney and Raiffa (1976, updated 1993) can perhaps be regarded as the culmination of the early development of the field, and their methods are not fundamentally different from those presented in this chapter. Von Winterfeldt and Edwards (1986) include an extensive discussion of multiattribute problems with special emphasis on behavioral aspects in addition to modeling. Watson and Buede (1987) include good material on multiattribute problems in their text on decision analysis. Kirkwood (1997) provides a classic text specifically on MADA. A recent mathematical treatment of the methodology is in Chapters 26–28 of Howard and Abbas (2016).

Professional decision analysts like to imagine themselves working for senior decision makers stumped by multi-billion-dollar life-or-death problems. People like that usually imagine themselves to be pretty good decision makers on their own. Usually they are right – they wouldn't have obtained the position without a record of good decision making, particularly in the military. And yet decision analysts do get hired, at least occasionally. Why is this? The US Army teaches a "Military Decision Making Process" or MDMP (United States, 2014, Chapter 9). This is "an iterative planning methodology to understand the situation and mission, develop a course of action, and produce an operation plan or order." The methodology uses the ideas of alternatives, decision criteria, and weighted measures in a way not fundamentally different from MADA. However, it puts more emphasis on thorough understanding of the problem and on careful and thorough planning and coordination than on actual decision making. It is more rough-and-ready and less careful about technical correctness than a decision analyst might like. This emphasis is appropriate, because in military operations it is more important to make a reasonable decision quickly and carry it out well and vigorously than to make the best possible decision. But there are other non-operational situations where there is enough time to look at the problem carefully. Perhaps there is a lot at stake in the decision and the best course of action is not obvious to all. This can be the case in system acquisition, base location, and strategy selection, for example. These situations call for a different method of decision making. A technically sound, defensible, and transparent decision making process can be of great help, especially when the decision must be publicly defended in front of stakeholders who may not be happy with the result. This is where the methods of formal decision analysis are most useful. Furthermore, knowledge of DA techniques can improve decision making even when the formal methods are not applied. They accustom one to clear and systematic thought when perplexed by a decision problem. The methods of DA are appropriate when the decision maker feels they are needed – and perhaps when an advisor to a decision maker feels that there is danger of making a wrong turn, and that some objective quantitative support will improve the chances of deciding correctly.

Howard and Abbas (2016) describe six linked elements in a good decision:

- the right frame, so that the correct problem is being addressed
- the right alternatives, so that we can be assured that the right one is among them
- the right values, so we base our decision on what's important
- the right information, so we are not misled

- the right reasoning, so we arrive at the correct conclusion
- the decision maker (DM), who accepts the result and carries it out

All six of these are important, maybe even equally important, but this chapter focuses on right reasoning, the model-building part of decision analysis that is most closely aligned with operations research as a field. Frame, alternatives, values, information, and DM will generally be assumed given for our purposes, even though in professional practice it is often difficult to pin down one or more of these. For an extended account of all these aspects of professional decision analysis, one could do far worse than consult the *Handbook of Decision Analysis* by Parnell et al. (2013), which is the product of a great deal of experience in the craft.

Since its goal is to improve decisions, the approach of this chapter is often called *prescriptive* decision analysis, i.e., describing how people ought to make decisions. It is distinguished from *descriptive* decision analysis, which describes how people really do make decisions. Needless to say, the two are often not the same. Nevertheless, an analyst applying DA in the military domain should understand the various biases and psychological traps that bedevil human decision making. So should anyone else. These logical errors are very easy to see in others once you understand them, and always very difficult to see in oneself. A very accessible account of descriptive DA has been provided by Kahneman (2011).

The importance of DA in defense applications has developed over the past 30 years. Corner and Kirkwood (1991) surveyed the OR literature up to 1989 and found no published defense applications of DA at all. Keeney and Raiffa (1993) included no military applications either. However, a later survey (Keefer, Kirkwood & Corner, 2004) found 13 between 1990 and 2001, amounting to 15% of all published applications. The very first one (Buede & Bresnick, 1992) used MADA.

An idea of the current prevalence and importance of decision analysis within military operations research can be had from the following statistic: of 338 technical articles in the journal *Military Operations Research* (*MOR*) for which keywords could be found, 75 of them (22%) included Decision Analysis or a synonym among them. (This covers 1997 through 2018.) A notion of the breadth and scope of the problems addressed by DA can be had from the following: in the most recent five years (mid-2013–mid-2018), 13 DA articles in *MOR* advanced the state of the art in the following areas:

- Jose and Zhuang (2013) investigated the dynamics between a defender and an attacker when considering the issue of technology in a multiperiod sequential game with uncertainty. They developed a modified dynamic programming algorithm to computationally analyze such problems
- Schroden et al. (2013) developed a new non-doctrinal assessment paradigm to assess progress in the complexities of the counter-insurgency in Afghanistan, using a two-tier structure and multiple criteria
- Yoho et al. (2013) combined wargaming with cost analysis to explore concepts of operation and resource allocation decisions for maritime irregular warfare
- Reed et al. (2013) developed a quantitative multiattribute measure of different types of harm to non-combatants or to physical infrastructure, with the idea that it could be used to assess proposed military courses of action
- Colombi et al. (2015) used Value-Focused Thinking and multi-objective decision analysis to evaluate different types of interfaces for open systems architectures
- Kuikka (2015) developed an analytical stochastic model of combat attrition that can provide rules for operational decisions

- Miner et al. (2015) developed metrics for system-of-systems adaptability and reported a method for combining the metrics into a single value that can be used to compare designs
- Dewald et al. (2016) applied an analytical Bayesian technique to combining information from operational tests and simulations to improve confidence in the assessment of the test results
- Coles and Zhuang (2016) introduced a new approach to developing dynamic profiles for terrorist organizations, and analyzing the evolution of terrorist organizations and estimating the likelihood of future attacks, building on aspects of Bayesian probability and multi-objective decision analysis
- Wall and LaCivita (2016) used ideas from microeconomics to develop a quantitative measure of affordability and cost risk and use it with multi-objective optimization to recommend procurement decisions
- Dreyfuss and Giat (2017) analyzed a multi-echelon repair system with the purpose of finding an optimal investment policy for spares, repair capabilities, and travel time, described an efficient algorithm, and bounded its distance from optimality
- Washburn (2018) defined a two-person zero-sum game to model the planting and clearing of land mines, and solved the game to find optimal policies for both sides
- Intrater et al. (2018) used linear regression and a zero-inflated negative binomial regression model to investigate what helps Navy recruiting

Note that six of these 13 articles (Schroden et al.; Reed et al.; Colombi et al.; Miner et al.; Coles and Zhuang; Wall and LaCivita) used multiattribute decision methods. Burk and Parnell (2011) surveyed 24 published military applications of portfolio decision analysis between 1992 and 2010; 18 of them (75%) used multiattribute DA.

The following section develops a technically sound yet reasonably easy to apply method for MADA modeling under certainty, from establishing one's value structure through calculating multiattribute value, along with methods of handling cost vs. value and of dealing with problem stakeholders. The section after that deals more briefly with MADA under uncertainty, in which the result of the decision will be in part the result of chance. Following that, a section discusses sensitivity analysis, an indispensable step for the analyst to show the robustness of his or her result, and another briefly discusses available commercial software for DA applications. The chapter concludes with a real-world example of the application of these methods.

Those who are interested in an applications-oriented text on these topics may find Clemen and Reilly (2001) helpful, particularly Chapters 15 and 16 for MADA. Each chapter also has a good list of references at the end.

3.2   Multiattribute Problems under Certainty

In business, most decisions come down to a single criterion: money (to include proxies for future streams of money, such as market share). This is appropriate because businesses exist primarily for the one purpose of making money. In the public sector, including the military, it is not so simple. There are multiple stakeholders with many different priorities and multiple incommensurable objectives. Money is certainly always an issue, but so is readiness for different kinds of conflicts, troop welfare, training, development of future capabilities, and sometimes national civil priorities like equal opportunity. It is common for a decision to be perplexing because there is no obvious alternative that is strong in all areas, so that some difficult tradeoff must be made. To weigh these incommensurables, the most straightforward technically sound approach is a value model (VM). This requires the following things: a value hierarchy, measures for each objective, ranges of variation for each measure, value functions to translate measure scores to single-attribute value, and relative weights to relate the different criteria to each other. By using this process, analyzing a difficult multiattribute decision can be broken down into a series of easily understood and accomplished steps.

This section starts with a discussion of the need to thoroughly understand what the DM is trying to achieve, recommending an approach called Value-Focused Thinking. Next is an account of how to gather all relevant information together in a consequence table. With these preliminaries accomplished, the following subsection presents the technical details of a value model that can calculate a best alternative, using a notional military example. Subsequent subsections discuss the special role of cost and the particular problems when one has multiple decision makers.

3.2.1 The First Step: Value-Focused Thinking

In a sense, decision analysis always starts with alternatives, since we have to be aware of at least two alternatives before understanding that we are in a decision situation at all. Nevertheless, Keeney (1992, 2008) has given us the insight that in a multiattribute problem, the first order of business is to thoroughly understand one's values, not the alternatives. After all, he reasons, our values are why we are concerned with the decision at all, why we would rather have one outcome rather than another. There may be many problems in which the values are clear and agreed to by all, but those are not generally the ones that get referred for careful analysis. If we have a thorough understanding of what we value in a decision, and make our estimation of value more precise by defining objectives and specifying desired outcomes, using engineering units if possible, we are more likely to make a good choice. A special issue of *Military Operations Research* (Vol. 13, no. 2, 2008) was devoted to Value-Focused Thinking, showing its fundamental importance to MADA.

Another reason to focus first on values is that they will actually help us develop better alternatives. For each objective, consider what type of alternative is likely to perform especially well in that particular area. This will result in a variety of possible alternatives that are different from each other and perform well, at least in one area. Remember that your outcome can be no better than the best alternative that you consider. In order to get better outcomes, it makes sense to develop alternatives that perform well in areas that you value.

The values, objectives, and measures for a decision problem do not come from the analyst. They have to be elicited from the DM and from other stakeholders in the problem. This is not always easy to do, especially when there are multiple stakeholders with different ideas of what is valuable. However, experience shows that it is not always as difficult as one might expect. Stakeholders rarely value opposite things; more commonly, they value the same things, but put them in different orders. It will help if they can be encouraged to move away from thinking about alternatives, where they often already have identified favorites, and toward thinking only of what performance outcomes would be of value.

For purposes of clear communication, it is often a good idea to organize the objectives into a value hierarchy, with related objectives grouped together. Figure 3.1 provides a realistic example of how this might be done for a decision about procuring an imagery surveillance system. There are three main objectives that provide value to the DM. One of the objectives is further broken down into subobjectives. Each subobjective has one or more measures that will be used to assess performance. There is a consistent numbering scheme for clear identification. Collectively, performance on the measures should tell the DM everything he or she needs to know to identify the best alternative – they should be *collectively exhaustive*. Care should also be taken that no two measures actually measure the same aspect of performance – they should be *mutually exclusive*. These ideals are relatively easy to approach, but hard to hit exactly, especially for a complex system. Some complex decision problems require one hundred or more separate measures to satisfy each stakeholder that their needs are represented (Burk et al., 2002; Parnell et al., 2004).

Objectives can be characterized as *fundamental* or *means*. A fundamental objective is something that is desired for its own sake, at least in the context of the problem at hand. A means objective provides a method of accomplishing a fundamental objective. The subobjectives shown in Figure 3.1 would qualify as fundamental; if the subobjectives under "Collect Data" were "Fast Slew Rate" and "High Quantum Efficiency," these would be means objectives. In general, fundamental objectives are preferable because they reflect what the DM really wants, rather than a way of getting it. On the other hand, it is often easier to identify natural and practical measures for means objectives.

The measures at the lowest level of the value hierarchy can be *natural*, *proxy*, or *constructed*. A natural measure is on a scale of engineering units of time, distance, or some other physical quantity, and it directly measures what is of interest, as "minutes" provide a direct natural measure of "Response Time." A proxy measure is also in engineering units, but measures something associated with the desired outcome instead of the outcome itself. For instance, "Weight" might be used as a proxy measure for "Cost," since the weight of a device is often closely associated with its cost (among devices of the same type). If there is no natural or proxy measure that is feasible and practical and cost-effective, recourse may be had to a constructed scale. In the simplest form,

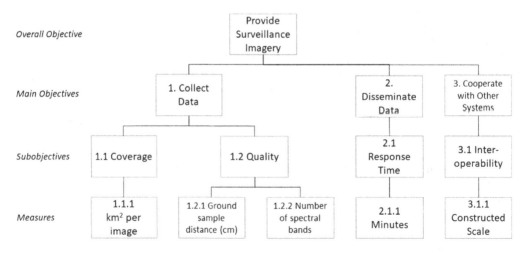

FIGURE 3.1: Example value hierarchy.

this can be no more than an expert's judgment that an alternative is "Excellent" or "Good" or "Fair" or "Minimally Acceptable" on the given objective. However, for public sector decisions with large consequences, a more complete, clear, and detailed scale may be necessary. Such a scale can be constructed by describing the best performance in a few sentences, then describing the minimally acceptable performance in a few sentences, then adding similar descriptions of intermediate points, for a total of perhaps five. Chapter 3 of Clemen and Reilly (2001) includes good discussions of both natural, proxy, and constructed measures, and of fundamental and means objectives.

Some objectives can be very difficult or impossible to measure objectively. This might include an objective like "User Friendliness" in a system acquisition, or "Future Development Potential" in a decision about research direction. There is an understandable tendency among decision makers, perhaps especially among practical-minded military ones, to focus on what is objectively measurable. This can be carried too far. Some things are important even if they are not easily measurable. It is better to have an imprecise measure of an important objective than to ignore it altogether. If no natural or proxy measure is available, then a suitable constructed measure, backed by the judgment of credible stakeholders and subject matter experts, should be developed.

The measures at the bottom of the value hierarchy are what count when it comes to calculating multiattribute value. The higher levels are there to provide organization, clarity, and communication. Needless to say, a sound set of measures, based on a cogent identification of lowest-level objectives, is crucial to the success of the analysis.

3.2.2   Generating Alternatives

An alternative is a possible decision, anything that might be done in response to a problem. As noted above, the decision outcome cannot be better than the best alternative, so it is sometimes worthwhile to spend more time finding a better solution than finding the absolute optimal among a set you already have. Do everything you can to generate a wide variety of good alternatives. Do not start with a group "brainstorming session" – those tend to be dominated by the first ideas presented by the most forceful personalities, even when people are instructed to suspend judgment. Instead, have people do independent thinking first, then combine the results. Perhaps a new and better alternative can be generated by combining features of two others. Give your subconscious time to operate. Challenge your constraints. Look for alternatives that perform particularly well on each objective. Set high aspirations, since you won't do any better than your aspiration level. Ask for suggestions and research what others have done. Do not stop until you have a range of distinctly different alternatives and if possible at least one alternative that does well on each objective, or until time spent on other aspects of the problem (or on other problems) would be more productive.

3.2.3   The Consequence Table

Let us suppose that we have a well-defined problem, a set of objectives for the problem, and a set of measures for the objectives to determine the value of an alternative that we might select. Suppose also that we have a set of alternatives, and that we have evaluated the alternatives on the objectives, using the scales of measure. This gathering of data can easily be the most difficult part of a decision problem, but we will assume it has been done. The next thing to do is to collect the results in a consequence

TABLE 3.1: Example Consequence Table

| | Alternatives | | |
Measures	Albatross	Bluebird	Canary
1.1.1 Coverage (km² per image)	10	20	20
1.2.1 Ground Sample Distance (cm)	11	18	23
1.2.2 Number of Spectral Bands (count)	6	3	3
2.1.1 Response Time (min)	15	20	20
3.1.1 Interoperability (constructed scale)	"Good"	"Fair"	"Minimally acceptable"

table, with one column for each alternative and one row for each measure, as shown in Table 3.1 for three hypothetical alternatives using the value model in Figure 3.1.

The consequence table will be invaluable for communicating results, but sometimes it can also enable us to simplify or even solve the problem. If one alternative is at least equal in performance to another in every measure, and strictly better in at least one, then it can be said to *dominate* the second, which can be eliminated from further consideration (after due diligence to ensure no important consideration has been left out). In Table 3.1, it can be seen that Bluebird does just as well as Canary in Coverage, Number of Spectral Bands, and Response Time, and better in Ground Sample Distance and Interoperability, and so Bluebird dominates Canary, which can be eliminated. Even if there is no strict dominance, it may be that one alternative does much better in some measures, and at most, only slightly worse in others. In the most favorable cases, the best decision can be found by looking at the consequence table, with no further formal analysis required.

3.2.4 Value Modeling

If a multiattribute decision is perplexing enough to require detailed analysis, especially if it is a public decision that must be explained and defended, then value modeling is likely to be the best approach. It offers an objective, quantitative, transparent, logical, and theoretically sound method to estimate the overall value for each alternative (Keeney & Raiffa, 1993; Watson & Buede, 1987; Kirkwood, 1997). It can also provide a lot of insight into the problem.

A value model starts with the attributes or measures identified in the lowest level of the value hierarchy (Figure 3.1). An additive value model has this mathematical form:

$$v(x) = \sum_{i=1}^{n} w_i v_i(x_i) \tag{3.1}$$

where v is the multiattribute value function, x is a multiattribute alternative, n is the number of measures, w_i is the swing weight of the ith measure, v_i is the single-attribute value function for measure i, x_i is the score of alternative x on measure i, and $\sum w_i = 1$, as described in the following sections. Thus, once the measures are identified and the alternatives all scored on them, only two things are necessary to calculate multiattribute value: the set of swing weights w_i and the set of single-attribute value functions v_i. For decisions under certainty with $n \geq 3$, when there is mutual preferential independence among the attributes, we can be assured that there is a set of weights that will reflect the decision maker's true values, so this additive form is very widely applicable. (Preferential independence means that preferences among levels in each attribute are the same regardless of the levels of other attributes. This condition is commonly met. See Chapter 3 of Keeney & Raiffa, 1993, for details.)

3.2.5   Ranges of Measure Variation

The proper first step in creating a quantitative value model from a set of measures is neither determining the swing weights w_i nor determining the form of the functions v_i, but instead, simply determining the range of variation of each measure. This is partly to establish the domain over which each v_i will be defined. More importantly, it is necessary in order to assign meaningful swing weights. In the context of multiattribute alternative comparison, measures and attributes in the abstract do not have value – only particular levels of measures and attributes have value. It is a fallacy to think that one can meaningfully attribute value to a particular measure without considering what range of scores on the measure one is referring to. Doing so may be the most common mistake in multiattribute decision analysis.

There are two ways to establish the range of variation of a measure. The first is to look at the range of alternative performance in the measure. This has the advantage of focusing attention on the real trade space. However, it encourages "Alternative-Focused Thinking" rather than the recommended Value-Focused Thinking. Also, if a new alternative comes to light that scores outside this range it may force you to redo some of your work quantifying the value model. The second method is to use the range from some minimally acceptable level to an ideally desired level. This has the advantage of encouraging development of new high-value alternatives. However, if the ideal level is set too high, it can divert attention to pondering the value of the unattainable. The recommended practice is to find an intelligent compromise between the two methods. One caution: the identified minimum-performance level must represent truly *acceptable* performance. A shift from unacceptable to acceptable within the range of the measure is not consistent with the additive form of Equation (3.1). An unacceptable minimum performance may be the second-most common mistake in multiattribute decision analysis.

3.2.6   Single-Attribute Value Functions

Single-attribute value functions for the n measures are necessary for two reasons. The first is that the different measures are generally in different units, as they are in Figure 3.1, and they must be converted into common units of value in order to be added to each other. The second reason is to account for returns to scale. Value may not increase at the same rate as the measure moves from its low-performance limit to its high-performance limit. For instance, the scale for Response Time in Figure 3.1 might run from 1 minute (best) to 60 minutes (minimum acceptable performance), but most of the value might be lost after 5 minutes, with a much more gradual loss of value after that. The value function for Response Time will capture this effect.

The domain (x-axis) of each function v_i is the range of the measure as identified above, and the range (y-axis) of the function is an interval scale that is conventionally taken to be 0 to 100. Let x_i^- be the minimum-performance score on the measure i, and let x_i^+ be the maximum-performance score. Then $v_i\left(x_i^-\right) = 0$ and $v_i\left(x_i^+\right) = 100$.

If more is better on measure i, then $x_i^- < x_i^+$ and v_i is monotonically increasing, but if less is better, then $x_i^- > x_i^+$ and v_i is monotonically decreasing. (Occasionally the stakeholders will identify an intermediate measure value that is best, so that x_i^+ occurs in the interior of the domain of v_i, which will then be non-monotonic and go to zero at both ends of its domain.)

In the simplest cases, it is acceptable to assume that value increases linearly as measure i goes from x_i^- to x_i^+, so that

$$v_i\left(x_i\right) = 100\,\frac{x_i - x_i^-}{x_i^+ - x_i^-} \tag{3.2}$$

More complex functions have been used to represent nonlinear returns to scale in a way that can be parameterized (e.g., Burk et al., 2002), but these functions put *a priori* constraints on the form of the value function that have no theoretical justification. It is usually just as easy to use the following *midvalue splitting* procedure to elicit a piecewise linear approximation of a nonlinear return to scale (Keeney & Raiffa, 1976). Using measure 2.1.1 in Figure 3.1 as an example, i.e., minutes of response time, suppose that we have established that $v_{2.1.1}\left(60\,\text{min}\right) = 0$ and $v_{2.1.1}\left(1\,\text{min}\right) = 100$. Then we can establish a midpoint in value by asking a suitable group of stakeholders and subject matter experts to tell us when half the value is lost as response time goes from 1 min to 60 min. If the answer is 4 min, we have $v_{2.1.1}\left(4\,\text{min}\right) = 50$. Further questions might ask when half the value is lost in going from 1 min to 4 min, and then in going from 4 min to 60 min, perhaps resulting in $v_{2.1.1}\left(2\,\text{min}\right) = 75$ and $v_{2.1.1}\left(10\,\text{min}\right) = 25$, respectively. These points can be used to define a piecewise linear function, as shown by the solid line in Figure 3.2. If desired, additional points can be elicited to provide additional precision. Of course, this method imposes its own functional form, which also has no theoretical justification, but the method has the flexibility to approximate any reasonable function to any desired precision. These functions measure value, which is inherently subjective, and it will not be possible to achieve engineering precision anyway. The analyst may feel an urge to replace the piecewise linear

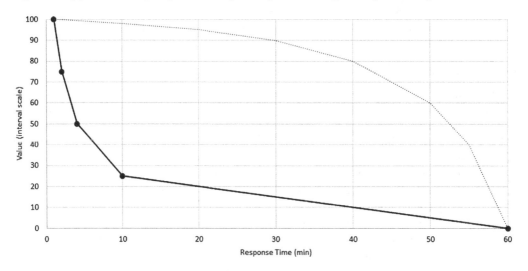

FIGURE 3.2: Example approximate single-attribute value function.

function with a smoother analytical approximation; this temptation should be resisted. The breakpoints are directly elicited and they should be respected. Also, a smooth function will create a spurious impression of precision. The straight lines on the graph will remind us that these results are not to be taken to three significant figures.

The solid line in Figure 3.2, which is concave from above, represents an "increasing return to scale" in which the rate of value accumulation increases as one approaches the best performance. The gray dotted line would represent a "decreasing return to scale." The distinction between increasing and decreasing return to scale should not be confused with the distinction between more-is-better (monotonically increasing) and less-is-better (decreasing) functions. Both the curves in Figure 3.2 show less-is-better. A more-is-better curve with an increasing return to scale would be increasing to the right and concave from above. It is also not uncommon to see curves that are not consistently either increasing or decreasing in return to scale, but rather have an S shape, with most of the value increase coming in the middle of the x_i range.

The single-attribute value functions $v_i(x_i)$ and the multiattribute value function $v(x)$ are sometimes improperly called "utility" functions. This comes from the seminal work of von Neumann and Morgenstern (1947), who used the term "utility function" for functions elicited using reference lotteries for decisions under uncertainty. Utility functions are used mathematically to evaluate alternatives in a way that is very similar to value functions, but they are elicited differently. They capture both return to scale and attitude toward risk. Value functions capture only return to scale.

3.2.7 Attribute Swing Weights

It remains to elicit the swing weights w_i. These relate the value to the stakeholders of the swings from x_i^- to x_i^+ on each of the attributes. Many good ways of doing this have been proposed, including using visual or tactile aids such as physical coins or tokens to distribute among the objectives, bar or pie graphs to partition, or imagined balance beams. The following method has been found to practical, easy to understand, and as precise as the domain allows. It is also technically valid.

The first step is relatively easy: rank order the n swings. This requires eliciting answers to questions of the form, "Which is of more value: a swing from x_i^- to x_i^+ on attribute i, or a swing from x_j^- to x_j^+ on attribute j?" By repeating this question with different pairs of attributes, one can rank all of them from the one with the most important swing to the one with the least. For an initial quick look, or when time is short and/or consequences small, approximate surrogate swing weights can be derived directly from the ordinal information. A number of ways to do this have been proposed, including rank sum (Jia, Fischer & Dyer, 1998), rank reciprocal (Stillwell, Seaver & Edwards, 1981), sum reciprocal (Danielson, Ekenberg & He, 2014), equal ratio (Lootsma 1999), rank order centroid (Rao & Sobel, 1980), and rank order distribution (Roberts & Goodwin, 2002). An empirical study (Nehring, 2015) found that the rank sum method is closest on average to matching weights derived from elicitation. Using this method,

$$w_{(i)} = \frac{n+1-i}{\sum_{j=1}^{n} j}, \quad i = 1 \ldots n \tag{3.3}$$

where $w_{(i)}$ is the ith largest swing weight and n is the number of objectives.

In most cases it will be better to go ahead and elicit the attribute swing weights themselves, rather than rely on surrogates. This will increase precision and reduce opportunities to question the results. To do this, start by assigning a relative non-normalized swing weight $\omega_{(1)} = 100$ to the most important attribute. Then ask the stakeholders to give the relative importance of the second-most important swing, yielding perhaps $\omega_{(2)} = 85$. Continue down the ordered list of attributes until a relative weight is elicited for $\omega_{(n)}$. Finally, normalize the weights so that they sum to 1:

$$w_{(i)} = \frac{\omega_{(i)}}{\sum_{j=1}^{n} \omega_{(j)}}, \quad i = 1 \ldots n \tag{3.4}$$

It is important to remember at each stage of the elicitation process that what are being compared are specific swings from low-performance values (x_j^-) to high-performance values (x_j^+) in each measure. We are not comparing the objectives in any abstract sense, or the values of swings from worst imaginable to best possible.

If the number of measures n is very large compared to the patience of the stakeholders, it may be necessary to simplify the elicitation process by asking for local weights at each level of the value hierarchy. For instance, in Figure 3.1, measure 1.2.1 would be weighted against 1.2.2 only, perhaps yielding $w_{1.2.2}^{\text{loc}} = 0.4$ (and consequently $w_{1.2.1}^{\text{loc}} = 0.6$). Then subobjectives 1.1 and 1.2 can be compared, perhaps yielding $w_{1.2}^{\text{loc}} = 0.3$, and main objectives 1, 2, and 3 can be compared, perhaps yielding $w_1^{\text{loc}} = 0.5$. Then the *global* weight for measure 1.2.2, i.e., the weight used in Equation (3.1), will be the product of the local weights up the hierarchy: $w_{1.2.2} = 0.4 * 0.3 * 0.5 = 0.06$. This procedure will ensure that the global weights sum to 1, as they are required to do, while simplifying the required elicitations by asking for comparisons between related measures only. A disadvantage is that it may be hard for the stakeholders to keep in mind what the local weight really means at a higher level in the hierarchy: the value of a simultaneous swing of all included measures from their minimum-performance levels to their maximum-performance levels.

3.2.8 Calculating Multiattribute Value

Once each w_i, v_i, and x_i is in hand, it is straightforward to calculate the multiattribute value of an alternative x using Equation (3.1). The result can be interpreted as the relative value of the alternative on a 0 to 100 scale, where 0 is the value of a hypothetical alternative that performs at the minimum acceptable level on each measure, and 100 would be the value of an alternative that performed at the highest level on each measure. (Note that a multiattribute value score of 0 does not mean that the alternative has zero value in the ordinary or absolute sense – it means that its increment of value over the minimum acceptable is zero.)

Figure 3.3 shows how calculations might be completed for the two nondominated alternatives in Table 3.1, based on the value model in Figure 3.1. The definition of the value function has been completed with notional value functions and attribute swing weights. The numerical results can also be presented insightfully in a stacked bar chart of weighted value, as shown in Figure 3.4. Note that Albatross scores better than Bluebird in four out of the five measures, but Bluebird still scores higher overall.

Attributes:	1.1 Coverage	1.2 Quality		2.1 Response Time	3.1 Inter- operability
Measures:	1.1.1 km² per image	1.2.1 Ground sample distance (cm)	1.2.2 Number of spectral bands	2.1.1 Minutes	3.1.1 Constructed Scale
Value functions:					
Swing weights:	0.35	0.09	0.06	0.3	0.2
Scoring for Albatross:	*10 km²*	*11 cm*	*6*	*15 min*	*"Good"*
	$v_{1.1.1}(10) = 33$	$v_{1.2.1}(11) = 95$	$v_{1.1.2}(6) = 50$	$v_{2.1.1}(15) = 23$	$v_{3.1.1}("G") = 75$
	$v(Albatross) = (0.35)(33) + (0.09)(95) + (0.06)(50) + (0.3)(23) + (0.2)(75) = 45$				
Scoring for Bluebird:	*20 km²*	*18 cm*	*3*	*20 min*	*"Fair"*
	$v_{1.1.1}(20) = 100$	$v_{1.2.1}(18) = 40$	$v_{1.2.2}(3) = 10$	$v_{2.1.1}(20) = 20$	$v_{3.1.1}("F") = 40$
	$v(Bluebird) = (0.35)(100) + (0.09)(40) + (0.06)(10) + (0.3)(20) + (0.2)(40) = 53.2$				

FIGURE 3.3: Example multiattribute value calculations.

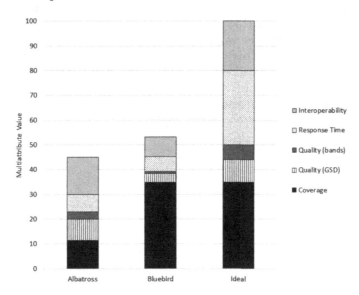

FIGURE 3.4: Example stacked bar chart of weighted value.

This is because its substantial advantage in the most important measure has been judged by the stakeholders (via the elicitation for value functions and swing weights) to be more important than all Albatross's advantages in all the other attributes put together. It is also noteworthy that neither alternative scored particularly well in Response time, which might suggest a search for another option that does well on that attribute.

Of course, a conclusion reported as "Bluebird beats Albatross by a score of 53.2 to 45" will not convince any decision maker. The conclusion has to be translated back into the terms in which the problem was posed. In this case, it might be phrased like this:

> Bluebird has an overwhelming advantage in coverage performance. Albatross has some advantages of its own, but they are all less important and together do not take the advantage away from Bluebird. Albatross's biggest advantages were in ground sample distance and interoperability, but the stakeholders found these differences in performance to be much less important than coverage for this decision.

3.2.9   Cost versus Value

Monetary cost is often an important consideration, in public as well as private sector decisions. This can be simply acquisition cost, as in an off-the-shelf procurement, or it can include various development, testing, acquisition, deployment, maintenance, upgrade, and finally, disposal costs over a long period of time. Often these can be combined into one net present value, using some annual discount rate for future costs. (In commercial applications, net present value would include future revenue streams.) To include these costs in a multiattribute decision, one common approach is simply to include cost as an attribute. However, it is usually more insightful to keep track of cost separately from performance value and to plot the alternatives on a cost vs. value scatterplot, as in Figure 3.5. This will reveal when a small increase in expenditure will yield a large increase in performance value, and when a large cost savings can be had with little loss of performance. Also, it will reveal when one alternative has better performance and lower cost than another, thus dominating it. The set of nondominated alternatives define the *efficient frontier* of the plot.

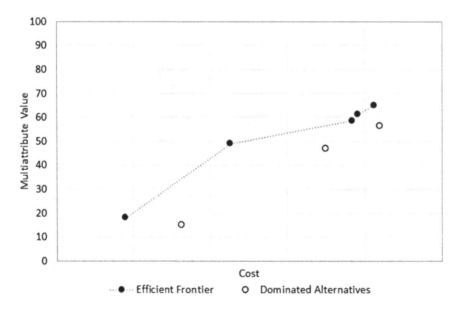

FIGURE 3.5: Cost vs. value scatterplot.

3.2.10 Value Elicitation, Multiple Decision Makers, and Problems with Voting

Decision analysis is about modeling preferences, and it requires eliciting preferences and values from the decision maker and from those with a legitimate stake in the outcome of the decision. The decision will only be as good as the information that the modeler can get from these stakeholders. This is not always an easy process. The stakeholders may not be accustomed to thinking in decision-analytic terms of measures, scales, weights, value functions, and so forth. However, these concepts are intuitive enough that people usually pick them up fairly quickly. The analyst's job is to lead them into the correct frame of mind while eliciting from them what they know. It is not always as hard as one might expect.

Sometimes the situation is complicated because more than one DM must agree on the decision. Even if there is one person with the formal authority to make the decision, in practice, he or she may have to get agreement, or at least acquiescence, from a group of stakeholders. Somehow the values of all these people have to be represented in one value model. The ideal approach is to get all the stakeholders together in one room with a trained decision analyst to lead them through the process, and have them develop and agree to a consensus value hierarchy, set of measures, value functions, and weights. The discussion is likely to lead to improved understanding and ultimately to a better value model. Unfortunately, people at the level where they can be credible sources for value on a major decision will be very busy and it will be hard to get them to commit the time required for this process. This may lead to repeated rounds of interviews, smaller meetings, drafts, comments, and revisions before consensus is reached.

One hopes that a consensus value model can be agreed on, just as one hopes the US Congress can agree on what laws the country needs. Sometimes it seems that consensus will be impossible because key stakeholders just have contrary values. In the military domain, this probably does not happen as often as it does in the public sector in general, since all stakeholders have the common ultimate goal of defending the country. They only disagree about the best means to that end. Stakeholders may differ about which objective is more important, but it is seldom the case that one thinks an objective is worth 95% of the problem and another thinks it is only 5%. More often it is more like 60% vs. 40%. If two conflicting stakeholders have the time to sit down and discuss their values and their justifications for them, there is hope that they will gain more mutual understanding and arrive at a compromise. This is especially true if the discussion takes place in the presence of a mutual superior. Sometimes initial estimates of value have some character of negotiating positions rather than carefully considered and firmly held beliefs.

One obvious approach to the problem of multiple stakeholders with different value structures is to simply ask them to vote on the alternative models. This approach should be viewed with extreme caution. The problem is that there is no way to aggregate votes on three or more alternatives that is without serious flaws. For instance, one might try "instant runoff," in which each voter ranks the alternatives from most preferred to least. Then the alternative with the fewest first-place votes is eliminated and the votes distributed to the voters' second choices. This is repeated until one alternative has a majority. But suppose 30% of the voters prefer A to B to C, 36% prefer B to A to C, and 34% prefer C to A to B. On the first ballot A is eliminated, and on the second B wins.

This violates majority rule, since 64% of the voters preferred A to B. Another approach using voter ranks would be to weight the votes in a Borda count: first place among n alternatives receives n points, second place receives $n - 1$, and so forth. This method can also violate majority rule. If there are five voters and two prefer A to B to C, two prefer B to C to A, and one prefers C to A to B, then B wins, though 60% of the voters prefer A to B. In fact, Arrow (1950) proved that there is no way to aggregate votes for three or more alternatives that does not have a flaw like this or one just as serious. This is known as "Arrow's Impossibility Theorem." The conclusion is that voting is a poor way to make a group decision if there are more than two alternatives.

The best approach when there are multiple stakeholders with apparently different values is to facilitate a discussion of what's important in the problem until they come to consensus on a value model. This can be difficult, but in a military application it is not hopeless because there is an underlying common goal. If agreement on one value model seems beyond reach, different value scores can be calculated based on the values of different stakeholders and the results compared. Sometimes there is little difference in the final results because the differences in multiattribute value are caused mostly by differences in performance measure score (x_i) and not by the weights (w_i) and value functions (v_i). Analysis of where the important differences lie may lead to better understanding and perhaps consensus. If time is short and a decision must be made (e.g., military operations, or captaining a ship at sea), a single decision maker must be identified. If all but two alternatives can be eliminated, voting is safe because Arrow's theorem no longer applies.

3.3 Multiattribute Decisions under Uncertainty

In previous sections we have assumed that the outcomes of all possible decisions are known with certainty. In defense applications that is often not the case. Chance reigns on the battlefield, and it is also often prominent before and after that point. Sometimes system acquisition decisions must be made before the technical outcome of system development is known with perfect confidence. Force development and training decisions have to be made before the future political state of the world can be known, and before the armed forces can know what kind of war they will be asked to fight. This aspect of DA is worthy of a chapter of its own, and indeed many chapters. This section will identify some of the main concerns when analyzing a multiattribute decision under conditions of uncertainty: expected value decisions, subjective probability elicitation, expected utility decisions, and limitations of the additive model. It will then give a recommended approach.

One crude approach to uncertainty is to simply include "Risk" or "Uncertainty" as one of the attributes, to indicate more value from alternatives that have less uncertainty. This does not do a very good job of capturing the effects of having multiple possible outcomes from a given decision. A better approach is the classical way to lay out a decision under uncertainty: putting it into a decision tree, such as the one in Figure 3.6. Squares in a decision tree represent choices and circles represent chance events. Outcome probabilities for the chances are given to the right of the corresponding circle. Final consequences are shown on the right. For simplicity, we will begin by supposing there is a single decision criterion that is to be maximized. This criterion could be additive multiattribute value as in Equation (3.1) (subject to some important

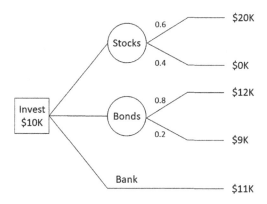

FIGURE 3.6: Decision tree for a simple investment decision.

caveats which will be explained below), but we will start by using dollar return for illustration. The decision tree makes it easy to determine the best course of action according to expected value (EV): starting from the right, attribute to each chance node the expected value of its possible outcomes, and to each decision node the value of the option that leads to the greatest EV. In Figure 3.6 that would be the Stocks option, since the expected value is $12K, versus $11.4K for Bonds and $11K for Bank.

Note that selecting the Stocks option in Figure 3.6 could result in losing everything because of bad luck. In ordinary speech one might say that picking Stocks was therefore a bad decision if that had happened. From an analytical point of view that is not correct. A good decision can lead to a bad result because of bad luck, just as a bad decision can lead to a good outcome because of good luck. This distinction between good decisions and good outcomes is one of the most important things to understand about decision analysis under uncertainty.

A complicated decision situation can involve a number of decisions under uncertainty made over time, with the options and probabilities at later decisions determined in part by the outcome of earlier chance events. For instance, a major system acquisition may involve a decision on what technologies to pursue, a decision on what kind of system to develop based on the result of technology development, a decision on which system configuration to acquire based on system development results, and so on. In financial matters, an outcome may include a regular stream of income or expenses for a period of time in the future (these are usually discounted by a fixed factor for every year in the future before they happen, to give a net present value, or NPV, to be used for current decision making). There may be an opportunity to structure a set of future decisions by buying a financial option, pursuing a certain research project, or otherwise incurring a certain but limited current cost in order to create the possibility of a very advantageous decision in the future. A complex situation like this can be captured in more complex decision tree, such as Figure 3.7.

However complex, a decision tree is always straightforward to evaluate to get the best decision based on expected value, starting from the right and working to the left, assigning an EV to every node along the way. This will also show the best decision to make at every square decision node, should the result of the various choices and chances result in arriving at that node. Decision trees are easy to draw, understand, and evaluate, and they accurately and clearly capture the effects of chance events and

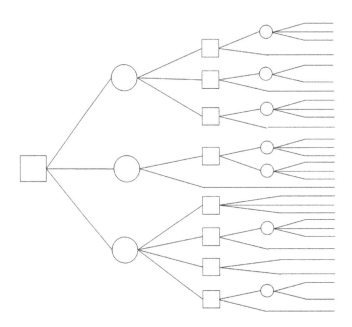

FIGURE 3.7: Complex decision tree.

decision interleaved over time. However, they can become huge and unwieldy as the situation becomes more complex (this can be ameliorated to some extent with software; see below). They also can include much repetition, e.g., when the same chance is encountered for any decision at a given node. Nevertheless, a decision tree is usually the first step in understanding a complex decision under uncertainty.

Once the structure of a decision tree is established, the most troublesome remaining task is usually determining the probabilities. Probabilities can be needed for things that haven't happened yet, or for things that have happened, but for which we don't yet know the outcome. The former might be a future political change or the outcome a research project; the latter might be the existence of extractable oil at a given drilling site. Both types of probability can be treated the same way. In the best situation, objective data may be available that allow one to estimate a probability. For instance, a geologist may know of 35 times when wells were drilled in certain geological conditions, and in 13 of them oil was found, giving an estimate of 37% probability of striking oil at a similar new drill site. More often, subjective probability estimates must be elicited from the decision maker or from acceptable subject matter experts. Chapter 8 of Clemen and Reilly (2001) provides some specific methods.

The method of deriving a best decision described above will result in getting the highest possible expected value, but when is it best to decide according to EV? For the sort of common decisions that are made every day or every week, it is usually best to decide according to EV. The Law of Large Numbers assures us that the sum of the outcomes of all the chancy decisions will tend to approach the sum of the expected values, so by maximizing EV we maximize our total result in the long run. However, for rare or once-in-a-lifetime decisions the Law of Large Numbers does not apply and expected value may not be the best criterion. For instance, in Figure 3.8 almost everyone would choose option B, though option A offers much better expected value. In situations like

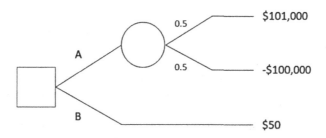

FIGURE 3.8: A decision best made not on expected value.

this, decision analysis cannot tell us which option to take without more information on the attitude of the decision maker toward risk.

Attitude toward risk can be elicited from a DM in the form of a utility function like those shown in Figure 3.9. The *x*-axis is the range of possible outcomes and the *y*-axis is an interval scale in arbitrary units that is traditionally termed "utility," and is usually scaled from 0 (for the least preferred outcome) to 100 (for the most preferred). A utility function looks very like a single-attribute value function (Figure 3.2), but it is elicited very differently. Reference lotteries must be used to capture attitude toward risk. For instance, to develop a utility function over the domain in Figure 3.9, one would ask, "What fixed amount would you be willing to exchange on an equal basis for a 50/50 chance of winning $101,000 or losing $100,000?" The answer to this question, called a *certainty equivalent*, provides the Utility = 50 point on the utility function. If the answer is $500, then the DM is going by expected value and the point falls on the solid line in Figure 3.9. If the answer is −$34,000 (i.e., the DM would pay $34,000 to be excused from the lottery), then the Utility = 50 point falls on the dotted line, and

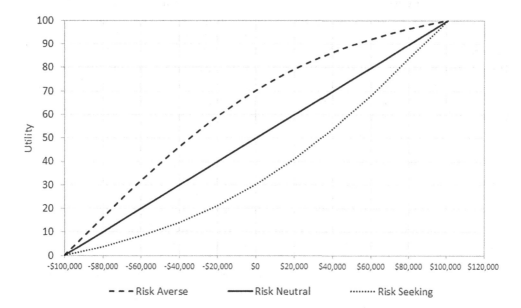

FIGURE 3.9: Example utility functions.

the DM is risk-averse. If the answer is \$35,000 (i.e., the DM would want that amount of money to forego a chance at the lottery), then the point falls on the dotted line and the DM is risk-seeking. Follow-up questions to find the certainty equivalents for 50/50 lotteries between outcomes of known utility can fill in more points on the utility function to any desired degree of precision. Needless to say, it can be difficult to get decision makers or stakeholders to commit the time required to carefully consider these theoretical and high-stakes lotteries and give truly well-thought-through certainty equivalents. Nevertheless, some such elicitation is required to capture the decision maker's utility function. It should also be remembered that utility functions are specific to a given decision maker and specific to a given decision situation.

Once the utility function is found, it is straightforward to determine the recommended decision in any situation represented in a decision tree. Simply replace each outcome with its utility, and evaluate the tree based on expected utility rather than expected value. For instance, if for the situation in Figure 3.8 we set U(−\$100K) = 0, U(\$101K) = 100, and elicited a utility function such that U(\$50) = 70, we would find that the utility of the sure thing exceeded the utility of the gamble, and so recommend taking the \$50, as almost any person would do.

It is perhaps astonishing that we can be sure that in principle there is implicitly in the mind of any coherent decision maker facing any chancy problem a utility function such as that shown in Figure 3.9, and that all we have to do is elicit it and put it on paper, and that the decision based on expected utility will be the correct one. This result is due to von Neumann and Morgenstern (1947). The proof depends only on a few axioms that almost any DM will accept: ranking (any two outcomes are either equally preferred or one is preferred to the other), transitivity (if lottery A is preferred to B and B to C, then A is preferred to C), and a few others more technical to state but equally hard to deny.

One might be tempted to create an additive multiattribute utility function in the same form as Equation (3.1), only substituting utility functions $u_i(x_i)$ elicited via reference lotteries for the single-attribute value functions $v_i(x_i)$. Unfortunately, this straightforward approach can only be used with extreme caution. Any multiattribute utility function of this form implies that the DM will be indifferent between (1) a 50/50 lottery between an alternative that is best on two attributes and another alternative that is worst on the same two, and (2) a 50/50 lottery between attributes that are best on one and worst on the other. Many decision makers would prefer (2) in a choice like this. It can be shown that an additive model like Equation (3.1) will accurately capture a decision maker's preference only if his or her preference structure shows *additive independence* in the situation, meaning that preferences between lotteries in one attribute are not affected by changes in lotteries in other attributes. This is much stronger than the preferential independence that is sufficient to justify the additive model under certainty. In fact, additive independence often seems not to hold in practice (von Winterfeldt & Edwards, 1986). There are functions more complex than the additive utility function that can be used to model DM preference without assuming additive independence, but they require more (and more difficult) elicitations from the DM. His or her patience will likely be taxed enough by the elicitation required for the additive model. Also, the results of modeling based on reference lotteries may well be harder for the DM to understand, accept, and be willing to take action on.

So what is a decision analyst to do to assist a decision maker with a multiattribute decision under uncertainty? In most cases it is probably best to start with the standard value model as developed under conditions of certainty, even though it implies dubious assumptions of risk neutrality and additive independence. Then if uncertainty seems to be a big part of the problem, ask a few test questions about reference lotteries to see what the situation is. Often risk-aversion, non-additive utility, and so forth are relatively small, at least compared to the main effects of value decomposition and alternative performance. In general, start with a simple model and only add complexity when it becomes clear that it is needed to capture the essential features of the problem. A complex model can quickly become too difficult to explain or understand, at least for non-specialists. If the decision maker cannot understand the model, he or she is unlikely to accept its results.

3.4   Sensitivity Analysis

The basic idea of sensitivity analysis (SA, also known as post-optimality analysis) is to re-accomplish a finished value calculation after changing one or more of the parameters that went into the model, in order to see if the result changes. Let us suppose that a decision analysis has resulted in a recommended decision. Almost inevitably, some of those affected by the decision will dislike the recommendation. They will examine the modeling that supports it, looking for aspects they can question so as to throw its soundness into doubt. Since decision analysis necessarily relies on elicited subjective values that can never be known with engineering precision, they will find them. For this reason, it is always a good idea to do some sensitivity analysis to see to what extent the final recommendation depends on debatable inputs. SA also provides insight into what really matters in the problem, and is usually worth doing for that reason alone.

In the MADA example that resulted in recommending Bluebird over Albatross (Figure 3.3 and Figure 3.4), the multiattribute value scores for the two alternatives depended on three types of inputs: the swing weights w_i, the single-attribute value functions v_i, and the performance scores x_i for the two alternatives. SA can be done on any of the three, but swing weights are the most common subjects because they can easily be seen to determine the answer and are wholly subjective in nature. The single-attribute value functions do not have so obvious an effect on the result, and the performance scores are generally based on engineering data. For decisions under uncertainty like Figure 3.6, elicited subjective probabilities are also prominent candidates for SA. This section will describe two methods for displaying SA results when one input varies at a time (*one-way* sensitivity analysis), using swing weights and probabilities as the subjects of investigation. For more detail on these methods, and other methods (including two- and three-way SA), see Chapter 5 of Clemen and Reilly (2001).

In the Albatross vs. Bluebird analysis, let us suppose that the Albatross program manager is distressed that his system was not selected. He protests that since it performed better on four of the five measures, it should be showing up better in the value model. He concludes that something must be amiss with the subjective swing weights. To address his concern, one can construct a *tornado diagram* showing the results of varying each of the swing weights by (say) ±0.1, as shown in Figure 3.10 (the name and concept of tornado diagrams are from Howard, 1988). This is constructed by recalculating the multiattribute value of Bluebird with each swing weight above and below its baseline value, and showing the results as horizontal bars from the lower resulting value to the higher, sorted with the longest bar at the top. For instance, the baseline swing weight of Coverage is 0.35, so the low value would be 0.25.

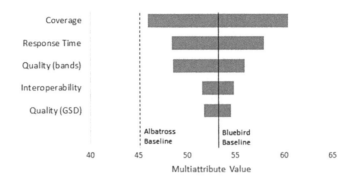

FIGURE 3.10: Tornado diagram example.

The other swing weights need to be adjusted to maintain the total of 1, while keeping the same proportions to each other, so the weights 0.09/0.06/0.3/0.2 become approximately 0.104/0.069/0.346/0.231. When these weights and Bluebird's single-attribute value scores are put into Equation (3.1), the resulting multiattribute value is 46, which is the left-hand limit of the Coverage bar in Figure 3.10. When all the bars are plotted with the longer ones on top, they form a tornado shape centered on the baseline value, in this case 53.2. When they are compared to Albatross's baseline value of 45, it is easy to see that the $w_{\text{Cov}} = 0.25$ case is the only one that comes close. This can focus the discussion on what Albatross's score is with those weights (it is ~46.8) and whether it is really reasonable to consider such a low weight on Coverage. The swing weights should have been determined by consensus in an elicitation process that included all stakeholders, including the Albatross program manager, and hopefully it will not be judged credible that the elicitation was wrong by so great an extent. This will justify the decision in favor of Bluebird.

A tornado diagram is an excellent way to show the results of sensitivity analysis on many parameters varied one at a time. To look in more detail at the effect of varying one parameter, a *sensitivity graph* is appropriate. Such a graph shows alternative scores as a function of a single parameter. Suppose in the problem shown in Figure 3.6 one wanted to explore the impact of uncertainty in the probabilities. One could plot the expected value of each alternative as a function of probability, as shown in Figure 3.11. This example shows the *x*-axis extended to cover all possible values of probability, which has the advantage of guaranteeing that no possibility will be left out. The baseline estimates of probability for both Stocks and Bonds are shown with vertical lines. It is easy to see that the selection of the best alternative is strongly sensitive to the probability estimates, particularly in the case of Stocks. Even if one were willing to decide based on expected value, this graph might make one pause to consider how much faith one really had in one's estimates of probability.

Tornado diagrams and sensitivity graphs are only two of a large number of ways to present sensitivity analysis. The method used should be selected to explore whatever is most controversial, most doubtful, or most likely to change the result in the analysis. The purpose of SA is to give confidence that the recommended decision is unlikely to change as the result of any reasonable change in the input parameters. Failing that, SA will show which input parameters should be investigated more closely to get firmer estimates of their correct values.

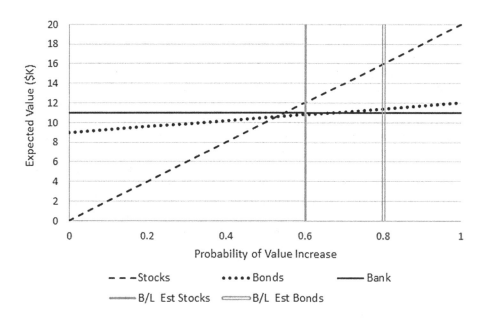

FIGURE 3.11: Investment decision expected outcomes as a function of probability of success.

3.5 Software

All the quantitative techniques discussed in this chapter can be fairly easily implemented using standard spreadsheet software. However, there are dozens of commercial software packages available that can make the implementation even easier, provided one is careful to get a package that has the capabilities needed for the problem at hand. The magazine *OR/MS Today* publishes a very useful biennial survey of decision analysis software in alternating October issues (e.g., Oleson, 2016; Amoyal, 2018). This provides capability, pricing, training, and vendor information on a wide variety of DA tools.

3.6 An Example: Army Base Realignment and Closure

The practical application of these methods was demonstrated in the US Army's development of a multiattribute military value model for Army bases in response to Congress's call for a round of Base Realignment and Closure (BRAC) in 2005 (Ewing, Tarantino & Parnell, 2006; Center for Army Analysis, 2004). This effort required an assessment of the value of Army bases that involved many disparate attributes: maneuver space, firing ranges, environmental impact, training facilities, expansion opportunity, mobilization and deployment capability, accessibility, logistics, local workforce, and so on. The decision to re-base Army units and close installations also had very significant impacts on many stakeholders, most notably all parts of the Army, the local communities near losing and gaining bases, and their political representatives. This made it a good candidate for formal multiattribute decision analysis. This example is interesting because of its scope and importance, and also because it extends the approach given in this chapter in a few ways to account for particular issues in BRAC. It is also of interest because research is underway now to generalize and improve the model in case Congress calls for another round of BRAC.

The project began with a research effort, including document reviews and interviews with senior leaders. Relevant Army, Department of Defense (DoD), Joint Service, and other documents were reviewed and summarized. Hour-long interviews were conducted with 36 senior Army leaders, following a carefully designed protocol. Based on this research, a qualitative value hierarchy was developed that divided overall military value into six capabilities that were important to an installation's value (Support army and joint training transformation; Maintain future joint stationing options; Power projection for joint operations; Support army materiel and joint logistics; Achieve cost-efficient installations; Enhance solder and family wellbeing). These were broken down into twelve capabilities at an intermediate level of the hierarchy, and finally into 40 value measures at the lowest level.

Of the 40 value measures, 26 were single-dimensional natural or constructed scales, and single-attribute value functions were elicited for them for value on a 0–10 scale. Some of these value functions were linear in form like Equation (3.2), but for others the experts consulted made a compelling argument that the function should be non-linear. For these an exponential functional form was assumed and a *midlevel splitting* approach was used (Kirkwood, 1997):

$$v_i\left(x_i\right) = 10\,\frac{1-e^{-\frac{x_i-x_i^-}{\rho_i}}}{1-e^{-\frac{x_i^+-x_i^-}{\rho_i}}} \tag{3.5}$$

where ρ_i is a parameter derived from the elicitations. The other 14 value measures had multidimensional constructed scales. The method of dealing with these can be shown by the example of Heavy Maneuver Area, which had two dimensions: Total Area and Largest Contiguous Area. The range of variation in these dimensions was partitioned into four intervals (<10, 10–50, 50–100, and >100, in 1000s of acres) and a 4 × 4 table built, then value on a 10-point scale was directly elicited for each entry in the table.

Swing weights for the 40 measures were elicited fundamentally as described in this chapter, but with the use of a *swing weight matrix*, as introduced by Trainor et al. (2004). A table was built with six columns for six levels of ability to change the measure, from immutable (e.g., Heavy Maneuver Area) to relatively easy to change with a modest expenditure (e.g., General Instructional Facilities). The table had three rows to represent high, medium, and low range of variation in the measures (x_i^- to x_i^+) among the alternatives. Each of the 40 measures was then assigned to one of the 18 table entries. Those measures that were judged immutable and had a high range of variation were given a non-normalized swing weight of 100, those in the opposite corner received a non-normalized swing weight of 1, and the others got appropriate intermediate weights. These weights were then normalized as in Equation (3.4).

With the measures, value functions, and swing weights determined, the military value of all Army installations could be calculated according to Equation (3.1).

A complete discussion of the results is available (DoD, 2005). The sound and convincing installation value model supported the Army's recommendations to the BRAC 2005 Commission, 95% of which were accepted.

3.7 Conclusion

Though it has been shown to be very useful at times, there is still an irony in the practice of decision analysis. If the method is resorted to, that indicates that the right decision is not obvious. But if the right decision is not obvious, the alternatives are probably pretty close to each other in terms of real value to the decision maker. Thus, DA is only necessary when it is not really very important. So why do we do it?

One answer is that sometimes the right answer becomes obvious only after some amount of analysis, problem structuring, and data collection. The complete model, additive multiattribute value function or elaborate decision tree, may not be needed. If that happens, and the DM comes to realize that he or she can be confident of the answer without further detailed modeling, that should be considered a great success for decision analysis. The analysis should only proceed as far as necessary to make the correct decision clear.

Another answer is that sometimes it is not simply a matter of convincing the decision maker. The chief requirement may be to make a public, transparent, and well-justified decision. It must be able to withstand criticism from a wide variety of stakeholders, who may be disappointed in the result and looking for grounds to question it. (This was certainly the case in the BRAC example.) In cases like this, one needs a sound decision modeling methodology, consistent with the best practices of decision analysis and also reasonably understandable by non-specialists. This chapter presents the basic methods of widest importance in multiattribute decision analysis in the defense sector. These will cover a large fraction of the problems that military decision analysts will face.

References

Amoyal, J. (2018). Software survey: Decision analysis. *OR/MS Today, 45*(5), 38–47.

Arrow, K. (1950). A difficulty in the concept of social welfare. *Journal of Political Economy, 58*(4), 328–346.

Buede, D. M., & Bresnick, T. A. (1992). Applications of decision analysis to the military systems acquisition process. *Interfaces, 22*(6), 110–125.

Burk, R. C., Deschapelles, C., Doty, K., Gayek, J. E., & Gurlitz, T. (2002). Performance analysis in the selection of imagery intelligence satellites. *Military Operations Research, 7*(2), 45–60.

Burk, R. C., & Parnell, G. S. (2011). Portfolio decision analysis: Lessons from military applications. In A. Salo, J. Keisler, and A. Morton (Eds.), *Portfolio decision analysis: Improved methods for resource allocation* (pp. 333–357). New York, NY: Springer.

Center for Army Analysis. (2004). *Decision analysis support for base realignment* (Report CAA-R-04-6). Ft. Belvoir, VA: United States Army.

Clemen, R. T., & Reilly, T. (2001). *Making hard decisions with DecisionTools*. Pacific Grove, CA: Duxbury.

Coles, J., & Zhuang, J. (2016). Introducing terrorist archetypes: Using terrorist objectives and behavior to predict new, complex, and changing threats. *Military Operations Research, 21*(4), 47–62.

Colombi, J. M., Robbins, M. J., Burger, J. A., & Weber, Y. S. (2015). Interface evaluation for open system architectures using multiobjective decision analysis. *Military Operations Research*, 20(2), 55–69.

Corner, J. L., & Kirkwood, C. W. (1991). Decision analysis applications in the operations research literature, 1970–1989. *Operations Research*, 39(2) 206–219.

Danielson, M., Ekenberg, L., & He, Y. 2014. Augmenting ordinal methods of attribute weight approximation. *Decision Analysis*, 11(1), 21–26.

Department of Defense. (2005). *Report to the Defense Base Closure and Realignment Commission* (Vol. III). Washington, DC: BRAC 2005.

Dewald, L., Holcomb, R., Parry, S., & Wilson, A. (2016) A Bayesian approach to evaluation of operational testing of land warfare systems. *Military Operations Research*, 21(4), 23–32.

Dreyfuss, M., & Gial, Y. (2017). Multi-echelon exchangeable item repair system optimization. *Military Operations Research*, 22(3), 35–49.

Ewing, P. L. Jr., Tarantino, W., & Parnell, G. S. (2006). Use of decision analysis in the Army Base Realignment and Closure (BRAC) 2005 military value analysis. *Decision Analysis*, 3(1), 33–49.

Howard, R. A. (1988). Decision analysis: Practice and promise. *Management Science*, 34, 679–695.

Howard, R. A., & Abbas, A. E. (2016). *Foundations of decision analysis*. Boston, MA: Pearson.

Intrater, B. C., Alt, J. K., Buttrey, S. E., House, J. B., & Evans, M. (2018). Understanding the impact of socioeconomic factors on Navy accessions. *Military Operations Research*, 23(1), 31–47.

Jia, J., Fischer, G. W., & Dyer, J. S. (1998). Attribute weighting methods and decision quality in the presence of response error: A simulation study. *Journal of Behavioral Decision Making*, 11(2), 85–105.

Jose, V. R. R., & Zhuang, J. (2013). Technology, adoption, accumulation, and competition in multi-period attacker-defender games. *Military Operations Research*, 18(2), 33–47.

Kahneman, D. (2011). *Thinking, fast and slow*. New York: Farrar, Strauss and Giroux.

Keefer, D. L., Kirkwood, C. W., & Corner, J. L. (2004). Perspective on decision analysis applications, 1990–2001. *Decision Analysis*, 1(1), 4–22.

Keeney, R. L. (1992). *Value–focused thinking: A path to creative decision making*. Cambridge, MA: Harvard University Press.

Keeney, R. L. (2008). Applying value-focused thinking. *Military Operations Research*, 13(2), 7–17.

Keeney, R. L., & Raiffa, H. (1976). *Decision making with multiple objectives: Preferences and value tradeoffs*. New York, NY: Wiley.

Keeney, R. L., & Raiffa, H. (1993). *Decision making with multiple objectives: Preferences and value tradeoffs*. Cambridge, UK: Cambridge University Press.

Kirkwood, C. W. (1997). *Strategic decision making: Multiobjective decision analysis with spreadsheets*. Belmont, CA: Duxbury.

Kuikka, V. (2015). A combat equation derived from stochastic modeling of attrition data. *Military Operations Research*, 20(3), 49–69.

Lootsma, F. A. (1999). *Multi-criteria decision analysis via ratio and difference judgement*. Boston, MA: Kluwer.

Miner, N. E., Gauthier, J. H., Wilson, M. L., Le, H. D., Melander, D. J., & Longsine, D. E. (2015). Measuring the adaptability of systems of systems. *Military Operations Research*, 20(3), 25–37.

Nehring, R. (2015). *Empirical study of rank order attribute swing weight calculation methods* (honors paper). West Point, NY: Department of Systems Engineering, U.S. Military Academy.

Oleson, S. (2016). Decision analysis software survey. *OR/MS Today*, 43(5), 36–45.

Parnell, G. S., Bresnick, T. A., Tani, S. N., & Johnson, E. R. (2013). *Handbook of decision analysis*. Hoboken, NJ: John Wiley & Sons, Inc.

Parnell, G. S., Burk, R. C., Westphal, D., Schulman, A., Kwan, L., & Blackhurst, J. L. (2004). Air Force Research Laboratory space technology value model: Creating capabilities for future customers. *Military Operations Research*, 9(1), 5–17.

Rao, J. S., & Sobel, M. (1980). Incomplete Dirichlet integrals with applications to ordered uniform spacings. *Journal of Multivariate Analysis*, 10, 603–610.

Reed, G. S., Tackett, G. B., Petty, M. D., & Ballenger J. P. (2013). A model for "evil" for course of action analysis. *Military Operations Research*, 18(4), 61–76.

Roberts, R., & Goodwin, P. (2002). Weight approximations in multi-attribute decision models. *Journal of Multi-Criteria Decision Analysis*, 11, 291–303.

Schroden, J., Thompson, R., Foster, R., Lukens, M., & Bell, R. (2013). A new paradigm for assessment in counter-insurgency. *Military Operations Research*, 18(3), 5–20.

Stillwell, W. G., Seaver, D. A., & Edwards, W. (1981). A comparison of weight approximation techniques in multiattribute utility decision making. *Organizational Behavior and Human Performance*, 28, 62–77.

Trainor, T., Parnell, G. S., Kwinn, B., Brence, J., Tollefson, E., Burk, R. K., ... Harris, J. (2004). *USMA study of the installation management agency CONUS region structure* (Report DSE-R-0506, DTIC ADA-427027). West Point, NY: United States Military Academy.

United States. (2014). *Commander and staff organization and operations: Field manual 6-0, Change 2*. Washington, DC: Headquarters, Dept. of the Army.

von Neumann, J., & Morgenstern, O. (1947). *Theory of games and economic behavior* (2nd ed.). Princeton, NJ: Princeton University Press.

von Winterfeldt, D., & Edwards, W. (1986). *Decision analysis and behavioral research*. Cambridge, UK: Cambridge University Press.

Wall, D. W., & LaCivita, C. J. (2016). On a quantitative definition of affordability. *Military Operations Research*, 21(4), 33–45.

Washburn, A. (2018). Ratio game on a network. *Military Operations Research*, 23(1), 23–29.

Watson, S. R., & Buede, D. M. (1987). *Decision synthesis: The principles and practice of decision analysis*. Cambridge, UK: Cambridge University Press.

Yoho, K. D., Bummara, J., Clark, W., & Kelley, C. (2013). Strategic resource allocation: Selecting vessels to support maritime irregular warfare. *Military Operations Research*, 18(3), 21–33.

Zeleny, M. (1982). *Multiple criteria decision making*. New York, NY: McGraw-Hill.

Chapter 4

Military Workforce Planning and Manpower Modeling

Nathaniel D. Bastian and Andrew O. Hall

4.1 Introduction

Human resources planning and workforce design decisions are an extension of traditional operations research (OR) problems, as workforce policies strive to place the appropriate and accurate numbers of the correct types of people in the right jobs at the right time. Traditional OR methods focus on how managers oversee, design, and redesign business operations in the production of goods and/or services in an efficient and effective manner. Comprehensive management methods explicitly consider personnel systems as part of these business operations.

Since the conception of military OR, manpower models study human resources in the military (Abrams, 1957). Military manpower models are properly employed whenever the personnel system has several distinct stages and is predominantly a closed system with a bar to lateral entry. In this manner, military manpower models must account for the need to grow the experienced personnel that are needed within the system. Wilson (1969) contains an early collection of papers and summarizes mathematical models and military manpower systems. Bartholomew, Forbes, and McClean (1991) describe statistical methods for manpower planning, and Vajda (1978) provides a mathematical modeling description of manpower planning. Edwards (1983) acknowledges the need for additional multi-disciplinary research in manpower planning, focusing first on the economics models and then on behavioral models. Charnes, Cooper, and Niehaus (1978) provides a good introduction to the management science approaches to manpower planning.

Military workforce planning and manpower modeling employs methods and techniques of OR to address problems of creating jobs and positions, as well as finding and assigning people to these positions. Advanced analytic techniques can be applied to the scheduling of soldiers or employees against shifts and optimally assigning the workforce against tasks. The variety of management issues within human resources provide opportunities to leverage simulation, optimization, economics, and advanced analytics against practical problems.

When employers maintain a large workforce, the cost of personnel is a large sunk cost. There is potential for an excess of employee inventory, possibly in the wrong skills, and a myriad of potential inefficiencies might emerge. When one overestimates the personnel requirements, there are too many people with not enough work to go around. When there are shortages in the labor pool, there are fewer employees available to complete the necessary work. Employee satisfaction, production, and task performance may suffer as managers pushes their employees to complete more work before authorizing overtime. Managers create inventory by hiring new workers, either in entry-level or more senior positions. Over time, managers promote workers to satisfy the demand for more advanced positions. Managers face the challenge of determining the number of people to hire into entry-level positions and the number of people already in the workforce to promote to senior-level positions.

Current hiring and promotion decisions not only affect the present-day workforce structure but also influence the workforce structure's future wellbeing. If organizations try to save resources by hiring fewer entry-level workers, this upstream policy will have an effect in the future, downstream, when the organization is trying to meet senior-level position requirements. Human resources planning policies may promote workers to eliminate shortages in senior-level positions. But promotion policies may also prohibit mass or large-scale promotions where all or many members of the workforce are promoted and there is no system to ensure that only qualified individuals are selected for promotion. The design of organizations across the services are as important as the compensation, retirement systems, and service obligations in forecasting the behavior of the workforce.

4.1.1 Terminology in Military Workforce Planning and Manpower Modeling

Some key terminology frequently used in military workforce planning and manpower modeling include the following: accessions, promotions, and authorizations. Accessions are new hires, which is when military personnel enter military service. Promotion is an increase in military grade (rank), which requires the selection from an approved board of leaders. Authorizations are military positions funded by the US Congress to carry out the mission of the US Department of Defense. A documented authorization is a funded position within an organization that identifies a specific specialty and rank required to meet a stated capability. These documented authorizations comprise the underlying workforce structure.

One method of analyzing the manpower model breaks the model into three components: recruiting, retention, and compensation. The US Army uses eight life cycle functions to describe human resource management: personnel structure, acquisition, distribution, development, deployment, compensation, sustainment, and transition (Hall, 2009). These eight functions are condensed into the three components included in much of the academic research. Issues of recruiting are addressed in the marketing domain, organizational science, and behavioral psychology. Retention is addressed in most manpower models, as the current workforce is the only source of experience within a military manpower model. Compensation is researched within labor economics, as well as in the interface between compensation, recruiting, and retention.

4.1.2   Chapter Overview

The remainder of this chapter provides an overview of military workforce planning and manpower modeling as it pertains to the military OR practitioner and researcher. Section 4.2 reviews previous work in the domain, addressing several models in the context of the issues facing the military at the time of their creation. Section 4.3 describes the four main classes of OR methods primarily used for workforce planning and manpower modeling. Section 4.4 investigates multi-disciplinary research supporting retention and talent management, while Section 4.5 provides two real-world case studies to highlight applications in practice. Finally, Section 4.6 provides a summary and directions for future work.

4.2   Previous Work in Military Workforce Planning and Manpower Modeling

There are many questions that arise when discussing the management decisions and policies that surround workforce planning and manpower modeling. This domain covers the spectrum of human resource management policy in which employees enter at an initial state and flow through the system until they are eventually a loss to the system. The military manpower system is one where there is limited lateral entry, and the system is constrained to hire entry-level workers and retain experienced workers.

Workforce planning and manpower models seek to answer questions on the number of personnel needed and what skills are required to optimize capability. This optimization is traditionally viewed as constrained to defense budgets. The US transition from conscription to the modern all-volunteer force over 45 years ago provided a rich time for OR in personnel systems within the Department of Defense. The military roots of OR, the immensity of defense budgets, the complexity of military systems, and the consequences of failure encourage research tackling rising questions concerning how to acquire, retain, and compensate the new all-volunteer military.

The transitions from peacetime overage to wartime shortages and conscription motivates early research concerning an all-volunteer system. In Daily's (1958) examination of US Navy re-enlistment rates, he finds the rates are amazingly consistent in total numbers but vary considerably with regards to the percentage of sailors that re-enlist each year. He proposes a model with an inverse relationship between the number of sailors inducted in any year and the re-enlistment percentage, which provides a predictive model of re-enlistment for retaining sailors past their first enlistment. Fisher and Morton (1967a, b) investigate re-enlistment rates for electronics personnel in the US Navy. Their model analyzes the value of additional incentives over the period of service to keep or retain experienced technical experts in whom a considerable training cost has been invested. McCall and Wallace (1969) address the issue of re-enlistments in the US Air Force, where they study the retention of electronic specialists. They develop a logistic regression model to relate the probability of re-enlisting and the difference in military and civilian pay, and their results indicate the importance of competitive initial pay levels for military personnel.

Flynn (1975a) creates a dynamic programming model to study the issue of retaining productive units. The model finds an optimal retention policy in units where the productivity of an element is determined with age and a predetermined retirement age for each element. Further, he finds conditions under which it is optimal to bring in just enough units at the entry level to satisfy linear production constraints. Jaquette and Nelson (1976) develop a mathematical model to determine steady-state wage rates and the force distribution by the length of service. The model adjusts the accession and retention policies to optimize the steady-state system, and they find lower accessions, higher first term re-enlistment rates, and slightly lower career re-enlistment rates are the optimal policy to maximize production, subject to existing budget constraints.

Bres et al. (1980) develop a goal programming model for planning officer hires to the US Navy from various commissioning sources. Using transition rates, the model projects the expected on-board flows between successive time periods in a Markovian fashion to add new officer hires entering the system. Gass et al. (1988) develop the Army Manpower Long-Range Planning System, which integrates a Markov chain and linear goal programming model to forecast the flow of an initial force (given by grade and years-of-service) to a future force over a 20-year planning horizon, and to determine the optimal transition rates (continuation, promotion, and skill migration) and accession values to obtain the desired end state force structure or the rates required to minimize the deviation from the desired end state force structure. Silverman, Steuer, and Whisman (1988) develop a multi-period, multiple criteria trajectory optimization system to help manage the enlisted force structure of the US Navy. Their workforce accession planning model employs an interactive augmented weighted Tchebycheff method while examining various recruitment and promotion strategies.

Gass (1991) describes network-flow goal programming models to provide the US Army with decision support tools to effectively manage its workforce. The transition rate models describe people going from one state to another during their life cycle in the workforce system. Weigel and Wilcox (1993) develop the Army's enlisted personnel decision support system, which combines a variety of modeling approaches (goal programming, network models, linear programming, and Markov-type inventory projection) with a management information system to support the analysis of long-term personnel planning decisions. In addition to these long-term workforce planning models, Corbett (1995) develops a workforce optimization model that assists personnel planners in determining yearly officer hires as well as transfers to functional areas as part of the branch detail program. The model employs a multi-year weighted goal program designed to maximize the Army's ability to meet forecasted authorization requirements.

Hall (1999) develops a mixed integer optimization to minimize the number of lost man-years resulting from inefficient scheduling of initial entry and advanced individual training. Yamada (2000) develops the infinite horizon workforce planning model using convex quadratic programming for managing officer hires, promotions, and separations annually to best meet desired inventory targets. Henry and Ravindran (2005) present both pre-emptive and non-pre-emptive goal programming models for determining the optimal hires cohort – the number of new Army officers for each of 15 different career branches. Shrimpton and Newman's (2005) network-optimization model designates mid-career level officers into new career fields to meet end-strength requirements and maximize the overall utility of officers.

Cashbaugh et al. (2007) use network-based mathematical programming to model the assignment of US Army enlisted personnel in a 96-month planning horizon. Kinstler et al. (2008) develop a Markov model using promotion and attrition rates to improve workforce management decisions in the US Navy Nurse Corps. Hall (2009) uses dynamic programming and linear programming techniques to model the optimal retirement behavior for an Army officer from any point in their career. McMurry et al. (2010) develops a set of non-linear mathematical programming models to provide the Army Medical Department (AMEDD) with a multiple criteria decision support mechanism for determining optimal hiring and promotion policies for medical officers. Coates et al. (2011) investigate the US Army's Captain retention program and use a chi-square and odds ratio analysis to determine whether the practice of providing financial bonuses to individuals agreeing to continue their service is an effective retention tool. Lesinski et al. (2011) uses discrete-event simulation to model the current flow process that an officer negotiates from pre-commissioning to the first unit of assignment. This model assists with synchronization of the officer accession and training with the Army Force Generation process.

Bastian et al. (2015) proffer deterministic and stochastic mixed-integer linear weighted goal programming models to optimize AMEDD workforce planning decisions. They use discrete event simulation to verify and validate the results of the optimizations. Hall and Fu (2015) explore the optimization of US Army officer force profiles, where an officer force profile describes the distribution of rank and specialty as well as age and experience within the officer corps. They propose a new network structure that incorporates both rank and years in grade to combine cohort, rank, and specialty modeling. Abdessameud et al. (2018) propose a technique to combine both statutory and competence logics for military manpower planning in the same integrated model to allow simultaneous optimization. They illustrate a model based on a flow network, which uses integer and goal programming to find optimal solutions. McDonald et al. (2017) use time series analysis and regression methods to develop more comprehensive forecasting methods which US Army recruiting leaders can use to better establish recruiting goals. Evans and Bae (2017) use discrete event simulation to estimate the limitations of a forced distribution performance appraisal system within the military to identify the highest performing individuals within an organization.

Given this review of the existing body of knowledge, the next section highlights the most commonly used OR techniques in military workforce planning and manpower modeling.

4.3   Mathematical Model Classes

Four types of mathematical models have been used in the preponderance of military workforce planning and manpower models. The early models predominantly solve using dynamic programming. Dynamic programming papers often cite manpower models as one of the sources of their application and motivation for the mathematical methods (Flynn, 1975b). The second type of model is the Markov decision model. Transition rate models require transition probabilities as an underlying assumption. Given an initial distribution, Markov models will answer the question of what attributes the force will have at each phase in the planning period. When modeling using transition rates, each individual follows the same Markov process and individuals are independent.

The third major type of model, goal programming, models the manpower system typically as a network flow problem, and policy decisions as constraints on the multiple competing components of the objective function. Network models ensure conservation of flow and have the advantage of integer solutions. The models are especially powerful when modeling systems with a limited number of characteristics at each time period. Goal programming expands upon the network structure to add multi-year goals to the modeling formulation. The different years will have constraints on the number of separations, promotions, grade targets, total strength, and operating strength targets. The objective function penalizes deviations from the goals but relies on the underlying network structure to simplify the optimization. Current Army manpower network models have a 96-month planning horizon, and aggregate on months of service, grade, term of service, and gender.

The final major type of model is statistical models. The statistical models of economics and OR are heavily relied on in shaping force structure, as well as attracting and retaining the required military and civilian manpower. Statistical methods in workforce planning and manpower modeling employ regression, time series analysis, forecasting, econometrics, and simulation-based approaches (discrete-event simulation and Monte Carlo).

Given this overview of the four main mathematical modeling classes, we next discuss each of them in greater detail in the following sub-sections.

4.3.1   Dynamic Programming

Dynamic programming is a multi-stage optimization approach that transforms a complex problem into a sequence of simpler problems. The objective of deterministic dynamic programming is to develop a steady-state policy, within the defined state space, action space, reward functions, and transition functions. The early works of Fisher and Morton (1967a, 1967b) and Flynn (1975a, 1975b) describe earlier dynamic programming models. Flynn (1975a, 1975b) first poses and solves the manpower models in management science, and then he extends the models with more detail mathematical foundations for steady-states policies (Flynn, 1979).

In dynamic programming, a state space $x_t \in X$, an action space $a_t \in A_t$, and a reward $R(x_t, a_t)$ for taking an action a in a state x are defined. The state space reflects the information required to fully assess the consequences that the current decision has upon future actions (e.g., number of soldiers in inventory), whereas the action space reflects the decision options to make within a state (e.g., number of soldiers to bring onto active duty). A transition function links x_{t+1} to x_t when you take action a_t at time t. The dynamic program can have a finite or infinite time horizon, but the formulation is deterministic.

The steady-state policy will be a decision rule that prescribes the action that a decision-maker should make upon reaching any particular state in a certain period. The policy may be a deterministic or a randomized policy and may be time-dependent or stationary. An optimal policy is a policy that results in the maximum (minimum) value of the stated objective function. The theory for solution is based upon the principle of optimality, due to Bellman (1954). The steady-state solutions for the discrete models provide the foundation for stochastic models which will leverage dynamic programming. For more details on dynamic programming, please refer to Bellman (1957).

4.3.2 Markov Decision Models

Markov decision processes (MDPs) are powerful analytical tools that generalize standard Markov models by embedding the stochastic sequential decision process in the model and allowing multiple decisions in multiple time periods. Dynamic programming models are a special type of discrete-state MDP. Gotz and McCall (1983) use MDPs to analyze US Air Force officer retention and study policy changes in promotion and retirement. Their model shows that retirement pay is the most important factor for officer retention between the 10- and 20-year marks. Markov manpower models in a discrete time setting are described and solved in Feichtinger and Mehlmann (1976), and Mehlmann (1977) extends the results to continuous Markovian manpower models. Davies (1982) considers a model with fixed promotions controlled by management.

The basic definition of a discrete-time MDP contains five components. The decision epochs, $T = 1,..., N$, are the set of points in time at which decisions are made (days, hours, etc.). The state space S is the set of all possible values of dynamic information relevant to the decision process. For any state $s \in S$, A_s is the action space, the set of possible actions that the decision-maker can take at state s. Transition probabilities, $p_t(. \mid s,a)$, are the probabilities that determine the state of the system in the next decision epoch, which are conditional on the state and action at the current decision epoch. Finally, the reward function, $r_t(s,a)$, is the immediate result of taking action a at state s. Thus, these five components collectively define an MDP.

A decision rule is a procedure for action selection from A_s for each state at a particular decision epoch, namely $d_t(s) \in A_s$. We can drop the index s from the expression and use $d_t \in A$, which represents a decision rule specifying the actions to be taken at all states, where A is the set of all actions. A policy π is a sequence of the decision rules to be used at each decision epoch and defined as $\pi = (d_1, ..., d_{N-1})$.

The objective of solving an MDP model is to find the optimal policy that maximizes a measure of long-run expected rewards. Note that future rewards are often discounted over time. In the absence of a discounting factor, if we let $u_t^*(s_t)$ be the optimal value of the total expected reward when the state at time t is s and there are N-t periods to the end of the time horizon, then the optimal value functions and the optimal policy giving these equations can be obtained by iteratively solving the following recursive equations (known as the Bellman equations):

$$u_N^*(s_N) = r_n(s_n) \quad \text{for all } s_N \in S$$

$$u_t^*(s_t) = \max_{a \in A_s} \left\{ r_t(s_t, a) + \lambda \sum_{j \in S} p_t(j \mid s_t, a) u_{t+1}^*(j) \right\} \text{ for } t = 1, ..., N-1, s_t \in S$$

where $r_n(s_n)$ denotes the terminal reward that occurs at the end of the process when the state of the system at time N is s_n, and λ represents the discount factor ($0 < \lambda < 1$) specifying the relative importance of rewards.

At each decision epoch t, the optimality equations given by the second equation chooses the action that maximizes the total expected reward that can be obtained for periods $t, t+1, ..., N$ for each state s_t. For a given state s_t and action a, the total expected reward is calculated by summing the immediate reward, $r_t(s,a)$, and future reward, which is obtained by multiplying the probability of moving from state s_t to j at time $t+1$ with the maximum total expected reward $u_{t+1}^*(s)$ for state j at time $t+1$ and summing over all possible states at time $t+1$.

A finite-horizon MDP model is appropriate for systems that terminate at some specific point in time. At each stage, we choose the following:

$$a_{s_t,t}^* \in \operatorname*{argmax}_{a \in A_s} \left\{ r_t(s_t,a) + \lambda \sum_{j \in S} p_t(j \mid s_t,a) u_{t+1}^*(j) \right\} \text{ for } t = 1,...,N-1$$

where $a_{s_t,t}^*$ is the best action maximizing the total expected reward at time t for state s.

In other situations, an infinite-horizon MDP model is more appropriate, in which case the use of a discount factor is sufficient to ensure the existence of an optimal policy. The most commonly used optimality criterion for infinite-horizon ($N = \infty$) problems is the total expected discounted reward. In an infinite-horizon MDP model, the following very reasonable assumptions guarantee the existence of optimal stationary policies: stationary (time-invariant) rewards and transition probabilities, discounting with λ, and discrete state and action spaces. Optimal stationary policies still exist in the absence of a discount factor when there is an absorbing state with immediate reward zero (such as retirement from the military in manpower models). In a stationary infinite-horizon MDP model, the time indices can be dropped for the reward function and transition probabilities, leaving the following Bellman equations:

$$V(s) = \max_{a \in A_s} \left\{ r(s,a) + \lambda \sum_{j \in S} p(j \mid s,a) V(j) \right\} \text{ for } s \in S$$

where $V(s)$ is the optimal value of the MDP for state s (the expected value of future rewards discounted over an infinite horizon). The optimal policy consists of the actions maximizing this set of equals.

Markov decision processes may be classified according to the time horizon in which the decisions are made: finite- and infinite-horizon MDPs. Finite-horizon and infinite-horizon MDPs have different analytical properties and solution algorithms. Because the optimal solution of a finite-horizon MDP with stationary rewards and transition probabilities converges to that of an equivalent infinite-horizon MDP as the planning horizon increases and infinite-horizon MDPs are easier to solve and to calibrate than finite-horizon MDPs, infinite-horizon models are typically preferred when the transition probabilities and reward functions are stationary. For more details on MDPs, please refer to Puterman (2014).

4.3.3 Goal Programming

Decision problems in military workforce planning and manpower modeling often exhibit the presence of multiple conflicting criteria for judging alternatives, as well as the need for making compromises or trade-offs regarding the outcomes of alternative courses of action. Multiple-criteria decision-making is a practical approach for helping make better, more informed decisions. The focus of multiple criteria mathematical programming (MCMP) problems is to fashion or create an alternative when the possible number of alternatives is high (or infinite) and all alternatives are not known in advance. MCMP problems are usually modeled using explicit mathematical relationships, involving decision variables incorporated within constraints and objectives.

For MCMP problems, one way to treat multiple criteria is to select one objective as primary and the others as secondary. The primary criterion is then used as the optimization objective function, while the secondary criteria are assigned acceptable minimum or maximum values depending on whether the criterion is maximum or minimum and are treated as problem constraints. However, if careful consideration is not given while selecting the acceptable levels, a feasible design that satisfies all the constraints may not exist. Fortunately, this problem is overcome using goal programming (GP); this method uses completely prespecified preferences of the decision-maker.

Price and Piskor (1972) provide a goal programming formation for officers in the Canadian Forces. Price (1978) shows how to re-formulate goal programming problems as capacitated network flow problems. Gass et al. (1988) describe a goal programming formulation of the Army Manpower Long-range Planning System. Hall and Fu (2015) create a goal program with a network structure to incorporate tenure within an optimal officer force profile. Bastian et al. (2015) present a mixed-integer linear weighted goal programming model to solve the workforce planning problem for the AMEDD, given both deterministic and stochastic continuation rates.

GP is a practical method for handling multiple criteria. It can be used in order to achieve the best compromise solution while considering many desires that the decision-maker would like to achieve. In GP, all of the objectives are assigned target levels for achievement and relative priority of achieving these levels. The aim is to minimize deviations between the specified targets of the decision-maker and what can actually be achieved for the multiple objective functions within the given constraints. In other words, GP attempts to find an optimal solution that comes as close as possible to the targets in the order of specified weights or priorities. Note that deviations can be either positive or negative, depending on whether we overachieve or underachieve a specific goal, respectively.

In pre-emptive GP, priorities are provided by the decision-maker, whereas in non-pre-emptive GP, numeric weights are provided by the decision-maker for each goal. In pre-emptive GP, there is a hierarchy of priority levels (as opposed to weights), so that the goal of primary importance is considered first when solving, followed by those of secondary importance, etc. In non-pre-emptive GP, however, the numeric weights are determined *a priori* such that the GP is solved all at once, rather than in sequence. For goals that are very important, high values are used as weights and for the other goals, relatively lower weights are used. In that way, there is a chance that high priority goals will be satisfied first. In other words, numeric weights are assigned to each deviation variable in the objective function, where the magnitude of the weight assigned to a specific deviation variable depends on the relative importance of that goal. Hence, the results of GP models provide a feasible, best compromise solution for the decision-maker.

The following non-pre-emptive (weighted) GP formulation is provided to reinforce the concept. For this GP formulation, the decision maker specifies an acceptable level of achievement (b_i) for each criterion f_i and specifies a weight w_i to be associated with the deviation between f_i and b_i. The weighted GP model looks like the following:

$$\textbf{Minimize } Z = \sum_{i=1}^{k}(w_i^+ d_i^+ + w_i^- d_i^-)$$

$$f_i(x) + d_i^- - d_i^+ = b_i \quad \forall i = 1, 2, \ldots, k$$

$$\textbf{subject to} \quad g_j(x) \leq 0 \quad \forall j = 1, 2, \ldots m$$

$$x, d_i^-, d_i^+ \geq 0 \ \forall i, j$$

The objective function seeks to minimize the sum of weighted goal deviations. The first set of constraints represent the goal constraints relating the multiple criteria to the goal/targets for those criteria. It should be noted that d_i^- and d_i^+ are the deviational variables, representing the underachievement and overachievement respectively. For more details on goal programming, please refer to Schniederjans (2012).

4.3.4 Statistical Models

In addition to the empirical analysis and modeling presented earlier, regression techniques are also used to solve manpower problems. Drui (1963) describes regression equations to study optimal manpower levels. Rizvi (1986) studies manpower demand using time series, regression, and econometric models. Hall (2004) and Cashbaugh et al. (2007) describe the use of statistical methods as a part of manpower planning systems that rely upon the use of historical data and time series methods alongside simulation and optimization techniques to address program objective memorandum (POM) planning tasks. Evans and Bae (2018) use a simulation optimization approach to explore US Army performance appraisals based on extensive statistical analysis on trends in officer performance appraisals. Further, statistical methods have formed the backbone of the budget and manpower controls necessary to maintain force levels under fixed authorizations (Rostker, 2006). For more details on statistical modeling, please refer to Bartholomew, Forbes, and McClean (1991).

Upon this review of the four mathematical classes commonly used in military workforce planning and manpower modeling, the next section discusses some approaches to retaining quality and talent management in the domain.

4.4 Retaining Quality and Talent Management

In a closed system with a bar to lateral entry, the organization must train its own leaders and build its own expertise base. This is especially true in the military, but this investment in employees has been reflected in the literature with a growing need to see a return on the training that occurs after employees have joined the organization. Much of the current research has appeared in the economics literature but is also found at the intersection of the business areas of management and operations.

Schneider (1981) explores military recruitment and retention from a social science approach during the height of the Cold War. He discusses the basis of military compensation in deferred compensation and the challenges of finding incentives for first-term enlistees that highly discount future income. He suggests longer terms of service to decrease the compensation needed to retain service members and reducing reenlistment opportunities to only 10–20% of the force. Cotton and Tuttle (1986) use meta-analysis to investigate the literature on employee turnover. They find that the research in turnover needs to be subjected to increased scrutiny in order to ensure improved academic rigor in empirical research and develop causal relationships. Lee and Mowday (1987) conduct an empirical test of Steers and Mowday's model of turnover. This paper is illustrative of the empirical research in organizations conducted to investigate the issue of why workers stay with a firm.

Campbell (1994) provides an example of an economic model of quit behavior. His work shows an example of an analytic model and obtains results showing that the change in wages is important in addition to the level of wages. Mitchell et al. (2001) investigate job embeddedness and its effect on voluntary turnover, finding that that highly embedded employees have lower quit rates. Warner and Goldberg (1984) investigate non-pecuniary factors that influence the labor supply through an empirical study of the re-enlistment of US Navy enlisted personnel. They find that pay elasticities are inversely related to the incidence of sea duty. Lakhani (1988) creates a model to study reenlistments in the Army comparing combat and non-combat specialties for US Army enlisted personnel. He poses that combat skills are less marketable outside the Army, and as a result, combat soldiers should have a stronger reaction to retention bonuses. Daula and Moffitt (1995) create a dynamic choice model with discrete choice variables. Their stochastic dynamic programming model explores the differences in military and civilian pay levels and the value of retirement. Mehay and Hogan (1998) use a probit model to study how much to pay service members to leave rather than stay. They find that incentives have only a moderate effect on voluntarily leaving the service, but the effect of allowing personnel to terminate their enlistment contract early is substantial.

Arguden (1988) addresses questions about changes in the military retirement system and the unintended effects on retention and workforce mix. Empirical results and forecasts are created using the Annualized Cost of Leaving (ACOL) model and the Dynamic Retention Model (DRM) calibrated to US Air Force retention rates. The overall attractiveness of the military with reduced compensation will alter the quality of personnel that are willing to stay in the military due to the reduced compensation. Congress has since reversed the policy after observing the retention shortfalls of the 1990s. It was changed as part of the National Defense Authorization Act of 2000 (Asch et al., 2002). Asch and Warner (2001) develop an individual decision model to explain why the military deviates from standard practice in terms of its flatter compensation structure, unusual retirement systems, and an up-or-out promotion system. Their result indicates that the current combination of up-or-out and generous retirement make the compensation consistent without an extreme skew in compensation between senior and junior personnel.

Talent management is a key research thread at the United States Military Academy, as Wardynski, Lyle, and Colarusso (2009a, 2009b) detail an alternative talent management paradigm for the US Army and Dabkowski et al. (2010, 2011) employ simulation modeling to model the flow of officers' careers through the creation of an officer career model. Wallace et al. (2015) explore implications of new talent management ideas to the military retirement system, and Colarusso et al. (2016) detail implications across military compensation. Hall and Schultz (2017) address talent management considerations involving direct commissioning of cyber officers and lifting the bar to lateral entry within the US Army officer corps.

Given this understanding of modeling approaches to retain quality personnel and manage talent, the next section details two real-world case studies for OR practitioners to get a better understanding of how to employ these types of mathematical models in practice.

4.5 Case Studies

In the context of military workforce planning and manpower modeling, this chapter has described the literature, highlighted the main mathematical modeling classes, and discussed modeling efforts around retention and talent management. This section proceeds by detailing two real-world case studies to discuss how these approaches apply in practice. In the first case study, a goal programming model helps determine optimal hiring and promotion policies for military medical workforce planning. In the second case study, a statistical modeling approach is used for grade plate roll down and the creation of an Army officer grade distribution.

4.5.1 Military Medical Workforce Planning

This case study from Bastian (2015) investigates long-term workforce planning and manpower modeling for military medical professionals within the AMEDD. The optimization model allows for better transparency of medical personnel for the AMEDD senior leadership, while effectively projecting the workforce skill levels (by grade) required to meet the demands of the current force. The goal programming model enables multiple criteria decision support for determining optimal hiring and promotion policies.

The AMEDD is a special branch of the US Army whose mission is to provide health services for the Army and, as directed, for other agencies, organizations, and military services. Since the establishment of the AMEDD in 1775, six officer corps (Medical Corps, Dental Corps, Nurse Corps, Veterinary Corps, Medical Specialist Corps, and Medical Service Corps) provide the organizational leadership and professional/clinical expertise necessary to accomplish the broad soldier support functions implicit to the mission (DA PAM 600-4 2007). Each corps is made up of individually managed career fields and duty titles called Areas of Concentration (AOC); there are a total of 100 officer AOCs in the AMEDD. The Medical Specialist Corps has the smallest number with four AOCs and the Medical Corps has the most with 41 AOCs.

The AMEDD manages medical officer personnel over a 30-year life cycle. Given the large number of AOCs in the AMEDD, determining the appropriate number of hires and promotions for each medical specialty is a complex task. The number of authorized medical personnel positions for each AOC varies significantly depending on the uniqueness of the career field and needs of the Army. Some AMEDD officers enter the Army and remain in the same AOC throughout their entire careers. There are others that start their career in one AOC but have the option to obtain additional education to qualify for a more specialized AOC. Finally, there are officers who enter the Army in one AOC but must obtain additional education and move to a more specialized AOC to stay competitive for promotion.

The rank structure of the AMEDD includes officers in the ranks of Second Lieutenant through General, but the AMEDD Personnel Proponent Directorate (APPD) is only responsible for managing officers below the rank of General, namely Second Lieutenant (2LT), First Lieutenant (1LT), Captain (CPT), Major (MAJ), Lieutenant Colonel (LTC), and Colonel (COL). The pay grades for these ranks are O-1 to O-6, respectively; note that there is a bijection between pay grades and military ranks. APPD is only responsible for managing these officers through the 30th year of commissioned federal service, because officers not selected for promotion to General are generally limited to a 30-year career in the military. The AMEDD's promotion policy is set forth by the Defense Officer Personnel Management Act (DOPMA) of 1980, which provides guidance on promotion selection target percentages (Rostker et al., 1993). For instance, the targeted promotion rate from LTC to COL is 50% by DOPMA, although this does not apply to physicians or dentists. In addition to the challenge of meeting federal mandates associated with promotion rates, uncertain officer continuation rates further complicate the workforce planning problem for the AMEDD, as officers may decide to leave at any time (given no active duty service obligation remains). Thus, uncertainty caused by attrition makes the officer accession and promotion decisions even more complex.

Although the Army Surgeon General has authority over the entire AMEDD, each corps has a Corps Chief who is responsible for making many decisions impacting the officers in his or her corps. Some of the key decisions include how many new officers to recruit and hire onto active duty each year within his or her corps, how many officers to promote to the next higher rank (grade) each year, and how many officers to train in each career field (or clinical specialty). The APPD seeks to provide workforce planning decision support to the Corps Chiefs by projecting the number of hires, promotions, and personnel inventory needed to support a 30-year life cycle within a corps' authorized officer positions. A 30-year life cycle allows for APPD to assess the availability of an officer throughout the anticipated lifespan. While the model is re-run year over year to re-assess, the number of hires must necessarily be based on the forecast for what is required based on attrition and promotion data.

Effective long-term workforce planning and personnel management of all medical professionals within the AMEDD is a complex problem. Prior to 2010, APPD used a manual approach to project the appropriate accessions (hires) and promotion goals for each medical specialty across the six separate corps. This case study describes the Objective Force Model (OFM), a deterministic, mixed-integer linear weighted goal programming model, to optimize AMEDD workforce planning for the Medical Specialist Corps (SP). The OFM allows for better transparency of AMEDD personnel for both the Corps Chiefs and the health services human resource planners at APPD, while effectively projecting the workforce skill levels (by grade) required to meet the demands of the current force. Note that a 30-year life cycle is considered due to Title 10 United States Code Chapter 36, which states that each officer who holds the grade of Colonel in the Regular Army and is not on the selection list to Brigadier General must retire the first day of the month after the month he/she completes 30 years of active federal commissioned service. Therefore, since officer ranks up to and including Colonel are modeled, 30 years constitutes the maximum life cycle and represents the target steady-state inventory of officers within each specialty, rank, and years of service.

AMEDD officers in the SP are hired (accessed) into the Army at the grade (rank) of either O-1 (2LT), O-2 (1LT), or O-3 (CPT). Unlike the more specialized and diversified corps, these officers remain in the same AOC throughout their entire careers. SP consists of four career AOCs: occupational therapists (AOC 65A), physical therapists (AOC

65B), dietitians (AOC 65C), and physician assistants (AOC 65D). Promotion decisions are made only for officers at the grade (rank) of O-4 (MAJ), O-5 (LTC), and O-6 (COL). According to APPD, a non-integer solution for promotions is an acceptable simplification. The non-integer structure is appropriate because of the concept of full-time equivalent employees, which may be fractional. While one may not hire a fractional person, one can augment any fractional requirement along the entire 30-year timeline. Next are some definitions and explanations of the military human resources terminology.

The set G represents the grade of the SP officers, which is indexed using {1, 2, 3, 4, 5, 6}. The set I represents the officer AOCs within the SP, which is indexed as {1 = 65A, 2 = 65B, 3 = 65C, 4 = 65D}. The set K represents the year of service for an officer, which is {1, 2, ..., 30} representing a full officer career. The set F represents the set of goals specified by APPD, which is {1, 2, 3, 4, 5}.

Authorizations are officer positions funded by the US Congress to carry out the mission of the US Army. A documented authorization is a funded position within an organization that identifies a specific specialty and rank required to meet a stated capability. There are two types of documented authorizations. The first is a career field specialty (or AOC) authorization that can only be filled by an officer specifically trained for that job (e.g., physical therapist). The second type of documented authorization is an immaterial authorization. Immaterial positions do not require an individual with a specific career specialty. Most immaterial authorizations are executive or leadership positions like commanders, directors, and administrators. The last type of authorization provides allowances for officers who are not assigned or contributing to the mission of an organization. This includes officers who are students, in transit between assignments, in long-term hospitalization or pending discharge (e.g., wounded warriors), or removed for disciplinary reasons (e.g., court-martial).

In the OFM, the parameter c_{ig} reflects the documented authorizations for each AOC and each grade, which reflect the requirements for SP officers to support both peacetime and wartime healthcare delivery for the Army. $D65X_g$ represents the SP immaterial documented authorizations for each grade; these are SP authorizations that SP officers can fill regardless of AOC. $D05A_g$ represents the AMEDD immaterial documented authorizations for each grade, which are AMEDD authorizations that SP officers can fill regardless of AOC. THS_g represents the transient, holdee, and student documented authorizations for each grade, which are authorizations that SP officers can fill regardless of AOC. Table 4.1 displays these data, which were provided by APPD. For example,

TABLE 4.1: Medical Specialist Corps Documented Number of Authorizations

Documented Authorizations	Area of Concentration							
	65A	**65B**	**65C**	**65D**	**Total**	**65X**	**05A**	**THS**
Total	75	255	122	780	1505	13	44	216
COL	3	6	5	3	28	4	5	2
LTC	9	23	19	28	103	1	15	8
MAJ	20	46	38	149	325	2	15	55
CPT	31	111	28	534	798	6	7	81
1LT	12	19	10	66	179	0	2	70
2LT	0	50	22	0	72	0	0	0
Company Grade	43	180	60	600	1049	6	9	151

for the 65C AOC, there are 122 total authorizations, which included five COL, 19 LTC, 38 MAJ, 28 CPT, 10 1LT, and 22 2LT.

The parameter Cap_i is the maximum allowable total structure distribution for each AOC (i.e., capacity). According to APPD, this upper bound only applies to AOCs 65A, 65B, and 65C. The parameter $Floor_i$ is the minimum acceptable total structure distribution for each AOC; this lower bound only applies to AOCs 65A and 65C. The parameter Cap_{ig} is the maximum allowable structure distribution for each AOC and each grade; this upper bound only applies to COL and LTC in AOC 65A. $Floor_{ig}$ is the minimum acceptable structure distribution for each AOC and each grade; this lower bound only applies to COL in AOCs 65A, 65B, and 65C as well as to LTC in AOC 65C. Table 4.2 displays these data, which are provided by APPD. For example, the 65A AOC has a maximum allowable total structure distribution of 96 and a minimum acceptable total structure distribution of 93.

Promotion rate is the number of officers selected for promotion divided by the number of officers considered. The number of officers selected is a variable in the OFM bounded by promotion rates usually based on DOPMA objectives ±10% when possible. In the OFM, the parameter pf_{ig} is the minimum promotion rate for each AOC and grade, which is not applicable to 2LT in each AOC. The parameter pc_{ig} is the maximum promotion rate for each AOC and grade, which is also not applicable to 2LT in each AOC. Table 4.3 displays these data, which were provided by APPD. For example, the minimum promotion rate for LTC 65B is 55%. Table 4.4 shows both the scheduled promotion evaluations by year and grade along with the DOPMA standard promotion

TABLE 4.2: Medical Specialist Corps Max and Min Structure Distribution

Structure Distribution	Area of Concentration			
	65A	*65B*	*65C*	*65D*
Total (Max)	96	295	154	
Total (Min)	93		149	
COL (Max)	4			
COL (Min)	4	8	7	
LTC (Max)	12			
LTC (Min)			20	

TABLE 4.3: Medical Specialist Corps Max and Min Promotion Rates

DOPMA Promotion Rate	65A					65B				
	1LT	*CPT*	*MAJ*	*LTC*	*COL*	*1LT*	*CPT*	*MAJ*	*LTC*	*COL*
Max	1	0.95	1	0.8	0.6	0.98	0.95	1	0.8	0.6
Min	1	0.95	0.7	0.6	0.4	0.98	0.95	0.7	0.55	0.3

DOPMA Promotion Rate	65C					65D				
	1LT	*CPT*	*MAJ*	*LTC*	*COL*	*1LT*	*CPT*	*MAJ*	*LTC*	*COL*
Max	1	0.95	1	0.8	0.6	1	0.95	1	0.8	0.6
Min	1	0.95	0.7	0.6	0.3	1	0.95	0.6	0.4	0.2

TABLE 4.4: DOPMA Promotion Standards

	Year				
Promotion Evaluation	*2*	*4*	*11*	*17*	*22*
2LT -> 1LT	98%				
1LT -> CPT		95%			
CPT -> MAJ			80%		
MAJ -> LTC				70%	
LTC -> COL					50%

rates, which APPD targets at ±10%. For example, the DOPMA standard promotion rate is 80% for CPTs being looked at for promotion to MAJ.

The continuation rate is the percentage of officers that stay in the Army from one year to the next year, categorized by specialty, rank and years of service. The rates are based on a 5-year average of actual data collected on every officer on active duty for each specialty. In the OFM, the parameter r_{igk} reflect the deterministic continuation rate for each AOC, grade, and year of service, which reflects both those SP officers who are selected for promotion when considered and those SP officers who are considered for promotion but not selected. Note that officers not selected for promotion are limited to a set number of years they may remain on active duty (rank dependent) before mandatory separation. These data provided by APPD come from the Medical Operational Data System, which is derived from multi-year averages. Finally, the parameter w_f reflects APPD's weight for each goal in the model.

The model decision variables p_{ig} represents the number of SP officers promoted in AOC i at grade g (for O-4 (MAJ), O-5 (LTC) and O-6 (COL) only). The model decision variables a_{ig} represents the number of SP officers accessed (hired) for AOC i at grade g (for O-1 (2LT), O-2 (1LT) and O-3 (CPT) only). The model decision variables d_{ig} represents the actual number of SP officers in the system for each AOC and each grade, whereas the model decision variables inv_{igk} represent the projected inventory of SP officers in the system by AOC, grade, and year. In terms of goal deviation variables, pos_f is the positive deviation for goal f and neg_f is the negative deviation for goal f.

The objective function of the OFM seeks to minimize the sum of the weighted goal deviations. The target for the first goal constraint is for the total structure distribution to equal the total documented authorizations. The target for the second goal constraint is for the COL structure distribution to equal the COL documented authorizations. The target for the third goal constraint is for the LTC structure distribution to equal the LTC documented authorizations. The target for the fourth goal constraint is for the MAJ structure distribution to equal the MAJ documented authorizations. The target for the last goal constraint is for the Company Grade (sum of 2LT, 1LT, and CPT) structure distribution to equal the Company Grade documented authorizations.

The hard constraints force inventory controls, promotion controls (floors and ceilings by AOC), and transition controls. These constraints are based on known promotion restrictions, transition data, and (primarily) decision-maker input. Some constraints apply to all AOCs; however, others are AOC-specific. Multiple constraints provide promotion floors and ceilings by year, AOC, and grade. These constraints are necessary to achieve decision-maker personnel requirements. In addition, inventory constraints are necessary to ensure proper roll-over from one time period to another by AOC and

TABLE 4.5: Promotion Percentage and Number Promoted by AOC and Rank

	CPT	MAJ	LTC	COL
65A	95%	83%	60%	43%
	7.6	3.9	1.8	0.5
65B	95%	70%	55%	39%
	18.4	10.2	5.1	1.5
65C	95%	89%	65%	41%
	9.5	5.2	3.0	1.1
65D	95%	61%	40%	20%
	85.8	37.1	7.0	1.2

by grade. Additional constraints ensure that promotions are considered only during those years when a promotion is feasible for each grade. A full description of the OFM formulation (sets, parameters, variables, objective function, and goal/hard constraints) is provided in the Appendix to this Chapter.

Upon formulating and solving the OFM, an objective function value of zero is obtained, indicating that all of the specified goal constraints were satisfied. In other words, the documented number of SP officer authorizations and the structure distribution determined by the model (for all the grades) are satisfied. The model solves in seconds.

The optimal number of accessions determined by the model are as follows: eight 1LT accessions for 65A, 20 2LT accessions for 65B, 10 2LT accessions for 65C, 85 1LT accessions for 65D, and five CPT accessions for 65D. In terms of promotions, the goal for the average promotion rate (across all ranks) for each of the AOCs is roughly 74% based on the DOPMA promotion standards (see Table 4.4), while the optimal solution achieves an average promotion rate of 70%. On average, 95% of the CPTs, 76% of the MAJs, 55% of the LTCs, and 36% of the COLs are promoted across all of the AOCs. Table 4.5 displays the promotion percentage and number promoted by officer AOC and rank (CPT through COL).

Similarly, on average 69% of CPTs, 75% of MAJs, and 81% of LTCs enter the rank (across all AOCs) and are remaining for consideration for promotion to the next higher rank. Table 4.6 shows officer continuation percentages for each AOC and rank (CPT, MAJ, LTC).

TABLE 4.6: Percentage of Remaining Officers Entering Rank for Promotion Consideration to Next Rank

	CPT	MAJ	LTC
65A	62%	75%	71%
65B	79%	91%	75%
65C	62%	88%	91%
65D	71%	47%	87%

The OFM is first solved with equal decision-maker weights (i.e., the documented authorizations at all of the grades are given equal weights). A sensitivity analysis is conducted to see if the adjustment of weights has any impact on the authorizations. Irrespective of the weights, all of the goals are achieved. However, the structure distribution for each grade depends on the weights that are chosen.

In addition to estimating the number of officer accessions and promotions, APPD uses the OFM as a workforce management decision-support tool for optimizing AOC grade structure, as well as assisting in workforce reduction/expansion program decisions. In particular, APPD uses the OFM results to generate histograms that visually represent the current workforce inventory (partitioned by the fiscal year the officer entered active duty) plotted against the 30-year lifecycle inventory as projected by OFM. For example, Figure 4.1 displays a generated histogram for the 65B AOC.

In Figure 4.1, the vertical bars of the histogram represent the actual inventory in terms of primary year group (cohort based on years of service) and rank. The black line represents the results of the OFM projections. Vertical bars that extend above the OFM line suggest that the AOC is over-strength (i.e., above the targeted number of officers) for that particular year group, and vertical bars that fall below the OFM line suggest that the AOC is under-strength (i.e., below the targeted number of officers) for that particular year group. APPD uses these results to provide the senior AMEDD leadership a quick reference to identify specific year groups that could require management focus and key personnel decisions. APPD periodically updates the histograms as personnel numbers change during the year and any time a new OFM is produced (usually annually).

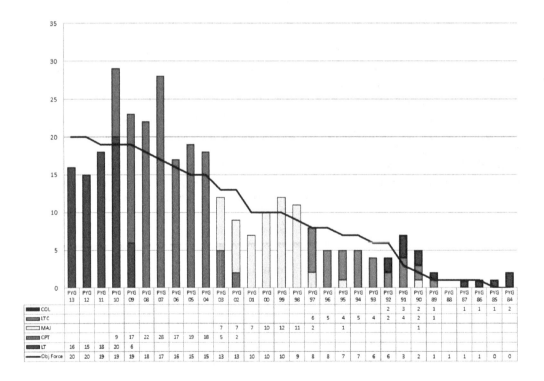

	PYG 13	PYG 12	PYG 11	PYG 10	PYG 09	PYG 08	PYG 07	PYG 06	PYG 05	PYG 04	PYG 03	PYG 02	PYG 01	PYG 00	PYG 99	PYG 98	PYG 97	PYG 96	PYG 95	PYG 94	PYG 93	PYG 92	PYG 91	PYG 90	PYG 89	PYG 88	PYG 87	PYG 86	PYG 85	PYG 84
COL																						2	3	2	1		1	1	1	2
LTC																	6	5	4	5	4	2	4	2	1					
MAJ									7	7	7	10	12	11	2		1						1							
CPT				9	17	22	28	17	19	18	5	2																		
LT	16	15	18	20	6																									
Obj Force	20	20	19	19	19	18	17	16	15	15	13	13	10	10	10	9	8	8	7	7	6	6	3	2	1	1	1	1	0	0

FIGURE 4.1: Objective force model vs. officer inventory.

In this case study, the Objective Force Model applied to the Medical Specialist Corps is described, which is currently in use by APPD for long-term workforce planning of medical professionals in the AMEDD. The expert workforce management decision support for estimating the proper number of accessions and promotions necessary to achieve the desired personnel structure under uncertainty is in itself of significant importance. The OFM provides tremendous value to APPD in terms of computation time, requiring only seconds to solve rather than months. This enables APPD to produce quick turn-around analysis in a transparent fashion that provides decision support that is invaluable, which was nearly impossible to do manually. One of the greatest benefits of the OFM is that APPD uses the optimization results to provide senior AMEDD leaders a quick reference to identify specific year groups that could require specific management focus as well as key personnel decisions in terms of hiring, promoting, or firing. As the US Army continues to change the size of its workforce structure, the use of the OFM will become vital for the AMEDD in its future medical professional reduction/expansion programs.

4.5.2 Grade Plate Roll-Down

The size of the U.S. Army has fluctuated during the time of the all-volunteer force, moving from active duty end strength of around 780,000 and 18 divisions to 480,000 with ten divisions. However, the force structure of the Army can vary wildly within potential end strength authorizations. As the size of companies, battalions, and brigades can change, so too can the number of soldiers needed to crew an armored fighting vehicle or radar system. The requirements for the US Army actually come from three sources: requirements for units, equipment, and personnel (Yuengert, 2018). The three sets of requirements are combined to detail the structure for the US Army.

The requirements for Army personnel come from two types of documents: tables of distribution and allowances (TDA) and tables of organization and equipment (TOE). The TOEs are developed for organizations of which the Army has multiple standardized units (e.g., brigade combat teams, infantry companies) and are developed by a proponent within the Training and Doctrine Command (e.g., infantry school). However, TDAs are developed for a particular unit (e.g., 780th Military Intelligence Brigade, 704th Military Intelligence Brigade) by their headquarters (e.g., US Army Intelligence and Security Command (INSCOM)). This puts the responsibility to create personnel documents both on the organization (e.g., INSCOM) as well as on the proponent (e.g., Intelligence Center of Excellence).

The statistical explanation of a system of possibly competing requirements generators set the stage for the project undertaken by Baez et al. (2015), a grade plate roll-down and creation of an Army officer grade distribution. Over the previous decade, the documented force had become significantly senior grade heavy, imposing pressure on the sustainability and feasibility of the personnel program. Continual force structure changes were adversely impacting the manpower program (i.e., grade distribution, leader-to-led ratios, promotion opportunity). Two distinct approaches to creating a sustainable and feasible program exist: adjusting the documentation or adjusting the inventory.

A sustainable grade structure drives the development and cost of the force. The number of officers and enlisted required at each grade impacts recruiting, training, retention, promotion, and development. The grade structure also determines competitiveness and thus drives the quality of the force. A misaligned grade structure can impact sustainability through career progression opportunities and selectivity rates, and it may require either a misalignment in inventory against requirements or unachievable retention rates (Figure 4.2). Figure 4.2 presents the composition of three alternative force levels (e.g., 482,400 Soldiers on Active Duty) and the percentage of officers, warrant officers, and enlisted soldiers in the requirements documents for each force level. The three force components are the Active, National Guard, and Reserve and the figure annotates the percentage of each manpower type in each component. For example, the Active Army (COMPO 1) proposed 450,000 force structure has 14.98% officers and the corresponding Reserve structure (COMPO 3) has 17.92% officers. The balance between grade adjustments (i.e., adjustments to inventory) and changing the grade plate (i.e., adjustments to documentation) had reached a tipping point, so a grade plate adjustment was needed.

A grade plate review is a deliberate effort to review and identify adjustments to documented positions based upon defined targets within the force to achieve a balanced grade structure. A grade plate reduction could be completed by the headquarters or through a consolidated effort by the document owners across the Army. Using considerations of grade distribution across each military specialty, targets were developed for proponents and units to guide the grade plate review. These targets required creating a template for officer requirements by career specialty and then applying this new requirement across the Army structure.

The enlisted force requirements are shaped by the Grade Cap Distribution Matrix (GCDM), and the warrant officer program is shaped by the Average Grade Distribution Matrix (AGDM). However, no such tool or guidance had been developed for the officer force, partially because of the number of specialty immaterial positions. After exploring the officer inventory data and the requirements data, a proposed officer GCDM was developed (Table 4.7). Table 4.7 details the number of positions by grade in a proposed 450,000 active component force structure and the proposed distribution of grades. For example, there are 22,183 requirements for O3/Captains in the proposed force structure document and the proposed GCDM would require 23,399 positions for Captains, an increase of 1,216 positions. As shown in Table 4.7, O2/First Lieutenants and O3/ Captains require more positions while O4/Majors, O5/Lieutenant Colonels, and O6/ Colonels require fewer positions than are currently in the force structure documents.

The key result of the analysis is the creation of an Officer Grade Distribution Matrix (OGDM) to complement the GCDM for enlisted soldiers and the AGDM for warrant

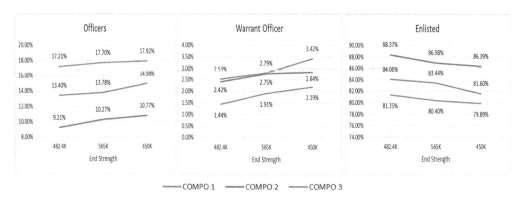

FIGURE 4.2: Force composition by Army component and grade.

TABLE 4.7: Officer Grade Distribution Matrix vs. Authorizations

	O2	O3	O4	O5	O6
450K Force	9,866	22,183	14,419	8,757	3,667
GCDM	11,083	23,399	13,287	7,651	3,472
Difference	−1,217	−1,216	1,132	1,106	195
450K %	16.8%	37.6%	24.5%	14.9%	6.2%
GCDM %	18.8%	39.7%	22.6%	13.0%	5.9%
Difference	−2.0%	−2.1%	1.9%	1.9%	0.3%

officers to provide guidance to the US Army commands, personnel proponents, and force management communities about the metrics of a feasible and sustainable force. This guidance provided new planning figures at the level of specialty and grade to maintain feasible career growth and opportunities. As future Army force structures are developed, this new tool guides force developers in adjusting to changing end strengths and potential future structures. This analysis also helped address concerns about the balance between generating force and operating force grade structures.

4.6 Future Work in Military Workforce Planning and Manpower Modeling

Military workforce planning and manpower modeling is an important area of interdisciplinary research. Many streams of research still await integration into decision-making models, for both military manpower planners and strategic business human resource planners. When examining the literature, several clear patterns are present. First, this area provides a rich field of application for mathematical and computational modeling. The early work in dynamic programming looked to the manpower problem for clear links to an application, and it generated papers both in the application areas as well as OR and applied mathematics journals.

The second observation is that changes in public policy require these previous results to be revisited. The early work was motivated by a need to move away from conscription. The Reagan-era buildups and policy changes necessitated better models to justify increasing budgets. The downsizing of the Clinton era provided another rich set of data to analyze the military members' personnel discount rate and the effectiveness of force shaping tools. Within the policy arena, Rostker (2006) provides a historical account of the development of the All-Volunteer Force and includes descriptions of the studies conducted to support the policy decision surrounding its creation. The *Handbook of Defense Economics* has an excellent chapter on the "New Economics of Manpower in the Post-Cold War Era" authored by Beth Asch, James Hosek, and John Warner (Hartley & Sandler, 2007). Changes in the 2020 National Defense Authorization Act (NDAA) will provide new opportunities for military workforce planning and manpower modeling.

The effects of several policy decisions have not yet been fully analyzed. For example, what is the full effect of the adjustments to the high three 40-year old military promotion system and the increased role of the Thrift Savings Plan with matching contributions? This has not yet been fully explored alongside direct commissioning and new talent management paradigms. What are the non-pecuniary effects of repeated combat employment on the all-volunteer force? The rates of re-enlistment in the post-9/11 environment are sure to exhibit unique characteristics. As the budgets move back toward balancing, how do we properly retain the enlisted force? What will be the most effective means of structuring a total compensation package for the next decade? Could the new research in dynamic contracting or multi-attribute auctions provide possible answers?

What should the structure and compensation of the officer corps look like? The behavior of the force is almost surely in flux, particularly with the increased operational tempo, perceived decreasing competition for promotion, and perceived military-civilian compensation difference. How have the most recent promotion trends affected the competition game, which has appeared in the past to keep the military from the necessitation of a pay skew as is common in the commercial sector?

In light of these timely and critically important questions, the field of military workforce planning and manpower modeling offers many opportunities for future research. There already exists a rich field of research, but there are many areas that need to be expanded and analyzed. The field offers many practical problems on which analytic and empirical modeling will produce interesting and policy-influencing research. These research challenges require the unique balance of qualitative and quantitative analysis that is aligned within the inherently multi-disciplinary purview of military OR.

Appendix

Objective Force Model (OFM) Formulation

Sets

G: index for officer grade with $g \in G$

I: index for officer career specialty (AOC) with $i \in I$

K: index for an officer's year in service with $k \in K$

F: index for each goal $f \in F$

Parameters

c_{ig}: documented authorizations for AOC i in grade g

r_{igk}: continuation (deterministic) rate for AOC i in grade g in year k

Cap_i: maximum allowable total structure distribution of AOC i

$Floor_i$: minimum acceptable total structure distribution for AOCi

Cap_{ig}: maximum allowable structure distribution of AOC i at grade g

$Floor_{ig}$: minimum acceptable structure distribution for AOC i at grade g

pf_{ig}: minimum promotion rate for AOC i at grade g

pc_{ig}: maximum promotion rate for AOCi at grade g

$D65X_g$: Specialist Corps immaterial authorizations for grade g

$D05A_g$: AMEDD immaterial authorizations for grade g

THS_g: transients, holdees, and students (THS) authorizations for grade g

w_f: decision – maker weight for goal f

Decision Variables and Goal Deviation Variables

p_{ig}: number of officers promoted in AOC i at grade g, for $g = 4, 5, 6$

a_{ig}: number of officers accessed (hired) for AOC i at grade $g = 1, 2, 3$

d_{ig}: structure distribution (officer quantity) for AOC i in grade g

Inv_{igk}: projected inventory of AOC i in grade g in year k

pos_f: positive deviation for goal f

neg_f: negative deviation for goal f

Objective Function

$$\text{Minimize} \, Z = \sum_f w_f(pos_f + neg_f) \tag{1}$$

Goal Constraints

$$\sum_g \sum_i d_{ig} - \left(\sum_g \sum_i \left(c_{ig} + D65X_g + D05A_g + THS_g \right) \right) - pos_{f=1} + neg_{f=1} = 0 \tag{2}$$

$$\sum_i d_{ig=6} - \left(\sum_i c_{ig=6} + D65X_{g=6} + D05A_{g=6} + THS_{g=6} \right) - pos_{f=2} + neg_{f=2} = 0 \tag{3}$$

$$\sum_i d_{ig=5} - \left(\sum_i c_{ig=5} + D65X_{g=5} + D05A_{g=5} + THS_{g=5} \right) - pos_{f=3} + neg_{f=3} = 0 \tag{4}$$

$$\sum_i d_{ig=4} - \left(\sum_i c_{ig=4} + D65X_{g=4} + D05A_{g=4} + THS_{g=4} \right) - pos_{f=4} + neg_{f=4} = 0 \tag{5}$$

$$\sum_{g=1}^{3} \sum_i d_{ig} - \left(\sum_{g=1}^{3} \sum_i \left(c_{ig} + D65X_g + D05A_g + THS_g \right) \right) - pos_{f=5} + neg_{f=5} = 0 \tag{6}$$

The objective function in (1) of the OFM seeks to minimize the sum of the weighted goal deviations. The target for the first goal constraint in (2) is for the total structure distribution over each grade and AOC to equal the total documented authorizations (over each grade and AOC as well as the SP immaterial, AMEDD immaterial, and THS). The target for the second goal constraint in (3) is for the total structure distribution of COLs over each AOC to equal the COL documented authorizations over each AOC as well as the SP immaterial, AMEDD immaterial, and THS. The target for the third goal constraint in (4) is for the total structure distribution of LTCs over each AOC to equal the LTC documented authorizations over each AOC as well as the SP immaterial, AMEDD immaterial, and THS. The target for the fourth goal constraint in (5) is for the total structure distribution of MAJs over each AOC to equal the MAJ documented authorizations over each AOC as well as the SP immaterial, AMEDD immaterial, and THS. The target for the last goal constraint in (6) is for the total structure distribution of Company Grade (sum of 2LT, 1LT and CPT) officers over each AOC to equal the Company Grade documented authorizations over each AOC as well as the SP immaterial, AMEDD immaterial, and THS.

Hard Constraints

$$\sum_{g} d_{i=1,g} \leq Cap_{i=1} \tag{7}$$

$$\sum_{g} d_{i=3,g} \geq Floor_{i=3} \tag{8}$$

$$\sum_{g} d_{i=1,g} \geq Floor_{i=1} \tag{9}$$

$$\sum_{g} d_{i=3,g} \leq Cap_{i=3} \tag{10}$$

$$\sum_{g} d_{i=2,g} \leq Cap_{i=2} \tag{11}$$

$$\sum_{g} d_{ig} \geq \sum_{g} c_{ig} \quad \forall i \tag{12}$$

$$\sum_{g=1}^{3} d_{ig} \geq \sum_{g=1}^{3} c_{ig} \quad \forall i \tag{13}$$

$$d_{i=3,g=5} \geq Floor_{i=3,g=5} \tag{14}$$

$$d_{i=3,g=6} \geq Floor_{i=3,g=6} \tag{15}$$

$$d_{i=1,g=6} \leq Cap_{i=1,g=6} \tag{16}$$

$$d_{i=2,g=6} \geq Floor_{i=2,g=6} \tag{17}$$

$$d_{i=1,g=5} \leq Cap_{i=1,g=5} \tag{18}$$

$$d_{i=1,g=6} \geq Floor_{i=1,g=6} \tag{19}$$

$$d_{ig} \geq c_{ig} \quad \forall \, i, g = 4,5,6 \tag{20}$$

The constraint in (7) is used to place a maximum allowable total structure distribution for AOC 65A, whereas the constraint in (8) is used to place a minimum acceptable total structure distribution for AOC 65C. The constraint in (9) is used to place a minimum allowable total structure distribution for AOC 65A, while the constraint in (10) is used to place a maximum allowable total structure distribution for AOC 65C. The constraint in (11) is used to place a maximum allowable total structure distribution for AOC 65B. The constraints in (12) ensure that the total structure distribution must meet or exceed the total documented authorizations for each AOC i. The constraints in (13) ensure that the total structure distribution for Company Grade (sum of 2LT, 1LT, and CPT) officers must meet or exceed the total documented authorizations for Company Grade (sum of 2LT, 1LT and CPT) officers for each AOC i. The constraint in (14) ensures that LTC structure distribution for AOC 65C must meet or exceed the minimum acceptable LTC structure distribution for AOC 65C. The constraint in (15) ensures that COL structure distribution for AOC 65C must meet or exceed the minimum acceptable COL structure distribution for AOC 65C. The constraint in (16) ensures that COL structure distribution for AOC 65A must be less than or equal to the maximum allowable COL structure distribution for AOC 65A. The constraint in (17) ensures that COL structure distribution for AOC 65B must meet or exceed the minimum acceptable COL structure distribution for AOC 65B. The constraint in (18) ensures that LTC structure distribution for AOC 65A must be less than or equal to the maximum allowable LTC structure distribution for AOC 65A. The constraint in (19) ensures that COL structure distribution for AOC 65A must meet or exceed the minimum acceptable COL structure distribution for AOC 65A. The constraints in (20) ensure that the MAJ, LTC, and COL structure distributions must meet or exceed MAJ, LTC, and COL documented authorizations for each AOC i.

$$p_{i,g=4} \geq pf_{ig=4} * Inv_{i,g=3,k=10} \quad \forall \, i \tag{21}$$

$$p_{i,g=5} \geq pf_{ig=5} * Inv_{i,g=4,k=16} \quad \forall \, i \tag{22}$$

$$p_{i,g=6} \geq pf_{ig=6} * Inv_{i,g=5,k=21} \quad \forall \, i \tag{23}$$

$$p_{i,g=4} \leq pc_{ig=4} * Inv_{i,g=3,k=10} \quad \forall \, i \tag{24}$$

$$p_{i,g=5} \leq pc_{ig=5} * Inv_{i,g=4,k=16} \quad \forall \, i \tag{25}$$

$$p_{i,g=6} \leq pc_{ig=6} * Inv_{i,g=5,k=21} \quad \forall \, i \tag{26}$$

The constraints in (21)–(23) ensure that the number of resultant Field Grade (MAJ, LTC, COL) promotions must be greater than or equal to the minimum number of promotions (product of the minimum promotion rate and pool) for each respective Field Grade, for each AOC i. The constraints in (24)–(26) ensure that the number of resultant

Field Grade (MAJ, LTC, COL) promotions must be less than or equal to the maximum number of promotions (product of the maximum promotion rate and pool) for each respective Field Grade, for each AOC i.

$$d_{ig} = \sum_k Inv_{igk} \quad \forall \; i,g \tag{27}$$

$$Inv_{i,g=1,k=1} = a_{ig=1}{}^*r_{i,g=1,k=1} \quad \forall \; i \tag{28}$$

$$Inv_{i=1,g=1,k=3} = Inv_{i=1,g=1,k=2}{}^*r_{i=1,g=1,k=3} \tag{29}$$

$$Inv_{i=1,g=2,k=3} = (Inv_{i=1,g=2,k=2} + a_{i=1,g=2}){}^*r_{i=1,g=2,k=3} \tag{30}$$

$$Inv_{igk=3} = Inv_{igk=2}{}^*r_{igk=3}, \quad \forall \; g,i = 2,3,4 \tag{31}$$

$$Inv_{i=1,g=2,k=2} = Inv_{i,g=1,k=1}{}^*r_{i,g=2,k=2} \tag{32}$$

$$Inv_{i,g=2,k=2} = ((Inv_{i,g=1,k=1}{}^*pf_{i,g=2}) + a_{ig=2}){}^*r_{i,g=2,k=2} \quad \forall \; i = 2,3,4 \tag{33}$$

$$Inv_{igk} = Inv_{igk-1}{}^*r_{igk}, \quad \forall \; k = 5-10,12-16,18-21,23-30 \quad \forall \; g,i \tag{34}$$

$$Inv_{i,g=1,k=2} = Inv_{i,g=1,k=1}{}^*(1 - pf_{i,g=2}){}^*r_{i,g=1,k=2} \quad \forall \; i \tag{35}$$

$$Inv_{i,g=3,k=4} = ((Inv_{i,g=2,k=3}{}^*pf_{i,g=3}) + a_{ig=3}){}^*r_{i,g=3,k=4} \quad \forall \; i \tag{36}$$

$$Inv_{i,g=2,k=4} = Inv_{i,g=2,k=3}{}^*(1 - pf_{i,g=3}){}^*r_{i,g=2,k=3} \quad \forall \; i \tag{37}$$

$$Inv_{i,g=4,k=11} = p_{ig=4} \quad \forall \; i \tag{38}$$

$$Inv_{i,g=3,k=11} = (Inv_{i,g=3,k=10} - p_{ig=4}){}^*r_{i,g=3,k=11} \quad \forall \; i \tag{39}$$

$$Inv_{i,g=5,k=17} = p_{ig=5} \quad \forall \; i \tag{40}$$

$$Inv_{i,g=4,k=17} = (Inv_{i,g=4,k=16} - p_{ig=5}){}^*r_{i,g=4,k=17} \quad \forall \; i \tag{41}$$

$$Inv_{i,g=6,k=22} = p_{ig=6} \quad \forall \; i \tag{42}$$

$$Inv_{i,g=5,k=22} = (Inv_{i,g=5,k=21} - p_{ig=6}){}^*r_{i,g=5,k=22} \quad \forall \; i \tag{43}$$

$$p_{ig} \geq 0 \; \forall \; i, \; g = 4,5,6; \quad a_{ig} \geq 0 \text{ and integer } \forall \; i, \; g = 1,2,3;$$

$$d_{ig} \geq 0 \text{ and integer } \forall \; i, \; g; Inv_{igk} \geq 0 \text{ and integer } \forall \; i, \; g; \; pos_f, \; neg_f \geq 0 \; \forall \; f \tag{44}$$

The constraints in (27) assign the structure distribution as the total projected inventory (over all years) for each AOC i and grade g. The constraints in (28) assign the inventory for year $k=1$ and grade $g=1$ for each AOC i. The constraint in (29) assigns the inventory for year $k=3$, grade $g=1$, and $i=1$, whereas the constraint in (30) assigns the inventory for year $k=3$, grade $g=2$, and $i=1$. The constraints in (31) assign the inventory for $k=3$ for all grades and $i=2, 3, 4$. The constraint in (32) assigns the inventory for $i=1$, $g=2$, and $k=2$, whereas the constraints in (33) assign the inventory for $g=2$, $k=2$, and $i=2, 3, 4$. The constraints in (34) assign the inventory for all years that are not year $k=1, 3$ or a promotion year, for each grade g and each AOC i. The constraints in (35) assign the inventory for year $k=2$ and grade $g=1$ for each AOC i. The constraints in (36) assign the inventory for year $k=4$ and grade $g=3$ for each AOC i. The constraints in (37) assign the inventory for year $k=4$ and grade $g=2$ for each AOC i. The constraints in (38) assign the inventory for year $k=11$ and grade $g=4$ for each AOC i. The constraints in (39) assign the inventory for year $k=11$ and grade $g=3$ for each AOC i. The constraints in (40) assign the inventory for year $k=17$ and grade $g=5$ for each AOC i. The constraints in (41) assign the inventory for year $k=17$ and grade $g=4$ for each AOC i. The constraints in (42) assign the inventory for year $k=22$ and grade $g=6$ for each AOC i. The constraints in (43) assign the inventory for year $k=22$ and grade $g=5$ for each AOC i. The constraints in (44) represent the non-negativity and integer constraints for the decision and deviational variables.

References

Abdessameud, O., Van Utterbeeck, F., Van Kerckhoven, J., and Guerry, M. (2018). "Military Manpower Planning: Towards Simultaneous Optimization and Statutory and Competence Logistics using Population based Approaches." Proceedings of ICORES 2018 - 7th International Conference on Operations Research and Enterprise Systems: 178–185.

Abrams, J. (1957). "Military Applications of Operational Research." *Operations Research* 5(3): 434–440.

Arguden, R. (1988). "There Is No Free Lunch: Unintended Effects of the New Military Retirement System." *Journal of Policy Analysis and Management* 7(3): 529–541.

Asch, B. J., Hosek, J. R., & Warner, J. T. (2007). New economics of manpower in the post-cold war era. *Handbook of defense economics, 2*, 1075–1138.

Asch, B., Hosek, J., Arkes, J., Fair, C., Sharp, J., and Totten, M. (2002). "Military Recruiting and Retention After the Fiscal Year 2000 Military Pay Legislation." Santa Monica, CA: RAND Corporation, MR-1532-OSD.

Asch, B. and Warner, J. (2001). "A Theory of Compensation and Personnel Policy in Hierarchical Organizations with Application to the United States Military." *Journal of Labor Economics* 19(3): 523–562.

Bartholomew, D., Forbes, A., and McClean, S. (1991). *Statistical Techniques for Manpower Planning.* Chichester; New York, Wiley.

Bastian, N. (2015). "Multiple Criteria Decision Engineering To Support Management In Military Healthcare And Logistics Operations." Published doctoral dissertation, University Park, Pennsylvania: Pennsylvania State University.

Bastian, N., McMurry, P., Fulton, L., Griffin, P., Cui, S., Hanson, T., and Srinivas, S. (2015). "The AMEDD Uses Goal Programming to Optimize Workforce Planning Decisions." *Interfaces* 45(4): 305–324.

Baez, F., Hall, A., Needham M., and Townsend, R. (2015). "Army Officer Grade Distribution." Presented at the 2015 INFORMS Annual Meeting in Philadelphia, PA.

Bellman, R. (1954) The Theory of Dynamic Programming." *Bulletin of the American Mathematical Society* **60**(6): 503–516.

Bellman, R. (1957). *Dynamic Programming*. Princeton, NJ: Princeton University Press.

Bres, E., Burns, D., Charnes, A., and Cooper, W. (1980). A Goal Programming Model for Planning Officer Accessions. *Management Science* **26**(8): 773–783.

Cashbaugh, D., Hall, A., Kwinn, M., Sriver, T., and Womer, N. (2007). "Manpower and Personnel." In A. L. L. Rainey, *Methods for Conducting Military Operational Analysis*. Military Operations Research Society.

Campbell, C. (1994). "Wage Change and the Quit Behavior of Workers: Implications for Efficiency Wage Theory." *Southern Economic Journal* **61**(1): 133–148.

Charnes, A., Cooper, W. W., & Niehaus, R. J. (1978). Management Science Approaches to Manpower Planning and Organizational Design. *Studies in the Management Science*, 8.

Coates, H., Silvernail, T., Fulton, L., and Ivanitskaya, L. (2011). "The Effectiveness of the Recent Army Captain Retention Program." *Armed Forces and Society* **37**(1): 5–18.

Colarusso, M., Heckel, K., Lyle, D. and Skimmyhorn, W. (2016). "Starting Strong: Talent-Based Branching of Newly Commissioned U.S. Army Officers." Strategic Studies Institute.

Corbett, J. (1995). "Military Manpower Planning: Optimization Modeling for the Army Officer Accession/Branch Detail Program." Master's thesis, Monterey, CA: Naval Postgraduate School.

Cotton, J. and Tuttle, J. (1986). "Employee Turnover: A Meta-Analysis and Review with Implications for Research." *The Academy of Management Review* **11**(1): 55–70.

Dabkowski, M., Huddleston, S., Kucik, P. and Lyle, D. (2010). "Shaping Senior Leader Officer Talent: How Personnel Management Decisions and Attrition Impact the Flow of Army Officer Talent Throughout the Officer Career Model." Proceedings of the 2010 Winter Simulation Conference.

Dabkowski, M., Huddleston, S., Kucik, P. and Lyle, D. (2011). "Shaping Senior Leader Officer Talent: Using a Multi-dimensional Model of Talent to Analyze the Effect of Personnel Management Decisions and Attrition on the Flow of Army Officer Talent Throughout the Officer Career Model." Proceedings of the 2011 Winter Simulation Conference.

Dailey, J. (1958). "Prediction of First-Cruise Reenlistment Rate." *Operations Research* **6**(5): 686–692.

Daula, T. and Moffitt, R. (1995). "Estimating Dynamic Models of Quit Behavior: The Case of Military Reenlistment." *Journal of Labor Economics* **13**(3): 499–523.

Davies, G. (1982). "Control of Grade Sizes in a Partially Stochastic Markov Manpower Model." *Journal of Applied Probability* **19**(2): 439–443.

Department of the Army Pamphlet 600-4 (DA PAM 600-4). (27 June 2007). "Army Medical Department Officer Development and Career Management." Washington, DC.

Drui, A. (1963). "The Use of Regression Equations to Predict Manpower Requirements." *Management Science* **9**(4): 669–677.

Edwards, J. (1983). *Manpower Planning: Strategy and Techniques in an Organizational Context*. Chichester [West Sussex]; New York: Wiley.

Evans, L. and Bae, K. (2017). "Simulation-Based Analysis of a Forced Distribution Performance Appraisal System." *Journal of Defense Analytics and Logistics* **1**(2): 120–136.

Evans, L. A., & Bae, K. H. G. (2019). US Army performance appraisal policy analysis: a simulation optimization approach. *The Journal of Defense Modeling and Simulation* **16**(2): 191–205.

Feichtinger, G. and Mehlmann, A. (1976). "The Recruitment Trajectory Corresponding to Particular Stock Sequences in Markovian Person-Flow Models." *Mathematics of Operations Research* **1**(2): 175–184.

Fisher, F. and Morton, A. (1967a). "Reenlistments in the U.S. Navy: A Cost Effectiveness Study." *The American Economic Review* **57**(2): 32–38.

Fisher, F. and Morton, A. (1967b). "The Costs and Effectiveness of Reenlistment Incentives in the Navy." *Operations Research* **15**(3): 373–387.

Flynn, J. (1975a). "Retaining Productive Units: A Dynamic Programming Model with a Steady State Solution." *Management Science* **21**(7): 753–764.

Flynn, J. (1975b). "Steady State Policies for a Class of Deterministic Dynamic Programming Models." *SIAM Journal on Applied Mathematics* **28**(1): 87–99.

Flynn, J. (1979). "Steady State Policies for Deterministic Dynamic Programs." *SIAM Journal on Applied Mathematics* **37**(1): 128–147.

Gass, S. (1991). "Military Manpower Planning Models." *Computers & Operations Research* **18**(1): 65–73.

Gass, S., Collins, R., Meinhardt, C., Lemon, D., and Gillette, M. (1988). "The Army Manpower Long-Range Planning System." *Operations Research* **36**(1): 5–17.

Gotz, G. and McCall, J. (1983). "Sequential Analysis of the Stay/Leave Decision: U.S. Air Force Officers." *Management Science* **29**(3): 335–351.

Hall, A. (2004). "Validation of the Enlisted Grade Model Gradebreaks". In Proceedings of the 36th Conference on Winter Simulation (WSC '04). 921–925.

Hall, A. (2009). "Simulating and Optimizing: Military Manpower Modeling and Mountain Range Options." Published doctoral dissertation, University of Maryland.

Hall, A., and Fu, M. (2015). "Optimal Army Officer Force Profiles." *Optimization Letters* **9**(8), 1769–1785.

Hall, A., and Schultz, B. (2017). Direct Commission for Cyberspace Specialties. *The Cyber Defense Review* **2**(2), 111–124.

Hall, M. (1999). "Optimal Scheduling of Army Initial Entry Training Courses." Master's thesis, Monterey, CA: Naval Postgraduate School.

Hartley, K. and Sandler, T. (2007). *Handbook of Defense Economics, Volume 2.* Amsterdam; New York: Elsevier.

Henry, T. and Ravindran, A. (2005). "A Goal Programming Application for Army Officer Accession Planning." *INFOR* **43**(2): 111–120.

Jaquette, D. and Nelson, G. (1976). "Optimal Wage Rates and Force Composition in Military Manpower Planning." *Management Science* **23**(3): 257–266.

Kinstler, D., Johnson, R., Richter, A. and Kocher, K. (2008). "Navy Nurse Corps Manpower Management Model." *Journal of Health Organization and Management* **22**(6): 614–626.

Lakhani, H. (1988). "The Effect of Pay and Retention Bonuses on Quit Rates in the U.S. Army." *Industrial and Labor Relations Review* **41**(3): 430–438.

Lee, T. and Mowday, R. (1987). "Voluntarily Leaving an Organization: An Empirical Investigation of Steers and Mowday's Model of Turnover." *The Academy of Management Journal* **30**(4): 721–743.

Lesinski, G., Pinter, J., Kucik, P., and Lamm, G. (2011). "Officer Accessions Flow Model." Technical Report: DSE-TR-1103. Operations Research Center of Excellence, U.S. Military Academy, West Point, NY.

McCall, J. and Wallace, N. (1969). "A Supply Function of First-Term Re-Enlistees to the Air Force." *The Journal of Human Resources* **4**(3): 293–310.

McDonald, J., White, E., Hill, R., and Pardo, C. (2017). "Forecasting U.S. Army Enlistment Contract Production in Complex Geographical Marketing Areas." *Journal of Defense Analytics and Logistics* **1**(1), 69–87.

McMurry, P., Fulton, L., Brooks, M., and Rogers, J. (2010). "Optimizing Army Medical Department Officer Accessions." *Journal of Defense Modeling and Simulation* **7**(3): 133–143.

Mehay, S. and Hogan, P. (1998). "The Effect of Separation Bonuses on Voluntary Quits: Evidence from the Military's Downsizing." *Southern Economic Journal* **65**(1): 127–139.

Mehlmann, A. (1977). "Markovian Manpower Models in Continuous Time." *Journal of Applied Probability* **14**(2): 249–259.

Mitchell, T. R., Holtom, B. C., Lee, T. W., Sablynski, C. J., and Erez, M. (2001). "Why People Stay: Using Job Embeddedness to Predict Voluntary Turnover." *The Academy of Management Journal* **44**(6): 1102–1121.

Price, W. (1978). "Solving Goal-Programming Manpower Models Using Advanced Network Codes." *The Journal of the Operational Research Society* **29**(12): 1231–1239.

Price, W. and Piskor, W. (1972). "The Application Of Goal Programming To Manpower Planning." *INFOR* **10**(3): 221–231.

Puterman, M. (2014). *Markov Decision Processes: Discrete Stochastic Dynamic Programming*. John Wiley & Sons.

Rizvi, S. (1986). "Manpower Demand Modelling." *The Statistician* **35**(3): 353–358.

Rostker, B. (2006). *I Want You!: The Evolution of the All-Volunteer Force*. Santa Monica, CA: RAND.

Rostker, B., Thie, H., Lacy, J., Kawata, J., and Purnell, S. (1993). "The Defense Officer Personnel Management Act of 1980, A Retrospective Assessment." RAND Research Report \#R4246. Retrieved 24 March 2014. Available from http://www.rand.org /pubs/reports/ R4246.html.

Schneider, W. (1981). "Personnel Recruitment and Retention: Problems and Prospects for the United States." *Annals of the American Academy of Political and Social Science* **457**: 164–173.

Schniederjans, M. (2012). *Goal Programming: Methodology and Applications: Methodology and Applications*. Springer Science & Business Media.

Shrimpton, D. and Newman, A. (2005). "The U.S. Army Uses a Network Optimization Model to Designate Career Fields for Officers." *Interfaces* **35**(3): 230–237.

Silverman, J., Steuer, R., and Whisman, A. (1988). "A Multi-Period, Multiple Criteria Optimization System for Manpower Planning." *European Journal of Operational Research* **34**: 160–170.

United States Code, Title 10, Chapter 36. "Promotion, Separation, and Involuntary Retirement of Officers on the Active-Duty List."

Vajda, S. (1978). *Mathematics of Manpower Planning*. Chichester, UK; New York: Wiley.

Wallace, R., Colarusso, M., Hall, A., Lyle, D. and Walker, M. (2015). "Paid to Perform – Aligning Total Military Compensation with Talent Management." Strategic Studies Institute.

Wardynski, C., Lyle, D. and Colarusso, W. (2009a). "Towards a U.S. Army Officer Corps Strategy: A Proposed Human Capital Model Focused Upon Talent." Strategic Studies Institute.

Wardynski, C., Lyle, D., and Colarusso, W. (2009b). "Towards a U.S. Army Officer Corps Strategy: Talent: Implications for a U.S. Army Officer Corps Strategy." Strategic Studies Institute.

Warner, J. and Goldberg, M. (1984). "The Influence of Non-Pecuniary Factors on Labor Supply: The Case of Navy Enlisted Personnel." *The Review of Economics and Statistics* **66**(1): 26–35.

Weigel, H. and Wilcox, S. (1993). "The Army's Personnel Decision Support System." *Decision Support Systems* **9**: 281–306.

Wilson, N. and NATO Science Committee. (1969). Manpower Research: The Proceedings of a Conference Held Under the Aegis of the N.A.T.O. Scientific Affairs Committee in London from 14th–18th August, 1967. London, English Universities P.

Yamada, W. (2000). "An Infinite Horizon Army Manpower Planning Model." Master's thesis, Monterey, CA: Naval Postgraduate School.

Yuengert, L. (2018). *How the Army Runs. A Senior Leader Reference* Handbook 2017–2018. Carlisle, PA: U.S. Army War College.

Chapter 5

Military Assessments

Lynette M.B. Arnhart and Marvin L. King III

5.1   Introduction

Assessments based on thorough analysis are crucial to underpinning good decisions. This is true not only in business but also in the military where the lives of service members, partners, and allies may be at stake. In an era of readily available data, the critical analysis of meaningful data that leads to the development of information and ultimately the imparting of knowledge is often overlooked in favor of the "art of war." As Gary Loveman notes in his foreword to *Competing on Analytics*, "Decision making, especially at high levels, not only fails to demand rigor and dispassionate analysis, but often champions the opposite" (Davenport & Harris, 2007). In this chapter, we focus on a particular type of military decision making known as assessments.

To introduce personnel to the topic of assessments, we outline the current state of assessments in the Department of Defense, we provide common definitions, cover the importance of objectives in assessment, and provide a generalized framework for conducting assessments. As there is no one formula or model for conducting an assessment, we then provide information on the best practices for three broad types of military assessment; campaign assessments, operation assessments, and training assessments. We conclude with data requirements, a topic common to all types of assessment, and a section on how to initiate an assessment process within a staff. The objective of the chapter is to help future assessors understand difficulties, nuances, and best practices of assessments and encourage them to be innovative and build cross-staff relationships that will enable the process.

5.1.1   Assessments in the Department of Defense

Assessments in the Department of Defense cover a wide variety of reporting mechanisms and military decision analysis tools. The authors have previously noted over a dozen different types of assessments employed at various levels by differing entities in the military.[1] These assessments range from annual evaluation of on-going campaigns to periodic reviews of current operations to quick turn analysis to provide near-immediate feedback to leaders.

However, there is little definitive doctrine, and almost no training or education to assist practitioners. This results in a gap in expertise. Therefore, many commands rely on operations research analysts to employ their quantitative training to lead the organization's efforts to design, collect data for, write, and present assessments. There are no existing definitive methodologies, frameworks, models, or process that are shown to be broadly successful. This chapter will provide advice and best practices for executing each type of assessment. Throughout this chapter, the authors will discuss assessments in the context of warfighting, field operations, and training. However, much of the thoughts regarding assessments can be adapted to garrison, staff work, and long-term projects.

5.1.2   What Assessments Are and Why Assessments Are Important

Assessments are periodic or continuous evaluations of situations, operations, or activities that support senior leaders' decision making by ascertaining progress toward an objective, creating an effect, or attaining an end state. Assessments also identify gaps and risks regarding accomplishing a task for the purpose of developing, adapting, and refining plans and for making strategies, campaigns, and operations more effective (adapted from JP 5.0). In this manner, good assessments enable the commander to see gaps, seams, shortfalls, and opportunities in current and planned operations. Gaps are newly identified facets of the problem which were not addressed in the plan. Seams are points of friction between missions, programs, or projects that an adversary may be able to exploit. Shortfalls are missions, programs, or policies that are not achieving intended effects due to failure, schedule, resourcing, or other causes. Opportunities are combinations of favorable events that provide an advantage if exploited in a timely manner. Based on this knowledge, the leader can make decisions to mitigate, modify, resource, leverage, and/or exploit based on the higher-level commander's intent. For this reason, while an assessment seeks to evaluate what has occurred in the past, it is in essence determining options for operations, activities, and actions that will occur in the future.

According to the Multi-Service Tactics, Techniques, and Procedures for Operation Assessment, assessments need to answer what happened and why, identify risks and opportunities, and make recommendations for future actions, activities, or operations (US Department of Defense, 2015). To do this, staffs require an assessment framework. Often, assessments are not considered until after the commander approves the plan and the critical questions of "How will we know when ...?" are not addressed until progress updates are demanded of the organization. Experienced assessors posit that it is ideal for assessments to be built into the plan rather than afterward. This means that assessors and planners work hand-in-hand to design and assemble a plan using objectives that the staff can realistically qualitatively and quantitatively evaluate. This is based on an assumption of clear intent and well-defined objectives.

5.1.3 The Importance of Objectives

In recent history, objectives have been fuzzy and obscure at the highest levels of military operations. For instance, Steven Metz, an expert in national security, claims that the United States no longer has a "discernable vision for world order" (Metz, 2018) and likens the objectives in Afghanistan as being akin to pushing the Sisyphean boulder up the hill (Metz, 2015). Contrary to this are the very distinct objectives in the 2018 National Defense Strategy (US Department of Defense, 2018). The difficulties come in how to assess objectives such as "Dissuading, preventing, or deterring state adversaries and non-state actors from acquiring, proliferating, or using weapons of mass destruction" or "Preventing terrorists from directing or supporting external operations against the United States homeland and our citizens, allies, and partners overseas" (US Department of Defense, 2018). While these are sound national defense strategic objectives, they are very difficult to assess and often result in questions such as "How do you measure deterrence and prevention?"

Fortunately, the military adapts national level objectives into military plans wherein military leaders outline their intent and objectives to achieve their directed tasks. These plans – whether campaigns for a combatant command, contingency operations, unit operations, or training activities – benefit from assessments based on the purpose outlined above. Each type of planned military activity results in a corresponding assessment; the most common types analysts encounter are Campaign Assessments, Operations Assessments, and Training Assessments. The authors will describe each of these Assessments based on experience in current named operations and as assessments leads in combatant commands.

5.1.4 A Generalized Framework for Assessments

During the war in Vietnam, the US military attempted to build an analytic assessment framework based on quantifiable data. Lacking in qualitative measures, the process failed to capture the complexities required to adequately describe the environment. Unfortunately, this experience led to the view that assessments are laborious, latent, and lacking. It is subsequent to post-September 11th counter-terror and counter-insurgency operations that this view has begun to dissipate. To ensure that assessments are relevant, timely, and useful, analysts must think critically about the structure of assessments and the framework before trying to conduct them. Unfortunately, most staffs hastily build assessments usually without considerable thought and after the plan is developed, approved, and implemented thus perpetuating misperceptions regarding assessment. Before delving into assessments themselves, it is worth discussing a generalized framework concept for developing assessments.

The first consideration the staff must consider when designing an assessment is to understand the needs of the leadership. This starts with understanding the strategic guidance and mission to the organization. The second consideration is understanding how assessment fits into the planning cycle. Figure 5.1 shows the military operations process, highlighting that assessment is part of every phase of the cycle. Assessment is continuous; reporting is periodic and depicts the understanding of the situation at a point in time.

With that understanding, the framework shown in Figure 5.2 can be used to develop an assessment schema. This framework outlines step by step questions that an assessment team can use to build their assessment mechanism. Organizational leadership, planners, and assessors should understand that there is not one "right" answer on assessment design any more than one single plan fits all missions. The team needs to work together to build what works for their leadership and their organization.

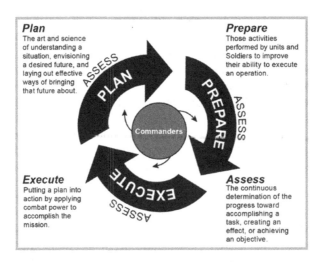

FIGURE 5.1: The military operations process (US Department of the Army, 2012).

Activity	Considerations
1. Identify the Ends	What is driving the requirement? How will the assessment be used? What questions are to be answered? Who is the consumer? What is the supported outcome?
2. Consider the Ways	How can an assessment be conducted and implemented in the organization? To what type assessment is leadership most amenable? Who are the stakeholders? What are their requirements? How can they best be involved? Is the assessment seeking to fill multiple requirements? What expertise exists in the organization or is otherwise accessible? What resources are available? What are the required timelines? Who will be the lead? Who will facilitate the process? Who will champion assessment? How can the assessment be nested with higher and lower organizations?
3. Determine Means	What metrics and data are needed and can be obtained? What best practices can be implemented and how? What are the mechanics of the proposed assessment? Who, specifically, will participate in the process? Do they need to be formally directed to do so? What is the periodicity? Do you need to communicate with with Allies and Partners? How will you communicate with them?
4. Design and Test a Prototype	What is the plan for collecting, analyzing, summarizing, presenting and back briefing? What worked? What can be improved? Are the technical means adequate? Did the communication channels work? What feedback do the stakeholders offer? How can it be integrated? Revise and update the assessment plan.
5. Conduct the Assessment	Establish a cross-staff assessment working group or similar cross-staff entity. Implement the initial plan. Establish a schedule. Launch a continuous process improvement plan to capture lessons learned and ensure that the assessment continues to meet the requirements of the organization.

FIGURE 5.2: Authors' suggested assessment framework.

In any assessment, the importance of a cross-staff team cannot be understated. Organizing a cross-staff team, including intelligence, operations, logistics, plans, and special staff improves the organizational understanding of linkages within the planned and on-going operations, activities, and actions and offers insights into anticipated and unanticipated primary, secondary, and tertiary effects. This team often becomes known as the assessments team or assessments working group, used interchangeably hereafter. Having discussed both assessments and building an assessment framework, the authors address the different types of assessments.

5.2   Campaign Assessments

5.2.1   Purpose of Campaign Assessments

Campaign Assessments use quantitative and qualitative methods to determine progress in a military campaign, normally at the Joint Staff, Combatant Command, Service Component, or Joint Task Force levels. These types of assessments have little standardization across the Department of Defense, with assessment staffs relying on trial and error to develop best practices and a supporting community of interest rather than doctrine, education, and training of assessors. The best practices for these assessments include strategic questions, theories of change, standards-based assessments, and written risk assessments, supplemented by traditional operations research data analysis and visualization techniques.

5.2.2   Process and Techniques

There are a number of techniques that can be used in the process of developing an assessment. This section outlines several that are considered best practice.

5.2.2.1   Theories of Change

A theory of change is a hypothesis that designates the if-then relationship between friendly action and result in the operating environment. There is no formal output for this best practice, but it helps to inform all other aspects of the assessment. A simple theory of change helps to refine the planner's objective, facilitates clear metrics, and catalyzes the staff around a common language and unity of thought in how they approach realizing an objective. Assessors may find list of relevant theories of change in "Theories of Change and Indicator Development in Conflict Management and Mitigation" by Susan Allen Nan and Mary Mulvihill, and a list of over 180 theories of change at the National Consortium for the Study of Terrorism and Responses to Terrorism (START) website (2018).

As any assessment that is useful to a commander articulates gaps and provides insight into real world problems, there are few relevant assessments that are unclassified. Through this section, we leverage a vignette to provide a simple example of the types of assessment described. The vignette is a simple counter-insurgency in steady state operations, such as Iraq or Afghanistan. In this vignette, some theories of change that may apply include:

- To what extent have military operations deterred the actions of terrorist groups
- To what extent have terrorist group members been effectively removed by counter-network actions

Assuming the assessments working group is well-versed in the environment and has translated strategic end states and objectives into military end states, the authors will assume for purposes of example that End State 1 of Campaign Plan A addresses reduction of terrorist influence in Country X. Authors further assume that End State 1 breaks down into a set of Objectives. The staff must ensure that each objective is clearly defined, decisive, and attainable. Using a theory of change, the working group develops logical if-then statements that help determine these specific objectives. For example, if terrorist violence is reduced, then terrorist influence has also been reduced.

5.2.2.2 Strategic Questions

Extending the example, assume that Objective 1-1 is terrorist violence in Country X is reduced. The working group can then implement strategic questions to determine what is measurable and how the organization can know if violence is reduced. For instance, strategic questions may include:

- What is the level of violence by region
- How has it changed from the prior evaluation period
- Does violence correlate to dates, events, or weather
- What constitutes "violence" for the purposes of this objective

Using questions such as these will help the organization determine what constitutes violence and how it measures violence. A broader discussion of the use of strategic questions can be found in Asking the Right Questions: A Framework for Assessing Counterterrorism Actions (Schroden, Rosenau & Mushen, 2016).

5.2.2.3 Standards-Based Assessment

The next step would be to determine what constitutes progress. The exercise involves thoughtfully considering the entire spectrum (within reason) of measure for the objective. In Objective 1-1 from above, it starts with defining what reduced means. This means asking, "What does success look like?" and "How will we know when we are there?" Prior to conducting an assessment, this benchmark should become a written standard that specifies the measurements that constitute success. For the purposes of example, assume the benchmark is that less than five violent incidents occur across all regions of Country X in any given week and no more than one major incident in the past rating period. This may still sound like a lot of violence; however, it should be considered in context. For illustration, Table 5.1 shows a partial listing of major mass casualty attacks by Islamic State from 28 February to 9 June in 2016 in Iraq (Iraq Violence, 2016). If the assessment period covers 90 days, a decrease to no more than one major incident constitutes a reduction of nearly 90%.

Other examples of scale can be found in the Global Terrorism Database (National Consortium for the Study of Terrorism and Responses to Terrorism, 2018). Drawing the example through, analysts need to ensure that leadership understand that a reduction

TABLE 5.1: Islamic State Terrorist Attacks in Iraq February 28 to 9 June 2016

Day, 2016	Location, Iraq	Type Attack	Deaths
28 Feb	Sadr City	Twin Suicide Bombs	70
6 Mar	Hilla	Fuel Tanker Bombed	47
26 Mar	Iskandariya	Suicide Bomb	32
1 May	Samawa	Two Car Bombs	At least 33
11 May	Sadr City	Car Bombs	93
17 May	Baghdad	Four Bombs	69
9 Jun	Baghdad	Suicide Bombs	30

Source: British Broadcasting Corporation (July 3, 2016).

in major incidents does not necessarily equate to less casualties. As analysts consider the relationship between incidents and types of incidents, they may find that as it becomes more difficult for adversaries to execute major incidents, adversaries may improve their planning skills and execute fewer but even larger attacks. Or they may offset their inability to effect major incidents by increasing their execution of smaller, more widespread attacks.

Once the benchmark is determined, the team must consider the current status of the environment as well as how much worse conditions could become. From these points, the working group can build bins or thresholds that clearly describe progressive states of the environment that can be used to demonstrate progress, no progress, or retrogression using the metrics explored earlier. Table 5.2 and 5.3. are examples of a standards-based binning with four mutually exclusive bins for Objective 1-1. Note that for the first assessment, both conditions must be met for the assessment to move from one bin to the improved bin, but only one condition not met can cause retrogression. Continuing the example from above, if there have been three major incidents in the period of assessment and a maximum of eleven minor incidents per week over the same timeframe, based on the bins in Table 5.2, the current assessment would be that the environment is in state C.

Alternatively, if sufficient data does not exist, or the nature of the objective does not lend itself to quantitative evaluation, a similar, but more subjective form may provide

TABLE 5.2: Example Standards Based Binning

Objective 1-1 Standard	# major incidents	# minor incidents
E (End state)	<= 1 per rating period	<= 5 per week
D	> 1 and <= 3	> 5 and <= 10
C	> 3 and <= 10	> 10 and <= 20
B	> 10	> 20
A	>15	>30

TABLE 5.3: Example Standards Based Binning (Qualitative)[2]

Objective 1-1 Standard	Criteria
E (End state)	Insurgent forces do not have freedom of movement, control territory, or the ability to conduct attacks.
D	Insurgents have freedom of movement only in some areas, and conduct attacks, but do not hold ground.
C	Some areas of the AO are controlled by insurgent forces, who have freedom of movement in most areas, and conduct attacks when targets of opportunity present themselves.
B	Insurgent forces have freedom of movement in all areas, except in the immediate vicinity of our bases, control multiple areas, and attack friendly forces regularly.
A	Insurgents have complete freedom of movement, control multiple areas within the AO, and attack friendly forces regularly.

the commander and staff a mechanism to visualize progress. Developing objective bins is a time-intensive process, requiring the participation of the staff to ensure the buy-in and creation of an evaluation "contract" to not change the bins when the assessment is completed, keeping the assessment objective. An example of standard-based bins with more qualitative criteria is shown in Table 5.3.

However, making an evaluation is insufficient. The assessment working group also needs to document the context of the assessment. The following questions may assist in the follow-on development of the assessment:

- Is the current evaluation an anomaly compared to the data from earlier rating periods
- Are there any events that influenced these numbers
- How did the organization expect the situation to change from the previous evaluation period
- Did the ratings change as expected or not? If not, why not
- What factors or actions impacted or affected these changes

These and similar questions help the assessors answer the questions of what has changed and why. When insufficient quantitative data exists, changes to the standards-based bins provide benchmarks to delineate change. In the context of the assessment, the answers to these questions should be preserved in a written narrative.

5.2.2.4 Written Risk Assessment

As part of this evaluation, assessors and planners must consider risk. Risks that need to be examined include risk to mission, risk to force, and risk to timelines (US Department of Defense, 2016). Risk also includes examination of authorities, policies, and resources necessary to successfully complete the campaign. The Joint Risk Assessment manual uses a four-step methodology as shown in Figure 5.3. The steps use questions to help the team develop a full understanding of the risk so that it can be communicated and decisions made to accept or mitigate the identified risk.

1. Problem Framing: define the type of risk – examples include the failure of the military mission, failure of the host nation government, or failure of another specific military objective;
2. Risk Assessment: scale the threats using the Joint Risk methodology;
3. Risk Judgment: determine how much risk exists and how much is acceptable;
4. Risk Management: decide what should be done to accept, avoid, transfer, or mitigate the risk?

Summarizing campaign assessments, the assessment is executed based on a well-designed framework by a cross-staff assessments working group. The assessment is developed into the campaign plan with objectives built using theories of change that specify measurable conditions. The assessment plan has benchmarked success criteria based on the conditions and defined evaluation levels to assist the team in determining progress, risk, and recommendations. Lastly, the context of the assessment is preserved in a written narrative for later reference.

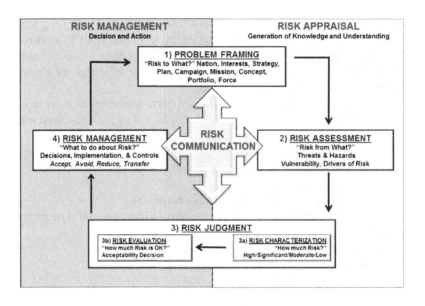

FIGURE 5.3: Four step risk methodology (US Department of Defense, 2016).

5.3   Operations Assessments

5.3.1   Purpose of Operations Assessments

Operations assessments (OA) measure progress of a named operation or exercise against mission success criteria outlined in an operations order at Corps or below. The objective of these assessments is evaluating success on the accomplishment of the decisive point, completion of key tasks, or supporting the commander's decision support template or commander's critical information requirements. While the Military Decision Making Process products are familiar to most officers serving on military staffs, the collection and display of progress in the military decision making process are not.

The NATO Handbook on Operations Assessment defines OA at the operational level as focusing on "the measurement of two distinct aspects of an operation; the generation of the intended effects (medium-term assessment), and the achievement of the objectives (long-term assessment)" (NATO, 2015). This definition clearly demonstrates the multiple purposes behind an assessment. What it does not show is how it should fit into the strategic or campaign level assessment. Recalling from above that the campaign assessment is also concerned with achieving objectives that support the desired end state, it can be seen how the nesting of assessments can benefit both staffs and commanders. In general, strategic end states are developed into campaign plans, portions of which are then parsed out to operational units. Therefore, the objectives the operational units are seeking to achieve should directly support the strategic campaign objectives. If that is the case, the assessment of operational objectives should feed up into the higher headquarters assessment and help reduce the data collection burden on lower level units.

The processes for Campaign Assessments also apply to Operations Assessments; however, they need to be brought down to a more local or immediate level. Regardless which definitions are used, it is easier to determine progress against well-written objectives than those with meanings that are open to interpretation.

5.3.2 Process and Techniques

The execution of an Operational Assessment differs from a campaign assessment in a number of ways. While campaign assessments are a quarterly or annual process, operational assessment occurs on a daily or weekly basis. The speed of decision making is also faster at the operational level than the strategic level. This changes the processes the staff leverages by demanding structured data collection processes that are well-specified and well-rehearsed. Sometimes, it may be helpful or necessary to conduct a longer-term focused assessment where the team examines a single objective about which the command needs information to facilitate a pending decision. At other times, leadership may request a quick turn assessment. Therefore, it helps to have well-developed and well-practiced assessment processes in place. Exercises where the organization practices its operations establish and practice these assessments.

In assessment of operational objectives, it is useful to consider Measures of Effectiveness (MOEs) and Measures of Performance (MOPs). MOEs measure how well tasks conducted contribute to achieving the desired conditions or objectives. MOPs measure how well an organization accomplishes its assigned tasks. Given these definitions, it can be seen that the two measures are not inherently linked; success in one does not imply nor ensure success in the other.

In building the assessment, the assessment team needs to consider specific sources for and types of data available. Is the available data quantitative, qualitative, objective, or subjective? Does the data rationally support assessment of effectiveness or performance? Despite the many possible uses, the implication behind each of them is that selected metrics need to be suitable and appropriate to what is being measured. Difficulties in the availability or timeliness of data can often lead to the use of proxy data which should be subjected to the same level of scrutiny to ensure it reasonably represents the desired measurements.

Assessors are often counseled to ensure that objectives have certain qualities to facilitate analysis, often referred to using the acronym SMART. However, as shown in Table 5.4, the acronym SMART can have multiple meanings (Farina, Williams & Kennedy-Chouane, 2013). Military doctrine ascribes the initialism RMRR: relevant,

TABLE 5.4: Typically Accepted Values for the SMART Acronym

Letter	Major Term	Minor Term
S	Specific	Significant, Stretching, Simple
M	Measurable	Meaningful, Motivational, Manageable
A	Agreed	Attainable, Actionable, Achievable, Appropriate, Aligned, Acceptable
R	Realistic	Results-oriented, Relevant, Resourced, Resonant
T	Time-bound	Timely, Trackable, Tangible, Time-oriented, Time-limited

Source: Processes for Assessing Outcomes of Multi-national Missions

measurable, responsive, and resourced (US Department of Defense, 2010). Additionally, objectives should be adequate (satisfy the tasking and will accomplish the mission), feasible (can accomplish the task within available resources and time), acceptable (proportional and worth the expected cost), and measurable (Mushen & Schroden, 2014). Measurable is not specified in planning doctrine but it is implied by the criteria outlined above. Adding "measurable" to these criteria helps ensure that end states and objectives are not ephemeral or ambiguous and are written in language that describes an observable condition. An observable condition is one for which evidence exists. While critical to operational assessment, the differences between effectiveness and performance, data quality, and SMART objectives also apply to campaign and training assessment. Having determined what to measure and availability of data, the assessment team should organize a data collection management plan. This plan specifies what data the staff collects, from where, and how often. It should also specify how the data is archived or preserved for future re-examination.

The briefing and display of an operational assessment takes many forms. Common examples include an objective-based evaluation, an assessment based on decision points and Commander's Critical Information Requirements (CCIR), and an orders-based decision process. Each provides the commander and staff with different types of information to make better decisions, and normally concludes with recommendations collected from the staff.

An objective-based assessment is conducted as a lower-level campaign assessment. The staff decomposes their mission into Lines of Effort, Objectives, and/or a list of end states. The staff evaluates these operational objectives in the same manner as a campaign assessment, although the staff may have more time constraints with an operational assessment. A Decision Point and CCIR-based assessment begins with the decision points the commander must make through the operation. The staff lists the Commander's Critical Information Requirements that he requires to make these decisions; a staff function performed as part of the Military Decision Making Process (MDMP). The assessment team, in coordination with the operations and intelligence staff, then makes a plan for the collection of data to inform the commander. Decisions may occur throughout the battle, such as committing the reserve, or at a specific point in time or space, such as having at least 50% of our combat power after achieving an objective. An orders-based decision-making process allows the staff to determine when to initiate branches, or sequels to the plan, as well as when to prepare a new operations order. These decisions possess the same links to CCIR are the previous type assessment.

Some operations assessments only differ slightly from campaign assessments; other staffs conduct different staff processes to produce different products, depending on the commander's desire and the staff's ability to internalize the results. While operations and campaign assessments share some commonalities, training assessments are much different.

5.4   Training Assessments

5.4.1   Purpose of Training Assessments

Training Assessments assist in the evaluation of concepts, new equipment, or the training of units at Division and below. These assessments leverage different research techniques than other assessments, including small sample surveys, focus groups, network analysis. Military units encounter these types of assessments at training centers and large exercises to assess their readiness but they are also important to the military services in determining the state and progress of doctrinal concepts, interoperability, and fielding new equipment.

Training Assessments are similar to other types of assessment in that they seek to inform leaders through observed insights and identification of gaps. The basis for these insights and gaps, however, is less clear than other types of assessment. While many military analysts are familiar with specific exercises and training events, especially the national training centers, the authors present training assessments as a holistic approach, a part of a broader campaign of analysis and assessment. This context allows some insights into how analysts can better conduct assessments for these events in the future. First, the authors describe the purpose of these assessments followed by some common analytic frameworks to decompose the problem. Next the authors provide several techniques for implementing this type of assessment and conclude with newer applications of how to data mine reports in training assessments.

A service's exercise directive, commander's training priorities, and/or service and joint processes external to the exercise provide the purpose of a training assessment. For the Army, training centers, such as the National Training Center at Fort Irwin, California, the Joint Readiness Training Center at Fort Polk, Louisiana, and the Joint Multinational Readiness Center (JMRC), at Hohenfels, Germany, train Brigade Combat Teams for combat. Warfighter exercises (WFX) train and certify multiple division staffs throughout the year. The Joint Warfighter Assessment (JWA), held to enhance training, enable interoperability with multinational forces, and inform the future force, is held at different locations each year. Each of these exercises is different in purpose and training audience but has common analytic methods.

Experienced assessors categorize the various purposes of training assessments into training observations, directed insights, gap assessment, and/or external process inputs. These four categories are not mutually exclusive, as assessment teams may reframe insights as gaps, and either may provide information for an external process. Training observations provide direct feedback to the unit on their ability to perform key tasks, usually a subset of their Mission Essential Task List (METL). Observer Controllers and Observer Analysts provide units training observations during exercises and training at a formal focus group dedicated to these insights named an After-Action Review (AAR). In experimentation exercises, observers identify gaps in a gap assessment process that provides input to changes to the force structure of the Army. While the training observations and insights cover the training of the unit, and gap assessment satisfies the Army and Joint requirements process, there are additional process inputs for NATO and other allies, and the America, Britain, Canada, Australia, and New Zealand (ABCANZ) Armies that improve interoperability between nations.

In some exercises, there are reports published providing information to multiple Joint or Service processes in addition to those designated in the exercise directive. These assessments often provide a multi-lens approach that attracts multiple observers and points of view observing the challenges presented in a given exercise. The training centers and the WFXs have specific cadre dedicated to training units for each of their exercises and have well-developed focus-groups and AAR-based methods for the execution of their analysis. While the authors identify some specific methods for these two types of training assessments, the rest of the article primarily addresses more open-ended analysis methods such as those used in the Joint Warfare Assessment.

5.4.2   Techniques for Training Assessments

Analysis begins with identification and decomposition of the problem. Most of the time, there are multiple study questions requiring analysis. Most Army training assessments use a common set of decomposition methods. For instance, training centers and WFXs use the warfighting functions as a categorical decomposition of the commander's training priorities. The analysis frameworks for training assessment include insights across the spectrum of Doctrine, Organization, Training, Materiel, Leadership and education, Personnel, and Facilities (DOTMLPF), Army's Warfighting Challenges, multinational organizational priorities, such as the North Atlantic Treaty Organization (NATO) or ABCANZ, and Capability Areas or Service Warfighting Functions.[3] While this is not an exhaustive list of analytic frameworks, it covers the main Joint and Army frameworks used for analysis.

A DOTMLPF analytic framework is the most holistic of the approaches. It begins with an input in one area, such as a change to material or doctrine, and determines how that change impacts the other aspects of DOTMLPF, such as Training or Organizational changes required for implementation. The Joint Staff process for gap development and mitigation (the Programming, Planning, Budget, and Execution Process, or PPBE) uses the Joint Capability areas. The Army aligns many organizations, such as the centers of excellence that are responsible for identifying gaps, to the Army Warfighting Functions. The Army also identifies specific gaps required for analysis in their Warfighting Challenges.

Techniques used to inform training assessments include observation reports, repeated focus groups and interviews, daily summary reports, formal surveys, summary report data, and data mining. Analysts, using the above frameworks, decompose the problem statement into subquestions, and then answer the sub-questions using these techniques. The authors provide a background on these types of military exercises and describe each of the techniques below.

TABLE 5.5:   Analysis Frameworks

DOTMLPF[1]	Joint Capability Areas (Joint)[2]	Warfighting Functions (Army)[3]	Army's Warfighting Challenges[4]
Doctrine	Force Integration	Mission Command	1. Develop Situational
Organization	Battlespace Awareness	Movement and	Understanding
Training	Force Application	Maneuver	2/3. Shape the Security
Materiel	Logistics	Intelligence	Environment
Leadership and	Command and Control	Fires	4. Adapt the
education	Communication and	Sustainment	Institutional Army
Personnel	Computers	Protection	and Innovate
Facilities	Protection		5. ...
	Corporate Management		...
	and Support		20. Develop Capable
			Formations

1. Chairman of the Joint Chiefs of Staff Instruction (CJCSI) 3170.01I, 2015.

2. https://intellipedia.intelink.gov/wiki/Joint_Capability_Areas. Accessed 18 December 2019.

3. Department of the Army, ADP 3-0: Unified Land Operations, 2011, p 13–14.

4. https://www.army.mil/article/38972/army_warfighting_challenges. Accessed 18 December 2019.

5.4.2.1 Observation Reports

The most common method for determining insights from a training exercise is individual observation reports. Personnel assigned to the collection of data are called "observer-controllers" if they have the ability to adjudicate tactical level engagement and "observer-analysts" if their main responsibility is to report and analyze the observations they record. The observations they collect are related to those writing the assessment through focus groups or consolidated written or typed reports. These observers normally serve eight- to 12-hour shifts if the exercise is continually running.

5.4.2.2 Focus Groups and Interviews

Social science methods may use specially selected subject matter experts (SMEs) in a focused session to elicit insights on various topics. In a military exercise assessment, SMEs by warfighting function (Mission Command, Movement and Maneuver, Intelligence, Fires, Sustainment, and Protection) interview the observers from each shift to consolidate and synthesize the observations on specific questions determined by the SMEs and supporting analysts. In a daily meeting, the SMEs report back to the intermediate exercise leadership (normally a Colonel and Sergeant Major) on the consolidated responses to the scheduled questions. The exercise staff then consolidates these responses into daily reports. Analysts also schedule other formal focus groups intermittently throughout the exercise to consolidate insights.

5.4.2.3 Daily Summary Reports

Daily summary reports provide the commanding general with the overall insights determined in the exercise. The format normally includes a daily briefing to the general, his staff, and his advisors, where he is provided a summary of processed reports. The general provides guidance for the rest of the exercise based on current knowledge and the staff's assessment of whether they have the data required to complete their analysis report. The staff then makes any adjustments to the planned events to ensure the analysis questions are met.

5.4.2.4 Surveys

Formal surveys provide responses to specific questions, usually in written form. Written surveys in exercises are normally limited to two pages, as the participants in the exercise are time constrained. Well-documented codes of best practice, examples, and properly trained personnel are required to develop surveys, as well as approval of the survey of ten or more personnel by the Army Research Institute for the Behavioral & Social Sciences.

5.4.3 Summary Report Data and Data Mining

Observers and analysts consolidate their reports in written form, but also in computer databases. When analysts collect more observations than they can read or process in a given time period, text analytics and summary statistics assist in the triage of observations by taking a more holistic view of the data, but also allowing analysts and observers to focus on key observations. The graphing and text analytic tools that best assist observers and analysts include frequency plots, sentiment analysis, word correlation, clustering, and topic models. While the mathematical techniques are widely available in the literature, the authors focus on how these techniques are best used in a training assessment.

5.4.3.1 Frequency Plots

The most commonly known frequency diagram is the word cloud. The purpose and the representations are fairly standard; the top fifty words in the database are arranged in a rough square, where the size of the word relates to the number of observations of the word. Analysts remove common stopwords, such as "a," "I," and "and," as well as military lexicon stopwords that do not provide relevance, such as the exercise's name or "operation." Word clouds generally are superior to frequency bar charts to quickly relate information since the exact number of times the word occurs is not as important as the relative frequency between words, and they are well-understood by all observers, not just analysts. Analysts and observers can easily pick out major themes not highlighted in the word cloud as well as spot outliers that lead to additional analysis. The word cloud in Figure 5.4 shows major themes in Chapter Two "Ends and Means in War" from Carl von Clausewitz' writings, *On the Nature of War*.[4]

5.4.3.2 Sentiment Analysis

Sentiment analysis scores an observation using a pre-determined list of words with associated positive and negative values. Observations selected by category or key word plotted by day over the exercise can reveal if a certain topic (from the list in Table 5.5 or other topic decomposition) is trending up or down, or if a certain observation or set of observations is out of the ordinary.

5.4.3.3 Word Correlations

Word Correlations provide context to the observations by linking common words together in a descriptive network map of words with a minimum number of connections. While these maps do not provide information that leadership can digest, analysts can use the information provided to spur investigation into certain topics. For example, if network and artillery have a correlation and negative sentiment score, then analysts should research additional observation reports to see if requests for indirect fire are hindered by network outages, if artillery missions are completed at the expense of other

FIGURE 5.4: Word cloud example for "Ends and Means in War" from the writings of Carl von Clausewitz (von Clausewitz & Graham, 2008).

bandwidth, or if there are multiple entries that possess an erroneous correlation, such as a callsign, operation, or location with the word "network" where artillery reside.

5.4.3.4 Clustering and Topic Models

Clustering and topic models provide alternate analysis for the same data that may lead to different conclusions. If analysts are using the Army Warfighting Functions for analysis and there is a challenge in identifying enemy artillery C2 nodes, analysts can generally find these reports distributed under Intelligence, Fires, Protection, or Mission Command. A clustering model may cluster these distributed reports under a common category, and then the list of topics, their frequency, and sentiment in that cluster reveal the underlying challenge.

5.5 Data Requirements

Data used in an assessment vary depending on the type of mission and the type of data available. In this section, we discuss the types of data that exist for assessment, the formulation of objectives that lend themselves to accurate measurement, the importance of quality control in data collection, and the benefits and pitfalls of direct command involvement in assessments.

For the purposes of military analysis, we may classify data into quantitative and qualitative data. Quantitative data implies personnel may apply statistical analysis to count or measurement data. The most common type of this data in military operations is count data, e.g., the number of attacks, the number of casualties, or the number of caches found in a given time period. Making assumptions on the quality of data, analysts may compare one time period to another to determine if there is a significant change in the operational environment over time or space using basic to advanced statistics. Control charts are also an effective method to leverage quantitative data to inform assessments to demonstrate change over time.

Qualitative data may take many forms, and analysts collect qualitative data in multiple ways. Daily or weekly situation reports, structured interviews, surveys, and staff input are some types of qualitative data. Situation reports provide an accurate and formatted report on the history of events but require either text mining or dedicated personnel to read the reports and form conclusions. Standardized formats and input controls, such as web-based drop-down options, assist in the aggregation and analysis of these reports. Dedicated analysts conduct structured interviews in both combat and in exercises to elicit opinions from subject matter experts or first-hand accounts from the field. Analysts and contractors conduct formal scientific surveys of exercise participants, personnel in combat situations, and host nation personnel. It should also be noted that in some cases, qualitative data recorded or observed over a span of time may become quantitative data.

As important as data is, not all leaders recognize that reporting creates data that is needed for effective analysis of operations and as a result data quality from recent conflicts is inconsistent. Analysts and operators in recent operations in Iraq, Afghanistan, and Somalia use a combination of qualitative and quantitative reporting with structured controls to maintain a web-based, online significant activities database. The quality of the data in these databases relies on two tenets of quality control: first, there are personnel dedicated to the quality control of the data, aggregating and reporting on the results on a regular basis to ensure that the data aligns with written situation and media reports and checking the inputs to ensure those that inputted the data follow a strict list of definitions and follow instructions diligently. Second, units must possess a mechanism to return reports that do not meet the quality standards to the originator for correction. The quality in historic data differ greatly between the military missions in Iraq and Afghanistan; military units in Iraq assigned personnel to conduct quality control on the database and provided dedicated operators to review and reject miscategorized or incomplete reports, while military units in Afghanistan did not dedicate personnel to these tasks. The differences in the quality control by both the analysts and operators demonstrated a wide disparity in the ability of the two missions to accurately report to their higher headquarters and in annual reports to Congress.

Further, leaders must acknowledge units collect on the parts of the mission on which they focus and focus on the portions of the mission on which they collect. As such, demands for reporting quantitative data can cause disproportionate attention and "market failures" in the ability of the staff to accurately determine the state of the operational environment. The most infamous of these failures is the collection of body counts in Vietnam, where commanders ascribed success to the number of enemy killed, and commanders attempted to increase the body count in accordance with this measure of success. The scars of this miscalculation remain today, resulting in some deployed commands refusing to release or brief the number of insurgents killed and others avoiding quantitative data in general. To ensure proper data collection, commanders and assessors need to ensure accurate and timely reporting, both quantitative and qualitative in the appropriate contexts.

5.6   Assessment Process Pitfalls and Best Practices

Assessments are broadly acknowledged to be difficult. As a result, many action officers shy away from involvement in the assessment process. However, like most complex processes, it helps to identify where the problems can lie and outline strategies to address them before they occur. Experience with the challenges of conducting assessments builds knowledge of best practices. In the final section, the authors offer their experience with challenges, pitfalls, and subsequently learned best practices.

5.6.1   Challenges and Pitfalls with Assessment

There are numerous challenges and pitfalls associated with the conduct of assessments. Whenever the assessment team encounters difficulty, it helps to take a step back to consider the resistance being encountered, the proximate cause, and how it might be resolved. Table 5.6 outlines some common issues, the underlying problems, and possible solutions. Familiarity with these challenges will help the assessment team proactively recognize and pre-empt many causes of frustration in the assessment process.

The frustrations with assessment can be reduced to four primary causes – lack of fully coordinated and integrated assessments, latency, poorly communicated findings, and inadequately developed recommendations. Instituting best practices from the start

TABLE 5.6: Commons Assessment Issues, Problems and Resolutions

Issue	Common Problem	Possible means of Resolution
Limited support for the assessment	Leadership and staff have not bought into the assessment or do see the value of the assessment	• Socialize the assessment with leadership, planners, and intelligence analysts • Work closely with planners and intel analysts to build the assessment and assessment process • Build the assessment into the plan; ensure that assessment is a routine part of the assessment process (See Figure 5.1)
Reliance on the Intelligence Estimate	Leadership does not see the complementary value of an operation assessment	• Clarify the difference between intel estimates and operation assessment; operation assessment focuses on how actions taken and the status of the environment contribution toward accomplishing the goals, objectives, or end states outlined in the plan
Assessment viewed as an additional burden on the staff	Assessment is not integrated into the planning cycle	• Build the assessment into the plan; ensure that assessment is a routine part of the assessment process (See Figure 5.1)
Staff resistance to improved assessments	General resistance to change or resistance toward fixing something that is not seen as broken	• Institutionalize consistent methodological processes; make incremental changes • If entire overhaul is needed, use a cross-functional staff team to build the assessment process for the future; ensure that the assessment process is "baked" into the planning process (See Figure 5.1)
Assessments viewed as evaluation of the staff	Planning, intel and other staff view the assessment as a "report card"	• Work closely with planners and intel analysts to build the assessment • Develop a reputation for fair and objective analysis.
Lack of staff consensus regarding assessment findings	Senior leaders often have a more nuanced or more informed strategic view than assessors and therefore may dispute findings	• Pre-brief senior stakeholders prior to briefing the assessment in a broader forum; discuss and fine tune findings • Develop an informed dissent process to facilitate discussion of findings
Poor visual presentation	Difficulty in concisely presenting findings results in use of poor graphics	• Refuse to use poor techniques (See Arnhart and King, 2018) • Study & implement data visualization techniques • Summarize in prose • Iterate with leadership to develop better charts for the organization

(Continued)

TABLE 5.6 (CONINUED):   Commons Assessment Issues, Problems and Resolutions

Issue	Common Problem	Possible means of Resolution
Latency of findings and recommendations	Assessments are backward looking; Operations are forward looking	• Some latency is unavoidable • Integrate major events that occur after the data cutoff (a timeline can be useful to showing major events that occurred post data cutoff but that have been informally integrated into the assessment) • Integrate predictive analysis techniques into the assessment • Ensure that all recommendations are appropriately future focused
Poorly developed recommendations	Recommendations are hurriedly developed and/or not thoroughly developed	• Recommendations should meet the same criteria as objectives and courses of action: feasible, acceptable, suitable, specific, measurable, achievable, realistic, and time-bound (See Table 5.4)
Approved recommendations not implemented	No process for tracking recommendation implementation and follow-up	• Assign responsibility for recommendation implementation as part of the approval process • Build a recursive function into the assessment that reviews the status of previously approved recommendations

of any assessment process is the best method to mitigate problems and set the conditions for successful assessment.

5.6.2   Best Practices

From the discussion above, it is clear that assessment processes begin either from scratch or evolve from earlier processes. In either case, assessment teams must develop and emplace a process embraced across a staff, ensuring buy-in while producing an objective assessment. The authors recommend four best practices for implementation based on the consolidated experience of multiple assessors (Arnhart, Carlucci & King, 2018): train the staff and educate leadership, staff a transparent assessment, leverage existing staffing processes including cross-functional working groups to own recommendations, and ensure the assessment is linked to higher and lower headquarters assessments. These best practices focus on fully coordinated and integrated assessments, as once this challenge is overcome, other issues tend to work themselves out. Nevertheless, we will also address latency, poorly communicated findings, and inadequately developed recommendations in this section.

As assessment teams rarely fall in on fully functioning assessment processes, the authors provide a suggested framework (refer to Figure 5.2) to help build an assessment. This implementation may use any or all of the methodological practices outlined above, depending on the staff, the mission, and the preferences of the commander. When implementing an assessment process, a paper and/or briefing that educates the staff on the meetings, briefings, expectations, and products helps to make the implementation successful. Multiple assessment staffs have maintained their process through leadership transitions by immediately briefing incoming leaders upon their arrival as part of their in-processing to the unit, sometimes referred to as a "no leader left behind" education policy.

Each staff implements assessments in a slightly different manner, complicating the formulation of a consistent recipe for as successful assessment. Assessments on smaller staffs find success in holding a separate meeting for assessing operations, while assessments for larger staffs succeed by leveraging existing bureaus, boards, and working groups. One best practice, now codified in doctrine, includes the use of cross-functional working groups (US Department of Defense, 2017). These working groups lead regular meetings on a particular line of effort, with the assessment of their line of effort a recurring topic. The cross-functional teams leverage the assessment team to provide expertise and guide the working group in their assessment, but the working group lead owns the recommendations resulting from the assessment. This allows the assessment team to focus on candid and objective assessment, avoiding friction between the assessment and planning staffs.

Staff and leadership coordination and buy-in on every part of the assessment is a key factor to success. The cross-functional team consists of assessors, planners, intelligence analysts, and other staff who all participate in building the process, conducting the assessment, developing the findings, and developing and implementing the recommendations, which is one of the best ways to coordinate and integrate assessments. In this manner, everyone on the team or working group builds a full understanding of the environment in which the challenge or crisis exists, end states and objectives under assessment, and what conditions constitute progress or success (or retrogression and failure). The team needs to emphasize that the intelligence estimate is not the same as an operations assessment. The operations assessment focuses on the impact actions undertaken and the status of the environment on achieving the objectives of the plan. The result of a cross-functional team building a coordinated, integrated assessment is enhanced quality, clarity, and transparency of the process and product.

Keeping the assessment transparent involves ensuring that every staff section feels their opinion is accounted for in the assessment, balanced by keeping the assessment honest. While assessors may reject some of the assertions of the staff claiming success, it is better to allow a public forum or staffing process to draw out inconsistencies between data collected and claims of success. Successful assessment staffs leverage multiple rounds of staffing and briefings before attempting to brief senior leadership to let the staff come to a consensus. Assessors that take on the role of judge, jury, and executioner of the assessment by briefing the commander behind closed doors incur the ire of staff sections that then become alienated, cutting off the sources of data required to conduct the assessment. Assessment staffs can augment transparency by successfully leveraging effective staff processes.

The assessment team can build consensus and credibility by institutionalizing consistent methodological processes. Consistent methodological processes assist the assessment staff with staffing inside the broader organization to come to a consensus on the results of the assessment. The authors' experience covers a broad range of assessments, from slides to written documents, and a common theme is that good assessments possess a written document to back up the results presented, as well as a process of vetting that provides transparent review from multiple stakeholders and perspectives. The act of writing and staffing a written document provides clarity that charts cannot convey. Often, producing only charts conceals the full depth, nuance, and meaning of the information presented. Charts usually contain summary bullet comments that do not stand alone. Despite local policy measures put in place to ensure clarity in presentation, this often translates to a paucity of discussion during the staffing process, since the reader of the slide is left to interpret unclear bullets in their own way. Well-written, active voice comments force clarity in the assessment, and allow the staff to come to consensus through common understanding. This applies equally to assessment findings and recommendations – the two key outputs of any assessment.

Assessment findings inform the leadership on how things are going – often referred to as "Are we doing the right things?" and "Are we doing things right?" Recommendations should answer the follow-on question, "Where do we go from here?" or "What should we do now/next?" Developing recommendations is hard for two reasons. First is the latency that is inherent in any product that is informed by the past. The second is that the future is, at best, only partially knowable. Therefore, as an assessment is developed, the assessment team needs to be integrating current events into the assessment. Using a timeline to inform leadership helps the assessors show that they are aware of the concern regarding latency. The timeline should show the cutoff date of the assessment with events prior to that date fully incorporated into the assessment and subsequent events that have also been considered, though less formally.

Recommendations need to be forward-looking. Keep in mind that recommendations may come at differing levels of decision authority. Often, some lower-level recommendations will be implemented before the staffing process is complete, as leaders at those levels take ownership of the actions they can implement. In the recommendation development process, some of the questions the cross-functional team needs to ask are:

- What actions need to be taken to further movement toward the desired objective
- What actions are already being taken
- What additional actions need to be taken
- What can go wrong and how can we take action to prevent that from occurring
- What actions need to be taken to shore up current success or to prevent retrogression
- What environmental (PMESII-PT[5]) conditions are necessary for progression/success? What can we do to create them

This partial list of questions should help assessors start thinking toward the future and indicate the need to integrate techniques like predictive forecasting and modeling into the development of recommendations. There are many methods of building an assessment – what is critical is to build one that works for the organization.

While the method is important, the presentation of the assessment is another critical component. Visual presentations of data in an assessment vary depending on the level of the assessment (tactical, operational, or strategic), and the type of assessment. While this is a good place for analysts and assessors to be creative, they should also be aware of, and avoid, a number of common pitfalls. Visualization techniques to avoid include thermographs, standard-less stoplights, one-hundred-point scales, color math, indices, arrows, and improper application of effects-based operations. Thermographs, the nominal plotting of a point on a rainbow bar, are similar to standard-less stoplights, which provide a red-yellow-green assessment without explanation or thresholds. One-hundred-point scales, an average of votes on a scale, are far too precise for the information presented and graphically present more detail than the underlying assessment data support. Color math and indices provide a convenient mathematical formula for obscuring background data and rolling it up into results more definitive than the supporting information. Arrows provide a direction of a given observation, such as the number of attacks increasing, without the proper context to determine if the change is statistically significant or operationally relevant or even valid for more than a single time period. One of the more common military application vices is following the outdated Effects-Based Operations (EBO) doctrine that consolidates numerical and categorical observations using a weighted average, similar to indices and color math (Arnhart & King, 2018).

Briefings should provide a standard format that adheres to the visualization methods described above. Expectations include the responsibilities for staffs' data collection and the timeline for staffing the written assessment and/or associated charts. Example products assist the staff in knowing the level of detail and the type of output that is expected for the assessment. Example assessment processes, charts, and reports are found in the NATO OA (Operation Assessment) Handbook, 1225 and 1230 reports (available online).[6]

Key to implementing a useful assessment process is communicating the assessment to higher and lower echelon staffs. Communicating assessments between staffs is best accomplished using the best practices outlined in the earlier section on methodological best practices. Most assessments fail to transmit or receive the information between higher and lower echelon staffs not because the assessment is bad or uninformative, but because the assessment lacks the processes to relate what the assessment learns. One method proven throughout multiple prior assessments is to initiate an assessment request from higher to lower with the mission, end state, and any specific strategic questions or data the higher headquarters requires for their assessment, provided the data comes with context and interpretation. In turn, the lower headquarters provides, given adequate time, a standards-based assessment based on the end state(s) for the mission(s), and the capability, capacity, and authority gaps that the lower headquarters observes; these are tempered by the risk these gaps pose to accomplishing the mission within the allotted time. Using the best practices in this manner allows staffs of multiple levels to communicate the learning they accumulate through their assessment processes. Assessment processes that fail to account for higher echelon demands are quickly overwhelmed with additional data requirements, and those that do not include lower echelons in their data collection are deemed out of touch.

A successfully implemented assessment process should result in a recurring assessment process that occurs regularly and becomes value-adding activity for the staff and commander. It is important to build the assessment with the assistance of a whole-of-staff approach, as well as properly communicating the assessment.

5.7 Conclusion

After reading this chapter, analysts will understand the best practices for a number of types of assessment and know how the assessments they conduct inform their command, their Service, or the Joint Force. Throughout the chapter, the authors demonstrated a generalized assessment taxonomy, lexicon, and framework to assist assessors in both discussing and conducting different kinds of assessments, as well as outlining data requirements and how to initiate and assessment process. The authors expect the emphasis on assessments to continue to grow and with it, the assessments' community of interest and the evolution of assessment doctrine, training, and education within the Department of Defense.

Notes

1. Special meetings were held by the Military Operations Research Society in 2012 and 2018: MORS Special Meeting: Assessments of Multinational Operations from Analysis to Doctrine and Policy, Report on Proceedings. (Arlington: Military Operations Research Society, 2012); Arnhart, Lynette, Renee Carlucci, and Marv King. "Advancing the Professionalism of Assessments." *Phalanx* 51(3), 2018. Also, the *Journal of Defense Modeling and Simulation* is publishing a special edition on assessments; relevant articles include: Jablonski, James A., Brian M. Wade, and Jonathan K. Alt. "Operation assessment: Lessons learned across echelons." *The Journal of Defense Modeling and Simulation*, 2019; Keith, Andrew, Darryl Ahner, and Nicole Curtis. "Evaluation theory and its application to military assessments." *The Journal of Defense Modeling and Simulation*, 2019.
2. AO is the military acronym for Area of Operations.
3. A further description of the Army Warfighting Challenges is found at: https://www.army.mil/article/38972/army_warfighting_challenges.
4. The text was used unedited for example purposes only. Normally, stopwords and other irrelevant words would be removed.
5. PMESII-PT is the military acronym for Political, Military, Economic, Social, Infrastructure, Information, Physical Environment, and Time conditions associated with planning and battlefield operations.
6. One example of the 1230 report is found at: https://dod.defense.gov/Portals/1/Documents/pubs/November_1230_Report_FINAL.pdf.

References

Arnhart, L., Carlucci, R. & King, M. (2018). Advancing the Professionalism of Assessments. *Phalanx*, 51(3), 26–31. Retrieved from http://www.mors.org/Portals/23/Docs/Publications/Phalanx/2018/Vol%2051%20N%203v2.pdf

Arnhart, L. & King, M. (2018). Are We There Yet? *Military Review*, 98(3), 20–29. Retrieved from https://www.armyupress.army.mil/Journals/Military-Review/English-Edition-Archives/May-June-2018/Are-We-There-Yet-Implementing-Best-Practices-in-Assessments/

Davenport, T. & Harris, J. (2007). *Competing on Analytics*. Boston, MA: Harvard Business School Press.

Farina, F. F., Williams, A., & Kennedy-Chouane, M. (2013). Civilian and Military Evaluation and Assessment: Synergies and Differences. In Williams, A., Bexfield, J., Farina, F.F., & de Nijs, J. (Eds.), *Innovation in Operations Assessment: Recent Developments in Measuring Results in Conflict Environments*, 117–144. Norfolk, VA: Headquarters Supreme Allied Commander Transformation. Retrieved from http://betterevaluation.org/sites/default/files/capdev_01.pdf

Iraq Violence: IS bombing kills 125 Ramadan shoppers in Baghdad. (2016, July 3). *British Broadcasting Corporation*. Retrieved from https://www.bbc.com/news/world-middle-east-36696568

Jablonski, J., Wade, B. & Alt, J. (2019). Operation Assessment: Lessons Learned Across Echelons. *The Journal of Defense Modeling and Simulation*. Retrieved from https://journals.sagepub.com/doi/abs/10.1177/1548512919826405

Keith, A., Ahner, D., & Curtis, N. (2019). Evaluation Theory and its Application to Military Assessments. *The Journal of Defense Modeling and Simulation*. Retrieved from https://journals.sagepub.com/doi/full/10.1177/1548512919834670

Metz, Steven. (2015, October 23). The U.S. Must Adopt Realistic Objectives for Afghanistan. Retrieved from https://www.worldpoliticsreview.com/articles/17029/the-u-s-must-adopt-realistic-objectives-for-afghanistan

Metz, Steven. (2018, July 13). Trump Has Hastened the Loss of America's Strategic Vision. Retrieved from https://www.worldpoliticsreview.com/articles/25054/trump-has-hastened-the-loss-of-america-s-strategic-vision

Mushen, E. & Schroden, J. (2014). *Are We Winning? A Brief History of Military Operations Assessment*. Arlington, VA: Center for Naval Analyses. Retrieved from https://www.cna.org/cna_files/pdf/DOP-2014-U-008512-1Rev.pdf

MORS Special Meeting: Assessments of Multinational Operations From Analysis to Doctrine and Policy, Report on Proceedings. (2012). Arlington, VA: Military Operations Research Society. Retrieved from http://citeseerx.ist.psu.edu/viewdoc/download;jsessionid=FB75CDCC8C9193772AFA2F59C3B24855?doi=10.1.1.641.4642&rep=rep1&type=pdf

National Consortium for the Study of Terrorism and Responses to Terrorism. (2018). College Park, MD: START. Retrieved from http://www.start.umd.edu/(START Theories of Change retrieved from http://start.foxtrotdev.com/)

North Atlantic Treaty Organization. (2015). *NATO Operations Assessment Handbook, Version 3.0*. Retrieved from http://www.natolibguides.info/ld.php?content_id=30192868

Schroden, J., Rosenau, W. & Warner, E. (2016). *Asking the Right Questions: A Framework for Assessing Counterterrorism Actions*. Arlington, VA: Center for Naval Analyses.

US Department of the Army. (2012). *The Operations Process, Army Doctrine Publication 5-0*. Washington, DC: Department of the Army.

US Department of Defense. (2010). *Joint Operation Planning, Chairman of the Joint Chiefs Staff Publication 5-0*. Washington, DC: Department of Defense.

US Department of Defense. (2015). *Operation Assessment: Multiservice Tactics, Techniques and Procedures for Operation Assessment*. Washington, DC: Department of Defense, Air Land Sea Application Center.

US Department of Defense. (2016). *Joint Risk Analysis, Chairman of the Joint Chiefs Staff* Manual 3105.01. Washington, DC: Department of Defense. http://www.jcs.mil/Portals/36/Documents/Library/Manuals/CJCSM%203105.01%C2%A0.pdf?ver=2017-02-15-105309-907

US Department of Defense. (2017). *Joint Planning, Chairman of the Joint Chiefs Staff Publication 5-0*. Washington, DC: Department of Defense. Retrieved from http://www.jcs.mil/Portals/36/Documents/Doctrine/pubs/jp5_0_20171606.pdf

US Department of Defense. (2018). *Summary of the 2018 National Defense Strategy of The United States of America*. Washington, DC: Department of Defense. Retrieved from https://dod.defense.gov/Portals/1/Documents/pubs/2018-National-Defense-Strategy-Summary.pdf

von Clausewitz, C. & Graham, J. (2008). *On War, Chapter 2*. Radford, VA: Wilder Publications.

Chapter 6

Threatcasting in a Military Setting

Natalie Vanatta and Brian David Johnson

6.1 Introduction

The intersection of digital and physical security is critical to the future of our military and national defense. Impending technological advances widen the attack plain over the next decade including cyber-social, cyber-physical, and cyber-kinetic attacks. Visualizing what the future will hold, and what new threat vectors could emerge, is a task that in the 21st century, traditional military planning mechanisms struggle to accomplish given the wide range of potential issues.

In February 2011, Secretary of Defense Robert Gates told West Point cadets:

> *When it comes to predicting the nature and location of our next military engagements, since Vietnam, our record has been perfect. We have never once gotten it right, from the Mayaguez to Grenada, Panama, Somalia, the Balkans, Haiti, Kuwait, Iraq, and more — we had no idea a year before any of these missions that we would be so engaged*

(Zenko, 2012).

Understanding and preparing for the future operating environment is the basis of an analytical process known as Threatcasting. Arizona State University's School for the Future of Innovation in Society, in collaboration with the Army Cyber Institute at West Point, use the Threatcasting analytical process to give researchers a structured way to envision and plan for risks ten years in the future. The Threatcasting analytical process assists and enables practitioners to imagine enemy innovations before they happen and identify actions that can disrupt or respond to these enemy innovations. For many organizations, the scope of this problem can seem overwhelming.

Threatcasting uses inputs from social science, technical research, cultural history, economics, trends, expert interviews, and even a little science fiction. These inputs allow the creation of potential futures. By placing the threats into an effects-based model (e.g., a person in a place with a problem), it allows organizations to understand what needs to be done immediately, and also in the future, to disrupt possible threats. The Threatcasting analytical process also exposes what events could happen that indicate the progression toward an increasingly possible threat landscape.

The Threatcasting analytical process draws strength from futures studies, a field that provides theoretical and applied tools designed to shed light on deep uncertainties and complexities that futures hold. Foresight tools are rooted in exploratory, rather than predictive methods of futures thinking, learning, and strategy in order to prepare and plan for long-term outcomes that are difficult to imagine and impossible to predict (Bell, 2009). Such methods often stand in contrast to causal, linear, "plan and predict" thinking that characterizes many contemporary practices of making and knowing futures.

As national security and technological possibilities change rapidly, new threats and opportunities become ever-present. Threatcasting is a means to make sense of potential military futures so that relevant institutions can anticipate, manage, and navigate both the uncertainty and complexity ahead. This chapter will explain the Threatcasting analytical process as well as use the weaponization of artificial intelligence (AI) in a supply chain setting to demonstrate how Threatcasting has been applied and used in the real world. Specifically, we will outline two case studies where the process was applied with specific results. One case study focuses on the digital and physical supply chain in private industry (Cisco Systems) and the second investigates similar threats to the military's supply chain (Military Logistics Officers).

6.2 Emerging Need for Foresight

In the last few decades of the 20th century, foresight and long-term strategic planning were introduced into corporations and private industry. The practice was pioneered and used for decades in the specific industries that needed to make decisions that might not pay off for five to ten years (e.g., energy, city planning, etc.). With the invention and proliferation of the personal computer and the internet, a wider range of organizations saw disruption and innovation happening at an increasing rate. This was especially apparent in the high tech or Information Technology (IT) industries (Popper, 2008). These companies' long lead-times in product development roadmaps means that they need to know what people would want to do with technology five to ten years in advance. Additionally, because of the complexity of these products, many of these companies needed to explain how a new product or service might be used years before it was ready for market. Ecosystems with multiple players needed to be convened around a vision for the future for the product to be successful.

Taking a cue from the high-tech industry, as companies prepared for the technological gains and advances in the 21st century paired with the seismic shock and loss of the Great Recession, a wider range of corporations and organizations began to look to strategic foresight to plan for the future, and to make sure that they were the disruptor in their markets.

As these corporations hired foresight professionals, they learned that the results of the process gave them a vision for possible futures/products and informed their current business strategy. Human Resources (HR) professionals began to use these long-term visions to prepare their hiring and training strategies. Mergers and Acquisition departments used the output to target early stage companies and startups for either investment and/or acquisition. Legal and Intellectual Property groups began to use the output to increase the company's patent portfolio and to develop "future proof" contracts. These are long-term contracts and agreements between companies that might span five to ten years and have language in them that prepare for possible innovations and technologies that will be released multiple years into the future.

Similarly, in the 21st century, the landscape and possible problems that will need to be addressed by the operations research community are growing more complex. Operations research, as an academic discipline, is about applying advanced analytical and mathematical methods to make better decisions in relation to complex problems. Within operations research, foresight could be used as a tool to provide richer data sets, as well as greater detail and definition to these possible and probable threat futures.

The threat landscape in the 21st century is agile, adaptive, fast-moving, and enhanced by evolving technology. These factors create a larger pool of threat adversaries that could have a massive effect on the United States (US) via cyber means, which were previously only seen via kinetic means, effectively lowering the barrier to entry for non-nation state actors to influence the US. Secretary of Defense Chuck Hagel articulated these threat concerns in 2014, with the creation of a third offset strategy. Offset strategies encourage innovation with an appropriate combination of technologies, and operational and organizational constructs, to achieve decisive advantage against our adversaries in peacetime to remain in a position of world power. A key piece of the third offset strategy is to develop cutting edge technologies in the field of robotics, autonomous systems, miniaturization, big data analysis, and advanced manufacturing to incorporate into military operations. Additionally, to ensure the US recruits and retains the individuals that are capable of these breakthroughs, to be able to respond to our adversaries.

Similarly, these agile and adaptive threat concerns are also reflected in the soon-to-be-published multi-domain operations (MDO) v1.5 which General Stephen Townsend states is the "evolution of a larger effort to develop and revise Army thinking and requirements to defeat multiple layers of stand-off in both competition and conflict" (Judson, 2018). MDO now recognizes that while we might have originally envisioned the future to be fighting across all domains – land, air, sea, space, and cyber – we really need to be prepared for a highly contested environment where the joint forces won't have dominance across the entire spectrum. This is a different future environment that our current military forces have not faced in their lifetimes – full of unknowns and evolving threats that are hard to define.

Current operations research processes and procedures are necessary, but do not identify the futures threat landscape. Therefore, the processes of foresight and Threatcasting can enable the traditional practice of operations research and fill gaps that currently exist. For example, Threatcasting can provide a broader range of "alternatives," a wider range of threats and futures to be analyzed, and probable futures. Furthermore, with these possible threats identified, practitioners can use backcasting (a process that defines time-phased alternative-actions) to imagine how to disrupt, mitigate, and recover from the threats. This provides greater clarity for possible actions and uncontrollable externalities for Optimization and the final data analysis after data-farming.

6.3   Definition of Threatcasting

The Threatcasting analytical process is a four-phase methodology that aligns with the body of academic work within the foresight and futures community. The process allows practitioners to approach military futures not in a vacuum nor with only an understanding of a small portion of the problem but instead with a systems-view to enable the user to grapple with complexity, uncertainty, and risk. The Threatcasting process begins with a research synthesis phase, which draws from the Delphi method (Linstone & Turoff, 2011). This is followed by the forecasting phase, which utilizes elements of scenario building and science fiction prototyping (SFP). Phase three is the time-phased, alternative-action definition (TAD) phase which generates multiple backcasts. The final phase consists of data analysis, technical documentation, and communication of both the future threats and the actions to be taken. A graphical depiction of the process is seen in Figure 6.1.

Ultimately, Threatcasting is a human-centric process. Practitioners' participation in the modeling session is essential. Bringing together individuals from the military, government, academia, and private industry, with the objective to envision possible threats ten years in the future, gives each group the ability to brainstorm what actions can be taken to identify, track, disrupt, mitigate, and recover from the possible threats in a way that is more comprehensive than if each group had done the modeling on their own.

6.3.1   Phase Zero

A fundamental component of the Threatcasting analytical process is selecting the appropriate research inputs to feed the process. These focus themes are selected to explore how their evolution from today contributes to the future, and how the intersections of the focus areas' growth modify each other. To select these themes, senior leaders inside the problem space and thought leaders outside the problem space are consulted on what "keeps them up at night" or what they feel no-one is focused on yet, in order to determine the severity and urgency of the proposed themes. These curated themes are then explored by Subject Matter Experts (SMEs) in a 10–15-minute recorded presentation to be used in Phase One.

When an organization is modeling possible threats, there is a tendency to try and "boil the ocean." Many groups attempt to comprehend and model all possible threats.

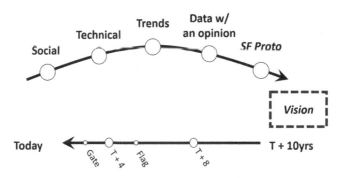

FIGURE 6.1: Threatcasting methodology (Johnson et al., 2017).

The Threatcasting analytical process ensures that groups are focused or "curated" only on specific threat areas. This enables the team to not only envision quality futures, but also get into designing potential disruption, mitigation, and recovery actions against these threats, as they are not attempting to solve all the problems in the world, just a curated set.

6.3.2   Phase One

Research synthesis is the first phase of the Threatcasting analytical process. The purpose of this phase is to allow each small group of practitioners to process the implications of the SME provided data while gathering the intelligence, expertise, and knowledge of the participants. The output of this phase becomes the raw data that is used to inform subsequent phases.

During this phase, all participants listen to each SME's presentation (curated from Phase Zero) and take notes. At the end of the presentations, they break into assigned small groups and using a specifically designed Research Synthesis Workbook (RSW) are led through an exercise to process and discuss the presentation material. Within the groups, they identify the key elements provided by the SMEs and discuss the larger implications of those elements in the future, based on their expertise. Additionally, the group characterizes each element as either positive or negative, and then lists ideas for what "we" should do about it. The "we" is purposely broad, as the input can be personal to the small group, the collected team in the room, the larger organization, or the entire human race. All of this information is captured in the RSWs.

The output of the research synthesis phase is a numbered list of these key points from the SMEs as determined by participants. Therefore, each circle in the top arc of Figure 6.1 is populated with a list of key considerations.

6.3.3   Phase Two

The core of the Threatcasting analytical process begins with phase two. The purpose of futurecasting is to model the future environment, based upon data compiled in the RSWs. These views of the future are effects-based models, meaning that the group is not modeling a specific threat or future first; they are exploring the layered effects that this threat will have on a single person, in a specific place. Threatcasting harnesses the futures wheel concept (Innatullah, 2008) for imagining and exploring, and further extends it beyond a single effect of a future event. This creates a more detailed effects-based model that ultimately explores the threat in greater depth.

Futurecasting is drawn as the upper arc in Figure 6.1, resulting in the "dashed" future at the far right of the figure. Each small group of participants generates this future in the form of a science fiction prototype (SFP). SFPs incorporate storytelling as a means of introducing detail into the future models and empowering the investigation into the human impacts, as well as scrutinizing the political, ethical, legal, and business impacts of these futures (Johnson, 2011, 2013). The SFP process follows a simple set of rules, as all stories have similar ingredients that drive the narrative, making them engaging enough for the reader to suspend disbelief with a structure to support potential plot resolution. Whether it is literature, motion pictures, or comic books, all stories or narratives contain a person, in a place, with a set of problems. Therefore, the output of phase two is a detailed outline for a specific future that the participants can then envision.

6.3.4   Phase Three

The third phase is the time-phased, alternative-action definition (TAD) process. TAD allows participants to explore multiple time-based futures and actions that can be taken to disrupt, mitigate, and recover from the future threats they have identified. Drawing from the practice of backcasting (Robinson, 1990, 2003), TAD provides multiple "backcasts," over a variety of timeframes and possible actions creating a multi-verse of options, plans, and strategies. Broadly speaking, Threatcasting engages the backcasting methodology by asking participants to work backwards in time from their one established future to identify what could be done to disrupt, mitigate, and/or recover from their defined threat. This is visualized as the backwards arrow in Figure 6.1. Participants are explicitly asked to imagine and place two types of indicators along their future trajectory: gates and flags.

Gates are actions (e.g., the use of technologies, capacities, systems) that defenders (government, military, industry, etc.) have control over that could disrupt, mitigate, and/or recover from the established threat. These are things that will occur along a concrete timeline from today (T) to T+10 years. Flags are events (e.g., economic, cultural, geo-political) or advances (e.g., technological, scientific) that defenders have no control over but once they occur, establish path dependencies with significant repercussions and consequence. Flags have an irreversible effect on the envisioned future and should be watched for as heralds of the future to come.

With the gates and flags established, the small groups then work from the future to the present to determine and timeline what specific actions (e.g., investments, organizational changes, technological development, security, policy) they might take to disrupt, mitigate, or recover from the threatcasted event. Thinking through concrete actions that would prevent their future threat gives participants the ability to understand how decision-making across time affects future outcomes. For the military, this provides a novel way to see how decisions to act today might help prevent tomorrow's threat.

One key benefit and output of the Threatcasting analytical process is its exploration of potential second- and third-order effects of these actions within the future. This is especially useful for large and complex military and business organizations. The SFPs craft a quick and easy way to understand the story, giving these organizations an efficient way to expeditiously understand threats and discuss what action(s) need to be taken.

6.3.5   Phase Four

Following the Threatcasting session, moderators use the RSWs as well as the small group future narratives (SFPs) as raw data for a synthesis session. Reviewing each workbook, the team of moderators look for patterns in the futures and for areas that were not explored.

The phases of the Threatcasting analytical process generate multiple futures and threats. In phase four, secondary research as well as the backcasting details from the practitioners give the moderating team the raw data needed to make specific recommendations for action in the near and long term. This post-analysis consists of multiple clustering and aggregation exercises to determine the patterns in all the recorded futures. These clusters are then examined in light of the SME presentations, looking for possible inconsistencies or areas that need more clarification. Additionally, the team highlights SME themes that the groups did not model but were strong components of the expert presentations. Combining all of these together, the team compiles a technical report with specific recommendations for next steps and areas of action, informed by the participants.

Additional details about how a practitioner can execute the phases of the Threatcasting analytical process can be found in the *Journal of Defense Modeling and Simulation* (Vanatta & Johnson, 2019). This chapter will now present case studies on the application of the Threatcasting analytical process and its importance to the field of operations research.

6.4 Supply Chains Defined by Industry and the Military

Stated simply, a supply chain is a network or system of companies, organizations, people, activities, information, and resources used to move products and sometimes services from single or multiple suppliers to an organization or customer. Over the last few decades, the subject of supply chain and the idea of supply chain management have become popular topics in both military and private industry.

In the private sector, this has been driven by globalization and the ability for a single organization to source different aspects of a product from multiple suppliers across the globe.

Corporations have turned increasingly to global sources for their supplies. This globalization of supply has forced companies to look for more effective ways to coordinate the flow of materials into and out of the company. Key to such coordination is an orientation toward closer relationships with suppliers. Companies in particular and supply chains in general compete more today on the basis of time and quality. Getting a defect-free product to the customer faster and more reliably than the competition is no longer seen as a competitive advantage, but simply a requirement to be in the market. Customers are demanding products consistently delivered faster, exactly on time, and with no damage. Each of these necessitates closer coordination with suppliers and distributors. This global orientation and increased performance-based competition, combined with rapidly changing technology and economic conditions, all contribute to marketplace uncertainty. This uncertainty requires greater flexibility on the part of individual companies and supply chains, which in turn demands more flexibility in supply chain relationships.

(Mentzer et al., 2011)

Although similar in many ways, corporate supply chains differ from military supply chains in several ways. The size of the military supply chain is considerably smaller than that of private industry. Part of this constraint comes simply from need. The size of the market that a military supply chain is addressing is just smaller, as the military is less than 1% of the entire US population. This means that the addressable market could be seen as 99% smaller than private supply chains. However, there is an inverse effect at work in a military supply chain. Due to the nature of the "business" of the military, if the supply chain breaks down or is weaponized, the possible effects are not 1% of a similar disruption in the private sector. The effects would be far more dangerous and potentially destabilizing.

Another difference between military and private sector supply chains is the barrier of entry to become a part of that supply chain. To receive a government contract and become a part of supply chain system, there are a greater number of requirements and associated regulations. Therefore, the makeup of the military supply chain is constrained by the number of organizations that can meet these requirements. The private sector supply chain is for the most part market-driven. Some private industries are regulated more than others, but these regulations come from governments. The military supply chain is constrained by itself, with a broader set of goals beyond its market-driven counterparts. The Federal acquisition process imposes regulations and laws upon the military to accomplish things like "spread the wealth" that are goals far different than a market-based supply chain.

Finally, military supply chains are also constrained by their inability to pull any manufacturing or ownership inside the organization. For example, global advertising and search giant Google, made the strategic decision to own their undersea cables to protect their business interests. Similarly, the company decided to manufacture their own hard-drives to meet their operating specifications in their data centers (both for efficiency and security). In another high-tech example, Apple started making its own microprocessors when they realized the risk they were taking on by allowing others to craft them. Unlike these examples, the military does not have the ability to resource their own supply chain and become manufactures of key components of systems. They must rely on a private, global industry to meet their needs.

The weaponization of any organization's supply chain and logistics systems poses a significant threat to national and global economic security. The very systems that are the engine of economies and the lifeline of goods and services to the world's population could, and most probably will, be turned against the very people and organizations that they serve. This new threat landscape and associated challenges will affect industry, militaries, and governments through loss of revenue and productivity and even loss of life. This weaponization will allow adversaries, whether they are criminal, state-sponsored, terrorists, or hacktivists to transform these systems from engines of productivity to enemies on the inside.

6.5 Threatcasting Applied in Industry

The Threatcasting analytical process has been used in industry for over a decade. Its use started in Silicon Valley with companies like microprocessor manufacturer Intel Corporation and software design tools company Autodesk. These organizations used the Threatcasting analytical process to build better products and solutions for customers. Much of this work is kept confidential because it contains company secrets, product strategies, patents, and other intellectual property. Therefore, there is little specific documentation or case studies of Threatcasting use within private industry. There are examples of Threatcasting use in academia, including by legal scholars (Bennett & Johnson, 2016) and the body of work is increasing each year. Additionally, multiple academic and military institutions (e.g., Georgetown University, Arizona State University, the United States Naval Academy, the United States Air Force Academy, the United States Military Academy) have begun teaching the Threatcasting analytical process as a tool for students to explore complex and uncertain futures. Recently, two industry organizations have publicly discussed their use of Threatcasting and their results.

Cisco Systems is an American technology company that develops, manufactures, and sells networking and telecommunications hardware equipment. Inside the Silicon Valley-based company is their innovation lab called CHILL. CHILL stands for Cisco Hyper Innovation Living Labs. It is

> *an innovation capability that aims to disrupt through the development of businesses and joint projects in 48 hours. It's a unique and new project in its third year for Cisco and the brain-child of Kate O'Keeffe, Senior Director. Each CHILL Lab tackles a different topic facing the globe.*
>
> (Bonime, 2018)

The CHILL event was titled "Securing the Digitized Supply Chain Powered by Blockchain." The team

> *aimed to drive disruptive innovation with and for customers ... which drive joint investment opportunities from multiple parties into projects or startups that get support. The partners for the Blockchain Lab included CitiBank, Intel, GE, and DB Schenker, among others.*
>
> (Wal-Aamal, 2017)

In preparation for the lab, O'Keeffe and team used the ASU's Threatcasting Lab report A Widening Attack Plain (Johnson, 2017a) to draw out specific threats and futures for the future of the digital supply chain. The authors of the report included experts from the government, military, private industry, trade associations, and academia. Specifically, the report explored the intersection between a global, digital, and automated supply chain, with the threats of cyber-attacks and terrorism.

Based upon the report, the CHILL team Threatcasted a future featuring a

> *state-sponsored terrorist attack using smart refrigerators and pantries that place excessive dairy and produce orders to a complex automated supply chain. With the roads and ports now clogged, the terrorists exploit a weakness in the Red Hook, New Jersey port system to sneak a dirty bomb into the country and detonate it in downtown New York.*
>
> (Johnson & Vanatta, 2017)

To better express this future in a short period of time, CHILL used the threat future as the basis for a science fiction prototype, "Two Days After Tuesday" (Johnson, 2017c). The goal was to develop and create a powerful narrative that would serve as a fact-based illustration of the future threats.

When asked about the effectiveness of the Threatcasting analytical process and the science fiction prototype, O'Keeffe said in an interview,

> *People aren't wired to imagine the future, ten or even five years out, which is a blocker to innovation ... We need to create that world for them, so they can immerse themselves in this future scenario, making it immediately apparent what kind of solutions we need to prepare for that future.*
>
> (Johnson & Vanatta, 2017)

The result of the CHILL event was success. Five new business or product concepts were generated by the lab based upon the Threatcasting futures and received funding at the event. When asked about the effectiveness of the labs in an interview O'Keeffe replied,

> *In our last three Labs we invested in six outcomes, of which 2 are still existing today. CHILL is an important part of our innovation engine as Cisco spends $6 billion in R&D, $2.2 billion in venture capital, and $250 million investment fund and has invested in 100 startups.*

> (Wal-Aamal, 2017)

Cisco used Threatcasting and its derivative deliverables as an input to their innovation process while other organizations have used it to identify future threats to markets and entire industries.

The American Production and Inventory Control Society (APICS) is a trade association that provides supply chain, logistics, and operations management research, publications, education, and certification programs. The organization provides Threatcasting as a key offering to their members who are concerned with digital and cyber-attacks. For example, they conducted a Threatcasting workshop during their annual conference in 2019.

In each seminar a presenter or speaker leads the group of supply chain professionals through the Threatcasting analytical process and works to apply the results specifically to the attendees' possible and probable future threats.

APICS specifically focused on the TAD component of the Threatcasting analytical process to identify areas of strategic focus for their organization and the entire supply chain industry. The results of the TAD mapped multiple areas that touched on "urbanization, Africa, the young and elderly, women in global society, technological autonomy and intelligence, data, transparency, fully conscious consumerism, and the speed of change" (Proctor, 2017)

These two different applications of Threatcasting in industry illustrate how the process can be used to identify future threats to an organization or systems. Once these threats have been identified, the organization can explore how to track, disrupt, mitigate, and recover from those threats.

Additionally, the process can be used to generate raw material for existing organization processes and procedures, giving practitioners a new or different viewpoint to better capture the current and future shifting landscape.

6.6   Threatcasting Applied to the Military

The Army Cyber Institute (ACI) is a national resource for research and engagement to enable effective Army cyber operations in the future. Put another way, this small thinktank exists to prevent strategic surprise in cyberspace and ensure the Army's dominance by scanning new developments on a three-to-ten-year horizon.

Similar to the previous section covering industry's concerns, ACI wanted to empower change within its larger organization (US Army) to prevent, disrupt, mitigate, and/or recover from these future threats and the impacts of the potential weaponization of Artificial Intelligence (AI). Like many challenges within the military domain, there is never a single solution. Many elements and factors are necessary to determine success. For instance, DOTMLPF (Doctrine, Organization, Training, Material, Leadership, Personnel, Facilities) is an acronym that Army leaders learn and adopt from their professional military education. The underlying idea is that multiple different facets and components go into developing a successful force. More importantly, that all aspects of DOTMLPF influence the end state and they also interact with each other (both negatively and positively). Threatcasting's Phase One builds on this military notion by harnessing the diverse SME domains to highlight how technology might evolve in the future, and how the other facts/domains might influence both the development and employment of the technology.

Exploring the future of military supply chains revealed several possible and probable scenarios. These included a future in which suppliers to the supply chain (such as a transportation company) embrace and implement advanced AI-driven automation. This high level of automation could include both the physical and the digital world. In the physical world, robots, self-driving cars, trains, and drones become the physical backbone for the movement of goods and supplies. In the digital world, AI is used to not only manage the physical autonomy but to make decisions and value calls that would have normally been taken on by humans.

This continued increase in automation will be driven by multiple factors. The first is a relentless drive for efficiency and cost-cutting measures. As budgets are tighter and the military is asked to do more with less, there is little choice but to automate as much of the process as possible. By removing the person from the loop, decisions can be made faster and the supply chain becomes increasingly optimized for speed and efficiency. These changes in society, culture, and economies influence the direction of the technology and its acceptance.

This relentless drive for efficiency exposes the supply chain to massive vulnerabilities. It is well known that a complex system, like a supply chain, cannot be designed for both efficiency and security (Johnson, 2017b). One needs to be sacrificed for the other. In this future scenario, security is sacrificed for efficiency. This becomes a direct threat to national security given the activities that are supported by commercial supply chains.

Therefore, the threats that the Threatcasting analytical process session produces, while not always focused on the military, can still be applied directly to military operations. A key component of Threatcasting is using a diverse group of participants to think about the future. Their diverse inputs can produce a wide variety of outputs and can be applied in a military setting. These future threats can be used even when their first instance was not in a military setting. SFP can take general Threatcasting results and make them more useful for a military audience. For example, the ACI used both the A Widening Attack Plain and The New Dogs of War: The Future of Weaponized Artificial Intelligence Threatcasting technical reports as the foundation of a targeted SFP workshop to create a tactically feasible, operationally correct vision of the future. The future operating environments outlined in these documents were then applied to military operations.

Over a two-day workshop, diverse military members (officers of all ranks, Marines and Soldiers, from each branch and many functional areas), with different experiences (both in terms of operational deployments, continents served on, and unit types) gathered to "turn the crank" on the Threatcasting results. Namely, to apply the threats from previous Threatcasting sessions to the context of military understanding and operational environments. A series of graphic novellas and animated movies were produced, with the intent of educating Army leaders at all levels about the contested domain of cyberspace. They demonstrate how the adversary can affect our ability to fight and win our nation's wars, at home and abroad. These graphic novellas were also designed to spark innovative thinking amongst the force and across society to help prevent, detect, mitigate, and recover from these possible futures.

"Our thought was to put these stories out, spread these ideas into the ecosystem, and build a community," said Colonel Hall, director of the ACI. "Our hope is to influence our partners in industry and academia, as well as the junior leaders within the Army we believe will have to deal with these issues in the future."

11.25.27 is the story of Lieutenant Jenkins and her skeleton crew working on Thanksgiving Day to supervise the loading of military equipment that arrived two days late to the port. Without a second thought one of them tweets "Finally ... Looks like I will get some turkey! #hatemylife" ... and the attack begins. Months before, the Army's highly automated supply chain and the deployment planning system had been breached, turning them into a weapon for a local terror cell. Small errors and minimal oversight sent a deadly payload to the docks of Seattle, WA. A pair of autonomous drones fly on a collision course with a specially loaded railcar ... millions will die. No one will ever forget 11/25/27. (Johnson, 2018)

"Graphic novellas are a medium with a rich tradition of use by the US Army," said Lieutenant General (Retired) Rhett A. Hernandez, former commander of the US Army Cyber Command.

> *For generations, the Army has successfully used this medium for conveying important messages across its force. Today, the Army Cyber Institute is continuing this tradition based on science fiction prototypes of cyber threats it may encounter on the future battlefields.*

6.7 Implications for Design

The Threatcasting analytical process and operations research intersect in some key areas: analysis of alternatives, wargaming, modeling and simulation, data analysis after data-farming and wrangling. For example, the process and output of Threatcasting can provide needed inputs to broaden the range of possible futures that are analyzed and eventually used for wargaming, while TAD can help clarify modeling and simulation impact decisions, as well as data analysis, after data-farming. Currently, the qualitative results of the Threatcasting analytical process cannot be mathematically validated; however, they are used as scenarios and inputs to quantitative models (e.g., agent-based threat modeling, risk analysis, gap analysis). With the increased use of Threatcasting within the discipline, future work is anticipated in this direction.

The primary goal of Threatcasting is to draw together a broad range of inputs to better define a wider range of possible threats in a specific area. For operations research, these inputs would provide the decision-maker more information and help screen out options when conducting an analysis of alternatives. These threat futures can also provide a wider range of potential futures that can then be used for the multiple criteria decision analysis process. Additionally, the effects-based models that are at the center of the Threatcasting analytical process can provide a new lens for the practitioner who is using value focused thinking, allowing them to build design options based upon an individuals' or organizations' values and desired outcome.

Threatcasting's broader range of threats provide practitioners more diversity of possible situations and landscapes to feed into the design of wargames. Often when designing a wargame, operations research practitioners focus more on the execution of the wargame and what the adversary might do. But in order to have truly effective results, it is also important to base the foundation of the wargame on a broader, more diverse set of potential scenarios. Threatcasting can provide a high degree of detail to these future scenarios, giving greater real-world complexity, as well as more specificity to the output.

Once the threat futures have been defined and the backcasting has taken place, this new dataset can provide practitioners valuable inputs to key areas of operations research. The data generated for how to disrupt, mitigate, and recover from these threats provide a more robust input for modeling and simulation. The powerful narrative component of Threatcasting can feed into this step and provide more detail in the multi-domain battle space. The TAD time-based step function of backcasting provide practitioners more detail for modeling and simulation, and gives practitioners multiple approaches and timeframes for the modeling and simulation.

In the post-analysis of data, the transparent nature of the Threatcasting analytical process allows practitioners to monitor the accuracy of their models. Practitioners can examine the driving factors for a threat, looking for key flags or indicators that the threat is on its way to becoming a reality. Original sources can be checked, thus allowing for assumptions to be challenged and better analysis to be derived.

Ultimately, operations research seeks to deal with uncertainty. Using Threatcasting to illuminate a broader range of possible threats and actions to be taken to meet those threats means that the areas of uncertainty will shrink. Additionally, the Threatcasting analytical process allows practitioners to process a wide range of disparate inputs to understand a wider range of possible futures. In this way Threatcasting can be seen as a way to actually deal with and process uncertainty.

6.8 Conclusion

In this chapter we reviewed the Threatcasting analytical process in detail so that it can be applied to multiple areas of interest for operations research. The Threatcasting analytical process assists and enables practitioners to imagine enemy innovations before they happen and identify actions that can disrupt or respond to these enemy innovations. It is a needed process that produces results that can provide a basis for operations research in areas such as a concept of operations (CONOPs) development or resource allocation decisions.

Using the analytical process, practitioners can explore and comprehend the close ties between the advancement of technology and the potential effect that it could have on society, economies, and national security. To illustrate this, we applied the Threatcasting analytical process to specific problems in military and corporate supply chains, by focusing on the weaponization of AI. As an output of this work, we showed the need for and motivation to create powerful narratives and fact-based illustrations that can provide leadership and decision-makers a concise and expeditious way to comprehend the results of the Threatcasting analytical process, and begin the initial steps of specific actions.

References

Bell, W. (2009). *Foundations of Futures Studies: History, Purposes, Knowledge. Volume I: Human Science for a New Era*. New York, NY: Taylor & Francis.

Bennett, M., & Johnson, B.D. (2016). Dark Future Precedents: Science Fiction, Futurism, and Law. In P. Novais and S. Konomi (Eds), *Intelligent Environments 2016*, (pp. 506–513). Washington, DC: IOS Press.

Bonime, W. (2018). Cisco CHILL to Tackle the Future of Work in 48 Hours. *Forbes*. Retrieved from https://www.forbes.com/sites/westernbonime/2018/02/04/how-cisco-chill-turns-ideas-into-companies-in-48-hours/

Inayatullah, S. (2008). Six Pillars: Futures Thinking for Transforming. *Foresight, 10*(1), 4–21.

Johnson, B. D. (2011). *Science Fiction Prototyping: Designing the Future with Science Fiction*. San Rafael, CA: Morgan & Claypool.

Johnson, B. D. (2013). Engineering Uncertainty: The Role of Uncertainty in the Design of Complex Technological and Business Systems. *Futures, 50*, 56–65.

Johnson, B. D. (2017a). A Widening Attack Plain. (Threatcasting Report: Army Cyber Institute). Retrieved from https://threatcasting.com/publications/

Johnson, B. D. (2017b). Efficiency is Easy to Hack. *Mechanical Engineering, 139*(08), 38–43. doi:10.1115/1.2017-Aug-2

Johnson, B. D. (2017c). Two Days After Tuesday. Retrieved from https://www.threatcasting.com/about/sci-fi-prototypes/

Johnson, B. D., & Vanatta, N. (2017). What the Heck is Threatcasting? *Future Tense*. Retrieved from https://slate.com/technology/2017/09/threatcasting-in-futurism-attempts-to-imagine-the-risks-we-might-face.html

Johnson, B. D., Vanatta, N., Draudt, A. & West, J. (2017). The New Dogs of War: The Future of Weaponized Artificial Intelligence. Retrieved from https://apps.dtic.mil/dtic/tr/fulltext/u2/1040008.pdf

Johnson, B. D. (2018). 11.25.2027. Retrieved from https://www.threatcasting.com/about/sci-fi-prototypes/

Judson, J. (2018). From Multi-Domain Battle to Multi-Domain Operations: Army Evolves its Guiding Concept. *AUSA*. Retrieved from https://www.defensenews.com/digital-show-dailies/ausa/2018/10/09/from-multi-domain-battle-to-multi-domain-operations-army-evolves-its-guiding-concept/

Linstone, H., & Turoff, M. (2011). Delphi: A Brief Look Backward and Forward. *Technological Forecasting & Social Change, 78*(9), 1712–1719.

Mentzer, J. T. et al. (2011). Defining Supply Chain Management. *Journal of Business Logistics*. doi:10.1002/j.2158-1592.2001.tb00001.x

Popper, R. (2008). Foresight Methodology. In L. Georghiou et al. (Eds), *The Handbook of Technology Foresight* (pp. 44–88). Cheltenham, UK: Edward Elgar.

Proctor, J. (2017). Futurecasting the Supply Chain. *APICS SCM Now Magazine*. Retrieved from http://www.apics.org/apics-for-individuals/apics-magazine-home/magazine-detail-page/2017/01/30/futurecasting-the-supply-chain

Robinson, J. (1990). Futures Under Glass: A Recipe for People Who Hate to Predict. *Futures, 22*(8), 820–842.

Robinson, J. (2003). Future Subjunctive: Backcasting as Social Learning. *Futures, 35*(8), 839–856.

Vanatta, N., & Johnson, B. D. (2019). Threatcasting: A Framework and Process to Model Future Operating Environments. *The Journal of Defense Modeling and Simulation, 16*(1), 79–88. doi:10.1177/1548512918806385.

Wal-Aamal, A. (2017). Cisco CHILLs about Securing Digitized Supply Chains on the Blockchain. [BLOG Post]. Retrieved from https://www.unlock-bc.com/news/2017-09-18/ciscos-chills-about-secure-digitized-supply-chains-on-the-blockchain

Zenko, M. (2012). 100% Right 0% of the Time. *Foreign Policy.* Retrieved from https://foreign-policy.com/2012/10/16/100-right-0-of-the-time/

Chapter 7

Analytical Modeling of Stochastic Systems

Roger Chapman Burk

7.1   Introduction

The purpose of this chapter is to promote and encourage the modeling of stochastic systems with analytical models. "War is the realm of chance" (von Clausewitz, 1976, p. 101), and discrete event simulation (DES) is a very common method to model the chances of war, not to mention the uncertainties of support and peacetime operations. DES has many advantages. It has flexibility that makes it adaptable to virtually any system that can be mathematically defined. Its relationship to the modeled system is easy to explain. On the other hand, DES also has some disadvantages, most basically that the output of a simulation is one sample from an unknown distribution, so multiple runs and statistical treatment of the results are required. DES models also have a seemingly inevitable tendency to grow in detail and complexity, making them harder and harder to interpret. Analytical models, such as Markov chains and queuing models, have their own strengths and weaknesses. They give instant and exact answers. However, they require that the system modeled have a certain mathematical structure, at least approximately, that is perhaps not always present. They are also more abstract and harder to explain to someone unfamiliar with the method. Despite this balance of advantages and disadvantages, it seems that simulation is much more widely used than analytical models in the practice of military operations research. This chapter will explore why that might be so, and propose some reasons why analysts might want to try to develop an analytical model for the problem at hand rather than turning at once to discrete event simulation. The reader is assumed to be broadly familiar with these methods, but not too familiar to find a review of the basics useful for this discussion.

Both DES and analytical models are in virtually all operations research curricula and are well described in standard textbooks (for instance, Denardo, 2002; Winston, 2004; and Hillier & Lieberman, 2001; from which the standard results presented below are taken). More depth on DES can be found in Law (2006) or Harrell, Biman, and Bowden (2012), among other texts. Good texts for analytics models include Pinsky and Karlin (2010) on Markov chains and Hall (1991) on queues. This chapter will provide a brief review of the basics of these methods, including a sample of some fundamental results, with particular attention to the particular advantages and disadvantages of each. This will be followed by a direct comparison, including some results from modeling a system that is easily modeled each way. Then examples will be presented of analytical modeling of military systems that one might expect to be modeled in a DES. Finally, there will be a discussion of the imprecision in results that might be attributed to perhaps unrealistic assumptions required to fit a real situation into an analytical framework; these turn out to be smaller than one might expect, and comparable to the imprecision one expects from any modeling method.

7.2   Discrete Event Simulation

A computer simulation is software that operates analogously to the system being modeled. For instance, a computer simulation of a simple queue would have a software module that generates arriving customers, another that handles the waiting line, and another that handles the service process. A simulation of the operation of a tank in the field would typically include modules for the movement of the tank, for the employment of its weapons, for its communication with other tanks and with higher echelons, and also for the terrain and for enemy units, in more or less detail according to the objectives of the model. Simulation can be deterministic or probabilistic, static or dynamic, and (if dynamic) discrete or continuous. Deterministic simulations like flight simulators are common in engineering applications, less so in operations research. A simulation that is probabilistic and static is often called a Monte Carlo simulation; this is the first type of simulation that had an important practical application, solving some problem in nuclear physics before the age of electronic computers, using tables of random numbers as input. A simulation that is probabilistic, dynamic (that is, changing over time), and discrete (changing state only at discrete intervals when particular modeled events happen) is a discrete event simulation, perhaps the most common method of modeling a stochastic system. Simulations that are probabilistic, dynamic, and continuous are harder to work with and much less common in analytical applications.

A DES obviously requires a great deal of computer code to keep track of everything that happens: creating and scheduling events, changing the state of modeled entities, finding and executing the next scheduled event, and tracking every parameter of interest as it changes during the course of the simulation run. Sometimes, and improperly, a large DES project comes to be regarded primarily as a computer programming problem. Fortunately, commercial simulation packages (e.g., Analytica, ExtendSim, ProModel; there is a biennial survey in *ORMS Today* (Swain, 2017)) are available to handle the common routine aspects of simulation and greatly reduce the programming effort.

Nevertheless, the problems of software development still exist for DESs: interfaces, versioning, documentation, testing, verification, validation, and so forth.

In order to simulate random events, DESs need a source of random numbers to represent random effects, time intervals, and so forth. The numbers should not be truly random (e.g., derived from some random physical process like throwing a die or particle decay) because a simulation may need to be repeated with the exact same random values, for instance, to track down a programming error. Instead, "pseudorandom" numbers are used that are derived from a deterministic algorithm but under statistical analysis look the same as true random numbers. A great deal of ingenuity has gone into generating pseudorandom numbers, and now reliable number streams from just about any distribution that can be described are readily available.

By its nature, the output of a DES run is a random number from an unknown distribution. It provides a point estimate for the parameter of interest, but no information on how good that estimate is. This means that multiple runs and statistical analysis of the results are almost always required. The wall-clock time of the computer runs is usually not as significant as it was in past decades, but the analytical work of deciding what runs to make and analyzing and interpreting the results remains undiminished. Care must be taken to account for correlation in the results. For instance, if one wanted to determine average wait time in a queuing system, and averaged the wait times of customers in a simulated queue to estimate the average, one would have to account for the fact that the wait times of successive customers are not independent. Also, if one is interested in the time-average state of a system, one has to account for the warm-up period. The simulation has to start in some particular state, e.g., the queue must start with some specific number of customers waiting. That initial state may not be very typical of the system – after all, you are likely running the simulation in order to learn what a typical state is. This is sometimes called the initial transient problem. These problems and others like them make interpreting simulation output more difficult than one might expect.

7.2.1 Advantages

For an analyst,, perhaps the most important advantage of discrete event simulations is that they operate analogously to the real system. That makes it easy to explain to the client and to other stakeholders how the model works and why its results can be trusted to give real information about the modeled system. Even a non-specialist can easily understand how software entities interact in ways that reflect how their prototypes interact. Many commercial simulation packages provide animations that show simulated jobs flowing through a factory, packages being routed and delivered, cars driving through a road network, and so forth. This makes it easy for the client to believe in your work.

Simulations also provide unsurpassable flexibility of application. They do not require any particular mathematical structure in the system being modeled. If it can be imagined and specified, it can be simulated. If changes need to be made to the simulated system, there is no danger that they will require a whole new approach to modeling – typically it can be handled by patching in some modified code at the point where some component behavior has to change (though system-level behavior sometimes changes in unexpectedly large ways for a seemingly small component-level change). This flexibility makes simulations very attractive for "what if" questions about system behavior.

Finally, DESs can provide valuable insight into how the modeled system will operate. It provides a time history of system functioning, at least under a given set of initial conditions. It provides not just steady-state averages, but a dynamic history of queue lengths, throughputs, backlogs, operating states, movements, idleness, downtime, and so forth. Animations can make this even more vivid. This kind of simulated operational history can give a very full understanding of the dynamic operations of the modeled system.

7.2.2 Disadvantages

There are three major disadvantages of discrete event simulation as a modeling technique, perhaps the largest of which is also the least obvious to the non-specialist. It is this: the very flexibility and adaptability of the method becomes a trap. Analysts' efforts get diverted into creating an ever-more-detailed model, in the hopes of better capturing reality, even when the added features have little effect on the result. Sometimes the simulation result conflicts with the expectations and/or organizational equities of a stakeholder, and the difference is attributed to poor modeling of some subtle feature of the system. Sometimes stakeholders unconsciously regard the reliability of the results as proportional to the number of lines of code in the simulation. Desire for more accurate answers leads to more complex simulations, which leads to poorly understood models, which leads to doubt about how to interpret the results, which leads to a desire for more accurate answers. Understanding the system becomes secondary to understanding the model. The model eventually becomes distrusted because it is too complex to understand, or perhaps worse, trusted because it is so complex. This is not a technical or theoretical limitation, but it is still very common and very real.

The second disadvantage is that, as already mentioned, DESs do not give simple answers. They must be carefully designed to account for correlation and initial transients. Care is required to ensure that the sample paths followed by the simulation are representative of the sample paths of the real system. Each individual result is then just one sample from an unknown distribution, and statistical analysis of multiple runs is required to get a confidence interval on any parameter of interest. Among other things, this makes it hard to adapt simulations to optimization of input parameters.

The third major disadvantage derives from the nature of DESs as large computer programs. The very size of the programs makes it more likely that there will be undetected errors in them. It is very hard to ensure that the programming is entirely correct. Testing every possible logical path through the code is a practical impossibility. One can never be totally sure that a surprising result is not just the result of some subtle bug lurking in the code.

7.3 Analytical Models

Analytical models develop equations describing system performance based on an abstract mathematical model of the system. Most optimization models fall into this category; for stochastic systems (i.e., systems that change randomly over time), such models include renewal processes, regenerative processes, and stochastic dynamic programming. Here we will confine the discussion to the simplest and most widely used stochastic analytical models, Markov chains and queuing models.

7.3.1 Markov Chains

There is a very large body of research on Markov chains; we will review here some of the fundamental and widely applicable results. The essence of creating a discrete-time Markov chain (DTMC) model is identifying a (usually finite) number s of states that the system can be in, a time interval over which the change of state is probabilistic (but independent of the prior history of the system), and the probabilities of transitioning from any state to any other in one time interval. (We will consider here only *stationary* Markov chains, i.e., those in which the transition probabilities do not change.) The probabilities p_{ij} of transitioning from state i to state j can then be assembled into the $s \times s$ probability transition matrix P, in which $\sum_{j=1}^{s} p_{ij} = 1$ for all i. If the initial probabilities (possibly 0) of being in each state are collected in the $1 \times s$ row vector q, then the probabilities of being in each state after n time periods are given by qP^n.

Many other properties of a DTMC can be directly calculated from P. If it is possible to move from any state to any state (possibly by way of other states), and if the states are aperiodic (that is, there is no period such that it is possible to return to the state only in multiples of that period), then the DTMC is called *ergodic* and there is a long-run steady state distribution π for the DTMC, where π is a $1 \times s$ row vector. The elements π_i of π represent the long-run probability of being in each state, and consequently they sum to 1. The probabilistic state of the DTMC will converge to π regardless of its initial state. This vector π can be can be calculated using the fact that one state transition cannot change the steady-state distribution: $\pi P = \pi$. Let I be the $s \times s$ identity matrix; let z be a $1 \times s$ row vector of all 0's; let z_1 be a $1 \times s$ row vector of all 0s except for a 1 in the last position; and for any matrix A, let A' be that matrix with the last column replaced by all 1s. Then

$$\pi P = \pi$$

$$\pi P = \pi I$$

$$\pi(P - I) = z$$

Since the elements of π must sum to 1, we have

$$\pi(P - I)' = z_1$$

$$\pi = z_1 (P - I)'^{-1} \tag{7.1}$$

In other words, π is found in the last row of $(P - I)'^{-1}$.

Suppose one is interested in the mean first passage times from a state i to all the other states; in other words, the average number of state transitions it takes to get from i to each possible target state. Let m_{ij} be the mean first passage time from i to j. Then $m_{ii} = 1/\pi_i$. The other m_{ij} can be found by conditioning on the first transition:

$$m_{ij} = \sum_{k=1}^{s} (m_{ij} \mid \text{first transition is to } k) p_{ik}$$

$$m_{ij} = \left(1 + m_{1j}\right)p_{i1} + \left(1 + m_{2j}\right)p_{i2} + \cdots + \left(1\right)p_{ij} + \cdots + \left(1 + m_{sj}\right)p_{is}$$

$$m_{ij} = \sum_{k=1}^{s} p_{ik} + \sum_{k \neq j} m_{kj} = 1 + \sum_{k \neq i} m_{kj}$$

$$m_{ij} - \sum_{k \neq j} m_{kj} = 1 \tag{7.2}$$

The set of all m_{ij} for any given j and for all $i \neq j$ can then be found by matrix algebra. Let $P_{\backslash j}$ be the state transition matrix with the jth row and column removed; let $M_{\backslash j}$ be an $s - 1 \times 1$ column vector containing all the m_{ij} except m_{jj}; and let E be an $s - 1 \times 1$ column vector of all 1's. Then the equations (7.2) for all $i \neq j$ can be represented in matrix form as

$$\left(I - P_{\backslash j}\right)M_{\backslash j} = E$$

$$M_{\backslash j} = \left(I - P_{\backslash j}\right)^{-1} E \tag{7.3}$$

In a non-ergodic DTMC, suppose some of the states are absorbing states (i.e., states i such that $p_{ii} = 1$) and the analyst is interested in time until absorption and in the state the DTMC ends up in. Let the number of absorbing states be a and the number of non-absorbing (or transient) states be t, so that $a + t = s$. If we order the states with the a absorbing states at the end, then we can rewrite the probability transition matrix in terms of submatrices:

$$P = \begin{pmatrix} T & R \\ Z & I \end{pmatrix} \tag{7.3}$$

where T is the $t \times t$ matrix of transition probabilities between transient states, R is the $t \times a$ matrix of transition probabilities from transient states to absorbing states, Z is the $a \times t$ matrix of transition probabilities from absorbing states to transient states (i.e., all 0's), and I is an identity matrix of the dimension the context requires. Then in the matrix $F = \left(I - T\right)^{-1}$, the element f_{ij} is the expected number of periods in state j given a start in state i. Furthermore, in the matrix FR, the element in the ith row and j^{th} column gives the probability of ending in the state j given a start in state i. (Winston, 2004, §5.6)

A continuous-time Markov chain (CTMC) has states like a DTMC, but it can change state at any point in time. The time until state change is an exponential random variable with a rate that depends only on the current state; thus, the future sample path of a CTMC depends only on its current state and not on its history. If the rate of transition from state i to state j is q_{ij}, then the probability of transitioning from i to a given state k is $q_{ik} / \sum_{j=1}^{s} q_{ij}$ and the total rate of transitioning out of i is $q_i = \sum_{j \neq i} q_{ij}$. We can find the steady-state distribution of the CTMC by observing that in steady state the total rate of transitioning out of i has to equal the total rate of transitioning into i:

$$\pi_i q_i = \sum_{j \neq i} \pi_j q_{ji} \tag{7.4}$$

If we define Q as an $s \times s$ matrix with $-q_i$ terms along the diagonal and q_{ij} elsewhere, then we can represent (7.4) for all i in one matrix equation.

$$\pi Q = z$$

$$\pi Q' = z_1$$

$$\pi = z_1 Q'^{-1} \tag{7.5}$$

and π is found in the last row of Q'^{-1}.

7.3.2 Queuing Models

There is a great deal of literature on queuing systems also; we will review some of the fundamental and commonly useful results. In a queuing system, items or customers (e.g., people, vehicles, job orders, messages, targets) come into the system at random intervals, wait if necessary for service at one of s servers, start service, and then depart after a random service time. Queues are very important in modeling because they occur often in real systems, because they are often very stochastic (the number of customers waiting for service, i.e., the state of the queuing system, can vary wildly and unpredictably over time), and because their structure allows for some very useful analytical results to be derived. Some common parameters of interest in a queuing system are

L: average number of customers in the system
L_q: average number of customers in the waiting line (not being served)
W: average time of a customer in the system
W_q: average time of a customer in the waiting line

Let λ be the average arrival rate and μ the average service rate. Then the four parameters above are related by "Little's Law" (Little, 1961; Stidham, 1974) and extensions thereof:

$$L = \lambda W = L_q + \frac{\lambda}{\mu} = \lambda W_q + \frac{\lambda}{\mu} \tag{7.6}$$

Thus, if one of these parameters can be found then the others follow immediately.

Usually interarrival times are modeled as exponential random variables that are identically and independently distributed (IID), and this is not an unreasonable assumption when the calling population is relatively large compared to the number of servers s. If the calling population is relatively small, then it is often reasonable to model the arrival rate as proportional to the number of customers not in the system. Service times are also often modeled as IID exponential, and while this makes the mathematics much easier, it is harder to justify. If both arrivals and services are IID exponential, and the calling population is finite, then the queuing system can be analyzed as a continuous-time Markov chain (CTMC). The same thing is true if the number of customers in the waiting line is limited (additional arrivals being rejected if the system if full). These are cases in which the number of states of the system is finite. There are also analytical formulas in most operations research textbooks that allow direct computation of L, π_i

(the probability of i customers being in the system), and other performance parameters, without going through matrix calculations.

Very often there is no natural limit to the length of the waiting line, so that the number of states in the system should be modeled as infinite. In these cases, the CTMC formulas given above can't be used, but analytical formulas are still often available. The simplest such queue is often called the M/M/1 queue, the two Ms indicating Markovian, that is, exponential, interarrival and service times, and 1 being the number of servers. Here the steady-state probabilities π_i can be found by considering the requirements of conservation of flow for state 0 in steady state, letting $\rho = \lambda/\mu$:

$$\mu\pi_1 = \lambda\pi_0 \rightarrow \pi_1 = \rho\pi_0$$

If $\pi_i = \rho^i\pi_0$ for all i up to j, then for $j + 1$, conservation of flow at state j gives

$$\mu\pi_{j+1} + \lambda\pi_{j-1} = (\lambda + \mu)\pi_j \rightarrow \pi_{j+1} = \rho\pi_j + \pi_j - \rho\pi_{j-1} = \rho\pi_i + \rho^j\pi_0 - \rho^j\pi_0 = \rho^{j+1}\pi_0$$

Then by mathematical induction, $\pi_i = \rho^i\pi_0$ for all i. Since all the probabilities π_i must sum to 1, we have

$$1 = \sum_{i=0}^{\infty}\pi_i = \sum_{i=0}^{\infty}\rho^i\pi_0 = \pi_0\sum_{i=0}^{\infty}\rho^i = \pi_0\frac{1}{1-\rho} \rightarrow \pi_0 = 1 - \rho$$

Then we can find L and hence the other three main performance parameters by using the identity $\sum_{i=0}^{\infty}i\rho^i = \rho/(1-\rho)^2$ for $0 < \rho < 1$:

$$L = \sum_{i=0}^{\infty}i\pi_i = \sum_{i=0}^{\infty}i\rho^i\pi_0 = \pi_0\sum_{i=0}^{\infty}i\rho^i = \pi_0\frac{\rho}{(1-\rho)^2} = \frac{\rho}{1-\rho} = \frac{\lambda}{\mu-\lambda} \tag{7.7}$$

Similar, but rather more complex, formulas have been developed for a similar queue with s servers working in parallel, each one having at most one customer at a time, which is commonly called the M/M/s queue.

All the queues mentioned here have had exponentially distributed service times, and as noted this often seems like a dubious assumption. There are results available for a single-server queue with Markovian arrivals but any IID distribution of service times. This is the M/G/1 queue (G for general). For this queue, $\pi_0 = 1-\rho$, and if the variance of the service time is σ^2,

$$L_q = \frac{\lambda^2\sigma^2 + \rho^2}{2(1-\rho)}$$

7.3.3 Advantages

Using these and other analytical methods, a large number of performance parameters can be calculated for many realistic systems, particularly regarding steady-state behavior. One obvious advantage of these models is that these parameters can be calculated immediately and there is no need for multiple runs or statistical analysis.

This makes it easy to use them for optimization. Also, analytical models do not require extensive computer programming and its potential for unsuspected errors. There is another indirect but perhaps more important advantage. To apply an analytical model properly, one must carefully analyze the modeled system and understand its mathematical structure. Analytical modeling encourages and even forces careful analysis of the system under study so that its major features can be distinguished and understood. It makes the analyst look more carefully at the system and think about the structure of the states, how long it stays in them, and how it moves from one state to another. It encourages him or her to see common structures like queues for which extensive analytical results are readily available. This promotion of careful and thoughtful analysis may be the most important advantage of analytical models.

7.3.4   Disadvantages

It is easy to see that analytical models require careful thought in their application, that a suitable structure must exist in the real system, and that modeling approximations might need to be made about distributions and probabilistic independence of which it would be hard to convince a skeptic. Many analytic techniques focus on steady states and long-term asymptotic averages, when transient effects may actually be more important. However, a bigger problem may be that analytical methods are hard to convey to a non-specialist. The structure of their formulas has a linkage to the real world that is not at all obvious and that takes a certain amount of specialized mathematical training to appreciate, unlike simulations, which have a structure based on that of the modeled system. When the analyst confesses the necessary modeling approximation, it is easy for the client to become dubious about the analyst's results, especially if he or she doesn't like them. If an analyst stands up in front of a client with much practical military experience but who knows little of operations research and says something like, "This system can be analyzed as an M/M/s queue with a limited calling population, and well-known formulas tell us that the average time from arrival to departure will be x," it just doesn't sound as convincing as saying, "We created a computer simulation of the system, ran 10,000 runs, and the average time in the system was x." This handicap in conveying results is a constant deterrent to using analytical models.

We have said that the exponential distributions often assumed in analytical modeling are often unrealistic (except for interarrival times, where they are not too bad). We should remember that the exponential has a relatively large variance, in proportion to its mean, compared to common empirical distributions. Since large variance generally leads to poor performance, other things being equal, modeling with exponential distributions is actually conservative – real performance is likely to be better in many situations.

7.4   A Comparison of Simulation and Analytical Models

A good way to see the relative merits of DES and analytical models is to use them on a simple system that is easily modeled either way. Let us take the venerable M/M/1 queue, with the service rate $\mu = 1$ and arrival rate $\lambda = 0.9$, and let us suppose we are interested in the average time a customer spends in the waiting line, W_q. Using the analytical models, we find at once from (7.7) that $L = 9$ and then from (7.6) that also $W_q = 9$. It is not difficult to model this in a spreadsheet with a DES that executes in a fraction of a second. The results of such a model are shown for the first 10,000 customers in Figure 7.1. Five replications

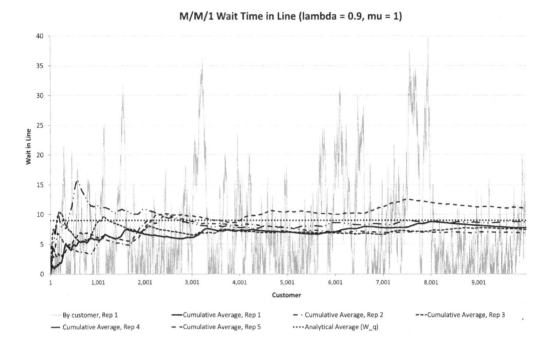

FIGURE 7.1: Simulation results for an M/M/1 queue.

are shown. The individual customer line waiting times are shown in gray for replication 1, along with the cumulative average in black. For the other replications, only the cumulative average is shown, in dashed black. The correct long-run analytical result (9) is shown in a dotted line. It is striking that most of the replications do not come particularly close to it, even after 10,000 simulated customers go through the system. It seems one would have to run the simulation much longer to reliably arrive at the correct result. Without multiple runs, the analyst would be unlikely to know he or she had not already arrived at the long-run average for several of these sample paths. On the other hand, the simulation shows the great variability of wait time, sometimes approaching 40, to which the long-term analytical average gives no clue.

Of course, the M/M/1 queue is not a typical stochastic system. It is perhaps more volatile than is common, despite (or because of) its simplicity. But one cannot know how typical a system one is dealing with *a priori*. It seems that analytical models have a great advantage in finding long-term averages and steady-state behavior. It might be objected that those models often require many simplifying assumptions about distributions, independence, structure, and so forth, but DESs typically require many more assumptions and approximations, in proportion as they are more complex and require more input parameters. Even if the effect of each single assumption is smaller, their cumulative effect can produce errors that are just as great. Analytical models have an advantage because their assumptions are made at the beginning of the analysis and are explicit, not buried in computer code with hundreds of others. If the assumptions can cause estimated parameters to be off, DESs can also be off because of stochastic variation, as four of the five replications in Figure 7.1 are. Simulations may also differ from reality for subtle reasons that are hidden in the code. Simulations can seem realistic because their output is strange and unpredictable, just like real life, but it is hard to be sure that they are strange *in the same way* that real life will be. It is their very complexity makes them seem lifelike and credible.

It is an advantage of DESs that they give automatic insight into how the system evolves over time, by looking at the sample paths from a few replications. If there are analytical ways to look at these, they are more involved than the methods we have discussed, and perhaps even harder to explain to the non-specialist.

If the problem is one of design, where a range of parameter settings need to be explored in order to find the best tradeoff, analytic models have a clear advantage. New results can be calculated by plugging the new values into existing formulas. Analytic optimization can sometimes be used. DESs have to be rerun, with all the replication repeated.

If a slight modification to the modeled system needs to be made, it is usually easier to see how to model it in a DES. A correspondingly small change in one subroutine may be all that is required. For an analytic model, a similar change might require rethinking of the whole approach. On the other hand, this rethinking may actually be an advantage, leading to a sounder understanding of the new whole. As described above, it is common for DESs to become more and more complex over time, as stakeholders demand including ever-more-subtle effects, to the point where the analyst loses track of the main important effects. The analyst himself may become so involved in the simulation that he becomes invested in it and reluctant to consider other methods of analysis.

Perhaps the most important difference between DESs and analytical models is not strictly a technical or modeling issue but an issue of how people interact with them. The two approaches have different strengths and weakness, and one or the other may be the best technical approach to a given problem. But the ready applicability of DESs and the ease with which they can be explained to non-specialists hide their tendency to grow in poorly understood complexity and gives their results an undeserved aura of believability. Analytical models require more careful thought and insight at the beginning, and more work in explaining them to the client. They should be used more often than they are. If they are harder to explain, that is partly because they are not used as often. If customers encountered them more often, they would become more easily accepted.

7.5 A Markov Model of a Tank Meeting Engagement

To see an example of how analytical modeling can be used, we will look at a discrete Markov model of a situation that might normally be thought more suitable for simulation: a meeting engagement in armored conflict. First, we will model a tank-on-tank duel, and define five states in the system, as shown in Table 7.1. Both tanks are modeled as belonging to a platoon advancing in a bounded overwatch fashion in rolling country. First contact will occur when a tank of one force moves into an overwatch position and sees an enemy tank, either moving forward or in its own overwatch position, waiting for its platoon to catch up. On contact, a tank in overwatch will fire, and moving tank will return fire while seeking improvised cover. Once in cover it will continue firing until the duel is resolved.

In real life many chance events can determine the outcome of such a duel, but a military analyst will want to concentrate on a few key parameters that will control the outcome in the great majority of situations. These are the quickness of acquiring the enemy target and firing, the probability of a kill or disablement (Pk) when firing in various circumstances, and the quickness with which a tank can move to improvised cover. We will represent these as shown in Table 7.2.

TABLE 7.1:   States in a Tank-on-Tank Duel

State	Blue	Red
OO	Overwatch	Overwatch
OM	Overwatch	Moving
OC	Overwatch	Improvised Cover
MO	Moving	Overwatch
CO	Improvised Cover	Overwatch

TABLE 7.2:   Input Parameters for a Tank-on-Tank Duel

Parameter	Blue	Red
Pk, overwatch to overwatch	p	P
Pk, overwatch to moving	q	Q
Pk, overwatch to cover or cover to overwatch	r	R
Pk, moving to overwatch	s	S
Rate of fire when stationary (per sec)	f	F
Rate of fire when moving (per sec)	g	G
Rate of moving to cover (per sec)	m	M

We are interested in the outcome of the duel, Blue Win (BW, meaning the Red tank is out of action) or Red Win, perhaps to evaluate the improvement to be expected from new armor or from new training that will improve Pk in given circumstances. Transition probabilities can be calculated from the parameters in Table 7.2; BW and RW are absorbing states that transition to themselves with probability 1. The transition probabilities are shown in Table 7.3.

TABLE 7.3:   Transition Probabilities for a Tank-on-Tank Duel

Initial State				Resulting State			
	OO	OM	OC	MO	CO	BW	RW
OO	0	0	0	0	0	$\dfrac{fp}{fp+FP}$	$\dfrac{FP}{fp+FP}$
OM	0	0	$\dfrac{M}{fq+GS+M}$	0	0	$\dfrac{fq}{fq+GS+M}$	$\dfrac{GS}{fq+GS+M}$
OC	0	0	0	0	0	$\dfrac{fr}{fr+FR}$	$\dfrac{FR}{fr+FR}$
MO	0	0	0	0	$\dfrac{m}{gs+FQ+m}$	$\dfrac{gs}{gs+FQ+m}$	$\dfrac{FQ}{gs+FQ+m}$
CO	0	0	0	0	0	$\dfrac{fr}{FR+fr}$	$\dfrac{FR}{FR+fr}$
BW	0	0	0	0	0	1	0
RW	0	0	0	0	0	0	1

The initial state will be OO with probability 0.5 and OM and MO with probability 0.25 each. With these probabilities, the transition probability matrix P can be constructed, and the probability of ending with a Blue or Red win calculated using the methods given above. The expected duration of the duel can also be determined. The time in each state will vary, but expected times can be calculated.

Using information on the rates and probabilities of tank-on-tank duels, a CTMC model can be constructed for a platoon-level meeting engagement. Let us assume that each platoon has four tanks and will withdraw from the battle if two of them are disabled. Then we can label the transient states 44, 43, 34, and 33, where the two digits give the number of surviving tanks for Blue and Red, respectively. The absorbing states can be labeled BW and RW. If we let λ be the rate at which one Blue tank achieves a win over one Red tank, and let μ be the win rate in the other direction, and the order of the states is $\begin{pmatrix} 44 & 43 & 34 & 33 & BW & RW \end{pmatrix}$, then the transition rate matrix will be

$$Q = \begin{pmatrix} -4\lambda-4\mu & 4\lambda & 4\mu & 0 & 0 & 0 \\ 0 & -4\lambda-3\mu & 0 & 3\mu & 4\lambda & 0 \\ 0 & 0 & -3\lambda-4\mu & 3\lambda & 0 & 4\mu \\ 0 & 0 & 0 & -3\lambda-3\mu & 3\lambda & 3\mu \\ 0 & 0 & 0 & 0 & 0 & 0 \\ 0 & 0 & 0 & 0 & 0 & 0 \end{pmatrix} \quad (7.8)$$

This will give information on the behavior of the system over time. The corresponding DTMC, which deals only with which states are entered, is

$$P = \begin{pmatrix} 0 & \dfrac{4\lambda}{4\lambda+4\mu} & \dfrac{4\mu}{4\lambda+4\mu} & 0 & 0 & 0 \\ 0 & 0 & 0 & \dfrac{3\mu}{4\lambda+3\mu} & \dfrac{4\lambda}{4\lambda+3\mu} & 0 \\ 0 & 0 & 0 & \dfrac{3\lambda}{3\lambda+4\mu} & 0 & \dfrac{4\mu}{3\lambda+4\mu} \\ 0 & 0 & 0 & 0 & \dfrac{3\lambda}{3\lambda+3\mu} & \dfrac{3\mu}{3\lambda+3\mu} \\ 0 & 0 & 0 & 0 & 1 & 0 \\ 0 & 0 & 0 & 0 & 0 & 1 \end{pmatrix} \quad (7.9)$$

This can be analyzed to find absorption probabilities in state BW or RW, given a start in state 44.

Of course, there is much more to a meeting engagement than is captured in these simple models. Some may want to add detail to them, perhaps much detail. But there is an argument that this model captures what is essential and universal for such engagements, and that this approach, or one like it, is appropriate for analysis.

7.6 A Queuing Model of Small Unmanned Aircraft Employment

This example of the use of queuing formulas for modeling a system that is not obviously a queue is taken from a study of different architectures for suppling unmanned aircraft support to companies in the field (Burk and Burk, 2007). The system modeled is an infantry battalion with three subordinate companies, each of which has a vehicle-based small unmanned aircraft system (UAS) in support. The UAS launches and recovers aircraft in response to the needs of the company commander. The parameter of interest is the average number of air vehicles in the air for the battalion. We can model this as an M/M/3 queue (exponentially distributed interarrival and service times, three servers) with a calling population of 3. In queuing terms, the average number of aircraft in the air is L, the average number of customers in the system. This can be calculated as follows, using standard queuing formulas (Hillier and Lieberman, 2001, §17.6):

t = average time required over the target

d = average distance from transporter vehicle to target

v = air vehicle dash speed

μ = queue service rate $= \left(t + \dfrac{d}{s} \right)^{-1}$

λ = arrival rate of mission requests per company

N = calling population = number of companies = 3

s = number of servers = 3

π_0 = probability of 0 customers in the system $= \left[\displaystyle\sum_{n=0}^{N} \dfrac{N!}{(N-n)!\,n!} \left(\dfrac{\lambda}{\mu} \right)^n \right]^{-1}$

π_n = probability of n customers in the system $= \dfrac{N!}{(N-n)!\,n!} \left(\dfrac{\lambda}{\mu} \right)^n \pi_0$

$L = \displaystyle\sum_{i=1}^{N} i\pi_i$

The power of this modeling approach is that it gives immediate information on the behavior of the system based on its structure and the input parameters. One limitation is that the calculations are based on an exponential distribution of interarrival times and service times, and it would be easy to call that assumption into question. However, there is some reason to think that the errors introduced by that assumption may not be too large, as will be discussed in the next section.

7.7   The Impact of the Exponential Assumption

One might reasonably be wary of the assumption of exponential distribution that is so common in analytical modeling. Of course, it can be expected that distributions with different means and variances will result in different system performance. However, higher-order moments may not have so great an effect. Here is a numerical experiment that illustrates this.

Three distributions were selected, all three with mean and variance both equal to 1. One was a bimodal beta, one exponential, and one lognormal. A single-server queue with a calling population of 3 was simulated using each of these for interarrival time distribution and for service time distribution in all nine possible combinations. Six replications were done for each combination and a 97% confidence interval (the min-to-max spread) for W_q generated for each. The analytic result for the exponential-exponential combination fell within the corresponding CI. The results are shown in Figure 7.2.

FIGURE 7.2: Simulation results for queues with different distributions.

Distribution of interarrival time had little effect on waiting line time (<1.6%). Distribution of service time was somewhat larger – as much as 5%. Of course, all these distributions had the same mean and variance. We should expect performance as measured by W_q, for instance, to worsen as variance increases. The good news here is that the exponential distribution has a relatively large variance, as empirical distributions go, and a distribution developed to have the same variance while matching the mean has to be somewhat contrived, especially if it is over a finite domain, as a beta is.

7.8 Conclusion

Discrete event simulation often seems to be the first tool an analyst reaches for when modeling a stochastic system. The thesis here is that instead it should be the last. DES has some advantages, the strongest one being its flexibility and ability to model any system that can be defined. But the analyst should remember its disadvantages also. That very flexibility tends to divert analytical attention from truly understanding the mathematical structure of the presented problem, and to sidetrack it into an endless cycle of ever-more-intricate model building. Analytical models have their disadvantages, including the need to identify particular structures and to make assumptions about distributions, but those disadvantages are the flip side of their strength, how they encourage and promote real understanding of the system. Besides, those assumptions often have no greater effect on the result of the investigation than is common in all operational modeling. Analytical models are harder to explain and justify to a client, but if the analytical community were less reluctant to use them, they would perhaps become more widely recognized for the insight they can bring to modeling a stochastic system.

References

Burk, R. C., & Burk, R. K. (2007). *Comparing organic vs. handoff UAV support to the maneuver company* (DTC Report ADA469275). West Point, NY: Operations Research Center of Excellence.

Denardo, E. V. (2002). *The science of decision making: A problem-based approach using Excel.* New York, NY: John Wiley & Sons, Inc.

Hall, R. W. (1991). *Queueing methods for services and manufacturing.* Upper Saddle River, NJ: Prentice Hall.

Harrell, C., Biman, G., & Bowden, R. (2012). *Simulation using ProModel* (3rd ed.). New York, NY: McGraw-Hill.

Hillier, F. S., & Lieberman, G. J. (2001). *Introduction to operations research* (7th ed.). New York, NY: McGraw-Hill.

Law, A. M. (2006). *Simulation modeling and analysis* (4th ed.). New York, NY: McGraw-Hill.

Little, J. D. C. (1961). A proof for the queueing formula $L = \lambda W$. *Operations Research*, 9(3), 383–387.

Pinsky, M., & Karlin, S. (2010). *An introduction to stochastic modeling* (4th ed.). Cambridge, MA: Academic Press.

Stidham Jr., S. (1974). A last word on $L = \lambda W$. *Operations Research*, 22(2), 417–421.

Swain, J. S. (2017). Simulation takes over: Reality is for sissies. *ORMS Today*, 44(5), 38–49.

von Clausewitz, C. (1976). *On War* (M. Howard & P. Paret, Trans.). Princeton, NJ: Princeton University Press.

Winston, W. L. (2004). *Introduction to probability models* (4th ed.). Belmont, CA: Brooks/Cole Cengage Learning.

Chapter 8

Modern Methods for Characterization of Social Networks through Network Models

Christine M. Schubert Kabban, Fairul Mohd-Zaid and Richard F. Deckro

8.1   Introduction

The measure of strategic success is ultimately the change in relevant actor behavior, in addition to physical results on the battlefield.

(Office of the Joint Chiefs of Staff, 2016, p. 1)

This chapter 1) provides a background for the usefulness of network models as a means to enable social network analysis (SNA), 2) discusses the importance of network characterization via a network model and 3) provides detail on the methods useful for such characterization, with an emphasis on newly emerging statistical methods. These methods are important to the operational analyst as a means to characterize and monitor networks of interest through the network model. The application of these methods extends beyond social networks and can be applied to any network of interest. Both students and early practitioners interested in SNA will benefit from the rich background provided within this chapter and may use the references and resources herein to learn more detail on any of the methods discussed. Further, data is used to demonstrate emerging statistical methods which may be used to monitor a network of interest.

An emphasis is placed on the statistical methods due to their flexibility, sensitivity, and ease of use. Students and practitioners should be able to follow, apply, and use these methods to both characterize and monitor the networks that are built to represent the structures of interest to them. Important features for the operations researcher to identify when characterizing a network of interest are whether or not the network has changed, whether the network behaves as expected, or whether the network does not exhibit characteristics as expected. The latter feature is of particular interest in that it represents emergent operational tasks intent on identifying fabricated, synthetic network structures which mask the true relations and intent within the network of interest. Identification of these features facilitates planning and response, both offensively and defensively. For instance, understanding and unmasking the real actors and relationships in a synthetic network may enable responders to pre-empt negative world events through network characterization and monitoring tasks.

8.1.1 Social Network Analysis Background: A Historical Perspective

The study of social, political, and military structures have been key intelligence considerations throughout the ages. Historically, the keys to understanding a foe's or an ally's strengths (and weaknesses) consisted of the lord/vassal, clans, and tribe structures, reinforced by familial, marriage, and financial links. Social network analysis (SNA) is a tool for the operations researcher to understand, monitor, and exploit these structures. Indeed, analysts have charted the alliances, coalitions, and military order of battles for centuries in order to derive factors providing operational advantages. Today, our interconnected world has created numerous additional considerations supplementing the traditional historical factors into which an analyst must be tuned in order to monitor and preserve his or her own network, while at the same time understanding and exploiting the weaknesses in a foe's network. SNA, with its associated mathematical tools enables the analyst to monitor, understand, and exploit network structures. These structures of our world abound today in complex ways.

> *The initial decade of the twenty-first century has been characterized by governments attempting to mitigate the effects of terrorist organizations, insurgent groups, organized criminal enterprises, drug cartels, human trafficking, piracy, and cybercrime. These entities utilize support networks composed of money laundering, weapons smuggling, illegal technology proliferation, and other illicit activities. Dealing with this myriad of interconnected organizations and activities has led to the development of nontraditional analytic techniques in support of strategies to cope with these threats to national security. One such analytic technique brought to bear on this problem set is social network analysis (SNA), not necessarily a new technique, but novel in its relatively recent application to the national security arena. As such, governments' initial unfamiliarity with SNA has now transitioned to various instantiations in levels of application and expertise in numerous organizations.*
>
> (Morris & Deckro, 2013, p. 70)

In addition, the nature of warfare has evolved over time. Von Clausewitz stated, "War is merely the continuation of policy by other means," implying wars were fought for political purposes to implement a far-reaching and/or overall strategy. While the classic image of war termination includes creating the conditions where a foe will acquiesce to unconditional surrender (whether it actually occurred or not), it has not generally been the outcome of modern conflicts. In addition, even when surrender is attained, winning the peace is problematic; the harsh terms of the Versailles Treaty ultimately lead to the Second World War. The Second World War termination evolved into the Cold War. The United States still maintains troops in Europe and throughout the world. While the US clearly demonstrated its military prowess in the Gulf Wars, ultimately toppling the Baathist regime, forces remain in Iraq attempting to stabilize the nation.

British General Sir Rupert Anthony Smith, in his text *The Utility of Force: The Art of War in the Modern World* (Smith, 2007), makes the point that warfare has shifted from "War between the people" (nations fighting to a distinct victory) to "War amongst the people" (non-nation states waging indefinite warfare) (Smith, 2007). The recent *Joint Concept for Human Aspects of Military Operations* (JC-HAMO) (Office of the Joint Chiefs of Staff, 2016) recognizes that these wars "among the people" coupled with the need to "win the peace" requires considering the effects on humans in planning *and* conducting military operations. It states that

> *Joint Force will enhance operations by impacting the will and influencing the decision making of relevant actors in the environment, shaping their behavior, both active and passive, in a manner that is consistent with U.S. objectives*

(Office of the Joint Chiefs of Staff, 2016, p. 1).

JC-HAMO goes on to point out that in planning conflict and building the peace, one must consider influence operations which include the social, cultural, physical, informational, and psychological elements that create a desired effect on behavior. Building on the use of SNA in counter-terrorism studies, it is the application of an analysis approach that can aid future joint planning. An understanding of SNA approaches will continue to be of value in the intelligence preparation of the battlespace, but with the adoption of the joint concept on human effects in planning operations, it will also be a planning analysis tool.

8.1.2   Emergence of Quantitative Approaches to Social Network Analysis

Humans remain the driving force at the center of technological advances and societal relations. As such, the study and analysis of human relations has expanded beyond the original interests of early sociologists. Over the last century, relations among individuals became represented by graphs in which individuals (nodes) were connected to each other through edges which represent the particular relations of interest; some of the earliest such graphs are attributed to the works of Jacob Moreno in the mid-1930s (Moreno, 1934). Dubbed sociograms by early sociologists, these graphical depictions of the social interactions among individuals are now a primary tool from which information and characteristics of the underlying social network are derived. Such information may be used to glean understanding of the potential implications that individual and collective group behavior may have on society or assets of interest.

Although initial analysis of social networks was conducted through qualitative assessments, the use of graphs enabled quantitative assessments through the use of mathematically based tools. Now these tools and assessments are paramount due to the growth of both the size and the application of social networks.

Today, the intelligence and operational communities widely use SNA in order to analyze relationships between individuals within groups of interest (Havig et al., 2012), and in many cases, to also monitor and possibly influence a network of interest. From a security and defense standpoint, individuals within the network may represent comrades or potential threats; or as applied to non-humans, could represent relations among assets or critical resources such as a network of computers, the power grid, or water sources. Irrespective of the network itself, network analysis, as a tool to understand, monitor, and potentially act upon a network of interest, is not possible without first characterizing the underlying structure of the network.

8.2   Characterization of a Network

Characterizing a network includes the compilation of operations and evaluations that comprehensively describes a specific network and facilitates the reproduction of the nodes and connections within a network by graphical or mathematical representation. These representations allow the analyst to modify, examine, and understand the relationships within the network in order to anticipate how the network may be affected by particular actions and to determine if the network is changing, e.g., if individuals are leaving or somehow now hidden from a network or if the network of individuals is growing.

There is not just one tool that can be used to accomplish network characterization. When possible, however, a mathematical model of the network should be constructed. Such network models may be specific to a network of interest or may be generated based upon common properties inherent within the network. As such, network models are efficient mathematical tools for the operations analyst. Not only can the network model be used to gain knowledge of network behavior, but it can be used to visualize the network. Network visualization methods are specific to the network structure and purpose for studying the network (Blaha, Arendt & Mohd-Zaid, 2014). However, through the use of the network model and, potentially, its visualization, an analyst may be able to study and respond to specific inquiries of interest regarding a real world network without the need to physically interact within the network. Network models, then, are the means by which planning and response is enabled in operational environments.

The next section describes graph properties and common network models that have formed the basis of many of the models that exist today. For further reading on the basic definitions and principles covered in the following sections, see Borgatti et al. (2009), Brass (1995), Newman (2010), and Wasserman and Faust (2009), among others.

8.2.1   Graph Properties

Many network models are based upon graphs which represent the relationships within the network of interest. Table 8.1 provides basic definitions and notation typically used for graphs and networks.

The notation is based upon network representation as graphs and is constructed from fundamentals in graph theory (West, 2001) and (Wasserman & Faust, 2009). A graph is defined by its nodes, V, and edges, E, and as the triple $G = (V, E)$. Simple

TABLE 8.1: Definitions for Common Graph Terms

Terms	Definitions
Graph, Network	A graph G is defined as a triple, $G = (V, E)$, with node set, V, and edge set, E.
Nodes, Vertices	A node set, $\{1,...,N\}$, is a set of points of size N that makes up a graph.
Edges, Links	An edge set, $E \subset \{V \times V\}$, is a set of ordered pairs, $e_{i,j} = (i,j)$, that connects two nodes in V.
Directed, Undirected	A graph is directed if $e_{i,j} \neq e_{j,i}$ and is undirected if $e_{i,j} = e_{j,i}$
Simple Graph, Multigraph	A simple graph is one such that there is only one instance of any particular edge, $e_{i,j}$, if it exists. A multigraph is a graph with multiple edges connecting the same pair of nodes.
K Regular Ring Lattice	A graph with N nodes where each node is connected to K neighbors on each side.
Degree	The number of edges connected to a given node.
Closeness	Inverse of the sum of pairwise distances between a node and other nodes in a graph.
Betweenness	The frequency that a particular node lies in the shortest path between all other paired nodes.
Clustering Coefficient	The proportions of local relationships amongst neighbors compared to the potential that all of the neighbors are connected.
Group Clustering Coefficient	The proportion of existing triangular relationships or triads in a graph over the number of all potential triangles.

graphs contain only a single edge between nodes and no edge which connects a node to itself. A graph is undirected if edges exist without associated direction between nodes. Special forms exists if nodes connect to neighbors in specific ways, such as being connected to exactly four neighbors as demonstrated in Figure 8.1 which is an example of what is called a 4-regular ring lattice on six nodes. Figure 8.2 gives an illustration of an undirected simple graph, $G = (V, E)$ with node set $V= \{1,2,3,4,5,6\}$ and edge set $E = \{(1,3), (1,5), (1,6), (2,4), (2,5), (2,6), (3,5), (3,6), (5,6)\}$.

Characteristics of a graph can be described through the relationships of the nodes and their edges. As defined in Table 8.1, five common measures of the characteristics of a graph can be summarized through nodal degree, closeness, betweenness, clustering coefficient and group clustering coefficient, in addition to many more. Table 8.2 provides the calculated values of these measurements for the graph example in Figure 8.2. The focus on these five, rather than others, is based upon the work of Guzman et al. (2014) in which an exploration of 24 different measures across many graphs could be divided into just five groups of highly correlated measures; the measures mentioned in Table 8.1 are single representations from each of these groups. Whereas the general definitions are provided in Table 8.1, there are various ways to compute these measures (Wasserman & Faust, 2009). For instance, closeness and betweenness may be standardized by dividing by the number of nodes or node pairs. When standardized, values for closeness or betweenness may be used to compare graphs of different sizes.

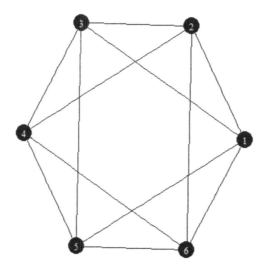

FIGURE 8.1: A 4-regular Ring Lattice on Six Nodes.

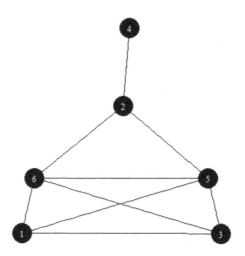

FIGURE 8.2: Undirected Simple Graph of Size n = 6.

8.2.2   Random Graph Generating Models

Random graph generating models are designed to model specific network properties of interest. As such, these models contain both fixed and random parameters so that properties, such as the size or connectivity within a network, could vary while other properties of the network could remain fixed. Therefore, families of possible network representations are generated by varying the random parameters while keeping constant those that are fixed. There are three basic random graph generating properties that form the foundation for many real-world network models: random behavior, scale-free behavior, and small-world behavior. Each of these are now described in turn.

TABLE 8.2: Nodal Measures for Graph in Figure 8.2

Node	Degree	Closeness	Betweenness	Clustering Coef.
1	3	0.1250	0	1.0000
2	3	0.1429	4	0.3333
3	3	0.1250	0	1.0000
4	1	0.0909	0	0.0000
5	4	0.1667	2	0.6667
6	4	0.1667	2	0.6667

Clustering Coef. – Clustering Coefficient

Further, algorithms that generate graphs from each of the models can be found in many software packages today.

Erdös and Rényi proposed the first commonly used random graph generating model (1959) in which a graph is generated by connecting any pair of nodes with an edge with probability p, and in which each edge is independent from every other edge. This results in a graph of N nodes and m edges having an equal probability of

$$p^m (1-p)^{\binom{N}{2} - m} \tag{8.1}$$

for all possible undirected simple graphs of N nodes and m edges. Although not mentioned earlier, the graph example in Figure 8.2 is an Erdös–Rényi graph with parameter p = 0.5.

Consequently, the graph has a group clustering coefficient of $C = k/N$ where k is the mean degree, and the statistical distribution of the degrees for all possible realizations of the Erdös–Rényi networks follows the Poisson probability distribution as the number of nodes grows large. One downside to the Erdös–Rényi graph generating algorithm is that it is not scale-free (Barabási & Albert, 1999), a property possessed by many real-world social networks such as the World Wide Web (Albert, Jeong & Barabási, 1999). A scale-free network is defined as one that has a power law degree distribution between nodes, which typically means that many nodes have few connections and only a smaller proportion of nodes have many connections. The Erdös–Rényi generated graph contains nodes that are randomly connected. However, given its history, the Erdös–Rényi algorithm is widely used as a baseline for comparisons of network metrics and classification.

Barabási and Albert proposed a model based on two mechanisms that govern the scale-free power law distribution of real world networks (Barabási & Albert, 1999). They defined the two mechanisms to be: (i) networks expand continuously by the addition of new nodes, and (ii) new nodes attach preferentially to existing nodes that are already well-connected. The model operates by first starting with an initial number of nodes N each having degree m. This is followed by an iterative process of adding one node with m edges where the edges are connected to an existing node i with degree d_i based on the following preferential attachment probability, denoted as $\pi(d_i)$:

$$\pi(d_i) = \frac{d_i}{\sum_j d_j} \tag{8.2}$$

which is the probability that node i will be attached to the new node. As developed, they claimed that their model indicates that the development of large networks is governed by robust selforganizing phenomena that is not specific to the domain, be it social, biology, or the World Wide Web (Barabási & Albert, 1999; Albert, Jeong & Barabási, 1999). Mathematically, the Barabási–Albert network has a group clustering coefficient (C) that is approximately $C \sim N^{-3/4}$ and the distribution of the degrees across all nodes follows a Pareto distribution with an exponent parameter of $\beta = 2$ (Barabási & Albert, 1999). Figure 8.3 is a representation of a Barabási–Albert graph of size $n = 6$ with $m = 2$. Note that the first and second nodes have considerably higher degrees in comparison to the other nodes due to preferential attachment. For completeness, the nodal measures for the example in Figure 8.3 are provided in Table 8.3.

The empirical degree distribution of the Barabási–Albert graph is different than what was originally derived, especially for relatively small graphs (Mohd-Zaid, Schubert Kabban & Deckro, 2017). Some have also suggested that the preferential attachment model causes biases on the connection of the high degree nodes in the Barabási–Albert graph (Li, Zhang & Small, 2011; Zhang, Small & Judd, 2015). Zhang, Small, and Judd proposed an optimal scale-free network model as an alternative to the Barabási–Albert model that accounts for the degree correlation in the Barabási–Albert model.

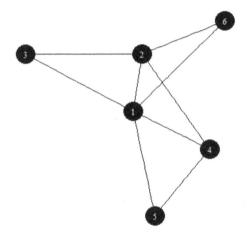

FIGURE 8.3: Barabási–Albert Graph of Size $n = 6$ with $m = 2$.

TABLE 8.3:   Nodal Measures for Graph in Figure 8.3

Node	Degree	Closeness	Betweenness	Clustering Coef.
1	5	0.2000	4.0	0.4000
2	4	0.1667	1.5	0.5000
3	2	0.1250	0.0	1.0000
4	3	0.1429	0.5	0.6667
5	2	0.1250	0.0	1.0000
6	2	0.1250	0.0	1.0000

Clustering Coef. – Clustering Coefficient

In addition, to date only Leydesdorff (2007) has studied the skewness and kurtosis of degree, betweenness, and closeness of various empirical networks.

A random graph generating model that produces small-world properties was introduced by Watts and Strogatz (1998). Small-world networks are networks where 1) the shortest path, L, between most pair of nodes in the network grows proportionately to the logarithm of the network size, N, such that $L \propto \log N$, and 2) the clustering coefficient is larger than that of a random network of the same size. The algorithm for this model functions by first starting with a ring lattice of size N (see an example of a ring lattice in Figure 8.1). This is then followed by rewiring each edge in the lattice with probability γ such that duplicates and self-loops are excluded. However, the nodes remain closely connected through short paths. Many real-world networks such as the neural network of the worm *Caenorhabditis elegans*, the power grid of the western United States, and the collaboration network of film actors are shown to possess small-world characteristics (Watts & Strogatz, 1998). That is, the average of the shortest paths between every pair of nodes is relatively small, and the ratio between the network's clustering coefficient is large compared to that of a random network. Further, the degree distribution for the Watts–Strogatz graph has been shown to follow a closed form distribution as described by Barat and Weigt (2000) and has a group clustering coefficient of

$$C \simeq \frac{3(k-2)}{4(k-1)}(1-\gamma)^3 \tag{8.3}$$

where k is the mean degree. Table 8.4 lists the average shortest path and the clustering coefficient characteristic of the real world networks listed above (Watts & Strogatz, 1998). Notice that despite the large size of these networks, the shortest path between nodes is on average as low as 2–3 nodes. Even the power grid has a relatively low shortest path as the average shortest path in a random graph of that size was reported as 12.7 (Watts & Strogatz, 1998). Further, the size of the clustering coefficient compared to a purely random network is large, demonstrating the clustering of nodes in these graphs. However, one disadvantage of the Watts–Strogatz algorithm is that it produces a network that does not have a scale-free power law distribution. Hence, neither the Barabási–Albert nor the Watts–Strogatz algorithm is capable of simultaneously modeling these real world properties in networks.

Other variations on these random graph generating models exist. Seshadhri, Kolda, and Pinar (2012) proposed a modified Erdös–Rényi model where community is defined as a subgraph that is internally highly connected and must contain a dense Erdös–Rényi subgraph. The proposed Block Two-Level Erdös–Rényi (BTER) model creates

TABLE 8.4: Average Shortest Path, $L_{average,}$ and Ratio of Clustering Coefficient with Respect to a Random Network of the Same Size, $C_{network}/C_{random,}$ for Some Real World Networks

Network	Size	$L_{average}$	$C_{network}/C_{random}$
Film actors	225,226	3.65	2925.9
Power grid	4,941	18.70	16
Caenorhabditis elegans	282	2.65	5.6

Data from Watts, D.J. and S.H. Strogatz, "Collective dynamics of 'small-world' networks", *Nature* 393, no 6684 (1998): 440–442.

connection via the Erdös–Rényi model within communities followed by cross-communities connections between nodes with degree one. They demonstrated that it accurately captures the scale-free and clustering properties of many real-world social networks. Morris, O'Neal, and Deckro (2014) created an algorithm for prescribed node degree connected graphs (PNDCG) that allows the user to define the scale parameter as well as the clustering coefficient of the network. The clustering coefficient is related to the transitivity concept in the social network literature where transitivity implies the idea of "a friend of a friend is a friend." Comparisons of the average clustering coefficient with those from the Erdös–Rényi and Barabási–Albert generated networks show that their algorithm is able to generate networks with a wider distribution of average clustering coefficients.

8.2.2.1   Mixtures of Random Graph Generating Models

Each of the random graph generating models described focus on a network property (i.e., scalefree or smallworld) and attempts to model that specific property through a single unified distribution for a network measure such as degree. More recent work has focused on using multiple distributions for a single network measure or a combination of network measures. Such models are called mixture models and are motivated by the lack of fit of any one individual property across similar real-world data sets. A mixture model is a single probabilistic model that combines several distributions, each from a different subpopulation, in which the distribution for each subpopulation is governed by different parameter settings. These distributions then are combined into one model through the use of a weighting term that weights each of the distributions; the sums of the weights must equal one. Parameter estimation for the mixture model is then accomplished through statistical techniques such as Expectation-Maximization or Bayesian methods.

With respect to networks, mixture models have been applied in two ways: either as a mixture of different distributions for the same network property, such as randomness, or more recently, as a mixture of varying properties, such as randomness with scale-free and small-world properties. Examples include the works of Thomas, Stoica, and Beuscart (2010) who modeled popularity with respect to social media as a mixture of Gaussian distributions, Zanghi, Ambroise, and Miele (2008) who created an Erdös–Rényi based mixture model to examine affiliation, and Daudin, Picard, and Robin (2008) who created a mixture of Poisson distributions to describe the Erdös–Rényi-based degree distribution within a bacteria network. Each of these examples focuses on mixtures of similar random variable distributions to model a network property. More recently, Durante, Dunson, and Vogelstein (2017) developed a nonparametric, Bayesian based method to model a network using a mixture of network properties. Applied to human brain data, their method demonstrates the feasibility and usefulness in creating a model that mixes several network properties in order to accurately model real-world data.

8.2.3   Network Model Summary

Network models are only useful if such models are able to characterize a network accurately enough for the intended analysis. This section discussed various network models and means by which network models are generated and combined (via mixtures) in order to reproduce network properties of interest. However, when we use specific network properties, such as small-world properties, in order to build a network model, how do we determine that the resulting model is accurate with respect to the real-world network of interest? In order to assess the accuracy of a network model, alternate techniques are used based upon mathematical or statistical methods. These methods for assessing the accuracy of a network model are discussed next.

8.3   Methods to Assess the Accuracy of a Network Model

There are several methods to determine whether or not the generated model of the network, a model based upon parameters, truly represents the real-world data of the network of interest to an acceptable level. These methods are based upon numerical and more recently, statistical techniques, and are categorized through the intent of the method, that is, via graph matching or graph classification methods.

8.3.1   Graph Matching

Graph matching methodology has developed over many decades and a comprehensive survey of methods and applications for graph matching over a span of three decades from the early 1970s to 2004 was conducted by Conte et al. (2004). In general, graph matching is categorized into two main groups: Exact Matching and Inexact Matching. Exact Matching methods consider edge preserving matches or edge mappings such as graph isomorphisms, graph homomorphisms, or maximum common subgraphs (MCS). Such methods are NP-complete with the exception of graph isomorphism which has not yet been shown to be NP-complete. Currently, MCS methods are only able to handle relatively small graphs due to their computational complexity. Exact Matching methods include other techniques such as Tree Searches which work to iteratively expand the set of paired matched nodes, group theory, subgraph decomposition, or decision trees on a library of known graphs.

Inexact Matching is an approximated matching technique that is more tolerant to slight differences between the graphs, and is achieved by finding a mapping that minimizes a stated matching cost. Here, cost is defined as either deformations or differences between the matched graphs or as a set of graph edit operations. Graph edit operations are operations where a new graph is formed from an original by adding or deleting a node or an edge. This can be performed through one of two ways such as optimal inexact matching, which finds a solution that gives the global minimum matching cost, or approximate matching that only guarantees that the solution is a locally minimized cost.

Optimal matching techniques are often more computationally expensive than Exact Matching itself, but approximate matching techniques are faster than Exact Matching, often exhibiting polynomial matching time. Like Exact Matching, Inexact Matching can also be broken into subcategories; specifically, Tree Searches, Continuous Optimization, Spectral Methods, and other techniques that do not fall into the former groups such as decomposition methods, neural networks, genetic algorithms, bipartite matching, and methods based on local properties. Regardless, the appeal of a faster computation time, which is usually polynomial, makes approximate matching techniques desirable if the application requires a timely analysis.

An updated survey was recently conducted by Livi and Rizzi (2013) with a focus on Inexact Matching methods that are divided into Graph Edit Distance (GED) based,

Graph Kernel-based, and Graph Embedding based techniques. GED based techniques fall in the family of tree searches whereas kernel and embedding based techniques fall in the continuous optimization family where the discrete problem is transformed into the continuous space in order to take advantage of the vast amount of metrics and matching techniques available in that space. Kernel-based techniques transform the graphs onto an induced feature space for evaluation whereas Graph Embedding based techniques transform the graphs to obtain a general vector representation and utilizes the metric space to compute dissimilarities. Livi and Rizzi (2013) suggested that all of the types of methods considered have polynomial complexity, but the computation cost is still dependent on a set of parameters that have to be tuned specifically to the application at hand.

8.3.2 Graph Classification

Graph classification methods classify a graph to the closest matched graph from a library of known graphs. Many such methods exist, and several of the more common methods are described here.

Frequent Subgraph Mining (FSG) is a method that takes a graph and a minimum support threshold $\varepsilon\%$ to generate all connected subgraphs that occur in at least $\varepsilon\%$ of the graph (Inokuchi, Washio & Motoda, 2000). Other methods, such as that proposed by Moonesinghe et al. (2007), uses the subgraphs generated by FSG to construct a binary feature vector which is then used to compute the maximum entropy of the graph in an iterative fashion until convergence occurs. This method was shown to perform comparatively to an AdaBoost and a support vector machine (SVM) classifier.

Ketkar, Holder, and Cook (2009) presented an empirical comparison of the major approaches for graph classification, namely SubdueCL (Gonzalez, Holder & Cook, 2002), FSG with SVM (FSG = flexible segmentation graph), a walk-based (direct product) kernel, FSG with AdaBoost, and DT-CLGBI which is a combination of FSG and decision trees (Nguyen et al., 2006). SubdueCL is the pioneering algorithm for graph classification, and functions by creating a decision list from subgraphs and performing an isomorphism test with a new graph for classification. FSG with SVM works by using FSG to create a feature vector that is then used as input for SVM classification. The walk-based kernel is created by taking the direct product of two graphs as a similarity measure. FSG with AdaBoost works by using FSG to create a list of subgraphs and AdaBoost to create a list of positive and negative examples from the subgraphs that result in the upper bound of the gain that is associated with the supergraph. DT-CLGBI combines aspects of frequent subgraph mining and decision trees. The algorithms were compared using a chemical compound dataset as well as artificial network data generated using an in-house data generating technique (Ketkar, Holder & Cook, 2009). The result showed that walk-based kernel performed poorly when the average degree is high and SubdueCL performed poorly when the graph is disconnected. Other methods performed similarly to one another.

Jin, Young, and Wang (2009) also proposed a graph classification method by deriving classification rules based on pattern co-occurrence from the subgraphs. The method only performs pattern mining once and as such, this method results in faster computation time than other methods that require multiple such iterations. The method can be integrated into any subgraph mining algorithm by organizing patterns into groups of co-occurrence rules to form a rule set. That is, whenever a pattern is generated, the discrimination score of every rule is calculated with the pattern's inclusion and then the pattern is inserted into the rule that yields the greatest increase in discrimination score. The algorithm then finds a co-occurrence rule set that maximizes the number of graphs that can be classified correctly. The authors compared their method against LEAP (Yan et al., 2008) with SVM (LEAP+SVM) and gPLS (partial least squares), showing that their technique performed comparably to the other techniques with faster computation time by magnitudes.

A technique for comparing graphs using subgraphs structures was introduced by Macindoe and Richards (Macindoe & Richards, 2010). This is performed by computing three summarizing features from the subgraphs, and then making a comparison using the earth mover's distance between the distributions of summarizing features to that of subgraphs from other graphs. The summarizing features used by the proposed method are Leadership (also known as Group Closeness) (Freeman, 1978), Bonding (also known as Group Clustering Coefficient) (Wasserman & Faust, 2009), and Diversity (Richards & Wormald, 2009) measures as previously defined. The results of Macindoe and Richards' analysis suggested that graphs can be shown to be similar based on the full graph structure but dissimilar by their local structures.

Employing the statistical knowledge obtained from nodal attributes, Gilbert, Valveny, and Bunke (2012) proposed four fuzzy graph embedding methods that utilize known statistical techniques, namely fuzzy k-means and Gaussian Mixture Models (GMM). These techniques were then applied to SVM for performing graph classification. These methods were compared against the k-Nearest Neighbor classifier as well as another GED-based embedding method on select datasets with the labels removed in order to illustrate the generalizability of their methods on unlabeled graphs. The results from their experiments showed that the proposed methods performed generally no better than the two reference methods; however, they claim that their methods were more computationally efficient.

By using already available network measures, Li et al. (2012) presented a graph classification technique that utilizes a feature vector of twenty graph measures which is then applied to SVM for classification. This approach did not have an overall better accuracy, but was consistently faster in comparison to other kernel-based graph classification techniques such as Random Walk, Shortest Path, Cyclic Pattern, Subtree, and Graphlet and Subgraph kernels. Further, this approach using the full graph measures was better than that using subgraph features and the most important features included: average clustering coefficient, number of nodes, number of eigenvalues, number of edges, energy (which is the squared sum of the eigenvalues of the adjacency matrix), and average degree. The smaller feature set was found to be sufficient to capture most of the important structural properties of the graph.

Ugander, Backstrom, and Kleinberg (2013) proposed a coordinate system based on triadic structure within subgraphs for characterizing possible sub-networks within a social network. This method begins by using a Markov Chain to model the frequency space of triadic evolution for a size $k = 3$, 4 subgraphs within a graph. They demonstrated their method on a Facebook® dataset by performing classification of sub-networks of various sizes into neighborhood, groups, and events through logistic regression. This was compared to the performance of using only global graph features such as size of k largest components, size of k-core, number of components in k-core, number of compositions in k-core, degeneracy, size of k-brace, and number of components in k-brace. The results showed that the proposed method performed much better than that which just uses global measures.

Lagraa et al. (2014) proposed a new distance measure for comparing graphs using modular decomposition for obtaining prime graphs. This is then used to compare with other network's prime graphs using probe distance, which measures the number of edit operations needed to transform one graph into a second graph. Modular decomposition is first used to obtain prime graphs which are graphs that have only trivial modules. The authors then used select datasets to perform comparisons and classifications. The resulting distance and computation time were then compared against those obtained using regular edit distance and star distance, which was proposed by Zeng et al. (2009). The results showed that the prime distance is only comparable to the star distance in terms of runtime and acts as an upper bound for the star distance.

8.3.3   Summary Graph Matching and Graph Classification

Graph matching and classification methods are tools that can be used to determine the similarity between a network of interest and a hypothetical, proposed network. Although these methods have been refined through the introduction of advanced methods as discussed in the previous subsections, these methods still contain difficulties, especially as the size of the network grows. Based on computational approaches, graph matching and classification methods suffer from scalability issues in that computational resources and time requirements grow exponentially as the size of the network grows. Further, these methods may provide estimates as to the ability of matching a particular network of interest, but do not formally determine if the associated graph characteristics follow those that are being modeled. That is, the characteristics may be misspecified even though the overall fit is adequate. This notion is similar to when a straightline regression model is fitted to data that contains some curvature. Even though the overall error of the straight-line model may be within acceptable bounds, the fit of the straightline model may be mis-specified as there is curvature in the data. In short, the fitted model, although containing an acceptable amount of minimum error, lacks overall fit to the underlying data. Similarly, although there may be an acceptable amount of mis-matching, there still may be mis-specification, and therefore, alternate models may be more appropriate. Recall, network models are only useful if such models are able to characterize a network accurately enough for the intended analysis. Therefore, methods verifying that network characteristics and measures are appropriately modeled are paramount to network analysis.

8.4   Statistical Testing Methods

Statistical, inferential methods overcome many of the issues presented by graph matching and classification by formally testing whether or not a feature of a particular network follows the statistical distribution associated with the proposed network model property of interest. For instance, if the real world network is believed to follow that as described by an Erdös–Rényi network model, then the degree distribution should behave statistically as a Binomial distribution. Therefore, the degree distribution for the network of interest could be tested to determine if it is distributed as Binomial with specific parameters.

Recent advances in statistical inference for networks can be found in works such as Blasio, Seierstad, and Aalen (2011) and Mohd-Zaid (2016) both of which examined the linear preferential attachment property. Whereas Blasio, Seierstad, and Aalen considered statistical testing for the linear preferential attachment property within a network in general, Mohd-Zaid directly developed a statistical test of hypothesis based upon the mechanisms governing the Barabási–Albert network model. The approach in this latter work provides a framework for setting up hypothesis testing based on the hypothesized distribution for network properties. This approach is outlined in detail for the degree distribution of a Barabási–Albert network model and may be used to derive statistical distributions for any network model.

Recall that under the property of linear preferential attachment governing the scale-free mechanism of the Barabási–Albert network model, Barabási and Albert developed equations that describe the probability that a new node would attach to an existing node (Barabási & Albert, 1999). Then, by fixing the minimum degree for any node (the minimum connections for any individual in the network), they derived the rate of change for the degree of each node over a fixed number of iterations (instances) at which a single node is added to the network on each iteration. It is from these equations that they determined that the asymptotic degree distribution follows a Pareto distribution with parameters m (the number of edges added on each iteration) and $\beta = 2$. The distribution for a random variable X which is distributed according to the Pareto(m, 2) distribution is given as follows (Casella & Berger, 2002):

$$f(x \mid m, \beta = 2) = \beta m^{\beta} x^{-\beta-1} I_{[m,\infty)}(x) = 2m^2 x^{-3} I_{[m,\infty)}(x), \tag{8.4}$$

where $I_{[m,\infty)}(x)$ is an indicator function taking the value of 1 if x takes the value of m or larger, and 0 otherwise. In this manner, the Pareto distribution is defined for (positive) values of m (assuming the number of edges added on each iteration is at least 1). Knowing that the degree distribution for a Barabási–Albert network model would have the Pareto(m, 2) distribution, Mohd-Zaid developed statistical hypothesis tests based on the Pareto distribution in order to test the fit of the network to the degree structure of a Barabási–Albert network model. Three tests were developed: one for the parameter m, one for the parameter β, and one that simultaneously tests the parameters m and β. The assumption (null hypothesis) is that the parameter of interest is equal to the value assumed under the Barabási–Albert network model. The alternative hypothesis varies depending on the parameter.

Consider the test for m assuming that the value of β is known. To test whether or not m is larger than that hypothesized for the particular Barabási–Albert network of interest (denote this hypothesized value as m_1), we would reject the null hypothesis and conclude that the number of edges being added on each iteration is more than that hypothesized if the observed minimum degree for the network (call this value $x_{(1)}$) is larger than

$$x_{(1)} = m_1 / a^{1/N\beta}, \tag{8.5}$$

where alpha is the Type 1 error rate, N is the total number of nodes, and β is the fixed parameter for the associated Pareto distribution. This is equivalent to testing the following:

$$H_o : m \leq m_1$$

$$H_1 : m > m_1$$

Similarly, we can test the value of β assuming that m is unknown. For this test, we would hypothesize that the value of β equals some value, (β_o), and we would reject this value for the Pareto distribution if there is any evidence that the parameter β does not equal this value. These hypotheses are given by:

$$H_o : \beta = \beta_o$$

$$H_1 : \beta \neq \beta_o$$

Similar to the test for m, when the null hypothesis is rejected in this statistical test, it establishes that the degree distribution is not following the expected hypothesized Pareto distribution, and therefore is not following that of a Barabási–Albert network model. The test statistic to test the value of β is

$$T = \ln\left(\left(\prod_{i=1}^{N} x_i\right) \bigg/ \left(x_{(1)}^{N}\right)\right), \tag{8.6}$$

where each x represents the degree for each of the N nodes and $x_{(1)}$ is the observed minimum degree for the network. This test will reject the null hypothesis, $H_o : \beta = \beta_o$, for

$$T \leq \frac{z_a\sqrt{N-1}+(N-1)}{\beta_o} \quad \text{or} \quad T \geq \frac{z_{1-a}\sqrt{N-1}+(N-1)}{\beta_o} \tag{8.7}$$

where z_a is the appropriate quantile of the standard normal (z) distribution.

If neither m nor β is known, then both parameters may be tested jointly such that if either one of the hypothesized values is rejected, then the conclusion would be that the network of interest does not have a degree distribution that follows the expected Pareto distribution associated with the Barabási–Albert network model and therefore, cannot be modeled as such. Such a conclusion is important in that, although the data may fit well enough (error-wise) to the network model, knowing that a particular feature of that network model is not being modeled correctly suggests that the assumed network model may be inappropriate and other models should be considered. The mathematical formulation of the hypotheses for this test is:

$$H_o : m \leq m_1 \cap \beta = 2$$

$$H_1 : m > m_1 \cup \beta \neq 2$$

To actually implement this joint test of hypothesis onto network data requires additional adjustments due to the non-independent nature of the nodes of the network (Mohd-Zaid, 2016). However, a simple version of the Pareto-based test as described above can be applied to the examples from Figures 8.2 and 8.3 to illustrate its use. First, consider the Erdös–Rényi graph in Figure 8.2. An application of the test on said graph should result in a rejection provided that the test has good power. Suppose the hypothesis that the graph is a Barabási–Albert graph is being tested on the data in Figure 8.2 with

$$H_o : m \leq 2 \cap \beta = 2$$

$$H_1 : m > 2 \cup \beta \neq 2$$

Then the test statistics for the graph based on its degree distribution given in Table 8.2 are $x_{\min} = 1$ and

$$T = \ln\left(\frac{3 \times 3 \times 3 \times 1 \times 4 \times 4}{1^6}\right) = 6.0684.$$

Assuming an acceptable Type-I error of $\alpha = 0.05$, the null hypothesis is rejected if

$$x_{\min} > \frac{2}{0.05^{1/(6 \times 2)}} = 2.5671$$

or

$$T \leq \frac{-1.96\sqrt{6-1} + (6-1)}{2} = 0.3087$$

or

$$T \geq \frac{1.96\sqrt{6-1} + (6-1)}{2} = 4.6913.$$

Here, since $T \geq 4.6913$, the null hypothesis is rejected and it is unlikely that the graph follows a Barabási–Albert network model which is true in this case. Applying the same hypothesis test on the Barabási–Albert graph in Figure 8.3 gives the test statistics $x_{\min} = 2$ and $T = 2.0149$. Since $x_{\min} < 2.5671$ and $0.3087 < T < 4.6913$, there is not enough evidence to reject the null hypothesis. Thus, an assumption that the graph in Figure 8.3 has a degree distribution that follows that of a Barabási–Albert network model, which is the ground truth, cannot be statistically disputed by the empirical evidence. With this knowledge, a Barabási–Albert network model may be used to represent this network, providing the analyst a means to study the characteristics of this network and to reasonably approximate what would happen to the network through the manipulation of the proposed Barabási–Albert network model. For instance, the analyst now can reasonably estimate how the network may expand, and whether or not the network is changing in structure if the network seems to be expanding in a way that is not anticipated.

Statistical hypothesis testing can be used to test the randomness property of the Erdös–Rényi network. Recall, we stated that if the network is believed to follow that described by an Erdös–Rényi network, then the degree distribution should behave statistically as a Binomial distribution. Therefore, the degree distribution for the network could be tested to determine if it is Binomial with specific parameters. Consider the same networks graphed in Figures 8.2 and 8.3. Suppose that we want to test the hypothesis that the networks in Figures 8.2 and 8.3 are random networks where each node connects independently to the other nodes with equal chance, thus $p = 0.5$. The hypothesis to be tested is then: H_o: $p = 0.5$ vs H_1: $p \neq 0.5$. For a network of size $n = 6$ where degree follows a Binomial distribution, the expected average degree is simply np, or in this case, 3. The test statistic (B) for the Binomial test is the number of instances where a node has a degree that is equal to the expected average degree. For the graph in Figure 8.2, three of the six nodes connect as expected with degree = 3 (Table 8.2). Thus, the test statistic for the Binomial test for the network graphed in Figure 8.2 is B = 3. Using a table of Binomial probabilities for $n = 6$ and a Type I error rate of 0.05, the rejection regions for this test are B > 5 or B < 1. Thus, we fail to reject the null hypothesis; there is no evidence to suggest that the graph in Figure 8.2 does not follow this property of the Erdös–Rényi network. That is, the network graphed in Figure 8.2

appears to follow the random property of the Erdös–Rényi network. The associated *p*-value for this test is 1.0. For Figure 8.3, B = 1 (see Table 8.3), and we once again fail to reject the null hypothesis; however, we acknowledge that the test statistic lies on the rejection region boundary. For this small graph and using the discrete Binomial distribution, the *p*-value for this lower bound is 0.1094 (the next possible *p*-value, if B = 0, is 0.0132). Given these values and the small size of these graphs, a Type I error rate of 0.05 may be too conservative for both tests.

8.4.1 Summary of Statistical Testing Methods

The use of statistical approaches to formally test network properties is a novel development in recent literature. Such approaches provide additional benefit to the network analyst in that these methods provide a means for characterization with a level of confidence that can be directly computed when network data is observed incompletely or with error. For these cases in which the error is within bounds, the real world data may still be generated and modeled using the hypothesized network model until such time that additional information is obtained. This provides the analyst with a means to observe network behavior on only a portion of the network. For example, if a portion of the network can be characterized as a Barabási–Albert network model, then the Pareto-based hypothesis test can be applied to this portion of the real world network in order to detect degradations of connections within the network (loss of nodes or edges) with good power (Mohd-Zaid, Schubert Kabban & Deckro, 2017).

8.5 Future Directions

Requirements for network characterization are rapidly evolving as SNA becomes a primary tool to passively observe and monitor the behavior within networks. As interest in network analysis and the size of the networks grows, tools that can rapidly model and generate networks, or portions of networks, are critical to time-sensitive responses. Novel methods such as the use of statistical testing to augment or supplant network classification or matching provide the analyst with confidence that the network model chosen for characterization is reasonably accurate. Further, and more critically to the national security analyst, it provides a sensitive tool to determine if a network has evolved or devolved. If the distribution of a network property has changed, then the network is also changing. Focusing analytical efforts on the regions of the network in which this change was experienced can provide insight into the social dynamics driving these changes. For example, an inquiry of interest may be whether certain agents in a real world network are disappearing or are growing connections to other agents. Such network changes may be indicative of future planned actions of the agents within the network.

Although the field of SNA is rooted in graph theory utilizing such techniques as graph matching and classification, the ability of statistical tests to scale accurately and compute more quickly make tools in this arena a viable option for future analyst needs. However, to use such tools, distributions must be traceable, or nonparametric testing techniques must be leveraged. The test of hypothesis approaches described in this chapter focused on statistical tests to determine if the degree distribution follows that expected for either a Barabási–Albert network model or an Erdös–Rényi network model. Although many real data networks exhibit the scale-free property of a Barabási–Albert network (Barabási & Albert, 1999; Zhao et al., 2013), recent findings suggest that real-data networks do not solely possess this property (Small, Judd & Zhang, 2014; Zhang, Small & Judd, 2015). However, with the onset of methods to describe network data as a mixture of network properties, statistical tests can be developed to test both the appropriate mixing proportion of these properties for a network of interest and the overall fit of the mix to the observed data. For instance, the degree distribution may be a mixture of Barabási–Albert network properties (Pareto distribution for the scale-free property) and Erdös–Rényi network properties (Binomial distribution for the random re-wiring property). Further, statistically based methods extend to multilayer models which may be used to describe higher and lower order functioning in a network of interest (Hamill et al., 2008). In short, these methods for testing may provide efficient solutions to assure adequate characterization of a network through a network model even for complex social networks. Ideally, by capturing all the facets of a social network, via a mixture of attributes or functional layers, a better understanding and representation of the network can be created for analysis and monitoring. The same type of approach can be applied to characterizing infrastructure networks for further analysis.

There are many outstanding questions that are of primary interest to those analyzing networks. The detection of fabricated synthetic networks is of particular importance within the operations research community as it identifies the need to uncover the true relations and intent with the network of interest. Knowing when and how a network may grow, change, or otherwise metamorphosize may facilitate planning and response from parties of interest. The statistical tests discussed previously are an efficient means to help answer these questions by identifying when real-world networks are exhibiting characteristics that are, or are not, expected.

Finally, all methods of network analysis help to shape how networks are visualized. Despite the growing use and dependence on data and analysis tools, human analysts require a means to visualize the behaviors of interest. Characterization of real-world network data to a network model is paramount for this visualization, and statistical tests on network properties help maintain that these properties are appropriately represented within the network model. Understanding the social network, how it is structured, and how it may change to shape the world around us are key tasks of SNA. Therefore, network characterization, monitoring, and the translation to an interpretable visualization are tasks of vital importance to the analyst. The tools discussed in this chapter are a means for the analyst to accomplish these tasks for the security of people, places, and resources vital to our survival.

References

Albert, R., Jeong, H., & Barabási, A. (1999). Internet: Diameter of the world-wide web. *Nature*, 401, 130–131.

Barabási, A., & Albert, R. (1999). Emergence of scaling in random networks. *Science*, 286(5439), 509–512.

Barat, A., & Weigt, M. (2000). On the properties of small-world network models. *European Physical Journal B-Condensed Matter and Complex Systems*, 13(3), 547–560.

Blaha, L.M., Arendt, D.L., & Mohd-Zaid, F. (2014). More bang for your research buck: Toward recommender systems for visual analytics. Proceedings of the Fifth Workshop on Beyond Time and Errors: Novel Evaluation Methods for Visualization, BELIV'14, 126–133.

Blasio, B.F, Seierstad, T.G., & Aalen, O.O. (2011). Frailty effects in networks: Comparison and identification of individual heterogeneity versus preferential attachment in evolving networks. *Journal of the Royal Statistics Society: Series C (Applied Statistics)*, 60(2), 239–259.

Borgatti, S.P., Mehra, A., Brass, D.J., & Labianca, G. (2009). Network analysis in the social sciences. *Science*, 323(5916), 892–895.

Brass, D.J. (1995). A social network perspective on human resources management. *Research in Personnel and Human Resources Management*, 13, 39–79.

Casella, G., & Berger, R.L. (2002). *Statistical Inference*. Duxbury, Thompson Learning.

Conte, D., Foggia, P., Sansone, C., & Vento, M. (2004). Thirty years of graph matching in pattern recognition. *International Journal of Pattern Recognition and Artificial Intelligence*, 18, 265–298.

Daudin, J.J., Picard, F., & Robin, S. (2008). A mixture model for random graphs. *Statistics and Computing*, 18(2), 173–183.

Durante, D., Dunson, D.B., & Vogelstein, J.T. (2017). Nonparametric bayes modeling of populations of networks. *Journal of the American Statistical Association*, 112, 1516–1530.

Erdös, P., & A. Rényi. (1959). On Random Graphs I. *Publicationes Mathematicae Debrecen*, 6, 290–297.

Freeman, L.C. (1978). Centrality in social networks conceptual clarification. *Social Networks*, 1(3), 215–239.

Gilbert, J., Valveny, E., & Bunke, H. (2012). Graph embedding in vector spaces by node attribute statistics. *Pattern Recognition*, 45(9), 3072–3083.

Gonzalez, J.A., Holder, L.B., & Cook, D.J. (2002). Graph-based relational concept learning. Proceedings of the Nineteenth International Conference on Machine Learning (ICML), 219–226. Morgan Kaufmann Publishers Inc., San Francisco, CA, USA.

Guzman, J.D., Deckro, R.F., Robbins, M.J., Morris, J.F., & Ballester, N.A. (2014). An analytical comparison of social network measures. *IEEE Transactions on Computational Social Systems*, 1(1), 35–45.

Hamill, J.T., Deckro, R.F., Chrissis, J.W., & Mills, R. (2008). Layered Social Networks. *IO Sphere*, pp. 27–33.

Havig, P.R., McIntire, J.P., Geiselman, E., & Mohd-Zaid, F. (2012). Why social network analysis is important to Air Force applications. Proceedings of SPIE, volume 8389, 83891E–83891E-9.

Inokuchi, A., Washio, T., & Motoda, H. (2000). An apriori-based algorithm for mining frequent substructures from graph data. Proceedings of the 4th European Conference on Principles of Data Mining and Knowledge Discovery (PKDD), 13–23. Springer-Verlag, London, UK.

Jin N., Young, C., & Wang, W. (2009). Graph classification based on pattern co-occurrence. Proceedings of the 18th ACM Conference on Information and Knowledge Management (CIKM), 573–582. New York, NY, USA.

Ketkar, N.S., Holder, L.B., & Cook, D.J. (2009). Empirical comparison of graph classification algorithms. Proceedings of the 2009 IEEE Symposium on computational Intelligence and Data Mining CIDM, 259–266.

Lagraa, S., Seba, H., Khennoufa, R., MBaya, A., & Kheddouci, H. (2014). A distance measure for large graphs based on prime graphs. *Pattern Recognition*, 47(9), 2993–3005.

Leydesdorff, L. (2007). Betweenness centrality as an indicator of the interdisciplinarity of scientific journals. *Journal of the American Society of Information Science and Technology* 58, 1303–1319.

Li, G., Semerci, M., Yener, B., & Zaki, M.J. (2012). Effective graph classification based on topological and label attributes. *Statistical Analysis and Data Mining*, 5(4), 265–283.

Li, P., Zhang, J., & Small, M. (2011). Emergence of scaling and assortative mixing through altruism. *Physica A: Statistical Mechanics and its Applications*, 390, 2192–2197.

Livi L., & Rizzi, A. (2013). The graph matching problem. *Pattern Analysis and Applications*, 16(3), 253–283.

Macindoe, O., & Richards, W. (2010). Graph comparison using fine structure analysis. SOCIALCOM'10 *Proceedings of the 2010 IEEE Second International Conference on Social Computing*, Minneapolis, MN, 193–200.

Mohd-Zaid, M.F. (2016). A statistical approach to characterize and detect degradation within the Barabási-Albert network, PhD diss., Air Force Institute of Technology (AFIT-ENC-DS-16-S-003).

Mohd-Zaid, F., Schubert Kabban, C.M., & Deckro, R.F. (2017). A test on the L-moments of the degree distribution of a Barabási -Albert network for detecting nodal and edge degradation. *Journal of Complex Networks*, 6(1), 24–53.

Moonesinghe, H.D.K., Valizadegan, H., Fodeh, S., & Tan, P-N. (2007). A probabilistic substructure-based approach for graph classification. 19th IEEE International Conference on Tools with Artificial Intelligence (ICTAI), 1, 346–349.

Moreno, J.L. (1934). *Who Shall Survive? A new Approach to the Problem of Human Interrelations.* Beacon House. ISBN 978-9992695722.

Morris, J.F., & Deckro, R.F. (2013). SNA data difficulties with dark networks. *Behavioral Sciences of Terrorism and Political Aggression*, 5(2), 70–93.

Morris, J.F., ONeal, J.W., & Deckro, R.F. (2014). A random graph generation algorithm for the analysis of social networks. *The Journal of Defense Modeling and Simulation: Applications, Methodology, Technology*, 11(3), 265–276.

Newman, M.E.J. (2010). *Networks: An Introduction.* Oxford University Press, Cambridge.

Nguyen, P.C., Ohara, K., Mogi, A., Motoda, H., & Washio, T. (2006). Constructing decision trees for graph-structured data by chunkingless graph-based induction. *Proceedings of the 10th Pacific-Asia Conference on Advances in Knowledge Discovery and Data Mining* (PAKDD), Springer-Verlag, Berlin, Heidelberg, 390–399.

Office of the Joint Chiefs of Staff. (2016). *Joint Concept for Human Aspects of Military Operations.* Washington, DC: Department of Defense.

Richards, W., & Wormald, N. (2009). Representing small group evolution. *Proceedings of the 2009 International Conference on Computational Science and Engineering, IEEE Computer Society*, Washington, DC, USA, 4, 159–165.

Seshadhri, C., Kolda, T., & Pinar, A. (2012). Community structure and scale-free collections of Erdos-Renyi graphs. *Physical Review E*, 056–109.

Small, M., Judd, K., & Zhang, L. (2014). How is that complex network complex? *IEEE International Symposium on Circuits and Systems (ISCAS)*, 1263–1266.

Smith, R. (2007). *The Utility of Force: The Art of War in the Modern World.* Alfred A. Knopf, New York.

Thomas, C., Stoica, A., & Beuscart, J-S. (2010). Online social network popularity evolution: An additive mixture model. *Proceedings of the 2010 International Conference on Advances in Social Network Analysis and Mining* (ASONAM), 346–350.

Ugander, J., Backstrom, L., & Kleinberg, J. (2013). Subgraph frequencies: Mapping the empirical and extremal geography of large graph collections. *Proceedings of the 22nd International Conference on World Wide Web*, 1307–1318.

Wasserman, S., & Faust, K. (2009). *Simple Distributions. Social Network Analysis.* New York: Cambridge University Press.

Watts, D.J., & Strogatz, S.H. (1998). Collective dynamics of 'small-world' networks. *Nature*, 393(6684), 440–442.

West, D.B. (2001). *Introduction to Graph Theory.* Prentice Hall.

Yan, X., Cheng, H., Han, J., & Yu, P.S. (2008). Mining significant graph patterns by leap search. *Proceedings of the 2008 ACM SIGMOD International Conference on Management of Data* (SIGMOD), 433–444.

Zanghi, H., Ambroise, C., & Miele, V. (2008). Fast online graph clustering via Erdös–Rényi mixture. *Pattern Recognition*, 41(12), 3592–3599.

Zeng, Z., Tung, A.K.H., Wang, J., Feng, J., & Zhou, L. (2009). Comparing Stars: On Approximating Graph Edit Distance. *Proceedings of the VLDB Endowment*, 2(1), 25–36.

Zhang, L., Small, M., & Judd, K. (2015). Exactly scale-free scale-free networks. *Physica A: Statistical Mechanics and its Applications*, 433, 182–197.

Zhao, Z.D., Yang, Z., Zhang, Z., Zhou, T., Huang, Z.G., & Lai, Y.C. (2013). Emergence of scaling in human-interest dynamics. *Scientific Reports*, 3, 3472.

Chapter 9

Process Optimization through Structured Problem Solving

David M. Bernacki, Robert E. Hamm Jr. and Hung-da Wan

9.1   Introduction

In its simplest form a process is nothing more than a series of steps executed to produce goods or services. The assembly of an automobile is executed through a series of interdependent steps. Those interdependent steps are supported by a host of sub-processes, also a series of interdependent steps. Some steps are short in duration, maybe just fractions of a second, while others take hours or even days. Many steps are complex while others are simple. Some steps in the process are performed by a human, others by a robot, and others by a computer executing a program designed by a human. Some steps are executed concurrently while others must take place before another can be executed. And do not forget that the bits of information necessary to trigger and execute a step are all part of the process. All the steps taken to produce a product or provide a service define the process. Sometimes all the steps that make up a process take place in a small room, while other processes include steps executed across thousands of acres of plant or miles of terrain. However, one thing is for certain, no process will last forever.

A process begins to die as soon as it is born due to constant change in organizations, technology, cost, competition, regulations, etc. Over time, unacceptable performance gaps begin to appear and negatively impact the organization's ability to accomplish its mission. Therefore, continuous improvement is essential for all organizations. In 2008, the US Department of Defense (DOD) institutionalized a Continuous Process Improvement

(CPI)/Lean Six Sigma (LSS) Program as a DOD-wide effort (Directive 5010.42, 2008). To reinforce the implementation of a CPI program, the US Air Force issued an Air Force Instruction in 2016 that laid out a detailed CPI plan, including roles, responsibilities, training, and certification (Instruction 38-401, 2016). In the instruction, an 8-step Problem-Solving method, also known as Practical Problem-Solving Method (PPSM), was adopted as a core method for CPI efforts in the US Air Force.

This chapter illustrates the use and impact of the well-structured 8-step Problem-Solving approach to process optimization. This approach begins with a charter, which describes the problem, process to be improved, performance gap, improvement target, and project scope. Following that, the eight steps of the approach are carried out to close the gap by finding and addressing the root causes. The 8-step approach does not require advanced mathematics or statistical analysis techniques, nor does it need complex computer programs. With its simplicity, the approach is deemed more effective than Value Stream Mapping or Rapid Improvement Events while implemented in the field, since it is less resource-intensive, easy to execute, and its results are better aligned with the organization's needs (Todd, 2008). Consequently, it has been used across the US Air Force.

In this chapter, the 8-step methodology is explained in detail via a case study involving reduced energy consumption on aircraft shelter lighting. The detailed procedures to develop a project charter and to carry out the eight steps are illustrated through the case study, along with practical techniques and tools commonly used in each step, as well as lessons learned (Do's and Don'ts) from numerous 8-step project experiences accumulated in the field. The objective is to share our insights and promote the use of this simple and effective method through this chapter, as a reference for practitioners in need.

9.2　The Life Cycle of a Process

Most processes were likely designed and first executed in a much different environment than the one in which they operate today. The organization looked different then; perhaps it was larger and the employees were more experienced. Technology taken for granted today may not have been present when the process was first introduced. Requirements critical to quality may have changed. Perhaps the operating cost of a piece of equipment was acceptable a decade ago but is no longer acceptable. It is not too much of a stretch to see that the environment we operate in today changes faster than in any time in recorded history. It is not reasonable to expect a process designed and introduced five or ten years ago to work today; thus the need for a structured approach to problem-solving to identify and then remove the waste, variation, and constraints in old processes, i.e., a structured approach to problem-solving that will result in an optimized process capable of producing products and services in today's world.

9.3　The Anatomy of a Process

In view of process optimization, processes consist of two kinds of steps: those that add value and those that do not. Those steps in a process that do not add value are considered waste and should be eliminated. It sounds straightforward. Should be simple right? Well, try telling an employee that any steps she accomplished as part of a process for the past two or three years is non-value-added and thereby is considered waste. That employee has grown accustomed to a very specific way of doing things. She is comfortable with the process. As a result of years and years of habit, she executes the process flawlessly. She feels good about her ability to make the process work for her. She does not think about each step; she thinks about the product or the service she provides and how many or how much of each she is assigned to produce; and it will be provided on time. She probably does not think of value, but she knows that every step in the process is essential to the product or service provided.

A value-added step in a process is defined by three characteristics. First, the step must be something that the customer is willing to pay for. Second, the step must directly change the form, fit, or function of something in order to produce a product or service. The final characteristic of a value-added step is that it is so important that it must be done right every time to successfully produce the intended product or service.

Every other step in the process is non-value-added and therefore renders the process less effective over time. Perhaps when first designed and introduced the waste was acceptable but as the environment changes both internally and externally, the waste is no longer acceptable and must be dealt with or the process grows old and dies. A structured approach to problem-solving works to optimize a process through the removal of waste, variation, and constraints (three types of root cause of sub-optimal processes) to the greatest extent possible. Here, variation refers to the deviation of process outcomes away from desired results, and it is often the target of Six Sigma projects (Breyfogle, 2003). As Walter A. Shewhart pointed out, a process may have two types of variations, i.e., chance variation and assignable-cause variation, and removing assignable causes will bring a process into statistical control (Rodriguez, 2010). Constraints, on the other hand, refers to bottlenecks of a series of processes that limits the performance. The Theory of Constraints (TOC) is a methodology to systematically identify and address the limiting factors to improve a system's performance (Naor, Bernardes & Coman, 2012; Sims & Wan, 2017). The concept was first introduced in a popular management-oriented novel, "The Goal" (Goldratt & Cox, 2004).

Waste comes in many forms. Over-production, over-processing, waiting for anything whether material or information, motion, transportation, excess inventory, and injuries are all examples of waste found in almost every process. Defects, whether the result of an employee's mistake or poorly operating machines, are waste. Constraints or bottlenecks in a process generally cause waiting and are thereby considered waste. It is generally considered impossible to remove all the waste, variations, and constraints from a process; there are no perfect processes. However, much of the waste can be removed through an incremental series of "passes" and the result is a redesigned process that will produce value in a new time. A simple acronym for the seven types of waste, anything that adds cost or time without adding value, is TIMWOOD. See Table 9.1 for a summary of the various forms of waste.

Probably the best way to define value is always to keep the organization's mission in sight. If a step in a process (1) isn't something the receiver of the goods and services hold dear, (2) doesn't change the form, fit, or function of something to ensure mission success, or (3) isn't so important to the mission that it has to be done right every time, it probably isn't value-added. However, the world changes continuously, including customer defined value. Continuous improvement assumes that at birth, (1) a process delivered a product or service correctly and economically, (2) the customer was satisfied with the value provided at the time but the definition of value can change, (3) our

TABLE 9.1: Seven Types of Waste

| | | 7 TYPES OF WASTE | | |
| | | TIM WOOD | | |
WASTE	DEFINITION	ISSUES	EXAMPLES	LOOK FOR:
TRANSPORTATION	Movement of people, materials, and information that does not add value	Adds time and takes up space; Increases handling damage; Adds potential for introducing more defects	Routing a hard copy form from one manager to another manager in another building; Moving supplies to warehouse that will later be sent to customer; Personnel having to drive > 200 miles to conduct training; Moving materials from one workstation to another; Retrieving or storing files	How is the information or work that is being transformed being delivered to other processes? Is work being delivered to the right place at the right time? Has work been consolidated where appropriate? How far is material being transported and how long does it take?
INVENTORY	More information, project, material on hand than is needed right now	Ties up money; Uses valuable working space; Risk of obsolescence and damage	Buying five cases of toner cartridges (12 per case and use about 1-month and toner starts drying out about 1-year from manufacturing date); Raw materials; Work in process; Finished goods; Consumable supplies	Are there boxes of material sitting on the floor? Are you using the hall for storage? Are there outdated materials or manuals in the area? Is material moving through the processes in batches or single-piece? Are you overstocking material? How much are you paying for inventory and do you use all of it?

(Continued)

TABLE 9.1 (CONTINUED): Seven Types of Waste

	7 TYPES OF WASTE			
		TIM WOOD		
WASTE	DEFINITION	ISSUES	EXAMPLES	LOOK FOR:
MOTION	Movement of people that does not add value	Takes extra time; Creates sense of frustration; Risk of injury	Reaching over for files in a file cabinet about 3 feet away from desk; Searching for wrench in a work cell (tools are placed haphazardly in area) to tighten bolt that is loose; Shuffling papers around on the desk in order to find a form that must be completed today	Can walking be reduced by repositioning equipment? Is information, tools, or material at point of use? Are there areas that impede the flow of work? How much repetitive human motion is involved in the work and how can such motion be mitigated? How much human reaching, stretching, bending, or kneeling is involved and how can such motion be mitigated?
WAITING	Idle time created when material information, people, or equipment are not ready	Increases lead time; Increases work in process; Slows delivery to your customer	Waiting on batch report to process payroll; Need eight approval signatures before item can be ordered; Waiting on the last of trainees to put on their uniforms before bus can depart	Are there delays in the delivery of work or information? Are there issues with punctuality with internal or external customers? Are there certain times when delays are more prevalent? Can you see a constraint or bottleneck in the process? Have delays been a problem or are they a recent development?

(Continued)

TABLE 9.1 (CONTINUED): Seven Types of Waste

			7 TYPES OF WASTE	
			TIM WOOD	
WASTE	**DEFINITION**	**ISSUES**	**EXAMPLES**	**LOOK FOR:**
OVER-PRODUCTION	Generating more than is needed right now	Ties up working capital; Hides process & quality problems; Creates and perpetuates all other forms of wastes	Built 100 engines (only sell 50 per year); Ran 50 different reports to analyze data (one of those was a summary report that had the information that was needed); avoid most over-production by building items on a build-to-order basis	Is work being performed ahead of schedule? Is this form a duplicate of some other form? Can information on a form be used in other areas? Is someone using all the information that is being provided? How many people approve the form?
OVER-PROCESSING	Efforts that create no value from the customer's viewpoint	Steps that don't add value; Creates through-put delay; Adds potential for introducing more defects	Creating reports; Repeated manual entry of data; Multiple cleaning of parts; Specialized machines; Overly tight tolerances	Is more effort put into the work than is required by internal or external customers? Has this work been done before? Is more information obtained than is required? Are there redundant phone calls and e-mails? Does the step add value?
DEFECTS	Work that contains errors, Rework, mistakes or lacks something necessary	Adds time; Wastes material and labor; Increases cost	Wrong forms used; Repair parts not to specification; Trouble ticket info not filled out correctly/missing info	Is there documented standard work? Does equipment have a maintenance schedule? Are there effective cross-training programs? Do employees have the proper amount of time to do their work?

environment constantly changes over time, and (4) that leaders have created an organizational culture that embraces change when necessary to improve the performance of critical processes (Hamm, Koshin & McSheffrey Gunther, 2017). When a process is first designed and put into place, it is put into place to produce value as defined by the customer. Over time, however, the environment in which the process must perform changes continuously. Changes in strategies, economies, regulations, and technology are just some examples that impact the customer's definition of value over time, and there are many others. This change in the customer's definition of value is called the "value proposition" (Hamm, 2016). It is hard for many to grasp the value proposition but the ability to see waste, variation, and constraints is the key to this structured approach to problem-solving and process optimization.

9.4 A Structured Approach to Problem-Solving

Over the past ten to 15 years, the United States Air Force has found success in optimizing processes with an 8-step structured and repeatable problem-solving model (Figure 9.1) that incorporates the best of the most popular process improvement methodologies in use in industry today, i.e., lean thinking, Six Sigma, business process engineering, and theory of constraints (AFSO21 Playbook, 2008). Teams that utilize this approach to process optimization work to identify the root causes of poorly performing process and then design solutions or countermeasures aimed at removing the waste, variation, or constraints slowing the process down. From AFSO21 to the current CPI and LSS certification program in the Air Force, the 8-step method (renamed to PPSM in 2016) has been the centerpiece of the project conduction and reporting (Herman, 2017). After the Air Force rolled it out in 2008, it has been taught in many professional military education courses as a disciplined approach to root cause analysis with sustained improvement (McAndrews, 2009).

The 8-step method is not the only problem-solving approach available for CPI. For example, Deming's Plan-Do-Check-Act (PDCA) problem-solving cycle and the DMAIC methodology in Six Sigma have been widely used. Mazur, McCreery, and Rothenberg (2012) introduced a lean implementation model involving a single loop and double loop of learning to carry out problem-solving practices that aims at finding root cause

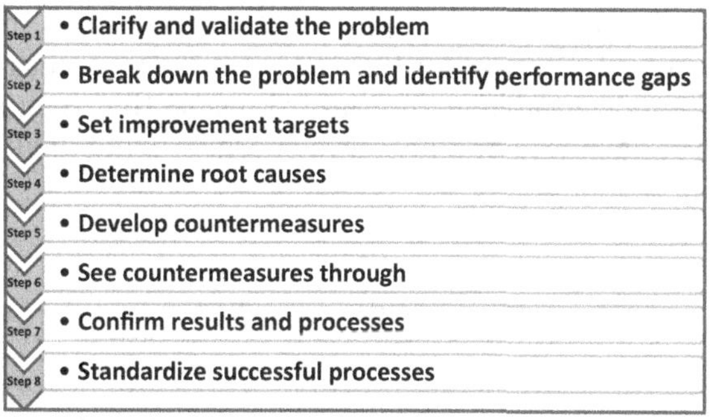

- Step 1 • Clarify and validate the problem
- Step 2 • Break down the problem and identify performance gaps
- Step 3 • Set improvement targets
- Step 4 • Determine root causes
- Step 5 • Develop countermeasures
- Step 6 • See countermeasures through
- Step 7 • Confirm results and processes
- Step 8 • Standardize successful processes

FIGURE 9.1: Eight-step process optimization model.

with sustainable results. The Eight Disciplines team-oriented problem-solving method (TOPS 8-D) is another example which was first developed by Ford Motor Company to address recurring engineering problems (Niggl, 2015). In general, the structured problem-solving techniques infused with CPI principles perform better than conventional methods, because it prevents the users from jumping to conclusion potentially resulting in only short-term results (Miller, 2018). The 8-step method is favored in the Air Force because it is simple and effective and fits well with the OODA Loop (Observe, Orient, Decide, Act) widely used in the Air Force (Todd, 2008).

The 8-step method is not a privilege to the Air Force. As a part of the famous A3 method in the Toyota Production System (TPS), the 8-step was developed by Toyota years ago as a practical problem-solving technique (Holland, 2013). It was evolved from Deming's Plan-Do-Check-Act (PDCA) problem-solving cycle, in which the Plan phase was divided into several steps (Goldsmith, 2014). In recent years, Toyota has all their employees, especially management level personnel, trained to be well-versed in using the 8-step method. Even co-op students and interns at Toyota present their projects in A3 format with 8-step embedded. Hogan, Huang, and Badurdeen (2015) also presented a case study of the 8-step method that optimized a service system at a university as an example in the non-manufacturing sector. In the Air Force, many success stories of process improvements and savings have been reported as a result of the CPI program, 8-step Problem-Solving classes, and the Green Belt program (McCully, 2016).

The key to enhancing an organization's performance is based on a culture of continuous improvement plus the development of the skills that allow all members to identify potential problems and visualize possible solutions (Markovitz, 2011). The ability to apply a structured problem-solving approach allows for the continual optimization of critical process necessary for mission success. Our experience has shown that learning organizations generally do not rely on random moments of inspiration; instead, the learning organization uses a repeatable structured approach to problem-solving to optimize critical processes.

The steps can be accomplished in hours, days, weeks, or months depending on the complexity of the problem and the resources dedicated to the development of solutions. We have found that there is a positive correlation between the effort put forth by the team and the quality of the solution. Problem-solving is hard work and requires teams of employees dedicated to making the process better than it was before. If the organization is not ready for change or does not embrace a culture of innovation/improvement or lacks real leadership commitment, you are wasting your time. In the pages that follow, we describe a structured approach to problem-solving that any organization can use to optimize processes. We demonstrate application of the model through a real-world case study involving an organization working to optimize a process designed to ensure aircraft shelter lighting is available at the right times so that technicians can work on aircraft during hours of darkness and preserve energy during daylight hours.

9.5   It All Starts with a Champion and a Charter to Improve

A common beginning point for most process optimization efforts is the development of a charter. This single one-page document is used to communicate to everyone that the organization's leader is committed to closing a performance gap that is standing in the way of success and he/she is willing to put the organization's energy against the challenges associated with optimizing the process. This is an important step in the problem-solving effort. Few solutions become a reality in organizations unless leadership is truly committed to implementing the solution developed by a problem-solving team.

Numerous studies indicate that solutions to problems are rarely implemented without commitment. For example, Talib, Rahman, and Qureshi (2011) reviewed 11 studies related to implementation of a structured approach to problem-solving, i.e., continuous process improvement. Among them, nine listed leadership commitment as the most critical factor. A survey of 265 plant managers working in a variety of manufacturing companies shows that management commitment is the most critical leadership competency necessary for successful implementation of continuous process improvement (Das, Kumar & Kamar, 2011). Why is this commitment so critical to effective problem-solving? Structured approaches to solving problems require the active and visible participation of committed leaders throughout the organization (Kotter, 1996; Mokhtar & Yusof, 2010; Venkateswarlu & Nilakant, 2005). Organizations must dedicate resources to the problem-solving tasks, monitor the progress of the improvement plan, and break down any barriers to implementation designed to improve process performance. Once the solution is implemented, organizations must verify that the performance gap has indeed been closed, and if not, determine the next step.

The basic elements of a charter include a problem statement, a brief (no more than ten high-level steps) description of the process, the performance gap, an improvement target, a scope that clearly describes what processes or sub-processes are in play, the key stakeholders, a team leader, a facilitator, and the names of employees assigned to the team. In this case study, the charter for a process optimization event aimed at saving the energy and money wasted by the illumination of facility lighting during daylight hours is presented in Figure 9.2.

Process Owner / Champion	Process Owner: Mr. David Long, 12th Flying Training Wing/Maintenance. Champion: Mr. Robert West, 12th Flying Training Wing/Maintenance. Signature: _____.
Problem / Opportunity Statement	The 12th Flying Training Wing aircraft shelter lighting is sometimes illuminated during daylight hours equating to an annual expenditure of $12,747 which is $5,555 above the projected baseline energy costs. During weekdays, 10% of the shelter lights are illuminated during daylight hours and on weekends 20% of the lights remain on most of the day. In addition, the annual bulb/ballast replacement cost is $3,500, 50% greater than required. VoC: Identify efficiencies to reduce aircraft shelter row lighting usage on JBSA Randolph.* *Cost figures are estimates.
Impact Statement	Inefficient energy usage creates a negative impact to the JBSA Randolph energy plan resulting in $5,555 and 129,169 kW hours wasted per year along with $3,500 bulb/ballast replacement costs.
Project Scope	Review the processes of JBSA Randolph aircraft shelter lighting operations.
Brief Description of Current Process	12th Flying Training Wing maintenance personnel turn on aircraft shelter lighting at their own discretion when proceeding to the aircraft rows to perform work.
Current Performance	The aircraft shelter energy costs are $12,747 per year which is $5,555 more than necessary, a 45% increase in energy costs and consumption. The bulb/ballast replacement is ~$3,500 annually or 50% greater than required.
Team Members	Team Lead: Mr. Michael Riddle. Facilitator: Mr. David "Naks" Bernacki (Green Belt candidate). Team members: Mr. Lincoln Sundman, Mr. Jaime Gomez, Ms. Sara Rodriguez, TSgt Arthur Rodriguez Jr., SSgt Larry B. Holms, Capt. Clint Waitcus , Mr. Dan Woolever (Black Belt monitor)
Potentially Affected Users	12th Flying Training Wing, JBSA Randolph.

Impact to Desired Effects	PRODUCTIVITY	ASSET AVAILABILITY	AGILITY	SAFETY & RELIABILITY	ENERGY EFFICIENCY
	Medium	High	High	High	High

FIGURE 9.2: The charter of lighting improvement project.

Charter development began with a conversation between the organization's leadership and a facilitator skilled in the art and science of leading teams tasked with process optimization. The champion must have the authority to implement solutions developed by the team. By developing and then signing the charter, the champion agrees to dedicate time, resources, and people to the problem-solving effort. It is important for the champion to understand that the success of the effort is his/her responsibility and taking the time and effort to clearly describe what the team is required to accomplish will help improve the chances of success. When the champion signs the charter, it becomes a powerful document that directs the team to develop solutions designed to optimize the process. But more importantly, the champion's signature guarantees commitment, follow-through, a willingness to break down barriers, and most importantly, a willingness to trust the team to develop solutions. Development of a charter may seem like a waste of time to some, but as Henry Ford wrote in *My Life and Work*, "Before everything else, getting ready is the secret of success" (Ford & Crowther, 2014). With charter in hand, the first step toward closing any performance gap is to clarify and validate the problem.

9.5.1   Step 1: Clarify and Validate the Problem

Simply put, a good problem statement identifies the, who, what, when, and where of the performance gap. Notice that a good problem statement does not include a "why." The problem statement is also void of potential solutions. Additionally, it is helpful if the problem statement includes "so what." In other words, why is it important that the organization put the time and energy toward solving this problem?

Good problem statements are the product of good data. Many organizations collect performance data. In these cases, the team's work is made easier. But in many cases the data necessary to write good problem statements has not been collected yet. Good data is the basis of any good decision, so if there is limited data, the team has some work to do and a trip to where the process is executed is a solid first step toward process optimization. The team will stand back and observe as the process is carried out. They will take notes, measure, ask questions. Making this trip is time well spent but it is too early to start solving the problem; the team is just gathering data so a workable problem statement can be written. It is okay at this step for members of the team to leave the place where the work is being done with data and ideas, but not solutions. It is not uncommon for members of the team to come away saying, "We need not move to the next step, we can see where the problem is right now." Trying to solve the problem at this point robs the organization of the creative thoughts of each member of the team. There is much more work to do; a seasoned team lead and facilitator will let this structured approach to process optimization work.

A good problem statement is supported by data that is clearly understood by the team. Look at the problem statement in Figure 9.3. It includes all the elements of a good problem statement, who, what, when, where, and so what. No solution is offered. A wise man once exclaimed, "Every time a leader offers a solution to a problem, he/she robs the employee of an opportunity to learn." With the problem statement in hand, the team is ready to learn, and solutions will jump from the knowledge they are about to gain.

In this case study, the team has clearly identified where the problem is occurring (aircraft shelter lighting), who is involved in the process (12th Flying Training Wing), what is happening that requires process improvement (lights are left on during daylight hours), when the problem is happening (weekdays and weekends), and the so what ($5,555 higher annual energy cost for the organization). The problem statement

Step 1 Clarify and Validate the Problem

The 12th FTW aircraft shelter lighting is sometimes illuminated during daylight hours. In 2014, the 12th FTW spent $12,747 for aircraft shelter lighting, $5,555 greater than required (if all lights were off during the day). During weekdays, 10% of the shelter lights are illuminated during daylight hours and on weekends 20% of the lights remain on most of the day. In addition, the annual bulb/ballast replacement cost is $3,500, 50% greater than required. 12 FTW/MX 2014 Strategic Plan objective #2: Ensure an Efficient and Operationally Safe Environment in which to work.

Voice of the Customer (VoC): Identify efficiencies to reduce aircraft shelter row lighting usage on JBSA Randolph.

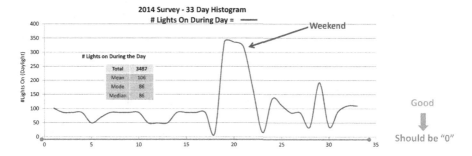

FIGURE 9.3: Step 1: Clarify and validate the problem.

provides a visual depiction of the data to confirm the existence of the problem. The problem statement provides no solutions, countermeasures, or possible root causes. The problem statement is the result of the team's collection of data pulled from existing metrics and observation. Understanding what objective data is needed and what the data, once it has been gathered, means is extremely critical to the root cause analysis and eventual development of solutions in later steps.

In this example, data was gathered by conducting a survey of facility lighting at a specific time for 33 days as shown in Figure 9.3. Through observation, team members were able to determine the number of lights that remained on after official sunrise and before official sunset and then calculated the cost associated with the current process. The graph and cost analysis show the problem clearly, and the performance gaps support the case for process optimization.

9.5.2 Step 2: Break Down the Problem and Identify Performance Gaps .

Now that the team has verified there is a problem, the next step is to break down the problem and present the performance gap as clearly and simply as possible. Visualizing the process by simply drawing the entire process on a "big map" can allow all individuals to identify problems which could include cost overages, man-hour expenditures, rework, and defects in the process. Useful tools for this step and related references for further reading include bottleneck analysis (Goldratt, 1999; Goldratt & Cox, 2004), value-stream mapping (Rother & Shook, 1999; Tapping, Luyster & Shuker, 2002), process map and supply-input-process-output-customer (SIPOC) diagram (Breyfogle, 2003; Pyzdek & Keller, 2018).

In this case study, the goal of Step 2 is to clearly describe current process performance in 2014 and compare it to the desired performance (VoC or voice of the customer)

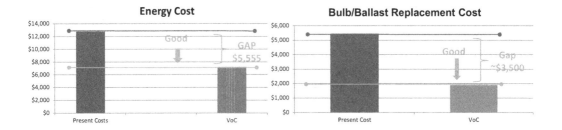

FIGURE 9.4: Step 2: Break down the problem and identify performance gaps.

in 2015, as shown in Figure 9.4. An effective way to break down the problem is to present it visually, in the form of a chart that represents the performance gaps.

Bar charts work well, but pie charts, line charts, histograms, scatter plots, and a variety of other formats also work well. In this case study, a team assembled to reduce the wasteful and unnecessary cost associated with leaving aircraft shelter lighting on during daylight hours has used bar charts to clearly frame and support the problem statement presented in Step 1.

Step 2 provides information that confirms and supports the problem statement in Step 1. Leaving the lights on when not necessary cost the organization approximately $13,000 in electricity and almost $5,500 in unnecessary bulb/ballast replacement in 2014. The charts are simple with green arrows indicating the performance gap to be closed. Performance gaps are generally the product of key performance indicators and associated metrics tied to an organization's strategy, goals, objectives, and key organizational drivers. In this case study, the problem handed to the team is aligned to an organizational goal to reduce energy costs.

9.5.3   Step 3: Set an Improvement Target

Armed with a clearly defined performance gap, setting improvement targets should be a straightforward exercise for the team. Organizations that get the most from their efforts to optimize processes are the ones that solve for waste, variation, and constraints, let the new process stabilize, compare current performance to production targets and if necessary, make another pass, and another and another until the process is optimized to the point that it efficiently and effectively produces value at the lowest possible cost.

In this case study, the organization's director is handing a problem to the team: cut the energy cost by $5,555 and bulb/ballast replacement costs by $3,500 as part of an effort to reduce energy cost throughout the organization. Therefore, in Step 3, the process optimization team set improvement targets designed to close this performance gap. The best way to communicate an improvement target is through a SMART statement (Table 9.2). The acronym SMART stands for specific, measurable, attainable, results-focused, and timely.

The improvement target should be aligned with the Voice of the Customer (VoC), which is the description of value defined by a customer. Determining VoC is a process

TABLE 9.2: SMART Targets

SMART PRINCIPLES	
SPECIFIC	Have desirable outputs that are based on subject matter expert knowledge and experience and are applicable to the process improvement activity
MEASURABLE	Includes time frames and have data that is obtainable from specific sources
ATTAINABLE	Resources are available; may have some risk, but success is possible
RESULTS ORIENTED	Link to the mission, vision, and goals and are meaningful to the user
TIMELY	Provide step-by-step views versus giant leaps and are measurable at interim milestones

to capture customer requirements (internal or external) to provide the best service or product at the lowest cost. There are numerous ways to capture VoC, such as direct discussion or interviews, surveys, focus groups, customer specifications, observation, and complaints (Found & Harrison, 2012).

The VoC in this project is to identify efficiencies to reduce aircraft shelter row lighting usage on Joint Base San Antonio (JBSA) Randolph. Using the SMART criteria, a specific target date is set to clearly address the problem statement described in Step 1. The statement does not introduce additional targets outside of the stated problem. In this case, the team has elected to develop a solution that will reduce electricity costs $463 per month and the cost of replacing ballast by $292 per month. As shown in Figure 9.5, the VoC and CY 2014 cost is depicted along with the CY 2015 target clearly indicated in green. Arrows indicate desired performance. Targets define the performance levels required to make the vision a reality.

Before the team moves to Step 4, it is a good idea to bring the champion back to the team to review the problem statement, performance gap, and improvement target. A quick vector check at this point in the event ensures the champion agrees that the team

Step 3 Set Improvement Targets

By 24 March 2015, reduce JBSA Randolph annual aircraft shelter energy costs to $463 or less per month and bulb/ballast replacement costs to $292 or less per month within 6 months of implementation.

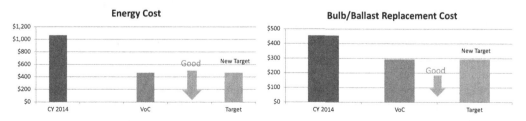

FIGURE 9.5: Step 3: Set improvement targets.

has a solid grasp of the problem to be solved, the gap to be closed, and the improvement target to be met. Getting concurrence from the champion at this point will save a great deal of frustration on the part of the team and the champion. There is nothing worse than expending effort solving the wrong problem or building solutions that the organization is not willing to support. Without the vector check, a team may wander off from the original problem. Once the vector check is complete, it is time for the team to move on to Step 4, determining the root causes for the problem.

9.5.4　Step 4: Determine Root Causes

Now the team is ready to start the work of problem-solving. This is where the tough work begins and where the decision to use a structured problem-solving model to optimize processes starts paying off. Caution: many times, members of the team will bring preconceived root causes to the table and attempt to move straight to developing solutions for these root causes. But this robs the organization of the collective innovative ideas of the entire team and must be avoided. Common root causes such as lack of funding, lack of manpower, and lack of training will undoubtedly surface, and easily become the recommended fixes. A good team leader will recognize what is happening and work to bring the team to instead identify waste, variation, and constraints in the current process – the true root causes of sub-optimized processes.

Teams skilled at problem-solving use cause and effect diagrams, 5 whys, brainstorming, value-stream mapping, or any number of tools to help identify the root causes of performance gaps. The team will break the process down step-by-step to determine if a step is value-added or a candidate for elimination. Further reading about LSS tools for root cause analysis can be pursued in references (Allen, Robinson & Stewart, 2001; Tapping, Luyster & Shuker, 2002; Breyfogle, 2003; AFSO21 Playbook, 2008; Snee, 2017; Pyzdek & Keller, 2018). Regardless of the tools used, it is during this discovery period that terms like "that's the way we've always done it" or "I'm not sure why we do it that way" are often heard. Waste, defects, or bottlenecks in the process (true root causes to the problem) begin to appear to the team. Rather than jump to conclusions or chase hunches and theories, the team will look at the entire process from a value perspective and seek out true root causes so they can develop solutions that will yield lasting effects on the process in Step 5.

Organizations find themselves addressing problems that have supposedly been "solved" many times before because the problem-solving focused on the symptom(s) of a problem and not the root cause of the problem. In this case study, the team worked to identify the most probable root causes of the high cost of shelter lighting through a structured and thoughtful analysis of the value stream. The team is looking for the high cost drivers, i.e., the costliest steps in the aircraft shelter lighting process. It should be noted that there is seldom a single root cause of any performance gap. Teams will most likely identify several root causes and typically attempt to develop countermeasures for each one later in Step 5. The reality is that teams are seldom given the time and resources necessary to solve for every root cause. As a result, the facilitator will lead the team through an exercise to prioritize the root causes. The root causes that the team believes will provide the greatest improvement in the process will be solved first using the Pareto principle or what most of us know as the 80/20 rule. Simply stated the Pareto principle asserts that a team can achieve as much as an 80% improvement in the performance of a process by providing countermeasures for the top 20% of the root causes of poor performance (Breyfogle, 2003). It is only a rule of thumb, but application

Step 4 Determine Root Causes

The team brainstormed and affinitized probable causes. The top three probable causes were chosen by multi-voting for root cause analysis using the "5 Whys" technique, cause and effect relationships.

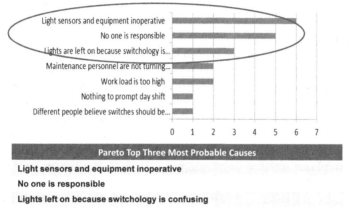

FIGURE 9.6: Step 4: Determine root causes.

of the 80/20 rule can help teams determine which root causes to address with the time and resources available.

In this case study, the team used several tools to conduct root cause analysis, including silent brainstorming (McFarland, 2014), affinity diagrams (Breyfogle, 2003), and multi-voting (BSC Institute, 1996). Cause and effect relationships were identified using a cause and effect (Fishbone) diagram (Breyfogle, 2003) and the 5 whys technique (Allen, Robinson & Stewart, 2001). The top three most probable root causes were chosen using the Pareto principle. The approach is documented, logical, and structured in determining the root causes. As shown in Figure 9.6, the team discovered the three probable root causes, (1) the sensors used to automatically turn the shelter lighting on and off were inoperative most of the time, (2) no-one was taking responsibility for the upkeep of the aircraft lighting system, and (3) no one fully understood the "switchology" necessary to manually activate/de-activate the lights on the shelters. Now it is time for the team to design solutions to address these root causes.

9.5.5   Step 5: Developing Countermeasures

In Step 5, the team develops countermeasures to address the root causes identified in Step 4. Various brainstorming techniques are used to garner innovative solutions from team members that execute the process every day. The best solutions come from the minds of those closest to the work. The trick is getting those great ideas out into the open so that the team can build on those ideas and develop countermeasures to address the root causes of poor process performance. Teams select the most practical and effective solutions using the SMART principles as shown in Table 9.2. All stakeholders should be involved in developing countermeasures to help develop a sense of ownership in the solution.

Typical countermeasures include but are not limited to development and implementation of new standard work, the elimination of non-value-added steps, creating flow by pulling instead of pushing materials through the value stream, error proofing, total productive maintenance, and visual management (see Table 9.3). There are many Lean Six Sigma tools as well as Theory of Constraints techniques and technical deployment methods to consider. To select appropriate countermeasures for better outcome with broader impact, a Lean RACE model is useful (Wan, 2018). As an acronym of four verbs, the RACE model suggests four different directions of possible improvements, including: (1) Reduce non-value-added (NVA) activities to decrease interruptions in flow, (2) Accelerate value-added processes to improve throughput and lead time, (3) Consolidate value-added processes into fewer steps to shorten the value stream, and (4) Enhance Value-Cost ratio of products and processes to improve competitiveness in the marketplace. After countermeasures are identified, a Force Field Analysis can help evaluate tradeoffs among them to prioritize the action items. The analysis helps to gain a comprehensive overview of the different forces acting for and against a potential change, in order to assess their strength. It is widely used to inform decision-making, particularly in planning and implementing change management (Breyfogle, 2003; Ramalingam, 2006).

In this case, to address root cause number one, i.e., the poor condition of the sensors designed to turn the lights on and off at sunset and sunrise, the team offered a solution that suggested the sensors be replaced totally or at least repaired as well as the implementation of a total production maintenance program designed to focus on keeping all equipment in top working condition to avoid breakdowns. Next, to address root cause number two, the fact that no-one was responsible for ensuring that the sensor/lighting system was maintained at peak efficiency, the team offered a solution calling for the appointment of a single point of contact to ensure a maintenance plan was put in place and carried out in accordance with a set schedule, a principle of total productive maintenance. Finally, to address the third root cause of this poorly performing process, confusion on the part of employees with respect to operation of the lights and sensor system, the team developed detailed easy-to-understand instructions or standard work for employees and made the standard work available through visual management.

Once countermeasures/solutions are developed, it is time to bring the champion back for another vector check to review the root causes and countermeasures developed by the team. The team will provide the champion with a list of potential solutions designed to optimize the process. The solutions are prioritized by the anticipated level of effort required to implement the countermeasure and the impact each will have on the performance gap using a PICK (possible, implement, consider, kill) chart as in Figure 9.7. A PICK chart is a visual tool for organizing and prioritize ideas. It is often used after brainstorming sessions to identify which ideas can be implemented easily and have a high payoff (Rouse, 2012; Graban, 2014). The chart is organized into four quadrants with two axes: Difficulty (X) and Payoff (Y). The solutions that provide the greatest return for the least effort are at the top left corner of the PICK chart and are implemented first.

The countermeasures are individually plotted on the PICK chart. The horizontal axis is used to determine the difficulty implementing the countermeasure while the vertical axis depicts the payoff from low to high. Countermeasures in the Implement (Just Do It) area are the easiest to implement with the most payoff; while the items in the Kill area are hard to implement with the least payoff. Most often, countermeasures in the kill zone are not implemented due to difficulty or are set aside for inclusion

TABLE 9.3: Lean Six Sigma Tools

ORGANIZATIONAL PROCESS IMPROVEMENT LEAN TOOLS		
	7 TOOLS	RESULT
6-S	Basic Lean tool that simplifies your work environment and reduces non-value activity while improving quality, efficiency and safety	Stabilizes work area and readies for improvement
Standard Work	The bedrock foundation of Continuous Improvement. Without standard work, it is impossible to tell if improvements are due to chance or due to our efforts. System of developing and documenting the best-known method for performing the work. Developed by the workers themselves.	Increases consistency and safety; Avoid unnecessary motion and wasted effort; Increase quality of output; Prevents equipment damage; Reduces cost; Provides workers a means to define their jobs; Maintains consistency between workers/shifts
Cell Design	Designing how workers are arranged relative to the work and to each other; characteristics (flow, pull, standard work, 6S, and visual management)	Creates flow
Variation reduction	The core of 6 Sigma	Defect reduction
Error Proofing	It is good to do something right the first time. It is better to make it impossible to do wrong; Basic premise is to prevent errors by placing limits on how a process or operation can perform	Reduces defects
Visual Management	Provide immediate, visual information that enables people to make correct decisions and manage their work and activities; Two components (visual display and visual control)	Increases understanding
Takt and Single-Piece Flow	Determines rate of production	Reduces inventory
Quick Changeover	Time spent adjusting the machine is time not spent adding value to the product	Quick changeovers decrease downtime; Well-designed changeovers decrease waste
Total Productive Maintenance (TPM)	In order to get the most productivity out of a machine, down time for maintenance must be planned. A machine run 90% of the time with maintenance during the other 10% will be more productive in the long run than a machine run 100% of the time	Reduces malfunctions

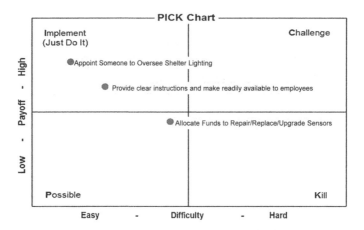

FIGURE 9.7: The PICK chart.

later. The scale payoff and difficulty as well as the breakpoint between the levels are determined by group consensus. In this case, the team offered a solution that called for appointing someone to oversee the shelter lighting/sensor system, followed by development of clear visible operating instructions for employees, while leadership works to find a way to fund the repair/replacement of the shelter lighting/sensor system and presented the solutions to the champion in an easy-to-understand table (Figure 9.8).

The impact of a solution is a combination of the quality of the solution and the acceptance of the solution by the people who must implement it. The relationship is like a mathematical formula: Q x A = I or [Quality of solution] × [Acceptance] = Impact. An excellent solution that receives no support has zero impact. On the other hand, an average solution that receives some support will have some impact. While the team has developed a solution designed to optimize the aircraft shelter lighting process, committed leadership is required to make the solution a reality. It is up to the organization's

Step 5 Develop Countermeasures

Used PICK Chart to prioritize countermeasures.

Priority	Root Cause	Countermeasures	LSS Tools
1	No One is Responsible	Appoint Someone to Oversee Shelter Lighting/Sensors	Total Productive Maintenance (TPM)
2	Confusing Switchology	Provide Clear Instructions	Standard Work/Visual Management
3	Inoperative Sensors	Provide Funding to Repair/Replace Lighting/Sensor System	Total Productive Maintenance (TPM)

FIGURE 9.8: Step 5: Develop countermeasures.

leadership to present the solution in such a way as to gain its acceptance by those who must implement it.

It is the job of the champion to validate that the countermeasures are realistic for the organization. Champions should express their concerns through open ended questions. Truly committed champions listen closely and ask, "What do you need from me; how can I help?" The most successful changes are the ones developed by those closest to the work and supported by the champion when the time comes to see the countermeasures through in Step 6. Committed, supportive leaders make changes stick. Although the team may have come up with countermeasures difficult to attain, the chances of optimizing the process using the solutions developed by the team are good if the champion will support the team's work. If a champion finds that a countermeasure just isn't realistic, now is the time to let everyone know because in Step 6 the team will build an action plan that will implement the countermeasures designed to optimize the process to deliver improved performance.

9.5.6   Step 6: See Countermeasures Through

The next step is to implement the countermeasures or, said another way, see the solutions designed to optimize the process through to completion. The team will build a detailed list of action items required to turn the countermeasures into action items that when fully implemented produce a new, optimized process. Action plans consists of specific actions, estimated completion dates and the names of those responsible for seeing the action items through to completion.

In the most successful efforts, committed champions take an active role by leading periodic action item reviews until all solutions are implemented. These reviews are easily added to regularly scheduled meetings and should move quickly. Active leadership is essential at this point. Failure to close performance gaps following an improvement effort is a direct result of the champion's failure to ensure the solutions developed by the team are carried out.

Failure to implement the ideas created by a team of employees is demoralizing, and employees will lose interest in problem-solving. If this occurs often enough, members of the organization will avoid process optimization events all together and learn to work around poorly performing processes. The result is shortcuts and workarounds that can lead to incidents, accidents, and processes that will continue to deliver poor performance. Remember, action plans mean change and leadership is essential if the changes developed by the improvement team are going to stick.

It is very easy to be excited about an effort designed to optimize a process immediately following a team event. But it takes discipline and committed leadership to keep the organization focused on implementing the solutions; follow-through is extremely important. Follow-through on action items not only ensures the process will perform better than before, but demonstrates leadership commitment, keeps employees interested and excited about process optimization, and more importantly, creates trust between problem-solvers and leaders.

An action plan to address the aircraft shelter lighting process is found at Figure 9.9. These action items should lead to a reduction in the energy and material cost associated with the aircraft shelter lighting process and when implemented, should close the performance gap. The first column lists the priority of the action items from the easiest to implement with the greatest payoff to the hardest to implement with the least amount of payoff. Columns two and three list the solutions and details actions. Column

Step 6 See Countermeasures Through

Priority	Countermeasures	Action Plan	OPR	Start	End	Status
1	Appoint someone to oversee shelter lighting/sensors	Assign via an additional duty	Dr. Hamm	24 Sep 14	24 Nov 14	G ↔
2	Provide Clear Instructions for operation of lights/sensors	Develop standard work and visual management	Mr. Sundman	24 Sep 24	24 Mar 15	G ↔
3	Provide funding to repair/replace lighting/sensor system	Insert funding for a contract to explore "new" technology and fix broken equipment into current budget	Dr. Hamm Mr. Sundman	24 Sep 14	24 Mar 15	G ↔

All action plans completed.

G On Track   Y Issues   R Off Track

FIGURE 9.9: Step 6: See countermeasures through.

four lists the person responsible for carrying out the action item. Columns five and six depict the start and stop dates. The last column provides a visual indicator of the status of the action item. This simple chart can be used to keep the champion informed of the status of the action plan.

Beware! As action items are implemented, there may be a temporary drop in the overall performance of the process. It may take some time for the new process to become a part of how work is performed in the organization. Equipment may need to be moved, employees trained, and leaders will need to break down barriers to change. Eventually, the organization will begin to see a drop in the cost of operating the aircraft shelter lighting system. Don't be surprised if some use the initial drop in performance to make the case that the solutions developed by the team will yield little benefit and the cost associated with operating the system was just the cost of doing business. How do we know the process is performing better now? Read on.

9.5.7   Step 7: Confirm Results and Process

Once the action plan is fully implemented, it is time to start measuring the performance of the new process. Watch the same key performance indicators that led the champion to charter the improvement event in the first place. Using these metrics, simply compare the results of the new process to the improvement target set in Step 3.

Process performance data is captured from the implementation of countermeasures until the target date, giving the new process time to stabilize. Look at Figure 9.10. In this case, a simple chart indicates the solutions developed by the team closed the performance gaps highlighted in Step 2. The energy and material cost associated with the aircraft shelter lighting system was reduced and operating costs now met the targets set in Step 3. If the improvement targets were not achieved, a detailed analysis would be provided and a plan to achieve the target should be presented.

Remember, failure to reach the target does not mean the entire effort was a failure. Be patient and take care to let the new process stabilize before making any changes.

Step 7 Confirm Results and Processes

By 24 March 2015, reduce JBSA Randolph annual aircraft shelter energy costs to $463 or less per month and bulb/ballast replacement costs to $292 or less per month within 6 months of implementation.

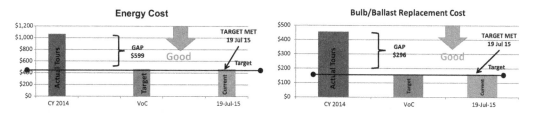

$9K savings a year for the 12 FTW/MX, JBSA Randolph

Target met as of 19 July 2015

FIGURE 9.10: Step 7: Confirm results and processes.

If the countermeasures developed by the team did not close the performance gap after a few months, it is possible they did not capture all of the root causes in Step 4. If this is the case, the champion can bring the original team, or a completely new team, back together to brainstorm for additional root causes and solve for these. Another option is to solve for root causes that may have been discarded on the first pass. A final option is to make another pass with a completely new team. This team should be given a new charter and conduct another improvement event, building on the work of the last team. Repeat as necessary until the performance gap is closed. Once it is determined that the countermeasures will yield the process improvement necessary to close the performance gaps, it is time to move on to the final step: standardize successful processes.

9.5.8 Step 8: Standardize Successful Processes

Once the new optimized process is stable and yielding the necessary performance the last step is to standardize the successful process and make it stick. Teams ask three important questions in this step:

- What's needed to standardize the improvements
- How should improvements and lessons learned be communicated
- Were other opportunities for optimization identified during the problem-solving effort

In this case, new standard work is distributed throughout the workplace as listed in Figure 9.11. First line supervisors ensure their employees are trained and periodic audits are conducted to make sure everyone understands and can execute the new way of accomplishing work. Champions visit the workplace periodically to answer questions, receive feedback from those performing the work, and identify barriers to improved process performance.

Keep in mind that old habits are hard to break; there will be constant pressure in the workplace to return to the old ways of doing business. Don't let it happen! Understand

Prior to Event

Post Event

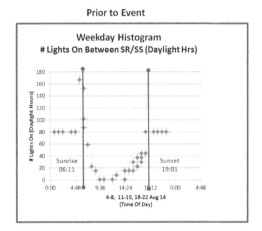

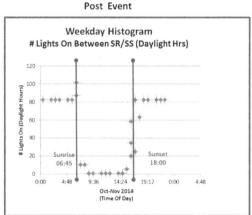

FIGURE 9.11: Step 8: Standardize optimized process.

that most employees were comfortable with the old way of accomplishing work even though the old process was not producing value at the lowest cost. Make it clear that the organization is going to give the new process a chance to stabilize and there will be another opportunity to adjust the process during a future pass. Training is completed and standard work is in place. Finally, the approved process with proven results is put forth for replication and benchmarking through newsletters, publication in journals, posters, and meetings.

9.6 Summary

In this chapter, the use of the well-structured 8-step problem-solving technique is introduced through a real-world case study to ensure aircraft shelter lighting is available at the right times and only when needed. The 8-step method does not require advanced mathematical optimization models. It is meant to be used by process owners from any background. In this case study, the effectiveness of the 8-step method was demonstrated in the data before and after a new standard process was established (Figure 9.12). It is just one of many cases which take place in the Air Force in an effort to achieve continuous improvement.

No process will last forever. Over time, processes designed in another time will not serve the organization well. Waste, variation, and constraints that may have been acceptable when the process was initially implemented just cannot stand up to constant change. Organizations need a structured problem-solving approach to process optimization. This simple 8-step process optimization model provides organizations with an easy to use quick way to optimize key processes through thoughtful analysis of data and presents solutions in an action plan that management teams can use to remove the waste, variation, and constraints that are negatively impacting process performance. This approach is most impactful when the organization is willing to learn, can embrace change, and is led by those truly committed to optimizing processes.

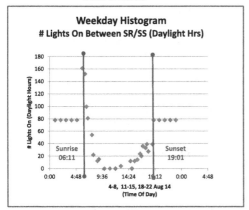
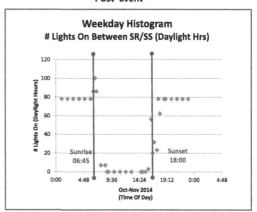

FIGURE 9.12: Before and after the 8-step problem-solving event.

References

AFSO21 Playbook. (2008). *Air Force Smart Operations for the 21st Century (AFSO21) Playbook.* U.S. Air Force. Retrieved from http://www.au.af.mil/au/awc/awcgate/af/afd-090327-040_afso21-playbook.pdf

Allen, J., Robinson, C., & Stewart, D. (2001). *Lean Manufacturing: A Plant Floor Guide.* Dearborn, MI: Society of Manufacturing Engineers.

Breyfogle, F. W. (2003). *Implementing Six Sigma: Smarter Solutions Using Statistical Methods.* Hoboken, NJ: John Wiley & Sons.

BSC Institute. (1996). Module 3: Decision-Making Tools. *Basic Tools for Process Improvement.* Cary, NC: Balanced Scorecard Institute.

Das, A., Kumar, V., & Kamar, U. (2011). The Role of Leadership Competencies for Implementing TQM: An Empirical Study in Thai Manufacturing Industry. *International Journal of Quality and Reliability Management*, 28(2), 195–219.

Directive 5010.42. (2008). *DoD-Wide Continuous Process Improvement (CPI)/Lean Six Sigma (LSS) Program.* Department of Defense Directive. Retrieved from http://www.esd.whs.mil/Portals/54/Documents/DD/issuances/dodd/501042p.pdf

Ford, H., & Crowther, S. (2014). *My Life and Work.* Scotts Valley: CreateSpace.

Found, P., & Harrison, R. (2012). Understanding the Lean Voice of the Customer. *International Journal of Lean Six Sigma*, 3(3), 251–267.

Goldratt, E. M. (1999). *Theory of Constraints.* Great Barrington, MA: North River Press.

Goldratt, E. M., & Cox, J. (2004). *The Goal: A Process of Ongoing Improvement.* Great Barrington, MA: North River Press.

Goldsmith, R. H. (2014). *Toyota's 8-Steps to Problem Solving.* Scotts Valley: CreateSpace.

Graban, M. (2014). Picking on the PICK Chart. *Mark Graban's Lean Blog.* Retrieved from https://www.leanblog.org/2014/07/picking-on-the-pick-chart/.

Hamm, R. (2016). *Continuous Process Improvement in Organizations Large and Small; A Guide for Leaders.* New York, NY: Momentum Press.

Hamm, R., Koshin, B., & McSheffrey Gunther, K. (2017). *Continuous Improvement; Values, Assumptions and Beliefs for Successful Implementation; It's All about the Culture.* New York, NY: Momentum Press.

Herman, R. L. (2017). *Economy of Force: Continuous Process Improvement and the Air Service* (Master's Thesis of the School of Advanced Air and Space Studies). Air University, Montgomery, AL.

Hogan, B., Huang, A., & Badurdeen, F. (2015). Wayfinding in the Graduate School: Structured Problem Solving in a Non-Manufacturing Environment. *Engineering Lean and Six Sigma Conference*. Atlanta, GA.

Holland, K. (2013). Eight Steps to Practical Problem Solving. *Kaizen News*. Retrieved from http://www.kaizen-news.com/eight-steps-practical-problem-solving/

Instruction 38-401. (2016). Continuous Process Improvement (CPI). *U.S. Air Force Instruction*. Retrieved from http://static.e-publishing.af.mil/production/1/saf_mg/publication/afi38-401/afi38-401.pdf

Kotter, J. P. (1996). *Leading Change*. Boston, MA: Harvard Business School Press.

Markovitz, D. (2011). *Factory of One: Applying Lean Principles to Banish Waste and Improve Your Personal Performance*. Portland: CRC Press.

Mazur, L., McCreery, J., & Rothenberg, L. (2012). Facilitating Lean Learning and Behaviors in Hospitals during the Early Stages of Lean Implementation. *Engineering Management Journal*, 24(1), 11–22.

McAndrews, L. (2009). AMC Office Seeks to Solve Problems in 8 Steps. *U.S. Air Force News*. Retrieved from https://www.af.mil/News/Article-Display/Article/119000/amc-office-seeks-to-solve-problems-in-8-steps/

McCully, S. (2016). Airmen Powered by Innovation, Continuous Process Improvement for 2016 and Beyond. *Desert Lightning News*. Retrieved from http://www.aerotechnews.com/nellisafb/2016/02/05/airmen-powered-by-innovation-continuous-process-improvement-for-2016-beyond/

.McFarland, M. (2014, March 26). When Silence Is Golden: The Benefits of Brainstorming without Talking. *The Washington Post*.

Miller, J. (2018). Top 10 Differences between Traditional and CI-Infused Problem-solving. *Gemba Academy*. Retrieved from https://blog.gembaacademy.com/2018/09/10/top-10-differences-between-traditional-and-ci-infused-problem-solving/

Mokhtar, S. S., & Yusof, R. Z. (2010). The Influence of Top Management Commitment, Process Quality Management and Quality Design on New Product Performance: A Case of Malaysian Manufacturers. *Total Quality Management & Business Excellence*, 21(3), 291–300.

Naor, M., Bernardes, E. S., & Coman, A. (2012). Theory of Constraints: Is it a Theory and a Good One? *International Journal of Production Research*, 51(2), 542–554.

Niggl, J. (2015). Problem Solving with the Eight Disciplines. *Global Sources News*. Retrieved from http://www.globalsources.com/NEWS/SIC-problem-solving-with-the-eight-disciplines.HTM

Pyzdek, T., & Keller, P. (2018). *The Six Sigma Handbook* (5th ed.). New York, NY: McGraw-Hill.

Ramalingam, B. (2006). *Tools for Knowledge and Learning*. London: Overseas Development Institute.

Rodriguez, R. N. (2010). It's All about Variation: Improving Your Business Process with Statistical Thinking. *SAS Global Forum 2010*. Seattle, WA.

Rother, M., & Shook, J. (1999). *Learning to See: Value-Stream Mapping to Create Value and Eliminate Muda*. Brookline, MA: Lean Enterprise Institute.

Rouse, M. (2012). PICK Chart (Possible, Implement, Challenge and Kill Chart). *TechTarget Network*. Retrieved from https://searchcio.techtarget.com/definition/PICK-chart-Possible-Implement-Challenge-and-Kill-chart.

Sims, T., & Wan, H. (2017). Constraint Identification Techniques for Lean Manufacturing Systems. *Robotics and Computer-Integrated Manufacturing*, 43, 50–58.

Snee, R. D. (2017). A Systems Approach to Root Cause Analysis and CAPA Investigations. *IVT Laboratory Week*. Philadelphia, PA.

Talib, F., Rahman, Z., & Qureshi, M. N. (2011). A Study of Total Quality Management and Supply Chain Management Practices. *International Journal of Productivity and Performance Management*, 60(3), 268–288.

Tapping, D., Luyster, T., & Shuker, T. (2002). *Value Stream Management*. New York, NY: Productivity Press.

Todd, J. (2008). AFSO21 Adopts 8-step Problem Solving Model. *Air Force Print News Today*. Retrieved from http://www.au.af.mil/au/awc/awcgate/af/afso21-8-step-prob-solv.htm

Venkateswarlu, P., & Nilakant, V. (2005). Adoption and Persistence of TQM Programmes: Case Studies of Five New Zealand Organizations. *Total Quality Management & Business Excellence*, 16(7), 807–825.

Wan, H. (2018). A Lean RACE Model for Different Directions of Continuous Improvement. *Engineering Lean and Six Sigma Conference*. Atlanta, GA.

Chapter 10

Simulation Optimization

Shane N. Hall, Brian M. Wade and Benjamin G. Thengvall

10.1 Introduction

Simulation and optimization are two powerful methodologies that are both widely used across the Department of Defense (DoD) (see Boginski, Pasiliao & Shen, 2015; Dirik, Hall & Moore, 2015; Kannon et al., 2015; and Hill, Miller & McIntyre, 2001 for specific military use cases of simulation and optimization). Simulations are used to understand complex system behavior at multiple levels of fidelity. For example, simulation can be used to perform detailed engineering, design, and testing of individual weapon system components; examine the interaction of components in a single advanced weapon system such as a fighter aircraft or nuclear submarine; analyze a tactical engagement between a few weapons systems such as two American F-22 fighters versus two Russian Su-57 fighters; evaluate and refine air operation plans by simulating multiple fighter engagements over a five-day mission to achieve air superiority; and, finally, estimate military risk for a full land, air, sea, and space theater-level military campaign against an adversary.

The simulation examples just described demonstrate the DoD modeling and simulation (M&S) hierarchy displayed in Figure 10.1. The resolution of the simulation models

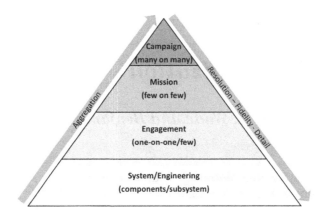

FIGURE 10.1: DoD modeling & simulation hierarchy.

increases as you move down the hierarchy, while the aggregation increases as you move up. For example, system/engineering models for the propulsion system on a submarine are high-resolution, detailed physics models. Results from the submarine's subcomponent engineering models are then aggregated to feed engagement models that evaluate how the submarine performs against an opposing enemy submarine. See Gallagher et al. (2014); Hill, Miller, and McIntyre (2001); and Hill and Tolk (2017) for additional details on the DoD M&S hierarchy.

Optimization provides a powerful way to determine the best option among many, sometimes infinitely many, options. For example, military analysts use optimization to maximize the amount of fuel and munitions delivered to an area of operations using the least number of ships and cargo aircraft; determine the best allocation of dollars to minimize the risk of failure in a future war; assign military personnel to bases in a way that maximizes personal preferences and professional development; and allocate weapons to targets in order to maximize the probability of damage while minimizing collateral damage. When resources are limited, and mission effectiveness is paramount, optimization offers military decision-makers keen insights and enables them to make the best choices.

Over the past two decades, increased computing power coupled with new techniques have contributed significantly to the usability of both simulation and optimization in military settings (see Fu, Glover & April, 2005; Lee et al., 2013; Swisher et al., 2000); Swisher et al., 2004; and Tekin and Sabuncuoglu, 2004). In fact, employing simulation and optimization together provides the military analyst with an extremely powerful methodology known as simulation optimization. The simulations used in DoD simulation optimization problems are typically large, complex models with hundreds or thousands of different inputs that result in multiple outputs of interest when executed. However, generally only a subset of the simulation inputs is varied for a particular simulation optimization study. For example, using a simulation that executes a military campaign, an analyst may vary inputs such as the number of tanks, ships, troops, aircraft, and missiles included and capture outputs such as the number of days required to degrade enemy forces to a desired level and the number of combat losses for tanks, ships, troops, and aircraft. The goal of a simulation optimization study is to answer the military commander's question: what combination of tanks, ships, troops, aircraft, and missiles will minimize the number of days to degrade enemy forces while also minimizing the number of combat losses?

This chapter proceeds as follows: Section 10.2 provides the basic concepts underpinning simulation optimization methodologies and a brief overview of the simulation optimization literature. Section 10.3 describes the relevant, real-world military problem of planning ballistic and cruise missile plans to maximize damage to an airfield. Section 10.4 discusses the analysis and solution of this military problem using a simulation optimization approach. Finally, Section 10.5 summarizes key simulation optimization concepts and analysis results.

10.2 Simulation Optimization Methodology and Literature

This section introduces simulation optimization by first discussing discrete-event simulation, then the general optimization, or math programming, problem, and, finally, their combination into a simulation optimization problem. Additional foundational concepts like black-box optimization and metaheuristics are also presented. Finally, military applications of simulation optimization from the literature are discussed.

To preface the methodology discussion, consider the role of these methodologies in providing analytic insights to an organization. Organizations often use multiple methodologies to perform different types of analyses that include descriptive, diagnostic, predictive, and prescriptive analysis (Ragsdale, 2015). Descriptive analysis seeks to understand what is happening in the system or process (e.g., equipment is not being delivered to a forward operating base). Diagnostic analysis seeks to understand why something occurs within the system or process (e.g., equipment is not being delivered because there is an insufficient number of C-130 cargo aircraft in the area of military operations). Predictive analysis seeks to determine what will happen to the system or process (e.g., future supply will not meet demand for equipment at some forward operating bases). Finally, prescriptive analysis seeks to determine what should be done to make something happen (e.g., in the context of the previous examples, increase planned C-130 aircraft missions to meet the expected demand at all forward operating bases). The ability of an organization to perform these types of analyses depends on the amount of available data, tempo of operations, availability of technical and operational experts, and access to software tools and hardware for computing. Simulation is a powerful method for performing descriptive, diagnostic, and predictive analysis while optimization is generally used to perform prescriptive analysis. Combining both techniques to perform simulation optimization is an advanced methodology that allows organizations to perform the full spectrum of analyses just described.

10.2.1 Simulation

"Simulation is the imitation of the operation of a real-world process or system over time" (Banks, Carson & Nelson, 1999). Think of your favorite fast-food restaurant as a real-world system to simulate. You could build a simple computer model of this restaurant that simulates the arrival of customers, the time it takes to submit food orders, and then the time it takes to process and deliver food orders. Such a simulation would allow the restaurant owner to analyze the best queuing system (many lines vs. a single line) or the number of cashiers needed to ensure a customer's wait time is reasonable. Since real-world military systems are often complex, laced with uncertainty, and exhibit non-linear relationships within the system or with external systems, a simulation is often the best method to ensure the system model sufficiently represents the realworld system. In many military settings, such as evaluating a future military operating concept or examining the effects of a new nuclear warhead, simulation is the only way to analyze the problem since the "real-world system" does not yet exist or cannot be observed.

There are multiple ways to categorize simulations. One way is by how the simulation models the state of the system of interest. Some simulations model the system using continuous functions, which means the simulation changes continuously over time such that there are an infinite number of states and the exact state of the system is represented at any point in time. These types of simulations are called continuous simulations. An example of such a simulation is one that simulates the path or trajectory of a missile over time. In contrast, a discrete-event simulation models the state of the system at discrete points in time resulting from a finite set of events and only represents the system at these enumerable points in time (Banks, Carson & Nelson, 1999). An example of a system that could be modeled as a discrete-event simulation is an air-to-air combat military scenario where the state of the system at any point in time is measured by the number of operational aircraft in the scenario. This number, or system state, only changes when specific events occur, such as when aircraft break, are shot down by the enemy, or deploy to the theater. Furthermore, each of these example events occur at discrete points in time. Thus, a discrete-event simulation steps through a list of finite events and evaluates the state of the system at each event. In military operations research analyses, discrete-event simulation is the most common type of simulation used.

Another way to categorize simulations is by live, virtual, or constructive (Defense Modeling & Simulation Coordination Office, 2014). Live simulations consist of real people operating real systems, such as a pilot flying an actual F-35 aircraft while simulating an attack. Virtual simulations are real people operating simulated systems, such as the same pilot sitting in a virtual reality flight simulator. Finally, constructive simulations are simulated people operating simulated systems. Keeping the same aircraft example, a constructive simulation would be a computer simulation that models the suppression of enemy air defenses using virtual F-35s against a virtual integrated air defense system. In practice, simulation optimization is almost exclusively used with constructive simulations since the required computations do not require a human-in-the-loop and, therefore, can be executed by a computer. The simulations discussed in this chapter are discrete-event constructive simulations.

10.2.2 Optimization

Conceptually, optimization seeks to prescribe decisions that produce the best (i.e., optimal) outcome for the real-world system. Military operations research analysts are usually first introduced to optimization with a course in linear programming (LP). Bazaraa, Jarvis, and Sherali (1977) define the basic LP problem as:

$$\text{Minimize} \quad f(\mathbf{x}) = c_1 x_1 + c_2 x_2 + \cdots + c_n x_n$$

$$\text{Subject to} \quad g_1(\mathbf{x}) = a_{11} x_1 + a_{12} x_2 + \cdots + a_{1n} x_n \geq b_1$$

$$g_2(\mathbf{x}) = a_{21}x_1 + a_{22}x_2 + \cdots + a_{2n}x_n \geq b_2 \qquad \text{(LP)}$$

$$\vdots \qquad \vdots \qquad \vdots \qquad \vdots \qquad \vdots$$

$$g_m(\mathbf{x}) = a_{m1}x_1 + a_{m2}x_2 + \cdots + a_{mn}x_n \geq b_m$$

$$x_1, \qquad x_2, \qquad \cdots, \qquad x_n \geq 0.$$

In LP, $f(\mathbf{x})$ is the objective function, $\mathbf{x} = (x_1, x_2, \ldots, x_n) \in \mathbb{R}^n$ is the real-valued decision vector, and $g_i(\mathbf{x})$, $i = 1, 2, \ldots, m$, are constraints. Additionally, all functions (i.e., $f(\mathbf{x})$ and $g_i(\mathbf{x})$, $i = 1, 2, \ldots, m$) are linear and all parameters (i.e., $\mathbf{c} = (c_1, c_2, \ldots, c_n)$, $\mathbf{b} = (b_1, b_2, \ldots, b_m)$, and $\mathbf{A} = \begin{bmatrix} a_{11} & \cdots & a_{1n} \\ \vdots & \ddots & \vdots \\ a_{m1} & \cdots & a_{mn} \end{bmatrix}$) are known constants. In LP, the mathematical problem is to find a nonnegative vector $\mathbf{x} = (x_1, x_2, \ldots, x_n) \in \mathbb{R}^n$ that minimizes the objective function, $f(\mathbf{x})$, subject to meeting all constraints, $g_i(\mathbf{x})$, $i = 1, 2, \ldots, m$. There are many algorithms to solve LP with the simplex method (Bazaraa, Jarvis & Sherali, 1977) being the most well-known and used algorithm.

LP can be generalized to the following math programming problem (MPP):

$$\text{Minimize} \quad f_1(\mathbf{x}), f_2(\mathbf{x}), \ldots, f_p(\mathbf{x})$$

$$\text{Subject to} \quad g_i(\mathbf{x}) = b_i \quad \text{for} \quad i = 1, 2, \ldots, m \qquad \text{(MPP)}$$

$$\mathbf{x} = (x_1, x_2, \ldots, x_n) \in \mathbb{R}^n.$$

In MPP, all model parameters are still assumed to be known constants and the mathematical problem is to find a vector $\mathbf{x} = (x_1, x_2, \ldots, x_n) \in \mathbb{R}^n$ that minimizes all p objective functions subject to meeting all m constraints; however, some important distinctions are worth noting. First, if $p > 1$, then MPP is a *multi-objective optimization problem* (Ehrgott, 2005). Second, if one or more functional relationships are non-linear, then MPP is a *non-linear optimization problem* (Nocedal and Wright, 1999). Finally, if one or more decision variables is required to be integer, then MPP is a *discrete (or integer) optimization problem* (Nemhauser and Wolsey, 1999).

As with LP, there are many different algorithms to solve MPP. In general, the theoretical and computational complexity of these algorithms increase for the multi-objective, non-linear, and discrete special cases of MPP. For example, integer programming problems can take significantly longer to solve than their corresponding linear programming relaxations (i.e., when the integer decision variable requirement is relaxed and the resulting problem is a linear program). Furthermore, the notions of global versus local optimality are important with nonlinear optimization problems. Most algorithms to solve non-linear optimization problems are gradient-based, which requires the functions in MPP to be well-behaved (i.e., continuous and differentiable) (Nocedal and Wright, 1999). However, even when non-linear functions in MPP are well-behaved, gradient-based algorithms generally converge to local optimal solutions, unless the functions are also convex; then the local optimal solution is also globally optimal.

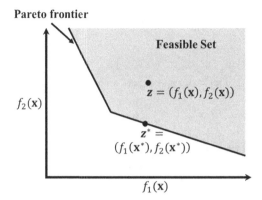

FIGURE 10.2: Objective space for multi-objective minimization problem with two objective functions.

Moreover, the definition of optimality is not as straightforward for a multi-objective optimization problem where two or more objective functions are simultaneously optimized. This situation presents the notion of Pareto optimality. A feasible solution **x*** is Pareto optimal if there does not exist another feasible solution that is better than **x*** for all p objective functions. The Pareto optimal solution **x*** is also called a non-dominated solution since no other feasible solution dominates (or is better than) **x*** for all p objective functions. To visually illustrate this concept, Figure 10.2 displays the objective space for a multi-objective optimization problem with $p=2$, where objective functions $f_1(\mathbf{x})$ and $f_2(\mathbf{x})$ are both being minimized (i.e., smaller values are better). The Pareto frontier indicated in Figure 10.2 is the set of all Pareto optimal solutions. For any point $\mathbf{z}^* = \left(f_1\left(\mathbf{x}^*\right), f_2\left(\mathbf{x}^*\right)\right)$ on this frontier, there is no other point $\mathbf{z} = \left(f_1\left(\mathbf{x}\right), f_2\left(\mathbf{x}\right)\right)$ in the feasible set that is smaller than $\mathbf{z}^*$ for <u>both</u> objective functions. Many algorithms for multiobjective optimization problems use simplifying strategies that sequentially solve single objective optimization problems.

Solving multi-objective, non-linear, discrete optimization problems is very challenging. This can be especially disconcerting when the MPP being solved doesn't adequately represent the realworld problem, which is often the case for complex systems where the limitations of representing the MPP as a closed-form collection of variables and constraints leads to simplifying assumptions. Therefore, not only is the model hard to solve, but its solution has limited applicability to the real-world. Another limitation of MPP is that all model parameters must be specified as known constants. This limits the ability of MPP to adequately model the stochastic nature of real-world problems. A methodology that allows for uncertainty in model parameters and enables high-fidelity representations of complex system behavior while maintaining the prescriptive abilities of traditional optimization approaches is needed. This is what simulation optimization methods provide and what makes them so powerful.

10.2.3   Simulation Optimization

Discrete-event constructive simulations are a flexible methodology to model complex, uncertain, and non-linear real-world military systems. These simulations can be viewed as "black boxes" where a given input vector, **x**, results in a random output vector, Y, as shown in Figure 10.3. The term black box is often used to describe simulations where the user does not have direct access to the model architecture or source code. Therefore, the user can only understand the inner workings of the simulation by observing how outputs change with changing inputs. From the user's perspective, the simulation model takes a set of inputs, performs a complex set of internal deterministic and stochastic calculations and then returns an output or set of outputs. The outputs are random for stochastic simulations because it is possible for the simulation to return many different outputs given the same input.

As an example, consider a discrete-event constructive simulation that models an engagement between two American F-22 fighters on the blue, or friendly, side versus two Russian Su-57 fighters on the red, or adversary, side. Suppose an analyst wants to vary the fighter location and weapon load on each fighter as inputs, simulate the air-to-air engagement, and collect the number of fighters shot down for both blue and red sides as output. This simulation models each fighter's radar sensors using a probability of detection and tracking in addition to each weapon having a probability of kill. Now, suppose the simulation is given an input: the F-22s are located at position P_1 and P_2 with weapons load L_1 and L_2 while the Su-57s are located at positions P_3 and P_4 with weapon loads L_3 and L_4. It is possible that a simulation of this same engagement returns one F-22 shot down and two Su-57s shot down on the first run and then zero F-22s shot down and one Su-57 shot down on the second run, given that each simulation run models the uncertainty in detecting, tracking, and killing an opposing fighter.

Figure 10.4 provides a visual representation of this example.

Returning to the MPP, if the value of at least one of the objective functions is determined by a constructive simulation, then MPP becomes a *simulation optimization problem*. Therefore, simulation optimization provides military analysts with a methodology to inject uncertainty into their optimization models. Reconsider the example in Figure 10.4 to define an MPP with two objective functions: 1) minimize the number of F-22s shot down (i.e., blue losses) and 2) maximize the number of Su-57s shot down (i.e., red losses). Suppose there are constraints on where the fighters can be located, and types of weapon loads they can carry, which are enforced using constraints $g_i(\mathbf{x}) = b_i$ for $i = 1, 2, \ldots, m$ in the MPP. This is a simulation optimization problem where the goal is

FIGURE 10.3: Simulation as a black-box.

FIGURE 10.4: Fighter engagement example.

for the blue side to determine the best F-22 locations and weapons loads that minimize its own losses while maximizing red losses. In this example, the objective functions are not explicit mathematical functions, but are evaluated using the engagement simulation. Therefore, the decision vector **x** is given to the simulation and the output vector Y provides the objective function values of interest.

10.2.4 Solving Simulation Optimization Problems

Recall that solving MPPs can be a mathematical challenge, particularly for multi-objective, nonlinear, and/or discrete models. Unfortunately, simulation optimization typically exacerbates this challenge with its black-box evaluations. Most algorithms used to solve non-linear optimization problems are gradient-based, but this requires functions that are differentiable. Simulations are not differentiable. Even if the simulation outputs were plotted on a graph, it's likely the resulting response surface would be non-linear with several peaks and valleys appearing very erratic and choppy. This is because the uncertainty that is typically modeled in the simulations yields statistical random variables as both intermediate and final outputs. Random intermediate variables can cause a compounding effect in a simulation, leading to apparent erratic behavior. Therefore, pseudo-gradient-based algorithms (using points close to each other to estimate a gradient) are not always a viable option for solving simulation optimization problems. Currently, the general concept of black-box optimization is an active area of research within operations research, computer science, and engineering (Audet and Hare, 2017).

Other approaches to solving simulation optimization problems include trial-and-error, enumeration, and design of experiments. For complex simulations, trial-and-error is akin to buying a lottery ticket and hoping you get lucky—success is highly unlikely and being able to justify or explain the quality of the solution is difficult. Enumeration is the brute-force way to attack an optimization problem with a finite number of input combinations. The amount of computing required to enumerate even small problems can quickly become intractable. To demonstrate, consider a simulation optimization problem with 80 decision variables. For the simple case, when each decision variable has only two possible values (e.g., binary decision variables), it would take a central processing unit (CPU) that can process 1 trillion solutions per second 38 millennia to process all 2^{80} possible solutions. Finally, design of experiments where a selective sample of solutions is evaluated is a reasonable method that many analysts use (Montgomery, 2017); however, no matter how robust the design, many of the best solutions may not be part of the sample.

An effective and efficient class of approaches to simulation optimization problems is metaheuristics (Fu, Glover & April, 2005). Effective in this context means the metaheuristic approach has a high probability of finding an optimal or near-optimal solution, and efficient means that the metaheuristic approach uses methods to explore a limited, but meaningful, portion of the possible set of solutions. Simply speaking, a heuristic is a clever and quick technique that strives to find a good solution without any assurance of feasibility, optimality, or closeness to optimality (Reeves, 1995). A metaheuristic, on the other hand, is a "master strategy that guides and modifies other heuristics" (Glover and Laguna, 1997).

Metaheuristics are designed to avoid getting stuck at local optimal solutions. For example, local (also called greedy) search is a widely used heuristic for many discrete optimization problems (Aarts and Lenstra, 2003). Local search begins with an initial solution, then iteratively searches the solution's neighborhood, moving to better and better solutions until no improving solution can be found. Local search can be improved by including an overarching heuristic that employs local search at many diverse starting solutions to search many neighborhoods in an attempt to find a globally best solution. One simple, yet effective, implementation of this idea is greedy randomized adaptive search procedures, or GRASP (Feo and Resende, 1995). Simulated annealing (Johnson & Jacobson, 2002) and tabu search (Glover & Laguna, 1997) are two well-used metaheuristics used to search for global optimum. While metaheuristics consistently show their effectiveness at finding optimal and near-optimal solutions for a broad class of problems, it is important to note that they are not exact algorithms. Exact algorithms will always find an optimal solution when given a feasible problem instance; however, metaheuristics do not come with this guarantee. The literature does provide some *theoretical* optimality convergence results for some metaheuristics; however, these results, while theoretically important, are not meaningful in a *practical* sense. For example, Johnson and Jacobson (2002) provide one such result for simulated annealing that shows convergence to an optimal solution as the number of local search iterations approaches infinity. While infinite search iterations are impossible, it does provide a guide for practitioners.

To illustrate the key metaheuristic concepts, consider the well-known Traveling Salesperson Problem (TSP) where a traveling salesperson will leave their home city and then visit $n - 1$ additional cities exactly once before returning home. A solution, called a tour, for the TSP can be viewed as an ordered sequence of integers $1, 2, ..., n$ that always begins at 1 (the home city). For example, $(1,3,4,2,5)$ is a tour where the salesperson leaves their home city 1, travels to city 3, then city 4, then city 2, then city 5, and finally back to city 1. The objective of the TSP is to find a tour that minimizes the distance traveled by the traveling salesperson. Realize there are $(n - 1)!$ possible tours to evaluate. To put this in perspective, if the traveling salesperson, say the President of the United States, wishes to leave Washington, DC and visit every state capital once before returning to the White House, then there are $50! = 3.04 \times 10^{64}$ ways to do this (how long would it take a modern CPU to explicitly evaluate all these tours?).

A simple heuristic for the TSP is the nearest neighbor heuristic that constructs a feasible tour by beginning at city 1, then selects the city nearest to city 1, say city i, and then selects the unvisited city nearest to city i, and so forth until all cities are visited. This nearest neighbor heuristic uses a greedy rule (i.e., choose the next city based on the immediate shortest distance) that can quickly construct a feasible tour. While greedy rules are intuitive and seek optimal decisions, they are myopic in nature and lead to sub-optimal solutions as is the case with a nearest neighbor tour, which is often a very poor solution. A simple metaheuristic for the TSP is to initialize the nearest neighbor heuristic with a different beginning city, iterate for all n cities and build n nearest neighbor tours, and then return the best nearest neighbor tour (note any tour can be adjusted to begin at city 1 (e.g., the tour $(4,2,5,1,3)$ is equivalent to the tour $(1,3,4,2,5)$ since the cities are visited in the same order)).

Local search could also be used to improve a feasible TSP tour by searching the neighborhood of the tour based on some defined metric, function, or move operator. Consider a feasible tour $\mathbf{x}'$ and neighborhood $N(\mathbf{x}') = \{\mathbf{x} \in X : \mathbf{x}$ is a 2-swap move away from $\mathbf{x}'\}$, where X is the set of all feasible tours and a two-swap move takes any two cities in $\mathbf{x}'$ and swaps their location to yield a neighboring tour. For example, if $\mathbf{x}' = (1,3,4,2,5)$ then $\mathbf{x} = (1,2,4,3,5) \in N(\mathbf{x}')$ since $\mathbf{x}$ is within one two-swap move of $\mathbf{x}'$ (i.e., swapping 2 and 3 in $\mathbf{x}'$ yields $\mathbf{x}$). Local search could examine the entire two-swap neighborhood of a given feasible tour (e.g., a nearest neighbor tour) and return the best tour found within the neighborhood (i.e., the local minimum for the defined move neighborhood).

Metaheuristics such as tabu search could then be applied to guide local search to explore many different neighborhoods. Tabu search uses memory throughout the search process to guide local search through a series of intensification and diversification steps (Glover and Laguna, 1997). *Intensification* and *diversification* are important general concepts for many metaheuristics that contribute significantly to their effectiveness. Intuitively, intensification involves strategies that aggressively search good areas of the solution space, whereas diversification involves strategies that encourage search in areas of the solution space yet to be explored. While intensification and diversification enhance the effectiveness of metaheuristics, they can also hinder the computational efficiency of metaheuristics because they require more elaborate strategies and explicit memory tracking throughout the search process. Furthermore, finding a balance between intensification and diversification within the metaheuristic is non-trivial, as too much intensification can lead to getting stuck in local optima while too much diversification can lead to near-random search. The design and analysis of metaheuristics for a broad set of problems is an active area of research. For example, McNabb et al. (2015) examine multiple neighborhood move operators used for intensification and diversification in many metaheuristics for a variant of the vehicle routing problem, which is an extension of the TSP.

Consider the concepts of intensification and diversification for a TSP metaheuristic. Suppose this TSP metaheuristic guides local search. While the search proceeds, the metaheuristic remembers two important items: first, it tracks an *elite list*, which consists of the ten best solutions found over the course of the search, and second, it tracks a *move frequency*, which records the number of times each city is moved (or swapped) over the course of the search. During intensification, the metaheuristic examines the neighborhood of each solution on the elite list using an aggressive move operator in hopes of further improving each elite solution. During diversification, the metaheuristic takes the current solution and then randomly swaps the ten cities with the smallest move frequencies to generate many new tours that are hopefully very different from the tours already evaluated (since the cities being swapped are the most infrequently moved). These new tours are then improved with local search.

While the TSP is not a simulation optimization problem, the metaheuristic concepts presented are directly applicable to solving simulation optimization problems – metaheuristic methods quickly and intelligently explore combinations of simulation inputs and return the set of simulation inputs that results in the best simulation outputs. The optimization tool discussed in Section 10.4 employs metaheuristics to solve the ballistic and cruise missile problem described in Section 10.3 via simulation optimization. This section now concludes by describing three specific military applications of simulation optimization from the literature.

10.2.5   Applications in the Literature

The first military application of simulation optimization comes from Melouk et al. (2014) who propose a simulation optimization approach to design sensor configurations for Air Force unmanned aerial systems (UAS). Sensors are the subsystems that detect and track environmental changes allowing the UAS to maintain situational awareness, gather intelligence information, and if armed, attack targets. In this problem, feasible UAS sensor configurations are passed to a simulation that determines UAS mission effectiveness (e.g., targets acquired, tracked, and engaged) for the given sensor configuration. This methodology allows military systems engineers to determine the optimal sensor configuration for future UAS. The authors compare their metaheuristic optimization approach to two other metaheuristics and a greedy heuristic.

Lobo et al. (2018) address the Army fuel planner problem, where daily load plans must be determined for a fleet of tankers (i.e., fuel tank trucks). Specifically, the problem is to identify which tankers to use, the type of fuel to load in each tanker, and then where to send each tanker. In this study, the authors use simulation optimization to determine optimal fuel distribution policies that provide a rule set for creating daily load plans. The simulation model determines the performance of an in-theater fuel supply-chain with uncertain supply and demand, given a specific fuel distribution policy. The optimization portion then determines the next policy to simulate and determine if the policy results in the performance required by the Army. Experimental results show the methodology effectively identifies a set of policies that meet the objectives of Army fuel planners and efficiently solves real-world scenarios.

Onal et al. (2016) use simulation optimization for scheduling the land restoration at degraded military training sites. The authors use simulation to determine the effects of training on military land for a given training schedule and set of financial resources. Their model considers weather forecasts, scheduled maneuver exercises, and the quality and importance of the maneuver areas. They apply their methodology to Fort Riley, Kansas and show: it is better to repair damaged land than to fix gullies; targeting a uniform distribution of maneuver damage increases total damage and adversely affects the overall landscape quality, implying that selective restoration is preferred; and, finally, sustaining the status quo requires a nearly $1 million annual restoration budget.

With a simulation optimization foundation established, the second part of the chapter presents a relevant and challenging military problem of determining a ballistic and cruise missile attack plan that maximizes damage to an airfield defended by an air defense artillery network using simulation optimization.

10.3 Military Problem – Missile Plans That Maximize Damage to an Airfield

This section defines a military problem that simulation optimization can be applied to as an example to reinforce topics described in the previous section. The problem involves a multiobjective optimization of the missile attack plan parameters to damage various parts of an airfield. Using the simulation hierarchy described in Section 10.1 and visualized in Figure 10.1, the overall architecture uses a mission-level simulation combined with two system-level simulations. The section begins with an explanation of the problem and scenario. It then outlines the simulation models used to adjudicate the engagements of the missile defense systems against the incoming missiles, and the simulation models used to describe the damage to the airfield caused by missiles that are not destroyed and impact the airfield. Finally, these simulation models are used with an optimization program that uses a metaheuristic approach similar to those described in Section 10.2 to conduct a multi-objective optimization to determine the missile fire plan parameters that cause the most, or maximum, damage to various parts of the airfield.

Most modern military forces maintain an arsenal of theater ballistic missiles (TBMs) and cruise missiles (CMs). These missiles provide a long-range strike capability that shapes the battlefield for future engagements. A TBM is a missile that follows a ballistic flight path (i.e., a parabola). This means that it spends a significant portion of its flight subject to only the forces of drag and weight. A CM is a missile that is self-propelled, guided, and sustains flight through aerodynamic forces for all or most of its flight, and whose mission is to deliver ordnance onto a target. A fire plan specifies the number, type, and attack parameters of each TBM or CM to shoot in a single raid against a specific target.

The defensive systems against TBMs and CMs are known as Air and Missile Defense (AMD) systems. These systems detect incoming threats with radar and/or electro-optical sensor systems and engage the threats with interceptor missiles, guns, or lasers. Incoming TBM and CM threats that are not shot down by the AMD platforms are called leakers. Leakers occur because either the detection systems failed to detect the threat, the AMD systems were unable to engage the threat, or because the interceptors failed to destroy the threat.

Historically, due to their high expense and complex nature, TBMs and CMs have been used in small numbers in uniform raids made up of only a few TBMs or CMs of the same type (Gormley, 2008). However, this trend will not continue. The proliferation of TBM and CM technology is accelerating, leading to growing stockpiles of these missiles (Feickert, 2005; Pollack, 2011; United States Air Force, 2013). Soon, many countries will be able to attack with large raids made up of many different types of missiles landing simultaneously onto a target area. The differences in accuracy, lethality, and vulnerability between TBMs and CMs of different types means that the capability of the AMD systems to defend against each missile type is different, and the effects of leakers from each missile type are also different (Gormley, 2002; Wade and Chang, 2015). These large and mixed raids open the opportunity for the attacking missile force to optimize missile raids since they can tailor the types, quantities, and attack parameters of the missiles in the raid to achieve certain effects. This optimization can be accomplished using the simulation optimization methods discussed previously.

10.3.1   Example Scenario

One of the primary targets for TBMs and CMs are airfields. The airfield and many of the targets on the airfield do not move making them more easily targeted by long-range missiles. The targets on an airfield can be grouped into three different categories: runway, fuel, and aircraft targets, as shown in Figure 10.5. For an attack on the runway, the missile force would like to damage the runway to such an extent that the number of takeoffs and landings by aircraft is reduced or eliminated for a period of time. For an attack on the fuel sites, commonly called Petroleum, Oil, and Lubrication (POL) sites, the attacking missile force wants to also reduce or eliminate the number of takeoffs by interrupting aircraft operations due to a lack of fuel or a bottleneck of aircraft trying to use the remaining undamaged POL sites. Finally, for an attack on the aircraft, the attacking missile force would like to destroy the parked aircraft to prevent their use in combat. While this is more advantageous than temporarily slowing or halting operations, it is more difficult because the aircraft are continually repositioning on the airfield due to missions, protection, or maintenance. Therefore, this simulation optimization problem can be stated as: what combination of missiles and their attack parameters will maximize the damage to the airfield's runway, fuel, and aircraft?

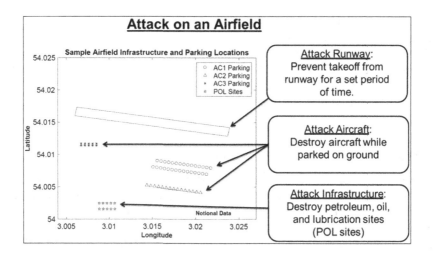

FIGURE 10.5: Notional airbase target architecture.

To show how this problem could be solved using simulation optimization, a hypothetical airfield was created with ten POL sites in two east–west rows of five sites each. The airfield also supports three different types of aircraft: AC1, AC2, and AC3. Aircraft type 1 (AC1) is arranged in two rows of 15 parking locations angled parallel to the runway. Parking for aircraft type 2 (AC2) is further south of the parking for AC1 in a single row of 15 locations, and parking for aircraft type 3 (AC3) is just north-west of the other parking locations in two east–west rows of five parking locations each.

A series of simulations are then required to adjudicate the results of the AMD defenses around the airfield as well as the damage to the airfield due to TBM and CM impacts. The sequence of these simulations is shown in Figure 10.6. Missiles specified by the enemy fire plan are first tested against defensive systems using an AMD simulation. Those that are not shot down, the leakers, are passed to one of the damage effects programs depending on their intended target. Leakers from the AMD simulation that were aimed at the runway are passed to the runway damage simulation model.

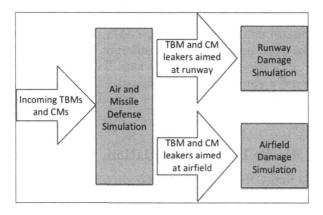

FIGURE 10.6: Airbase damage simulation models.

This model calculates the impact location and crater size for each munition, and then determines the amount of runway that is still useable after the attack. Leakers from the AMD simulation that were aimed at the fuel or aircraft are passed to the airfield damage simulation model. This model determines the location of all assets on the airfield and the impact location of all missiles. It then calculates which airfield assets were within the lethal radius of each impact and counts the damaged or destroyed aircraft and POL sites.

10.3.2   Warhead Types and Accuracy

TBMs and CMs typically attack a target with two different types of payloads: unitary warheads or submunitions. A unitary warhead has a single explosive payload. These warheads deliver all the explosive power to a single location on the ground. A submunition warhead dispenses many smaller "bomblets" that disperse as they fall to the ground. This type of warhead delivers the explosive power dispersed over a larger area.

In an attack on a runway, a specific type of submunition is often used called a runway penetrating munition. To create the largest possible crater, the detonation of the runway penetrating munitions is subsurface (Chabai, 1965; Kiger and Balsara, 1983). Therefore, these munitions are designed to maintain a higher velocity as they fall and are shaped to burrow into the concrete on impact. A delay-fuse is used to detonate the explosive once the penetrator has burrowed to a certain depth, defined by a time after impact with the surface.

The release altitude of the submunitions affects the spread of the submunitions around the mean impact point. The release altitude also serves as a defense measure against the AMD systems. Generally, the submunitions are too small and numerous to be engaged by the AMD systems. If they are released before the AMD system can engage the missile carrier, then the submunitions propagate to the ground unimpeded. This must be balanced with the desired concentration of the submunitions. If they are released too high above the target, the resulting spread of the munitions can dilute their effectiveness.

The accuracy of a ballistic or cruise missile is quantified by its circular error probable (CEP). The CEP is the radius of a circle, centered at the aim point, in which 50% of the weapons are expected to land (Driels, 2004). For both guided and unguided missiles, the CEP along with the probability of different terminal guidance states can be used to calculate the impact location as explained in Driels (2004).

The example scenario used three notional TBM and two notional CM types. TBM types 1 and 2 are the same missile with two different warheads: TBM1 is unitary and TBM2 contains 70 runway-penetrating munitions. TBM3 is a more accurate unitary missile. The CMs used a similar missile body with different warheads: CM1 carried 40 submunitions and CM2 carried a unitary warhead. The CEP (in meters) of all five missiles is shown in Table 10.1.

10.3.3   Air and Missile Defense Simulation

Before the TBMs and CMs can impact their targets, they must first pass through the AMD defenses that can shoot down some or all of the incoming missiles. The AMD simulation used in this work is a complex and widely used DoD simulation called the Extended Air Defense Simulation (EADSIM). EADSIM is a stochastic computer simulation of air and missile defense warfare that can model scenarios ranging from

TABLE 10.1: Notional Missile Guidance Parameters

Missile	Precision / Non-precision	CEP (m)	Unitary / Submunitions	Number of Submunitions
TBM1	Non-precision	200	Unitary	Not Applicable (NA)
TBM2	Non-precision	200	Submunitions	70
TBM3	Non-precision	50	Unitary	NA
CM1	Precision	20	Submunitions	40
CM2	Precision	20	Unitary	NA

system-on-system engagements to many-on-many theater-level engagements (Teledyne Brown Engineering, 2015). It includes robust flight, detection, command and control, and engagement models. The defending AMD forces are composed of two different notional systems: an Upper Tier (UT) system and a Lower Tier (LT) system. The UT system has a single interceptor type that can engage TBMs only. The LT system operates at altitudes below the UT system and has three different interceptor types: Interceptor A, B, and C that can engage both TBMs and CMs.

The AMD simulation uses a probabilistic radar model to determine which missiles are detected and when they are detected. For those missiles that are detected, the simulation uses a command and control model to determine which AMD systems to engage which threat and when those engagements will occur. Finally, the simulation uses an engagement model to probabilistically determine the success or failure of each intercept attempt. If the AMD system fails to destroy an incoming missile, it may attempt to re-engage that missile if there are time and resources available.

The number and type of AMD systems and their assigned tactics must be set before optimizing the red fire plan. These settings should be based on the expectations of the attacking missile forces. For this work, the notional simulation setup for the AMD systems and tactics are shown in Table 10.2. An enemy missile force would base these on

TABLE 10.2: Notional Base Case AMD System and Tactics Values

System	Variable	Set Value
Upper Tier	Number of Launchers	6
	Maximum Engagement Altitude	60,000 meters
	Minimum Engagement Altitude	25,000 meters
	Shot Doctrine	1
Lower Tier	Number of Interceptor A Launchers	2
	Number of Interceptor B Launchers	2
	Number of Interceptor C Launchers	4
	Maximum Engaging Altitude – TBMs	20,000 meters
	Minimum Engagement Altitude – TBMs	10,000 meters
	Maximum Engagement Range – CMs	5,000 meters
	Minimum Engagement Range – CMs	2,000 meters
	Shot Doctrine – TBMs	2
	Shot Doctrine – CMs	1
	Priority Interceptor – TBMs	A
	Priority Interceptor – CMs	C

known doctrine. The shot doctrine is the number of interceptors that are shot in rapid succession at a single inbound missile threat. The priority interceptor is the interceptor that the LT system will use against the given missile threat until exhausted. It will then use the interceptor with the shortest time to intercept. Note that there is a gap in engagement space between the upper and lower tier systems. This is common to account for time to conduct an assessment of the effects of any upper tier engagements that occurred before the lower tier system begins to engage those threats.

10.3.4 Runway Damage Simulation

TBM and CM leakers from the AMD simulation that were originally aimed at the runway are passed to the runway damage simulation. This is a stochastic simulation that calculates the amount of runway that is damaged from a series of TBM and CM impacts with different types of warheads. The runway damage simulation first stochastically determines the impact point relative to the aim point of the missile based on its CEP and guidance parameters. It then finds the speed and orientation of the TBM or CM at the impact point for unitary munitions, or at their release point for missiles with runway-penetrating munitions. For missiles with runwaypenetrating munitions, it then propagates the submunitions to the ground, and determines their impact location relative to the missile carrier impact point and their depth of burial. Using this depth of burial for runway-penetrating munitions or a depth of zero for unitary munitions, the simulation determines the crater size for each unitary or runway-penetrating munition.

Once the crater size and location are specified for each munition, the runway damage simulation determines the amount of usable runway that is left undamaged after the attack. Aircraft typically do not need the entire length and width of a runway to take off and land. The smallest strip of runway that an aircraft needs to either takeoff or land is called a minimum operating strip (MOS). A MOS is rectangular in shape and its dimensions are different for each aircraft type, operating conditions, and loading (AFPAM 10-219, Volume 4, 2008). The MOS must be completely clear of craters for its full length and width and must completely lie within the bounds of the runway. However, the MOS does not have to start or end at the beginning or ending of the runway. The simulation places as many MOSs of a given size on the undamaged portion of the runway as shown in Figure 10.7. This figure shows MOSs for nimble fighter aircraft, but the size of the MOS would likely increase for larger and heavier aircraft, wet or icy conditions, or changes in weather that reduce the air density (lower pressure and/or higher temperature). Once the MOSs are placed, the simulation then calculates the fraction of runway area that is still usable by dividing the sum of the areas from all available MOSs following an attack by the total area of the runway. Reducing this

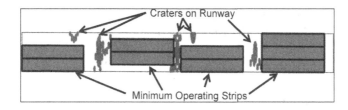

FIGURE 10.7: Sample results from the runway damage simulation.

fraction of runway that is usable is the overall objective of the attacking missile force. See Wade (2019) for more information on the models and processes of this runway damage simulation.

10.3.5 Airfield Damage Simulation

Referring to Figure 10.5, the second set of targets on an airbase lie off the runway and inside the rest of the airfield. Referring to Figure 10.6, leakers from the AMD simulation that were aimed at the POL sites or aircraft are passed to the airfield damage simulation. This stochastic simulation determines the impact location of all missiles and the damage from those impacts.

The POL sites are fixed locations that are known at the start of the attack. The aircraft, however, can reposition on the airfield to new parking locations. Typically, the aircraft are parked such that like aircraft are grouped to better facilitate maintenance and inspections; however, spreading the aircraft apart as far as possible helps to minimize the damage from missile impacts. This simulation assumes the forces operating on the airfield will try to balance these competing demands by spreading the aircraft as far as possible given the constraint that aircraft remain in each aircraft's designated parking area. The simulation does this by randomly selecting an initial aircraft to place on the airfield from the total number of aircraft. It then places that aircraft in a randomly selected parking spot in its assigned location. For the parking locations of all subsequent aircraft, an aircraft is randomly drawn (without replacement) from the remaining unplaced aircraft in the simulation. That aircraft is placed onto the airfield by examining the available parking locations for that aircraft type and selecting the location that maximizes the minimum distance between all aircraft already on the airfield.

For the attacking force, the selection of aim points follows a different process. There is typically a time delay from when a decision is made to attack an airfield, the missiles are loaded and transported to their launch location, and the missiles travel to their target. During this time, the airfield is continually operating, and aircraft are repositioning on the airfield. TBM launch crews will often operate without radio contact to conceal their location and many cruise missile types cannot be dynamically re-tasked once launched, therefore the simulation chooses the aim points based on all parking locations as opposed to occupied parking locations. To account for this dynamic, the simulation selects the aim points for a given attack such that the expected damage is maximized given the missiles in the raid, the number of each aircraft type, and the parking locations for each aircraft type.

When calculating the actual damage, however, the true leakers of the raid will only impact a proportional, but random, subset of the original optimized set of aim points. This is because the simulation generates a set of N aim points that would maximize damage given the N missiles. However, only M leakers ($M \leq N$) get through the AMD defenses (the other $N - M$ missiles were destroyed by the AMD defenses). The simulation then randomly selects M of the N aim points for impact. This is based on a random draw from a uniform (0,1) distribution where the likelihood of selecting an aim point is proportional to the total number of missiles targeting that aim point (for a given missile type and the number of that missile types that leaked through the defenses). For example, assume an attack includes five missiles of a given type aimed at two aim points, three missiles to the first aim point and two missiles to the second, and of the five missiles, two leak through the AMD defenses. For the first leaker, there is a 0.6 (3/5) chance the missile was aimed at the first aim point and 0.4 (2/5) that it was aimed at the second aim point. So, if the random draw from the uniform (0,1) is between 0 and 0.6, then the missile impacts the first aim point, otherwise it will impact the second. For the second leaker, the process is repeated, but the probabilities are adjusted based on which aim point the first missile impacted. For example, if the first random draw determined that the first leaker impacted the second aim point, then there is a 0.75 (3/4) chance that the second missile is aimed at the first aim point, and 0.25 (1/4) chance that it was aimed at the second aim point. If the missiles had the ability to adjust their trajectory based on knowledge of other missile's fate, that capability could be reflected by using a different type of distribution.

Once the aim points are determined for the leakers, the airfield damage simulation stochastically determines the impact point for each missile relative to its aim point based on its CEP and guidance parameters from a Rayleigh distribution as explained in Driels (2004). Finally, the simulation determines the damage that results from the blast and fragmentation of each missile for all nearby targets within the missile-target lethal radius. See Wade (2019) for more information on the models and processes of this simulation.

10.4 Solving the Military Problem Using Simulation Optimization

This section uses the AMD simulation, runway damage simulation, and airfield damage simulation described above with an optimization program to perform a multi-objective optimization of the TBM and CM fire plan parameters to maximize the damage to the airfield assets and runway for three fire plans of increasing size. The section begins by introducing the optimization program, then describes the test cases for the missile fire plan optimization, and concludes with a discussion of the results of the optimization of the missile fire plan.

10.4.1 Optimization Program

The analysis of the red fire plan requires that the three complex black-box stochastic simulations described in Section 10.3 be linked to create the overall damage function (see Figure 10.6). Each of the simulations is a discrete-event constructive simulation with discrete input variables and non-linear behavior. Section 10.2 discussed how gradient-based optimizers typically do not perform well with these types of simulations. Therefore, this work utilized an optimization program that employs metaheuristic techniques such as those described in Section 10.2. The optimization program that was selected is called OptDef, which interfaces with an optimization engine called OptQuest, both of which are developed by OptTek Systems (2018).

The OptQuest optimization engine uses a variety of metaheuristic techniques and employs tabu search and scatter search as primary methodologies (Glover and Laguna, 1997; Fu, et.al, 2005). The engine searches for global optima by iterating between diversification (exploring broadly) and intensification (finding local optima) phases. It then slowly increases the precision for each simulation input variable over the course of the optimization search in order to explore finer levels of variable value changes and identify further unique solutions. For problems with multiple objectives, the engine searches for the best solution in each objective direction and attempts to push out the Pareto frontier by exploring gaps between estimated frontier points found earlier in the search process.

10.4.2 Simulation Optimization Test Cases

To demonstrate how the simulation optimization results can change with different inputs, three examples of fire plans are presented: a small, medium, and large fire plan. The small fire plan was made up of five TBMs and five CMs impacting the target simultaneously. These TBMs and CMs could be of any type (TBM1, TBM2, TBM3, or CM1 and CM2) detailed in Table 10.1. The medium fire plan was made up of 15 TBMs and 15 CMs, of any type, impacting the target at the same time. Finally, the large fire plan was made up of 30 TBMs and 30 CMs, of any type, impacting the target simultaneously.

The three test cases were each optimized independently. The optimization criteria were to maximize the fraction (percentage) of aircraft destroyed, maximize the fraction of POL sites destroyed, and minimize the fraction of available MOSs. This multi-objective optimization creates a Pareto frontier of non-dominated solutions for each fire plan. Each point on the Pareto frontier represents a combination of fire plan inputs that achieves a non-dominated solution in at least one of the optimization goals.

All fire plans included an identical setup for the input variables. The only difference between the fire plans was a constraint that ensured the number of TBM and CM missiles matched the correct size for that fire plan. The fire plan input variables and ranges are shown in Table 10.3. Even though the largest fire plan allowed for 60 missiles (30 TBMs and 30 CMs), only a maximum of 20 missiles of any single type was allowed to account for the limitations in the number of launch platforms. For each type of missile, the fraction aimed at the runway was calculated as [100 − (percent aimed at the airfield)]. Of the missiles aimed at the airfield, the number of those missiles aimed at the POL sites were calculated as [100 − (percent of airfield missiles aimed at aircraft)]. Additionally, an attack sequencing and timing variable was included called "CM impact delay from TBM impact" to allow the CMs to impact before (negative delay) or after (positive delay) the TBMs. Changing the impact time of the CMs relative to the TBMs can affect the vulnerability of both missile types by altering the amount of time each threat is within engagement range of the AMD systems and the amount of time both TBMs and CMs are simultaneously within engagement range of the AMD systems.

Figure 10.8 provides a visual representation of the process used to solve this simulation optimization problem. The optimization program provides a set of simulation inputs to the AMD and airbase damage simulations. A set of simulation inputs includes one value within the minimum and maximum range for each variable shown in Table 10.3. These simulations execute and return the three optimization criteria (i.e., the fraction (percentage) of aircraft destroyed, the fraction of POL sites destroyed, and the fraction of available MOSs) as output. The optimization program then determines a new set of inputs using the optimization metaheuristics described in the previous section. For example, the optimization program may combine two distinct set of inputs that have yielded the best simulation outputs found so far in the search into a new distinct set of inputs to simulate next. This new set of inputs for the airbase damage simulations then results in another set of outputs. This process is repeated iteratively until

TABLE 10.3:   Red Fire Plan Optimization Input Variables

Input Variable	Type	Minimum	Maximum
Number of TBM1	Integer	0	20
Number of TBM2	Integer	0	20
Number of TBM3	Integer	0	20
Number of CM1	Integer	0	20
Number of CM2	Integer	0	20
TBM1 – Percent Aimed at Airfield v. Runway	Float	0	100
TBM2 – Percent Aimed at Airfield v. Runway	Float	0	100
TBM3 – Percent Aimed at Airfield v. Runway	Float	0	100
CM1 – Percent Aimed at Airfield v. Runway	Float	0	100
CM2 – Percent Aimed at Airfield v. Runway	Float	0	100
TBM Release Altitude for Submunitions (meters)	Float	10,000	30,000
CM Release Altitude for Submunitions (meters)	Float	50	500
Percent of Airfield Missiles Aimed at Aircraft v. POL sites	Float	0	100
CM impact delay from TBM impact (seconds)	Float	–60	+60

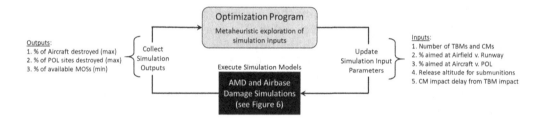

FIGURE 10.8: Red fire plan simulation optimization problem.

some stopping criteria is reached. For example, the total number of simulation runs to execute, a specified time to search, or a given number of search iterations without finding an improving solution are common stopping criteria. Once the stopping criteria are met, the optimization program returns the red fire plan input values that maximize the percentage of aircraft destroyed and the percentage of POL sites destroyed and minimize the percentage of available MOSs. Because these are competing objectives, there will be several red fire plan input values that result in a set of non-dominated output values, which estimates the Pareto frontier.

10.4.3   Simulation Optimization Results

The results of the optimization for the small, medium, and large fire plans are shown below in Figure 10.9 and the left side of Figure 10.10. These images show the estimated Pareto frontier of non-dominated solutions for each test case. Since the optimization method is a metaheuristic and only a subset of all possible input parameter combinations have been simulated, the estimated Pareto frontier is not guaranteed to be the

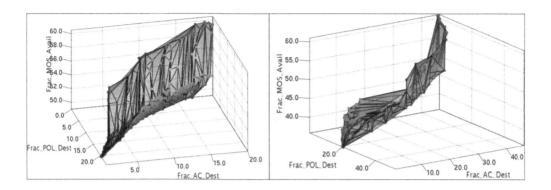

FIGURE 10.9: Optimal frontier for small fire plan (left) and medium fire plan (right).

true Pareto frontier. The right side of Figure 10.10 shows the Pareto frontier along with all of the dominated tested points for the large attack. In Figures 10.9 and 10.10, the two horizontal axes are the fraction of aircraft destroyed and fraction of POL sites destroyed. The vertical axis is the fraction of MOSs available on the runway for use. The missile force is trying to maximize the fraction of aircraft and POL sites destroyed and minimize the fraction of MOSs available. The points on the Pareto frontier represent the best tradeoffs that can be made between the three objectives. Thus, points on the lower-right of each graph represent solutions that perform well against all three objectives.

As the size of the fire plans increased from small to large, the non-dominated front's curvature also increased. The frontier also moves toward the lower-right of the graph indicating that the increased number of leakers from the larger fire plans caused greater damage to all parts of the airfield. The points on the frontier of all three fire plans show the tradeoffs between damaging various parts of the airfield. Certain fire plans caused a greater number of specific types of leakers, which in turn, caused greater damage to different parts of the airfield. This is illustrated in Table 10.4.

Table 10.4 shows two points on the Pareto frontier for the medium fire plan. Each of these points used a fire plan composed of 15 TBMs and 15 CMs; however, the outputs from each were different based on the type of missiles, targets, and attack

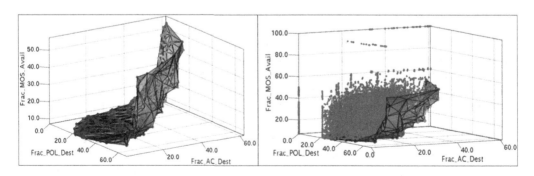

FIGURE 10.10: Optimal frontier for large fire plan (left) and large fire plan with both dominated and frontier points (right).

TABLE 10.4: Sample Points on Estimated Pareto Frontier for Medium Fire Plan

Input/ Output	Variable		Sample Frontier Point	
			1	2
Input	Number of Missiles in Fire Plan	TBM1	0	0
		TBM2	13	0
		TBM3	2	15
		CM1	9	15
		CM2	6	0
	Percent of Missiles Aimed at Airfield v. Runway	TBM1	0%	0%
		TBM2	0%	0%
		TBM3	50%	94%
		CM1	11%	100%
		CM2	0%	0%
	TBM Release Altitude for Submunitions (meters)		30,000	10,000
	CM Release Altitude for Submunitions (meters)		491	105
	Percent of Airfield Missiles Aimed at Aircraft v. POL sites		66%	85%
	CM impact delay from TBM impact (seconds)		−3	7
Average Output	Fraction of MOSs Remaining After Attack		40.7%	60.7%
	Fraction of POL Sites Destroyed in Attack		5.3%	11.6%
	Fraction of Aircraft Destroyed in Attack		10.9%	40.9%
	Number of Missiles that Impact the Ground (Leakers)	TBM1	0	0
		TBM2	3.9	0
		TBM3	0.3	5.5
		CM1	3.0	1.9
		CM2	2.9	0

parameters chosen. Sample frontier point 1 in Table 10.4 was composed of a mix of TBMs 2 and 3 and CMs 1 and 2. All but one of the TBMs was aimed at the runway and the TBMs that carried submunitions released them at the maximum allowable altitude. This reduced their vulnerability to the AMD systems. CM1 released its submunitions at a higher altitude that increased its spread around the runway, but this made it more vulnerable to the AMD systems since flying at a higher altitude causes the missiles to crest the radar horizon further away from the AMD systems which leads to a longer engagement window. To counter this disadvantage, the CMs were timed to arrive three seconds before the TBMs. This meant that the AMD systems were likely engaging the TBMs when the CMs first crossed the radar horizon, reducing the AMD system's engagement window against the CMs. This resulted in a smaller portion of usable runway following the attack compared with sample point number 2 in Table 10.4.

Contrast this with sample frontier point 2 in Table 10.4, which was a uniform attack consisting of only a single type of TBM and CM: TBM3 and CM1 missiles. All of the CM1 missiles were aimed at the runway and released their submunitions at a very low altitude, arriving seven seconds after the TBMs. Most of the TBM3 missiles were aimed at the airfield and were concentrated on the parked aircraft. This resulted in a larger portion of usable runway still available, but with a greater number of aircraft and POL sites destroyed in the attack, compared to sample frontier point 1 in Table 10.4.

Table 10.4 demonstrates the value of multi-objective optimization. Sample frontier point 1 is a fire plan that is more effective if the military commander's intent is to inflict more damage on the runway so that opposing aircraft have limited capability to take off; whereas sample frontier point 2 is a fire plan that is more effective if the commander's intent is to destroy aircraft and POL sites. Real-world complexities and uncertainties nearly always lead to competing decision criteria, and multi-objective optimization using simulation models provides decision-makers with the information required to begin to consider the trade-offs.

10.5   Conclusion

Simulation optimization is an advanced analytic technique that offers a robust modeling paradigm for complex, uncertain, and non-linear real-world problems. Simulation optimization also allows organizations to perform the full spectrum of analyses encompassing descriptive, diagnostic, predictive, and prescriptive analysis. Unfortunately, traditional algorithms for solving general optimization problems like MPP are not applicable to optimizing the simulation models that capture the stochastic, complex nature of the real world. Fortunately, metaheuristic techniques are effective and efficient for solving simulation optimization problems (Feo and Resende, 1995). An effective use of simulation optimization is demonstrated using three discrete-event constructive simulations (i.e., an AMD simulation, a runway damage simulation, and an airfield damage simulation) and a metaheuristic optimization program to address the military problem of determining a ballistic and cruise missile attack plan that maximizes damage to an airfield defended by an air defense artillery network. Results showed how the attacking missile force could adjust the fire plan parameters in order to maximize the destruction of various parts of the airfield.

References

Aarts, E., & Lenstra, J. (Eds.). (2003). *Local Search in Combinatorial Optimization*. Princeton, NJ: Princeton University Press.

AFPAM 10-219, Volume 4. (2008). *Airfield Damage Repair Operations*. Washington, DC: Department of the Air Force.

Audet, C., & Hare, W. (2017). *Derivative-Free and Blackbox Optimization*. Springer International Publishing.

Banks, J., Carson, J. S., & Nelson, B. L. (1999). *Discrete-Event System Simulation* (2nd ed.). Upper Saddle River, NJ: Prentice Hall.

Bazaraa, M. S., Jarvis, J. J., & Sherali, H. D. (1977). *Linear Programming and Network Flows* (2nd ed.). New York: John Wiley & Sons, Inc.

Boginski, V., Pasiliao, E. L., & Shen, S. (2015). Special issue on optimization in military applications. *Optimization Letters, 9*(8), 1475–1476.

Chabai, A. J. (1965). On scaling dimensions of craters produced by buried explosives. *Journal of Geophysical Research, 70*, 5075–5098.

Defense Modeling & Simulation Coordination Office. (2014, March 19). *DoD Modeling and Simulation Glossary*. Retrieved August 8, 2018 from https://www.msco.mil/MSReferences/Glossary/MSGlossary.aspx

Dirik, N., Hall, S. N., & Moore, J. T. (2015). Maximizing strike aircraft planning efficiency for a given class of ground targets. *Optimization Letters, 9*(8), 1729–1748.

Driels, M. R. (2004). *Weaponeering: Conventional Weapon System Effectiveness*. Reston, VA: American Institute of Aeronautics and Astronautics.

Ehrgott, M. (2005). *Multicriteria Optimization* (2nd ed.). Berlin-Heidelberg: Springer.

Feickert, A. (2005). *Cruise Missile Proliferation*. Washington, DC: Congressional Research Service, The Library of Congress.

Feo, T. A., & Resende, T. A. (1995). Greedy randomized adaptive search procedures. *Journal of Global Optimization, 6*(2), 109–133.

Fu, M. C., Glover, F. W., & April, J. (2005). Simulation optimization: a review, new developments, and applications. *Winter Simulation Conference Proceedings* (pp. 83–95). Orlando, FL.

Gallagher, M. A., Caswell, D. J., Hanlon, B., & Hill, J. M. (2014). Rethinking the hierarchy of analytic models and simulations for conflicts. *Military Operations Research, 19*(4), 15–24.

Glover, F., & Laguna, M. (1997). *Tabu Search*. Kluwer Academic Publishers.

Gormley, D. M. (2002). The neglected dimension: controlling cruise missile proliferation. *The Nonproliferation Review, 9*(2), 21–29.

Gormley, D. M. (2008). *Missile Contagion: Cruise Missile Proliferation and the Threat to International Security*. Westport, CT: Praeger Security International.

Hill, R. R., Miller, J. O., & McIntyre, G. A. (2001). Simulation analysis: applications of discrete event simulation modeling to military problems. *Winter Simulation Conference Proceedings* (pp. 780–788). Arlington, VA: IEEE Computer Society.

Hill, R. R., & Tolk, A. (2017). A history of military computer simulation. In A. Tolk, J. Fowler, S. Guodong, & E. Yucesan (Eds.), *Advances in Modeling and Simulation* (pp. 277–299). Springer, Cham.

Johnson, A. W., & Jacobson, S. H. (2002). A class of convergent generalized hill climbing algorithms. *Applied Mathematics and Computation, 125*(2–3), 359–373.

Kannon, T. E., Nurre, S. G., Lunday, B. J., & Hill, R. R. (2015). The aircraft routing problem with refueling. *Optimization Letters, 9*(8), 1609–1624.

Kiger, S. A., & Balsara, J. P. (1983). *A Review of the 1983 Revision of TM 5-855-1 "Fundamentals of Protective Design" (Nonnuclear)*. Vicksburg, MS: USAE Waterways Experiment Station.

Lee, L. H., Chew, E. P., Frazier, P. I., Jia, Q., & Chen, C. (2013). Advances in simulation optimization and its applications. *IIE Transactions, 45*(7), 683–684.

Lobo, B. J., Brown, D. E., Gerber, M. S., & Grazaitis, P. J. (2018). A transient stochastic simulation-optimization model for operational fuel planning in-theater. *European Journal of Operational Research, 264*(2), 637–652.

McNabb, M. E., Weir, J. D., Hill, R. R., & Hall, S. N. (2015). Testing local search mover operators on the vehicle routing problem with split deliveries and time windows. *Computers & Operations Research, 56*, 93–109.

Melouk, S. H., Fontem, A. F., Waymire, E., & Hall, S. N. (2014). Stochastic resource allocation using a predictor-based heuristic for optimization via simulation. *Computers & Operations Research, 46*, 36–48.

Montgomery, D. C. (2017). *Design and Analysis of Experiments* (9th ed.). John Wiley & Sons, Inc.

Nemhauser, G. L., & Wolsey, L. A. (1999). *Integer and Combinatorial Optimization*. New York: John Wiley & Sons, Inc.

Nocedal, J., & Wright, S. (1999). *Numerical Optimization*. New York: Springer-Verlag.

Onal, H., Woodford, P., Tweddale, S. A., Westervelt, J. D., Chen, M., Dissanayake, S. T., & Pitois, G. (2016). A dynamic simulation/optimization model for scheduling restoration of degraded military training lands. *Journal of Environmental Management, 171*, 144–157.

OptTek Systems, Inc. (2018). *OptDef Government Solutions.* Retrieved September 1, 2018 from www.opttek.com/optdef

Pollack, J. (2011). Ballistic trajectory: the evolution of North Korea's ballistic missile market. *Nonproliferation Review, 18*, 411–426.

Ragsdale, C. T. (2015). *Spreadsheet Modeling & Decision Analysis* (7th ed.). Stamford, CT: Cengage Learning.

Reeves, C. R. (Ed.). (1995). *Modern Heuristic Techniques for Combinatorial Problems.* London: McGraw Hill.

Swisher, J. R., Hyden, P. D., Jacobson, S. H., & Schruben, L. W. (2000). A survey of simulation optimization techniques and procedures. *Winter Simulation Conference Proceedings* (pp. 119–128). Orlando, FL.

Swisher, J. R., Hyden, P. D., Jacobson, S. H., & Schruben, L. W. (2004). A survey of recent advances in discrete input parameter discrete-event simulation optimization. *IIE Transactions, 36*(6), 591–600.

Tekin, E., & Sabuncuoglu, I. (2004). Simulation optimization: a comprehensive review on theory and application. *IIE Transactions, 36*(11), 1067–1081.

Teledyne Brown Engineering, Inc. (2015, July 3). *EADSIM Overview.* Retrieved from https:// eadsim.teledyne.com/Overview.aspx

United States Air Force. (2013). *Ballistic and Cruise Missile Threat.* Wright-Patterson AFB, OH: National Air and Space Intelligence Center.

Wade, B. M. (2019). A Multi-objective optimization of ballistic and cruise missile fire plans based on damage calculations from missile impacts on an airfield defended by an air defense artillery network. *Journal of Defense Modeling and Simulation: Applications, Methodology, Technology, 16*(2), 103–117.

Wade, B. M., & Chang, P. (2015). New measures of effectiveness for the air and missile defense simulation community. *Phalanx: The Magazine of National Security Analysis, 48*, 49–53.

Chapter 11

Analytical Test Planning for Defense Acquisitions

Darryl K. Ahner and Gina Sigler

11.1 Introduction: Why Test and Why Test Using Scientific Test and Analysis Techniques?

As weapon systems have become more complex, the ability to discern performance by simply building systems correctly and placing them in operational use has diminished greatly. This complexity arises from the interaction of more components and subsystems. People are generally poor at predicting patterns formed from the interactions of many elements (e.g., rules and computing artifacts). The only way that we may fairly elicit patterns of performance over the full range in which we want these complex systems to perform without introducing bias is to develop useful metrics and apply the scientific method to understand operational, technical, and systemic interactions. This is the focus of this chapter.

Before digging into the more technical aspects of this chapter, it is important to understand how defense acquisitions usually occur. First, an element of the Department of Defense decides that a new or improved warfighting capability is required. Performance characteristics or mission requirements are defined on what this capability should be able to achieve. A contract is typically awarded for a contractor to develop that capability. Evaluations are conducted by the contractor to ensure it meets these requirements, then by the government to ensure development is proceeding, and finally by the government again at the end of development to ensure it can accomplish the mission for which it was designed and built.

For these evaluations, Scientific Test and Analysis Techniques (STAT) are used to efficiently and effectively generate the data for evaluation. STAT are the statistical methods and processes used to enable the development of efficient, rigorous test strategies that will yield quality information to inform these evaluations. STAT enables efficient and effective development of structured data for decision-making. STAT is the essence of operations research in that it is an application of scientific and mathematical methods to the study and analysis of problems involving complex systems, in this case, the development of, and performance of, complex systems in operational military environments. STAT encompass techniques such as design of experiments, reliability growth, and survey design, among others. The suitability of each method is determined by the specific objectives of the test. A STAT process is described later in Figure 11.2 that guides the use of these techniques. Through STAT, we develop defendable and traceable evidence to make technical, engineering, and acquisition decisions throughout the development of the system. We combine this quantifiable knowledge with systems engineering and subject matter expertise to fully inform the decision-maker. The goal of the application of STAT is to assist acquisition program leadership and systems engineers in developing rigorous, defensible test strategies and plans throughout the acquisition life cycle. These test strategies and plans result in the ability to more effectively quantify and characterize system performance.

One might ask, is STAT really necessary? Several sources indicate it is needed to avoid system development delays and future costs.

In August 2014, the Director, Operational Test and Evaluation (DOT&E) published Reasons Behind Program Delays, a document which listed 115 programs on the DOT&E oversight list with delays of at least six months in duration. In 49 of these cases, performance problems were identified during the system's operational test (OT). By Service, 15 of 35 Army programs experienced delays where problems were discovered in OT, 20 of 43 Navy programs experienced delays where problems were discovered in OT, nine of 29 Air Force programs experienced delays where problems were discovered in OT, and five of eight non-Service specific programs experienced delays where problems were discovered in OT. In other words, over 42% of programs on oversight experienced delays of more than six months due to performance problems not found in developmental testing. While all of these problems in operational testing may not be discovered in developmental testing, these statistics run counter to the thought that effective developmental tests were completed and resulted in no major surprises in operational testing. An effective developmental testing program should be every program's goal.

The 2014 General Accounting Office (GAO) Report DEFENSE ACQUISITIONS Assessments of Selected Weapon Programs highlights the need for more effective developmental testing to provide a knowledge-based acquisition approach. The report states that many programs continue to commit to production before completing developmental testing. It also states that programs can improve the stability of their design by conducting reliability growth testing and by completing failure modes and effects analyses during development. The report only indicates, through surveying the program offices, whether developmental testing was conducted, but *not* whether these tests were planned effectively and efficiently. Furthermore, the report does not indicate whether testing was used to inform a risk assessment. The Deputy Assistant Secretary of Defense for Developmental Test and Evaluation (DASD(DT&E)) has advocated use of a developmental evaluation framework for this purpose since late 2014.

In the DOT&E FY 2015 Annual Report, 64% (48/75) of operational tests conducted in Fiscal Year 2015 (FY15) revealed problems significant enough to adversely affect evaluation of the system's operational effectiveness, suitability, or survivability. Almost 39% (29/75) of these operational tests discovered significant problems that were unknown prior to operational testing.

Finally, the 2016 GAO Report DEFENSE ACQUISITIONS: Assessments of Selected Weapon Programs states that high levels of concurrency between development testing and production may add risk to the acquisition portfolio. With 16 of the 43 current programs assessed in production, 11 of these programs plan to complete 30% or more of their developmental testing concurrent with production and three of these 11 programs plan to place more than 20% of procurement quantities under contract before testing is complete. The GAO states that as concurrency increases, so does total acquisition cost growth. It is this cost avoidance which is the end goal for the application of STAT.

Clearly, the completeness of developmental testing, as of 2015, is not adequate for program success across the acquisition portfolio. But even if testing occurs as prescribed, a lack of STAT-based efficient and effective test designs decreases the probability of meeting desired outcomes for an effective, affordable, and suitable system. We must also realize that STAT is not sufficient by itself. To determine why these performance problems are not uncovered in developmental testing, we must take a more holistic approach and can turn to root cause analysis. Root cause analysis is a systems engineering approach that decomposes the defect – in this case developmental test performance failure – into its components. In other words, we focus on "why" we did not find faults before operationally testing. Some reasons may be:

- testing is not scheduled
- unavailability of representable test items for developmental testing
- untested integration issues
- inadequacy of test ranges or facilities
- poorly planned tests
- faulty or incomplete test evaluation

STAT focuses on the rigor of ongoing test planning so that effective evaluation can be achieved. In its simplest form, the STAT process begins with a clear understanding of requirements and the capabilities that are desired from the acquisition program and ends with quantifiable data that informs evaluations.

The remainder of this chapter addresses why testing is needed for understanding the performance of complex systems and how the analysis of testing is used to manage performance risk. Beginning with system capabilities and requirements, the flow down to actual test design demand signals is explained.

The second section focuses on developing an effective STAT test strategy. This section addresses a well-defined Department of Defense system test & evaluation (T&E) strategy: (1) well-defined, end-to-end, mission-oriented objectives within the scope of the decision to be informed; (2) mission-oriented, traceable measures of capability and readiness; (3) complete coverage of the test space; (4) designs with good properties and test design structures; (5) disciplined test protocols; (6) statistical data analysis and assessment techniques and standardized evaluation criteria; and (7) reliable products to inform decisions.

The third section addresses the application of effective test planning using design of experiments applied to a specific example of a weapon system.

The fourth section concentrates on how STAT addresses development risk and the specific testing efficiencies often realized by employing rigorous statistical engineering through STAT.

The last section provides conclusions from lessons learned drawn from the implementation by the Scientific Test and Analysis Techniques Center of Excellence and other defense analytical organizations employing STAT.

11.2   The Demand Signal for Testing

The primary challenge in applying STAT to Department of Defense (DoD) testing is the broad scale and complexity of the systems, missions, and conditions. This challenge is best addressed by breaking down the requirement, system, and/or mission into more manageable pieces. At a lower level, these pieces are more readily translated into statistical designs that provide rigor, efficiency, and definitive, quantifiable analysis.

The leader's role is to understand if the proposed testing will be sufficient and ensure program risk is quantified and minimized. This process can generally be described as shown in Figure 11.1. The process starts with capabilities and requirements development. These desired capabilities and requirements are then decomposed using the systems engineering process into subsystems and components. The development of these subsystems and components carry with them a level of risk. For instance, a subsystem or component that has a well-established architecture and has been produced in the past may only have risk associated with its integration with other subsystems or components. Meanwhile, a subsystem or component that has a new design may have performance risk associated with it meeting its performance requirement, in addition to its integration risk as part of a larger subsystem or system. Evaluating and mitigating these risks are addressed in a performance evaluation framework that lays out the timing or conditions that must be met to plan test events. Finally, rigorous test designs are developed for each test event so that they are both effective in obtaining the required information to inform the risk assessment and overall system development success, and also efficient so that time and funding resources are not wasted.

While testing can be expensive, continued development that arises from a system unable to perform to the required capability and requirements after full-rate production has begun is exponentially more expensive. It is imperative that a careful risk and cost tradeoff occur and when testing is required, an effective STAT test strategy must be used.

FIGURE 11.1: Demand signal for test events/designs.

11.3 Developing an Effective STAT test strategy

STAT encompasses deliberate, methodical processes and procedures that seek to relate requirements to analysis in order to inform better decision-making. All phases and types of testing strive to deliver meaningful, defensible, and decision-enabling results in an efficient manner. The incorporation of STAT provides a rigorous methodology to accomplish this goal.

Although academic references and industrial examples of STAT are available in large quantities, many do not adequately capture the challenges posed by complex systems. Case studies, such as those that are used in test planning course curricula, are more relevant since they are application-oriented. However, even these rarely contain enough detail to be useful in practical settings. Another potential source of confusion arises from test documentation that not only fails to adequately define system requirements, but also omits critical conditions for the employment of these systems, such as a contested cyber environment. Such difficulties must be overcome, usually at a high cost in both time and effort, to produce efficient and highly effective, rigorous test plans that instill a sufficient level of assurance in the inferred performance of the system under test.

As systems become more complex, the application of STAT becomes simultaneously more necessary and more challenging. This challenge is best addressed by breaking down the requirement, system, and/or mission into subsystems or components, which can then be readily translated into rigorously quantifiable statistical designs. This point is expressed in the following excerpt from Montgomery (2017):

> *One large comprehensive experiment is unlikely to answer the key questions satisfactorily. A single comprehensive experiment requires the experimenters to know the answers to lots of questions, and if they are wrong, the results will be disappointing. This leads to wasting time, materials, and other resources, and may result in never answering the original research question satisfactorily. A sequential approach employing a series of smaller experiments, each with a specific objective ... is a better strategy.*

To support such a breakdown, STAT drives an iterative procedure that begins with the requirements and proceeds through the generation of test objectives, designs, and analysis plans, all of which may be directly traced back to the required capabilities or requirements. Critical questions at every stage help the planner keep the process on track. At the end, the design and analysis plan is reviewed to ensure that it supports the objective that began the process. Figure 11.2 provides a concise process flow diagram that summarizes the application of STAT.

We summarize the STAT Process and review each of these topics in more detail throughout the remainder of this section (Burke et al., 2017).

- Requirements
  - the information, purpose, and intent contained in the requirements drive the entire process
  - all subsequent steps support selection of test points that will provide sufficient data to definitively address the original requirement
- Identify/clarify/quantify STAT candidates

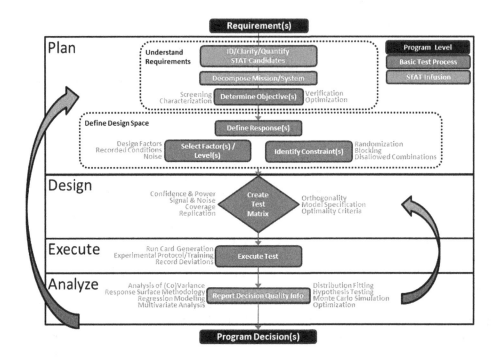

FIGURE 11.2: Scientific test and analysis techniques process.

- identify what systems or tests may benefit from STAT (not all tests require STAT)
- clarify the type of results the STAT candidate tests should produce;
- STAT candidates need to be associated with quantifiable requirements/metrics
- Mission/system decomposition
  - break the system or mission down into smaller segments
  - smaller segments make it easier to discern relevant response variables and the associated factors
- Determine objectives
  - derived from the requirements and reflect the focus and purpose of testing
  - serve to further define the scope of testing
  - should be unbiased, specific, measurable, and of practical consequence (Coleman & Montgomery, 1993)
- Define response variables
  - the measured output of a test event
  - the dependent outputs of the system that are influenced by the independent or controlled variables (known as factors)
  - used to quantify performance and address the requirement
  - should be quantitative whenever possible
- Select factors and levels
  - design factors are the input (independent) variables for which there are inferred effects on the response (dependent variable(s))
  - levels are the values set for a given factor throughout the experiment. Between each experimental run, the levels of each factor should be reset, even when the

next run may have the same level for some factors. Failing to do so can violate the assumption of independence of each result and therefore introduce bias into the analysis
- tests are designed to measure the response over different levels of a factor or factors
- statistical methods are then used to determine the statistical significance of any changes in response over different factor levels
- uncontrolled factors contribute to noise and are referred to as nuisance factors. The design should account for these factors
- Identify constraints
  - anything that limits the design or test space
  - may be resources, range procedures, operational limitations, and many others
  - limitations affect the choice of design, execution planning, execution, and analysis
- Create test design/matrix
  - provides the tester with an exact roadmap of what and how much to test
  - provides the framework for future statistical tests of the significance of factor effects on the measured response(s)
  - allows you to quantify risk prior to executing the test
  - all the aforementioned considerations combine to inform the final test design
- Execute test
  - the planning accounts for, and requires, certain aspects of the execution to be accomplished in a particular manner
  - since variations from the test plan during execution may potentially impact the analysis, they must be approved (if the change is voluntary) and documented as data is collected during test execution
- Analysis
  - begins when all of the data is collected and you can perform statistical analysis
  - must reflect how the experiment was actually executed (which may be different from the original test plan)
  - conclusions as to the efficacy of the system during test are documented, organized, and then reported to decision-makers
  - the information collected from the previous test phases may either influence the design of the next phase of testing or inform future program decisions
- Program decisions
  - the final step to decide whether to proceed into the production phase or to continue development of the test article
  - represents the fulfillment of the entire test process
  - the program utilizes data collected on the system to quantitatively comment on performance and adjudicate the acquisition process

Each one of these elements is required for effective analytical test planning for defense acquisitions.

11.3.1   Plan: Understanding Requirements

Requirements are both the starting and ending point for effective test planning and evaluation. Planning starts with mapping out a path to report on the requirement. Many choices must be made during test planning to address concerns such as operating conditions, resource constraints, and range limitations. These choices help focus the process toward the test design and execution methods so the analysis will provide the right information to decision-makers to make an informed decision about the requirement. If the requirement is not understood clearly at the beginning of the test-planning process, the test team may find that it cannot produce the data needed to report on the requirement. Two aspects are critical to understanding the requirements. The first is the source of the requirement from the mission decomposition within the systems engineering process. This may be thought of as the operational view of the derived requirement. The second is the systems engineering or technical view of the requirement. This aspect of the requirement involves a full understanding of the system under test and is best accomplished collaboratively with system technical experts.

According to Buede and Miller (2016) the requirements for a system set up standards and measurement tools (metrics) for judging the success of the system design. Furthermore, Buede and Miller (2016) call out four requirements categories:

1. Specification Level-Stakeholders', Derived, Implied, and Emergent: stakeholders' requirements are derived from the operational need. Derived requirements are those requirements defined by the systems engineer in engineering terms during the design process. Implied requirements are those that can be inferred from systems engineering documentation. Emergent requirements are those not known before development. Implied and emergent requirements should be avoided;
2. Performance Requirements versus Constraints: performance requirements are often defined as a threshold which is the minimum acceptable level and objective which is the desired level of performance above which the stakeholder sees no substantial additional value. Constraints rule out certain possible designs;
3. Application – Systems versus Program: systems requirements relate to the characteristics of the system's performance in the broadest sense. Program requirements relate either to the programmatic tasks that must be performed, programmatic tradeoffs among cost and schedule, and programmatic products associated with the systems engineering process;
4. Functional, Interface, or System-wide Requirements: functional requirements relate to specific functions that the system must perform. An interface requirement is a system requirement that involves an interaction with another system. System-wide requirements are characteristics of the entire system; examples include availability, reliability, maintainability, durability, supportability, safety, trainability, testability, extensibility, and affordability.

Beginning testing at the mission level is ill-advised. It is often difficult to determine how individual system components influence performance. Even if one could identify a particular component as being the specific cause of an observed behavior during test, the exact mechanism of this causal relationship may not be well understood. "Systems are becoming so synthesized or fused, complex and interdependent that they can, even without taking into account human agency, have emergent properties or exhibit behaviors that vary to an extent that is not easily predicted" (Joiner & Tutty, 2018, p. 4). Systems that exhibit complex and interdependent behavior are often referred to as systems of systems and can be classified as complex systems. Complex systems that exhibit dynamic, unpredictable, and multi-dimensional relationships are the subject of a relatively new research area referred to as complexity science.

> *Complexity can exist in the problem being addressed, in its environment or context, or in the system under consideration for providing a solution to the problem. The diagnoses made will allow the systems engineer to tailor his/ her approaches to key aspects of the systems engineering process: requirements elicitation, trade studies, the selection of a development process life cycle, solution architecting, system decomposition and subsystem integration, test and evaluation activities, and others.*

(Sheard et al., 2015)

This issue is mitigated by a sequential, progressive test strategy that starts with low-level test designs and culminates in top-level operational testing. This issue can also give rise to emergent behavior in development that may be explored through prototyping and holistic testing (Sheard et al., 2015). Box and Liu (1999) emphasize the importance of sequential testing and learning as follows:

* the fundamental starting point for any test should be to understand the performance of each critical component. The type and the extent of the component testing required is a function of criticality and previous knowledge
* the next level of testing is to evaluate individual functions. To the greatest extent possible, functional characteristics should be evaluated over the entire range of operational conditions. Testing over a subset of possible conditions greatly increases the risks of delaying failure discoveries
* next, the combination of all of the functions should be evaluated in the system. The goal is to discover failures caused by the integration of all of the functions
* finally, full system testing should be conducted in order to evaluate the highest-level system-of-systems measures and to validate all previous testing

In order to accomplish low-level test designs that culminate in top-level operational testing, a mission systems engineering V-model decomposition, as seen in Figure 11.3, is usually required. This process begins in the upper left with the engineering requirements for the mission success of the system. Requirements analysis encompasses the tasks that determine the needs or conditions that the new system must meet and outputs the operational, functional, and physical systems engineering views of the system. Functional analysis is a top-down process of translating system level requirements from the previous step into detailed functional and performance design criteria which are also used during developmental testing and analysis. Critical component analysis is where critical design elements occur, resulting in preliminary and, through iteration, detailed designs. An effective test strategy will "build back up" the other side of the systems engineering V so that the program manager will understand the technical baseline of the development and be better able to inform re-engineering decisions and manage developmental risk. This is done early on through contractor testing (CT) and early developmental testing (DT) applied to component specifications, then through DT applied to functional analysis on functional specifications, then DT, operational assessments and assessment of operational test readiness applied to critical test parameters which are measurable critical system characteristics that, when achieved, enable the attainment of desired operational performance capabilities, and finally, to operational testing (OT) applied to critical operational issues. Through this testing and evaluation

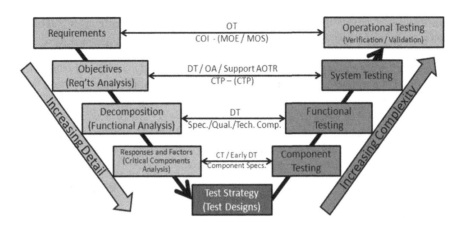

FIGURE 11.3: Systems engineering V-model decomposition.

process it is the goal of applying STAT, along with input from systems engineering architectures and documentation, and available subject matter expertise that acquisition programs understand and manage their technical baseline.

Planning cannot proceed without a clear set of test objectives. When writing test objectives, precise wording is critical. The action verb in the objective is especially important. It is also extremely important that objectives are stated clearly to remove any ambiguity and lessen the likelihood of misrepresentation. Montgomery (2017) provides example action verbs for objectives for just such a purpose, as shown in Figure 11.4. Once clear objectives are established, the test planner can progress to defining the test design space.

11.3.2   Plan: Define Design Space

As discussed in the previous section, the key to a good test design is to fully understand the goals of the test. A critical characteristic of a good objective is that

Objective	Action	Example
Characterize	To measure the response across a design space.	Characterize the effects of solder temperature and conveyor speed on circuit board defect rates.
Screen	To learn which factors have the most influence on the response.	Screen solder temperature, solder depth, or conveyor speed to see whether they affect the defect rate.
Optimize	To find the factor levels that result in a desired response.	Optimize solder temperature and conveyor speed to achieve the lowest defect rate.
Confirm	To verify the system behavior is consistent with theory or experience.	Confirm that the defect rate during full-rate production is the same as that during LRIP.
Discover	To determine what happens when factors are added/removed or the factor ranges are increased.	Discover the effect a new soldering material has (if any) on the defect rate.
Robustness	To find the factor levels that both provide desired response, AND reduce the variance of the response.	Find the solder temperature and conveyor speed that result in the lowest defect rate AND lowest variability from batch to batch.

FIGURE 11.4: Example action verbs that facilitate clear objectives.

it is measurable; the systems engineering V-model decomposition and using clear action verbs assists us to accomplish this. Responses are the measured outputs of a test event and are used to assess the objective of the test. There may be several responses measured for a given test or in support of a requirement. Responses which capture system performance measures should trace back to key performance parameters, measures of performance, measures of effectiveness, or measures of suitability that inform critical operational issues and critical technical issues which may have arisen from an understanding of the system architecture or flowed from previous testing.

Factors are inputs to, or conditions for, a test event that potentially influence the value and/or the variability in the response. Controllable factors are those factors that the tester may actively manage while uncontrollable factors (sometimes referred to as noise factors) cannot be managed, but still must be documented. An example of an uncontrollable factor would be the weather during the day of an outdoor test; it cannot be chosen, but it should be recorded since it could still impact the response. The goal of considering all potentially relevant factors is best accomplished through focused brainstorming sessions with system engineers and experienced testers. Additionally, factors must be clear and unambiguous and may have priorities assigned to them in case not enough resources are available to explore all factors in testing.

Lastly, constraints may exist that impact test design or execution. Constraints affect which combinations of factors may be considered. Common constraints are the test budget, restrictions to the experimental factor region due to safety or other factors, and hard to change factors effecting randomization in the execution of test points.

11.3.3 Test Design

Many considerations affect the choice of design. In this section, we consider design of experiments as the gold standard of test designs when factors can be controlled. The type of test objective is a key factor in choosing an appropriate design. For instance, Box, Hunter, and Hunter (2005) and Montgomery (2017) are two excellent references for expanding knowledge about design of experiments in general and understanding differences among various test designs. Burke et al. (2017) is also an excellent guide in choosing an appropriate design as seen in Table 11.1.

TABLE 11.1: Selecting a Design Type

Test Objective	Suggested Designs to Consider
Screening for Important Factors	Factorial, Fractional Factorial, Definitive screening
Characterize a System or Process over a Region of Interest	Factorial, Fractional Factorial, Response surface, Optimal
Process Optimization	Response surface, Optimal
Test for Problems (errors, faults, software bugs, cybersecurity vulnerabilities)	Combinatorial, Orthogonal Arrays
Analyze a Deterministic Response	Space Filling, Optimal
Reliability Assessment	Sampling Plans, Sequential Probability Ratio Test

In addition to the type of design selected, design metrics are used to determine performance. Among these metrics are the statistical model supported (e.g., main effects, two-factor interactions, quadratics), power, confidence, correlation between statistical model terms, design resolution, prediction variance, design efficiency, and others. An appropriate text addressing the specific type of design should be consulted. Myers et al. (2016) discuss 11 desirable properties of designed experiments which should be balanced to achieve the priorities of the test. A designed experiment should:

1. result in good fit of the model to the data;
2. provide good model parameter estimates;
3. provide a good distribution of prediction variance of the response throughout the region of interest;
4. provide an estimate of pure experimental error;
5. give sufficient information to allow for a test for curvature or lack of fit;
6. provide a check on the homogeneous variance assumption;
7. be insensitive to the presence of outliers;
8. be robust to errors in the control of the design levels;
9. allow models of increasing order to be constructed sequentially;
10. allow for experiments to be done in blocks;
11. be cost effective.

Too often in the acquisition process, the 11th property dominates the other properties. By using efficient designs developed using modern software, this situation should be avoidable.

11.3.4   Test Execution

The three principles of design – randomization, replication, and blocking – should be considered when planning the execution of a design. Randomization is the act of executing the chosen test runs in a random order. Randomization helps protect against unknown conditions affecting the analysis of the system. For example, suppose while executing a test, the temperature steadily increased. Without randomizing the factor conditions, we might accidentally attribute changes in the response to a controlled factor, when temperature was actually the driver. By randomizing, the effect of a non-controlled factor is averaged out over the controlled factors. If a design has hard-to-change factors, a split-plot design may be required (along with appropriate analysis) to account for restrictions in randomizing all test runs. Replication is the repeating of some test design points to estimate pure error, i.e., the natural variation in the response. Blocking is a planning and analysis technique where experimental runs are arranged in groups (blocks) that are similar to one another. A blocking factor is a source of variability that is not of primary interest to the experimenter. However, by accounting for the variability due to a blocking factor, it is easier to identify and estimate the effects of factors of interest on the response. A common blocking factor is the day of the week. If all the runs of a test cannot be executed on the same day, blocking the test design by day allows us to remove any potential affects the test day has on the response. Further reading of these design principles can be found in Box, Hunter, and Hunter (2005) and Montgomery (2017).

11.3.5   Test Analysis

All the test planning is for naught unless a good test analysis plan exists and is followed. This analysis should capture metrics (statistics) that are used to form a narrative around which the test objectives are viewed. For example, after executing a designed experiment, we can build empirical models of the response. The statistical analysis can, and should, be much more than reporting averages. Software plays a critical role in determining the statistics of interest. Test analysis is conducted against system requirements. In the next section, we look at an application of effective test planning using design of experiments applied to a weapon system.

11.4   Design of Experiments Application

In a nutshell, design of experiments focuses on breaking down a system or process into appropriate factors and responses that can be measured to provide data. This data then helps answer questions about the system or make decisions. There are many different types of designed experiments that vary with designated factors and responses. These can include full factorials, fractional factorials, screening experiments, response surface analysis, randomized block designs, split-plot designs, etc. For more information on any of the designs mentioned, please see Box, Hunter, and Hunter (2005) or Montgomery (2017). The choice of the design used in an experiment depends on the objectives of the test, the number of factors to be explored, and any test constraints (particularly the test budget).

Consider a weapon system called the ABC missile, a new missile based on legacy systems. The program office has stated that the goal of this new missile is to take the place of one of the old missiles that has an outdated targeting system. The new missile will have many of the same specifications as the legacy system, but it will include a new and updated targeting system. In order for the new missile to be acceptable, it must perform either the same or better as the old system in terms of the requirements.

Examples of some of the requirements are found below:

- the missile shall have a range of X meters
- the missile shall be usable on stationary targets of varying size
- the missile shall be usable during different times of day and under various states of cloud cover
- the missile shall have a maximum impact time of X seconds from time of launch
- the missile weight shall not exceed X pounds

The missing X values would be replaced by the values known for the legacy system. Note that the requirements have been constructed around quantifiable measurements. Complex or ambiguous requirements will make the application of testing much more difficult. Consider instead a requirement that stated "The missile shall be user-friendly." What does it mean for something to be user-friendly? There are no standard metrics to compare user-friendliness to, so we instead purposefully construct easily quantifiable requirements. Using these requirements as the basis for possible experiments, the test team discussed which requirements would require testing and which requirements might only require demonstration. It was determined that the weight requirement could be satisfied using only a demonstration, but that the remaining requirements need to be rigorously tested. The weight, while accompanied by some small missile to missile variation, should not have any changing factors that impact it. The other requirements (e.g., range and impact time) can be referred to as scientific test and analysis (STAT) candidates, requirements with quantifiable metrics that benefit from scientific and structured testing to best understand how they change across the operating space.

After reviewing the requirements, a team of subject matter experts was called in to discuss and analyze the system with the help of a STAT expert. Since most of the missile is comprised of legacy subsystems, the requirements tended to map to the new targeting system. The program objectives are two-fold. First, they want to show that the missile can hit the target (with some degree of accuracy) at a distance of at least X meters. STAT experts recommend stating all objectives in the form of a testable question that can then be directly answered by a response in the test. Using the following format, the testable question becomes: does the missile hit the target at a distance of X meters or less? The test team still needs to determine what the degree of accuracy is for a "hit" to the target. Next, the test team wants to show that a missile impact can be accomplished in X seconds from time of launch to impact. The testable question here becomes: does the missile spend X seconds or less in the air from time to launch to time of impact? Showing these two objectives in all possible conditions should satisfy the requirements. The test team determined that both of these requirements could be evaluated in the same test. The next step for the test team was to determine how to measure each of the requirements in a measured response in the test.

Responses should always be chosen to directly support the objective or answer the testable question. A continuous response will provide more information than a binary response because continuous measures allow better characterization of the variability in the response. For the first testable question, it might be intuitive to use a response variable of hit or miss, but this will lead to less overall information. Instead, the test team decided that the radial distance from the target to the point of impact would be measured as accuracy and that if any part of the target has been damaged by the missile, accuracy will be considered to be a distance of 0. The team went through a similar discussion for the second testable question and settled on measuring the time in seconds from missile launch to impact.

Finally, the team of subject matter experts and a STAT expert walked through possible factors that could affect the chosen responses. Just as was true with the responses, it is better to have continuous or numeric factors than binary or categorical factors. Numeric values can be interpolated so that predictions of values in the operating space that are not present in the test can still be made. Continuous factors, therefore, provide more information than their categorical counterparts, which do not allow for interpolation. Factors for this test were driven by a combination of the requirements and previous experience. The first factor is distance from the launch point to the target in meters. If the missile is required to have a certain range, then it should be tested at different distances for effectiveness. The team also determined that the launch angle and launch point height might play key roles as these have been significant factors on the legacy missile systems. The requirements state that the targets would be stationary and of varying size. After discussing possible targets, it was determined that the targets could be grouped into three different size categories of small, medium, and large. This was necessary since measuring the size of each target was infeasible due to the multitude of target shapes. Another requirement states that the missile must be usable at different times of the day. While it was suggested that lumens be measured to create a continuous measure of sunlight, the capability to measure this was not possible during testing. Instead, a time of day factor was created with only day or night as the recorded levels. The same requirement states that the missile must be used in different amounts of cloud cover. Cloud cover was chosen as a factor to be measured in oktas using a high-resolution image sensor. The team suggested that operator experience could play a role in the overall performance of the missile. Therefore, a factor of operator experience was created with new and experienced operators recorded. The last suggested factor was terrain. While not explicitly specified in the requirements, the missile might have additional uses in various terrains such as mountains, desert, jungle/forest, or the coast. All factors described above and appropriate levels are listed in Table 11.2 below.

As seen in Table 11.2, distance to target, launch angle, launch point height, and cloud cover are all to be varied as continuous variables. Values of −1 or 1 are set as placeholders for actual low and high values to be used as the minimum and maximum in the test. Categorical factors include target type with three different size targets, time of day, operator experience level, and terrain. It was also decided that terrain would be a difficult factor to change as there would have to be a change in physical location in the test. All other factors were determined to be relatively easy to change, although varying cloud cover could present a potential problem if weather does not cooperate. The main issue with hard to change factors is the impact on randomization, which in turn impacts the information provided. Hard to change factors should be minimized whenever possible. In this case, it is not in the budget to keep randomly moving around to the different terrains for every run.

The next step in testing is choosing a design for the test objectives, the desired factors, and responses. The team decided to start with a screening design to help narrow down the factor space. Their first choice would be conducting a full factorial design, but this would be difficult because terrain is a hard-to-change variable, which would interfere with the randomness of the runs. A full factorial also turns out not to be a very efficient design due to all the multi-level categorical factors included. Instead, they investigated split-plot designs because terrain cannot be randomized throughout the entire test. The test operators will need to do as many test points as possible at the same terrain location. The group thinks the budget will stretch to include two trips to each terrain location for a total of eight locations; this will create eight whole plots in

TABLE 11.2: Factors and Levels for ABC Missile Test

Factors	Levels			
Distance to target	−1	1	−	−
Launch angle	−1	1	−	−
Launch point height	−1	1	−	−
Target type	Small	Medium	Large	−
Time of day	Night	Day	−	−
Cloud cover	−1	1	−	−
Operator experience	New	Experienced	−	−
Terrain	Mountains	Desert	Jungle/forest	Coastal

the experiment. The main issue with split-plot designs is that the lack of randomization decreases the power on the hard to change variable. This means that we won't be able to estimate the effects of terrain very well, but other main effects should have decent power. Also, any interaction effects with terrain will be easy to detect, which should help support any decision about terrain alone. Split-plot designs are created using computer-generated optimal designs. When selecting a design to run in an experiment, STAT experts recommend creating several designs with varying user-supplied inputs to compare the pros and cons of each design. For this example, the test team wanted to investigate differences in designs for different total test size, number of whole plots, etc. Table 11.3 shows summary information for several designs considered by the test team which were created by changing the user-supplied input information to the design generator, particularly the test size. The metrics in the design comparison include model supported, power, prediction variance, and aliasing. Each design was created using JMP (V. 13) and has eight factors. Model supported explains which factors are in the model such as main effects or two-factor interactions. Ideally, the model would support both main effects and two-factor interactions. Signal-to-noise ratio is the desirable difference to detect over the expected natural variation. This should be decided by the subject matter experts before running any tests. Alpha is also referred to as the significance level or 1-confidence level; typically, this value is set to be between 0.01 and 0.2. Power is the ability to correctly identify when an effect is present; it is typically desired to be 0.8 or higher. Fractional design space (FDS) prediction error informs the test how precise predicted responses will be; typically, values below 1 are considered ideal.

TABLE 11.3: Potential Designed Experiments for ABC Missile Screening

Design #	1	2	3	4	5	6
Software Package	JMP	JMP	JMP	JMP	JMP	JMP
Name/Design Type	Full Factorial	Split Plot 1	Split Plot 2	Split Plot 3	Split Plot 4	Split Plot 5
Factors	8	8	8	8	8	8
Model Supported	ME, 2FI	ME, 2FI	ME, 2FI	ME, 2FI	ME, 2FI	ME
Signal-to-noise Ratio	2.0	2.0	2.0	3.0	3.0	3.0
Alpha	0.05	0.05	0.05	0.05	0.1	0.05
Total Runs	768	80	72	72	72	36
Power for ME @ S/N	1	0.16	0.15	0.29	0.46	0.15
Power for 2FI/Q @ S/N	1	0.81	0.67	0.88	0.95	–
FDS Pred Err @50%	0.05	1.12	1.31	1.31	1.31	0.77
FDS Pred Err @95%	0.06	1.50	1.92	1.92	1.92	0.81
Aliasing	none	low	low	low	low	medium

ME: Main effect; 2FI: Two-factor interaction; Q: Quadratic; S/N: Signal-to-noise; FDS: Fractional design space; Pred Err: Prediction error

Obviously, all test teams would like to have fewer runs in the test to minimize cost, but the number of test turns must be balanced by desirable design properties including power and prediction variance. Higher values of power are more desirable since this indicates a high probability of detecting a factor effect. Prediction variance is more desirable at lower values since this indicates smaller confidence bound on prediction of future values of the response. In an ideal situation, the two FDS prediction variance values would both be below 1 and power would be above 0.8 for all main effects and two-factor interactions. However, this is rarely possible in real life test cases.

Design 1 in Table 11.3 is the full factorial that is infeasible due to the total number of runs and hard to change factor (terrain). Designs 2 through 6 are all split-plot designs with different model terms supported, signal-to-noise ratio, alpha values, and number of total runs. After discussing the first three design options, the test team noted that the factors used in tests of the legacy system tended to have large effect sizes relative to the noise of the system, indicating that a signal-to-noise ratio of 3 would be adequate for the ABC missile test. This was taken into account for the final three design options. The team has also stated that it is not possible to devote more than 72 runs to the initial screening. Ultimately, it was decided that an alpha value of 0.1 would be used; the test team was more willing to accept a slightly higher risk of including a factor in a potential future test than they were to accept not correctly identifying a present effect. Even though the prediction variance is a bit high, design 5 was the one that was selected by the team. Since the primary purpose of this test is to screen factors, prediction is a less critical design capacity. The test team evaluated each of the designs and chose the one that best met their objectives while balancing test cost.

The design matrix for design 5 was constructed using JMP, and the 72 runs were later executed in the randomized order by the test team. Preliminary analysis showed that distance to target had the largest impact for both time in the air and impact distance, as expected. An empirical model was created for each of the two responses using linear regression. Figure 11.5 shows the JMP effect summary for the response of impact distance. Non-significant terms, terms with a p value lower than 0.1, have been removed from the model unless a main effect is included in a higher-order interaction. Launch point height and cloud cover have both been screened out of the model. This means that future tests do not need to include these two factors, which will save on testing time and the number of runs required in future tests.

Effect Summary

Source	LogWorth		PValue
Distance to Target(20,700)	7.937		0.00000
Operator experience	5.190		0.00001
Operator experience*Terrain	2.282		0.00522
Launch angle*Operator experience	2.008		0.00983
Launch angle*Time of day	1.822		0.01507
Target type	1.821		0.01510
Target type*Time of day	1.814		0.01533
Target type*Terrain	1.708		0.01959
Terrain	0.280		0.52495
Time of day	0.025		0.94305
Launch angle(5,35)	0.011		0.97550

FIGURE 11.5: Model effect summary for response of distance to target.

Effect Summary

Source	LogWorth		PValue
Distance to Target(20,700)	73.364		0.00000
Launch angle(5,35)	8.784		0.00000
Distance to Target*Launch angle	7.415		0.00000

FIGURE 11.6: Model effect summary for response of time from launch to impact.

Figure 11.6 shows the JMP effect summary for the second response of time in seconds from missile launch to impact. Again, non-significant terms have been removed. In this case, there are only two main effects and one interaction term remaining in the model. The only factors that have an impact on the time are the distance to the target, launch angle of the missile, and the interaction between the two. This is also not surprising to the test team, although they will still conduct their test with all of the factors that were significant for the other response variable in future tests to better refine the empirical models for these two responses.

The information from the screening design can now inform further testing. The non-important factors of cloud cover and launch point height can now be removed from the next test. If any additional factors had been found during this phase, those factors could also be added in to the next phase of testing. This is a primary part of the reason for sequential testing to be performed; it allows for real-time changes to be made to better improve testing by making it more efficient. Because terrain was found to be an important factor in missile performance, the next test was set up as another split-plot with terrain remaining as a hard-to-change variable. This would be done similarly to the first screening design, but would have an end goal of a confidence interval around the response variables. While the goal of the first test was to screen out unimportant factors, the goal of the follow-on testing is to make tighter predictions on the responses for observations not tested. After completing this testing and modeling, it was concluded that the new missile did indeed pass the requirements. Confidence intervals constructed around the two response variables confirmed that the missile was able to adequately hit the target at a distance of X meters and spent less than X seconds in the air.

If the system did not adequately pass the requirements, the factor breakdown would allow the test team to determine what was contributing to the deficiency so that it could be individually addressed. Without the application of the design of experiments, there would not be a way to track any potential issues. Design of experiments allowed the test team to rigorously explore the performance of the ABC missile. The test also provided the program leaders with defensible data to make informed decisions on the performance of the missile.

11.5 Why STAT?

Many ask why STAT is important to the Department of Defense. The short answer is that without STAT, there are often mistakes in testing (which are sometimes easy to fix) that then require expensive fixes. With STAT, deficiencies in a system can be found earlier because of the use of rigorous testing methods. Consider a program where STAT has recently had a major impact. This program had already set up all their testing, but had not yet consulted with a STAT expert. The program has now determined that the amount of testing is actually infeasible for the program, so they have decided to bring in a STAT expert to overhaul the test design.

The introduction of rigor can take many forms. One of the basic forms is simply understanding all of the moving parts of a designed experiment. As presented in the previous section, most experiments will already have considered factors and responses in some form. However, many test teams are not informed of the benefits of being able trade off number of runs, confidence level, power, and term aliasing.

This particular program is attempting to integrate a new missile onto a ship. The missile itself has already gone through rigorous testing of its own, but it has never been used before on this particular vessel. Upon looking at the test designs, the STAT expert notices two major opportunities to manage the overall size of the test design. First, the factor space for the original design was large and included factors that would explain both the integration of the missile into the new vessel and the overall performance of the missile. The performance of the missile is, of course, nice to know; however, missile performance will not ultimately affect the program decision because the missile has already been tested on its own. While the program team did a good job with factor scoping, it is actually a waste of time and resources to include performance factors for a missile that has already been tested. This means that the factor space can be greatly reduced for the current test, which focuses on integration of the missile into the vessel. Understanding the true objectives of the test led to a more informed and efficient test design because removing the unnecessary factors reduces the total number of runs required in the experiment. The other opportunity identified by the STAT expert arose by discussing the implications of adjusting the thresholds of the sources of risk. As discussed in a previous section, test designs can be compared and assessed prior to testing using various metrics including power, type I error, the signal-to-noise ratio, the level of model term aliasing, and the sample size. In the DoD, common thresholds for power and alpha (type I error) are 0.8 or greater and 0.2 or less, respectively. Testers are able to actively control the alpha level by specifying it upfront; however, power is not actively controlled and is dependent on alpha: decreasing alpha will cause power to decrease. In other words, enforcing a lower type I error rate results in a higher type II error rate (since power is equal to one minus type II error). In addition, many other design metrics have an effect on power including the sample size, the signal-to-noise ratio, and even the type of design used. Knowing the relationships among power, alpha, sample size, signal-to-noise ratio, and aliasing, allows better understanding of the tradeoffs in risk when changing the value of one metric versus the other. Of course, adjustments made to these values must be made carefully and reflect current assumptions known about the system. In this case, the program was vying for simply reducing the number of runs and accepting a test with lower power. However, the STAT experts were able to instead point out options for aliasing terms and increasing the type I error rate slightly. In the end, the new test design had fewer runs, providing a test design at a manageable level and saving over 100 days of testing.

Another major impact that STAT had on this project involved the idea of sequential testing. The original test plan contained two different tests. The first would be a tracking-only test where targets would be tracked and a firing solution would be developed. The second test would be a live fire exercise that actually fires off a real missile. Obviously, the second test is a more costly than the first test since no physical objects would be destroyed during the first test. The same argument was made that the performance of the missile had already been tested in a previous program and so the live fire test was not as important for the current program as the tracking testing. Ensuring that the tests match the objectives of testing (assessing the integration of the missile with the new vessel) allows for efficient and effective testing. The STAT expert recommended moving more of the budgeted runs to the tracking test from the live fire test in order to reduce overall test cost and place emphasis on the testing with higher risk. Finally, the program had been planning on executing a split-plot design, similar to the ABC missile described previously. As discussed, a split-plot design greatly sacrifices power of the hard-to-change factors in order to reduce some of the randomization factors that are hard or expensive to change after every run. In the ABC missile, terrain was truly a hard-to-change factor and would lead to a much longer test time. For this program, however, it was thought that not changing the factors between every run would be a simple way to reduce the time to complete testing, although no factor was hard or expensive to change after each run. Not fully randomizing the factors in this program's test would have led to more difficult analysis in the end with less clear results. The STAT expert explained the importance of a truly randomized design, which changed the both the implementation and possible analysis of the test.

By bringing STAT into their test plan, this program was able to mitigate risk, reduce cost, ease program burdens, and execute a test design that would provide the right data to support decision-making. Scientific and statistical methods can be one of a program's greatest assets. It is never too late to introduce STAT to a program.

11.6   Conclusions and Bringing it all Together

This chapter addressed test planning within the acquisition of Department of Defense programs. While studying statistics and statistical models as a student and later as a new practitioner when these are used, it is important to understand that application of analysis techniques often need frameworks in order to be applied effectively. When addressing evaluation of complex weapon system development, Scientific Test and Analysis Techniques and their use within the STAT process yield quality data for analysis.

The complexity of acquisition programs requires rigorous processes and techniques to gain understanding of their performance in order to mitigate risk. As we previously mentioned, many DoD systems had significant failures during operational testing that ideally would have been identified during development testing where fixes are less expensive. The DoD uses STAT as an overarching methodology to incorporate rigorous processes and techniques into system testing. STAT, however, is not a replacement for scientific knowledge and subject matter expertise. STAT is best employed through collaboration with systems engineers, range operators, and test engineers to understand the requirements, test objectives, the system under test, and all constraints (particularly cost and schedule) related to testing. This collaboration among the program's subject matter experts and a STAT expert results in a rigorous and defensible test strategy that will provide decision-quality information to the program leaders.

Analysts should follow the Plan – Design – Execute – Analyze process in Figure 11.2, paying close attention to each of the following elements of effective implementation of STAT:

- requirements
- identify/clarify/quantify STAT candidates
- mission/system decomposition

- determine objectives
- define response variables
- select factors and levels
- identify constraints
- create test design/matrix
- execute test
- analysis
- program decisions

Each one of these elements is required for effective analytical test planning for defense acquisitions. In addition to following these steps for a rigorously planned individual test, testing should ideally be a sequential process where one test builds upon the other in order to inform acquisition and programmatic decisions as the system maturity progresses. When assessing a system over the operational space and those factors are controllable, design of experiments is the gold standard for testing. Design of experiments provides the test planner with an efficient test that is also effective for gaining insight into whether system requirements have been met or where in the operational space the system fails. These decisions and insights learned after each test are identified in the performance evaluation framework to improve knowledge as a program moves from developmental testing to operational testing. However, while information from prior test events can be beneficial, great care must be taken not to rely too much on prior results if significant engineering changes are made to the system under test.

Due to the increasingly complex nature of acquisition systems, the DoD must continue to improve the effectiveness and efficiency of test and evaluation in order to deliver decision-quality information for systems development and acquisition decisions. While the implementation of rigorous STAT is not sufficient for gaining all knowledge for technical, programmatic, and acquisition decisions, STAT is a necessary component to ensure our armed services have the systems with the capabilities they need. STAT provides a structured test development process that results in the appropriate data to assess system performance.

The use of STAT has been codified in DoD Instruction 5000.02, making it a required element in DoD acquisition test and evaluation. However, STAT, in many cases, remains an afterthought during test planning. The consistent and effective use of STAT requires a change in DoD culture, beginning with increased education in STAT knowledge by our military and government workforce. This change in culture, one where STAT is embraced as a necessary and effective testing methodology, will lead to continuous improvement DoD testing for the increasingly complex acquisition systems.

References

Box, G. E., Hunter, J. S., & Hunter, W. G. (2005). *Statistics for Experimenters: Design, Innovation, and Discovery*. Hoboken, NJ, Wiley-Interscience.

Box, G. E., & Liu, P. Y. (1999). Statistics as a Catalyst to Learning by Scientific Method Part I—An Example. *Journal of Quality Technology*, 31(1), 1–15.

Buede, D. M., & Miller, W. D. (2016). *The Engineering Design of Systems: Models and Methods*. Hoboken, NJ, John Wiley & Sons.

Burke, S., Harman, M., Kolsti, K., Natoli, C., Ortiz, F., Ramert, A., Rowell, B., Truett, L. (2017). Guide to Developing an Effective STAT Test Strategy V5.0. https://www.afit.edu/STAT/statdocs.cfm. Accessed October 15 2018.

Coleman, D. E., & Montgomery, D. C. (1993). A Systematic Approach to Planning for a Designed Industrial Experiment. *Technometrics*, 35(1), 1–12.

Director, Operational Test and Evaluation. (2014, August 26). DOT&E Presentation Reasons behind Program Delays - 2014 Update. http://www.dote.osd.mil/news.html. Accessed October 09 2018.

Director, Operational Test and Evaluation. (2016). FY 2015 Annual Report. http://www.dote.osd.mil/pub/reports/FY2015/. Accessed October 15 2019.

Joiner, K. F., & Tutty, M. G. (2018). A Tale of Two Allied Defence Departments: New Assurance Initiatives for Managing Increasing System Complexity, Interconnectedness and Vulnerability. *Australian Journal of Multi-Disciplinary Engineering*, 14(1), 4–25.

Montgomery, D. C. (2017). *Design and analysis of experiments*. Hoboken, NJ, John Wiley & Sons.

Myers, R. H., Montgomery, D. C., & Anderson-Cook, C. M. (2016). *Response Surface Methodology: Process and Product Optimization Using Designed Experiments*. Hoboken, NJ, John Wiley & Sons.

Sheard, S., Cook, S., Honour, E., Hybertson, D., Krupa, J., McEver, McKinney, D., Ondrus, P., Ryan, A., Scheurer, R., & Singer, J. (2015, July). A Complexity Primer for Systems Engineers [INCOSE White paper]. https://www.incose.org/docs/default-source/ProductsPublications/a-complexity-primer-for-systems-engineers.pdf?sfvrsn=0&sfvrsn=0. Accessed October 14 2019.

U.S. Government Accountability Office. (2016, March). DEFENSE ACQUISITIONS: Assessments of Selected Weapon Programs. (Publication No. GAO-16-329SP). https://www.gao.gov/assets/680/676281.pdf. Accessed October 15 2019.

Section II

Soft Skills and Client Relations

Chapter 12

Why Won't They Use Our Model?

Walt DeGrange and Wilson L. Price

12.1 Introduction

What if someone were to furnish you with a Ferrari and pay for all the fuel, tires, insurance, and maintenance? How would you drive the car? Surprisingly this is what happens in the United States (US) military. The services (Army, Navy, Air Forces, and Marine Corps) provide forces to regional commanders and finance the fuel, maintenance, personnel, and supplies required to operate these forces. For example, in the US Navy, commanders decide the policy for sequencing replenishment missions to combat ships. Then, a group of 38 personnel manually produces a sequence of supply missions for the next 1–3 weeks. Anomalies such as the inadvertent creation of multiple supply missions to the same combat ship or the inadvertent omission of a combat ship from the schedule are identified through discussions among the schedulers, supply ships, and combat ships. Any schedule shortcomings are made up for in execution by increasing the speed of the supply ships and increasing fuel usage.

Usually, this tactic works but increases mission cost significantly. One cannot, however, fault regional commanders for basing scheduling decisions on mission completion rates and neglecting fuel costs. The mathematical difficulty of the computations required by the inclusion of the cost of fuel, indeed of almost any additional variable or objective, is too high for a manual system.

The US Navy attempted to address this situation with the development of the Replenishment at Sea Planning (RASP) tool (Brown et al., 2017). RASP is an analytic optimization model of the scheduling process, capable of producing thousands of feasible solutions in minutes and of treating costs as an objective to be minimized. The regional commander would have the ability to make significant savings in resupply operations and the door would be open for other innovations, such as the creation of global links among the Navy regions. Three years of extensive work by a multidisciplinary team yielded a computation engine that we expected to be quickly adopted. Unfortunately, rather than a quick and painless adoption, we had to ask ourselves, "Why won't they use our model?"

12.2   What Went Wrong?

Many authors have addressed the difficulties that arise in the implementation of new technologies such as the decision support systems, optimization models, and scheduling tools that are the basis of much operations research. While some of the major issues, such as linking to legacy systems and design of the user interface are primarily technological in nature, in many cases, it is in dealing with the human component that an implementation team finds the greatest challenges.

Analytics literature offers many descriptions of the pitfalls encountered in implementation and some prescriptions for avoiding them. In an OR/MS Today article (Cokins, 2012) referring to analytics, Cokins asks, "So what continues to obstruct the adoption rate of analytics?" He answers that the major causes are "social, behavioral and cultural issues, including people's resistance to change, fear of knowing the truth (or of someone else knowing it), reluctance to share data or information and a 'we don't do that here' mindset." Note that all of these issues are related to organizational and individual concerns rather than technical matters.

In another case, Leclerc and Thiboutot (2003) report their use of a model-based software suite for the routing and scheduling of a truck fleet in contrast to the policy of many organizations which still use manual or simple legacy systems. Leclerc and Thiboutot are, respectively, director of logistics and Information Technology (IT) manager at a leading furniture company in Eastern Canada. They observe that

> *There is strong resistance to change. Adoption rates should accelerate as resistance to change is overcome by the evidence of results driven by accurate modeling environments, clear demonstration of the scalability and deployability of newer software architectures, and proven return on investment and efficiency gains demonstrated by initial implementations.*

Why then are so many vehicle fleets still using primitive legacy systems?

It is not only analytics that is affected by the rejection of technological innovations. In their 2006 paper, Lapointe and Rivard (2006) report on attempts to implement a medical information system in three hospitals. They report that many physicians "are reluctant to use IT tools." In all three cases, there were strong negative reactions, and only one of the implementations was successful. Why was the success rate of these implementations so low? In this chapter, the authors analyze the issues raised in each case and are able to relate them not only to the attitudes adopted by the putative users of the systems but also to the decisions and actions of the management and the system implementation team. They argue that change agents, such as IT innovators and analytics professionals, can affect outcomes and implementation success rates.

Most analytics practitioners have implementation anecdotes to tell and cases to describe. Where things did not go as hoped, many of them are asking, "What went wrong?" In the case of RASP, reflection and analysis suggested several answers to this question:

- Most of the schedulers using the scheduling model did not, in general, have analytics or IT experience
- Only on-the-job (OJT) training in the use of the model was planned
- Senior officers responsible for ship scheduling were wary of IT-based scheduling, which they perhaps saw as limiting their control over the process and the results
- IT issues, in particular, the connection to legacy systems that contained much of the data required by the model, forced the double entry of much of the data, which was a significant increase in workload

12.3 Conceptual Models of the Reaction to Technological Change

A conceptual model of reaction to information technology implementation may be thought of as a lens through which an implementation may be viewed. It appears that no single model covers all situations and to obtain as complete a picture as possible of the situation, the analyst in search of relevant insights should consider each available model in turn. Lapointe and Rivard (2005) identify four such conceptual models and develop a fifth themselves.

Markus (1983) points out that there may be existing patterns in the organization that interact negatively with an implementation that is under way and these mismatches may create resistance. In particular, she refers to the interactions that a technology implementation may have with existing power structures within the organization. For example, a new IT system might give better access to data (e.g., monthly sales or production figures) that were previously restricted to a few, and this may be seen by the former gatekeepers as diminishing of their power and influence.

Joshi (1991) views the appearance of a negative reaction to change as arising from a user's perception of equity in the balance between inputs that the change requires versus the outcomes that it produces. The user may be measuring inputs in several ways, such as data entry effort, stress level, the difficulty of learning the system, and manual and cognitive effort required to perform the multiple process tasks. Different users may have different measures. Similarly, outcomes include the service level to customers, the change in power level of the user, job satisfaction, salary change, and promotion possibilities. The list of possible metrics for measuring equity is complex and asymmetric. If a user perceives that an increase in inputs, such as the level of effort required to learn the new system, is not balanced by the outcome of better service offered to the customer, a negative reaction to the change may arise. The identification of equity issues requires careful observation and discussion with users.

Marakas and Hornik (1996) identify a user reaction that has its source in fear and stress stemming from the intrusion of new technology into a previously stable environment. They refer to the inflexibility of some individuals faced with changes in the ways of work that leads them to use covert means to procrastinate, slow their performance, or otherwise delay and obstruct the implementation.

Martinko, Henry, and Zmud (1996) suggest that individuals perceive a new technology according to both internal (e.g., the users' perception of their ability to master the new technology) and external influences (e.g., experience with other similar technologies). These perceptions then lead the user to develop expectations as to the outcome and efficacy of the technology. Of course, unfavorable expectations can engender a negative reaction.

Lapointe and Rivard (2005) take a longitudinal perspective and propose a multilevel model that seeks to explain resistance behaviors at the individual and workgroup levels. When a system is introduced, users will assess it starting with their individual initial conditions (e.g., knowledge, experience, power) and of the characteristics of the proposed system. If the users' expected consequences of implementation of the system are unfavorable, this triggers a negative reaction. This analysis takes a longitudinal perspective of the implementation, tracking it through its different phases, and covers the reactions both of individuals and of work groups.

Ford, Ford, and D'Amelio 2008) recommend prudence in the analysis of implementations because change agents may not always deliver unbiased accounts of users' actions. Some cases of failure due to "resistance" should rather be attributed to the actions or omissions of the implementation team.

12.4 Are There Resistance-Prone Changes?

The principle of *continuous improvement* is at the heart of Toyota's highly successful strategy for the management of industrial production (Shingo, 1989). Supervisors and factory workers often meet, every day in some cases, to discuss ways of improving quality and production flow. Examples of continuous improvements that occur on the shop floor include the enhancement of tools and adjustments to inter-workstation conveyors. In the area of operations management and control, installing a faster computer to speed up solver computations is an example of continuous improvement. Toyota managers feel that they obtain better productivity and quality through the constant application of continuous improvement rather than from occasional radical changes. A great advantage of continuous improvement is that it is less likely to engender negative reactions ("resistance") to the proposed changes, particularly since workers are involved from the very beginning of the development.

It is not always possible to avoid radical change, however. Consider two cases from the past:

- The replacement of manually controlled machine tools by multi-axis computer-controlled machining centers
- The introduction of personal computers into a workplace wedded to mainframe computing and the IBM Selectric typewriter

For these two examples, one sees that the change required was great: the new devices that actually perform the work are completely different, new computer-based systems are required, workers must be extensively retrained, and the range of work that can be performed is both broadened and deepened. When these changes occurred in business and industry, they were radical, and there were many occurrences of negative reactions. In some environments, the change was chaotic, but in others, the transition was adroitly managed.

Is it possible to identify which analytics projects are likely to engender negative reactions, foot-dragging, and resistance before implementation commences? The System-Task-Planner framework may be used in the analysis of an analytics solution implementation to determine if it is likely to cause potential users to display negative reactions.

12.5   The System-Task-Person (STP) Framework

In our lexicon, a "framework" is a structured way of visualizing, recording, and analyzing observations of the people and events involved in technology implementation. It is an analysis tool.

The STP framework shows the links among three elements of a technological implementation: the system, the task, and the person who is the user of the technology.

The *system* is a purposeful arrangement of instruments and tools that aid the worker in the production of goods or services. The instruments and tools in question may include workstations, networks, models and solvers, information systems, and other software. One may still find a manual "system" composed of paper worksheets and forms, blackboards, flip charts, and a manual of procedures.

Campion and Medsker (1992) propose a definition of the task which, while general, is a good fit for factory work. They define the task as a set of actions to be performed by a worker in order to transform various inputs into an appropriate output by means of instruments, tools, and methods. Meister (2003) offers a complementary definition well-suited to knowledge work. He sees the task as an arrangement of behaviors of the worker (perceptions, cognitions, motions, actions) linked and ordered in time and organized so as to attain a goal.

The "person" is the individual responsible for the execution of a set of operational or planning tasks using the system provided. The "person" might be part of a work group performing essentially identical tasks (for example, work in a call center), or one of a series of tasks that are carried out in a fixed sequence (as in a formal business process).

The STP-Triangle (Price & Rousseau, 1994) in Figure 12.1, is a framework and a tool for the analysis of a model's implementation that illustrates the interconnection of these three elements. The triangle represents human involvement in producing work output.

Figure 12.1 displays text balloons that identify the roles of various actors involved in establishing and maintaining the execution of the Task. The three nodes of the framework involve the following roles:

- The **Task** is established and maintained by a task designer who may be an analytics professional, a business process designer, an experienced operator, or a manager. It is executed by a person referred to as a "user."
- The **System** is managed by an IT team that deals with the evolution of the hardware and software and with system security and with ensuring the compatibility of the various software modules. The System includes hardware (workstations, computers, networks, storage devices) and software (information system, databases, software tools, analytics models) designed to assist the user, in the execution of the task
- The **Person** (user) who executes the task is supported by a human resource manager who is responsible, among other duties, for hiring, training, performance evaluation, and workplace safety

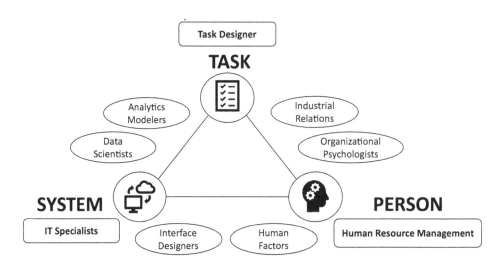

FIGURE 12.1: The System-Task-Person (STP) framework.

The three links of the framework may also be associated with specific roles:

- The **System-Task** link is where analytics professionals are usually found implementing models. The modelers work at formulating a representation of all or part of the task while the solvers devise algorithms (or choose them from among existing algorithms) for obtaining appropriate solutions that will assist the users in the execution of the task. The modelers will work nearer to the Task end of the link, while the solvers will need to work closely with the IT specialist
- The **System-Person** link requires the skills of the interface designer and the human factors specialist. The interface designer will ensure that display screens are clear and logical and are easily navigated. The human factors specialist will contribute knowledge of the cognitive efficiency of various interface choices
- The **Person-Task** link may involve industrial relations issues. For example, a proposed modification to the task may change job descriptions that are embedded in a collective agreement or require retraining of the users. Organizational psychology may contribute to the evaluation of personnel or to identifying potential health issues

Note that while nine roles have been identified, fewer actors may be required. For example, one analyst may redesign part of the task, model the revised task, design the model solvers, write the code to implement the model, and integrate it into the existing IT installation in a total of four roles. The human resources manager may cover issues with industrial relations, psychology, and human factors, and so this one actor will take on three roles. It is important for the analyst to note that for a model to be successfully implemented, each of the nine roles must be covered. In a given application, some roles may be "walk-on" bit parts – for example, there may be little or no industrial relations impact or health and safety issues. However, all roles must be covered.

The model implementation team may be "rich" in expertise. Some roles may be covered by a "resident" incumbent. For example, there may be an IT systems manager, a human resource manager, or a human factors person who has both the expertise and the organizational authority over a role. If there is no such resident incumbent or if that person is unavailable to the implementation team, there is a need for the analysts to develop minimal expertise in the field in question. There is no doubt that most analysts are more comfortable working along the System-Task link. The talents required to work effectively on the other two links are the "soft skills" mentioned so often in talks and publications.

To reduce project risk and to facilitate implementation, the analyst would do well to identify which actor will play each of the roles and to specify what script – the desired actions – the actors must follow for the project to be a success.

The manager who "owns" the STP triangle framework of Figure 12.1 is a tenth actor. This manager is the owner of the task, controls access to many elements of the triangle, approves the project, and judges progress and results. The analyst should be aware of what script this senior actor will follow and what constraints this script may add or remove on the implementation.

The situations outlined in Figure 12.1 are examples of how the nodes and arcs of the STP framework may be populated in a given project and not normative ideals. In practice one is likely to find a wide variation in the training and experience of the individuals playing a given role. For example, the Person role may be in the hands of a board-certified industrial psychologist in a firm with a depth of such professionals. Such a professional could conceivably be the overall team leader. This might occur in a firm developing software for public use, such as a touch-screen for an automobile dashboard which must transmit data to the driver without being an intrusive distraction. In other circumstances, the Person role could be managed by an experienced and respected engineer with a deep knowledge of the firm's Human Resource policies.

12.6 Supply Ship Scheduling in the US Navy

Let us return to the US Navy's ship scheduling model, RASP, that by 2017 had been implemented in two US Navy sites overseas. This experience in implementing the model enriched the framework and contributed to fashioning its present form.

Even now, in some locations, supply ship scheduling is being done manually. Schedulers seek a "good" feasible solution such that all operational ships receive timely resupply, supplies delivered are adequate in quantity, and that costs, particularly supply ship fuel costs, are controlled. A legacy information system records data concerning each operational vessel: its present position, time and place of the last resupply, and the current levels of onboard commodities and ordnance. It may also show a target time and location for the next resupply at sea. For the present time period (a specific number of days) manual scheduling assigns a specific supply ship to the task of resupplying one or more operational vessels at known times and locations. In the manual scheduling process, the schedulers must ensure that each supply ship can complete its assignments in the time required and supply the required material to the operational vessels.

As one would expect, some schedulers produce better supply ship assignments than others. Coordination among the schedulers ensures that the overall solution is feasible, but for several reasons, the manual solution may be far from what could be considered as optimal. For example:

- Manual scheduling is time-consuming and the schedulers may not have the time to improve the first feasible solution found

- Coordination among schedulers may be imperfect and details overlooked may render the proposed solution infeasible
- Human schedulers will often favor "greedy" heuristics in scheduling (e.g., first do the job with the shortest processing time, go to the closest city in the Traveling Sales Person) and non-intuitive, but high-quality solutions may never be found
- In order to render a trial schedule feasible, it may be necessary that supply ships travel at high speed, requiring more fuel, and driving up the cost of the supply operations
- The turnover rate for naval personnel assigned to the scheduling task is a problem because they are given only on-the-job training and are in post for relatively short periods (three years or less) which causes a loss of expertise

Naval Postgraduate School developed an operational planning solution that generates optimal replenishment schedules for the resupply ships. The primary focus of the tool is to minimize supply ship fuel consumption while servicing combatant customer ships with adequate supplies. The secondary focus shifts to the planning staff itself, seeking to reduce task force staff workload through the automation of data capture and standardized report generation.

Let us now look at a few of the elements that a prior analysis with the STP framework could have revealed.

On the Task node: At the Task node, the STP triangle owner is the supply ship scheduling organization Commanding Officer (CO). The CO owns the task and negotiates a rendezvous with the commanding officers of the customer ships to ensure timely resupply and will be held responsible for any failures. While the manual scheduling system has its faults, it functions nonetheless. The CO understands the system and is able to "tweak" it in unusual situations or when faced with unforeseen events. The CO may be nervous about accepting the work of what may seem to be a "black box." What can the CO do should the current model solution be unsatisfactory? The CO of the first implementation site actually had some of these concerns and informally revealed them to the analyst at a social occasion. Further compounding issues are that each supply ship command is independent of one another and have completely different business processes due to geographic differences. Each CO has a different opinion on how the scheduling model fits into their scheduling and planning processes.

System node: The legacy system was installed on a secure computer network, due to the ships' schedules being classified, and was managed by an IT specialist on site. Initially, the installation of the new model on the secure system was not authorized. This created an increase in overall workload because data from the legacy system had to be manually copied to the computer on which the model was installed. A heuristic of the model was developed to allow full implementation into a Microsoft Excel spreadsheet.

Person node: Schedulers are assigned to this task in the same manner as for any other shore assignment. They receive only on-the-job (OJT) training, and there is no specific computer literacy requirement. Not all schedulers fully commit to using the model because, in the short term, it involves an increase in workload as well a commitment to climbing the learning curve. The short-term assignment (one to three years) of the schedulers does not encourage investment in learning the new process. The new scheduler may or may not get adequate training on the model or have the background to immediately absorb the training.

Person-Task axis: An increase in workload was caused by the need for manual re-entry of a large volume of data. A junior reserve officer with an appropriate background was identified and trained for several days in the use of the model. He was assigned, as part of the implementation team, to carry out the day-to-day OJT of the schedulers, to shoulder the temporary increase in the overall workload, and to identify desirable modifications to the model and its computer implementation.

System-Person axis: Early in the implementation, it became evident that interface design was an important issue. Schedulers were keen to have screens and printed reports that exactly reproduced those of the legacy systems. Numerous requests for reprogramming the user interface were dealt with as expeditiously as possible. Many formatting issues such as font sizes and box colors were addressed to satisfy customer requirements in this area.

System-Task axis: Implementation of the new model modified the Task to some degree. Rather than producing one feasible solution, the schedulers would be able to explore some high-quality solutions and to modify them to some degree. The model also allows for exploration of "What-if" scenarios. Ten-page scheduling messages were automatically formatted and produced and that reduced production time and reduced errors from manually typing the schedule into a Microsoft Word document.

12.7   The State Space Framework

A Google search for the term "resistance to change" will yield approximately 75 million "hits." Resistance to change is real and observable, but merely saying that an implementation failure was caused by resistance to change leaves many questions unanswered.

Because *resistance* is so frequently blamed for failures, a project implementation team is sometimes asked to include dealing with negative reactions to change in the project plan. From the professional literature, many implementation teams have chosen to use the work of Kübler-Ross (Kübler-Ross, 1969) on death and dying as a model of "resistance to change." Kübler-Ross proposed a model of the emotional stages that will be experienced by a person who is told that death is approaching. The five stages are denial, anger, bargaining, depression, and acceptance. In adaptations of this model to represent resistance to change, more stages are sometimes added, and the terms changed to better fit a technology implementation project, but the principle is the same. Direct adaptations of the Kübler-Ross model assume that resisters display denial when the project is announced and then go through a progression of emotional states that end with acceptance.

Adopters of this interpretation of the Kübler-Ross model seek to move users through the intermediate stages and to overcome resistance through an appropriate communications plan. If the users continue to "resist," communications are repeated or augmented. This may actually help, and it is indeed important to have the best communications plan possible and to ensure that participants are not led to unhelpful positions through a lack of information. However, while there is value in the adaptation of the Kübler-Ross death and dying model to technology implementation, the two situations have characteristics that are quite different.

If careful medical diagnosis determines that a disease is fatal and untreatable, it may lead the person affected to *acceptance.* A technology implementation project simply does not have this weight of inevitability. Users know that the implementation is not preordained and that they are not obliged to accept it passively. The technology implementation situation is closer to the model depicted in the state space framework in Figure 12.2.

When a technology implementation project is announced, the participants will have different emotional reactions. Some will see the new technology as desirable progress while others may be more cautious.

It is useful to see the state space framework as a Markov chain in which a user may move from any state to any other state with some non-zero probability. For simplicity's sake, Figure 12.2 illustrates what are the most common states and transitions. Each of the states in this model will be familiar to practitioners who work in the area of technological innovation and organizational change, although others may carve up the state space in different ways.

One finds reference to such state space frameworks in Lapointe and Rivard (2005) and elsewhere. Lapointe and Rivard use a finer definition of the states of resistance, referring specifically to states of apathy, passive resistance, active resistance, aggressive resistance, and implicitly to those called trial acceptance and satisfying.

Some users will, for their own reasons, move directly from *Awareness* to *Trial Acceptance*. They may conclude that their knowledge and previous experience will allow them to use the new technology and benefit from it without great difficulty. Alternatively, they may feel that the new technology will bring advantages such that

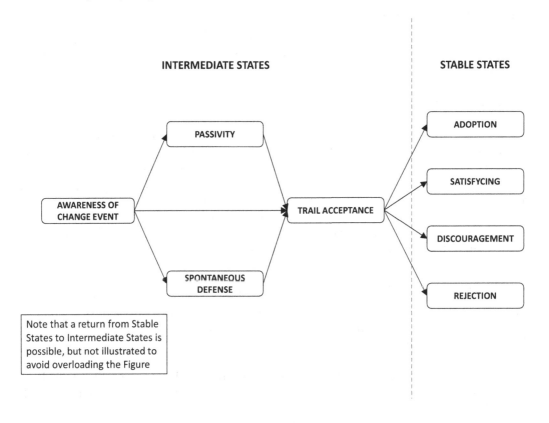

FIGURE 12.2: The state space framework.

even if a sustained adaptation effort is required, the benefits will outweigh the costs. They will refrain from the immediate and permanent adoption of the technology, so as to see if the supposed advantages materialize and if the barriers to adoption are as expected.

Others will exhibit *Passivity.* They feel that the new technology will not affect them. It may be that their work, as they see it, will not be touched by the new technology. Others expect to be transferred or to retire before the technology is implemented or are distracted in their work by issues in their private lives. These participants may move to *Trial Acceptance,* but perhaps not all of them will. Some, as a result of further thought, discussions with others, or due to new information received, might move to the intermediate state, *Spontaneous Defense,* and refuse to accept the new technology.

Spontaneous Defense is a state of aggressive and vocal objection to the technological implementation. Signs of this behavior may appear immediately following the announcement of the project. There are a number of possible causes for the adoption of this behavior. Experienced users who are adept and efficient with whatever legacy system is in place may feel downgraded by a new technology which will allow any new user to perform at their level. They see the new technology as part of a "de-skilling" process that devalues the work they perform. Others feel that the new technology is too difficult to master and that they will be left behind, or that mastering the new technology will require a considerable effort for which they will not be compensated.

There are also instances where vigorous objections to the proposed implementation are based on objective reality. The new technology may not, in fact, fit the job, and some users may fear that they will be obliged to invent a work-around and attempt to use both new and legacy technologies in parallel.

Possible transitions among the intermediate states that are not illustrated include passages from *Passivity* directly to *Discouragement* or *Rejection.* Where a user moves from *Trial Acceptance* to *Defense,* a good first impression may be shattered as a deeper understanding of the technology is gained through experience with its use as in Case Two of the Lapointe and Rivard (2005) report.

The four final states are:

- *Adoption* – A state describing users who have mastered the new technology and use it to the best of their ability
- *Satisfying* – A state describing users who have a basic knowledge of the new technology but who use it without enthusiasm and without actively seeking to improve their performance. They "get by"
- *Discouragement* – A state describing users who have been unable to master the new technology and who, at least for the moment, do not feel that they will be able to do so
- *Rejection* – Some users will definitely reject the technology and will seek to have it withdrawn, sometimes through the governance structures of the organization (committees, grievances, technical analyses) and sometimes outside of them (attempts to influence decision-makers). Others may use the technology but will vocally find fault or object to it at every opportunity and blame it for delays and errors

The state space framework of reactions to the implementation of new technology offers a richer representation of the sequence of emotions that a user may experience than the adaptations of the Kübler-Ross model, which was developed for a very different set of circumstances. It will be useful as a lens through which the reactions of users may be observed and interpreted. Rather than rely only on a common communications approach to helping users through the implementation of new technology, it encourages the analysis of users' behaviors and allows each individual to be situated within the state space framework. An individualized approach may then be crafted to address the concerns of the users and to better identify those unable or unwilling to embrace the change.

12.7.1 Motivating Users to Move to Acceptance and Adoption

There are actions that the implementation team can adopt to encourage users to adapt to the changes required by the proposed implementation. Stone, Deci, and Ryan (2009) offer a guide to actions that favor the reinforcement of a sense of competence, relatedness, and autonomy among users. They include the following list of actions in their recommendations:

- *Invite active user participation* in the implementation of change by asking users open questions
- *Avoid direct confrontation with users* who express emotional reactions to the implementation, for example, by formulating questions that invite clarification of their position
- *Offer choices if possible* (e.g., software systems, packages or platforms, team composition, implementation schedule) and clarify the roles and responsibilities expected of users
- *Provide positive praise* of specific contributions to the project but avoid routine formulaic praise which will be seen as insincere
- *Minimize coercive controls*, such as the use of compensation and benefits as a motivational strategy, as this may actually be counter-productive
- *Go beyond formal training* to develop talent and promote the learning of new skills and the sharing of knowledge. User groups, internet-based communities of practice, or professional and scientific societies may be useful forums

---

12.8 Anticipating Negative Reactions to the Implementation of a New Model

12.8.1 Use the STP Framework to Understand the Change

Radical technological change projects can present greater difficulties than more routine implementations, no matter how well they are designed and how great their benefits. If implementation of the analytics model and associated IT system also requires changes to the task process or to the user's job, this must be taken into account at the earliest opportunity. While the project plan may have no activity or resources for this analysis, if it is set aside, project risk will increase. The decision-makers may hesitate to engage in a revision of the task process or the users' ways of work if the analytics model is seen as "simply" an update to whatever legacy system is in place, but if the change is radical, these steps will indeed be required. From the analytics point of view, the change is most often centered on the system models and algorithms but may involve modifications to the task, in the tradition of industrial engineering. Will the changes to one of these elements be great enough to require changes in the other two?

Using the STP framework as a guide, one may ask if a proposed system implementation affects the power relationships (Markus, 1983), the perceived equity balance (Joshi, 1991), or the stress and fear level (Marakas & Hornik, 1996) of the users or if their previous experience with technology (Martinko, Henry & Zmud, 1996) may lead them to a negative reaction to the change. Answering these questions will require careful observation and interpretation of the ways users interact with the new technology, the task as currently defined, and the attitudes and intentions of decision-makers who are involved.

12.8.2   Use the State Space Framework to Observe Stakeholders

Most analysts have little or no formal training in the so-called soft-skills for dealing with human behavior. However, the implementation of a new technology cries out for such skills. The analyst can use the State Space framework to note current thinking of various users, follow the evolution of their points of view, and detect at the earliest opportunity signs of negative reactions. Typical mitigating actions are improved communication with stakeholders, changes to the project timetable and, if necessary, to project content and even to the group of users. This activity will remain ongoing throughout the implementation period and will require careful observation of stakeholders' verbal statements, work habits, and non-verbal communication. It also demands that the analyst develop the self-awareness and humility required to ensure that claims of user "resistance" are not used to mask the effects of shortcomings of the new system.

12.9   Summary

The STP triangle reminds us to ensure that the model and associated IT system are tuned to the needs and skills of the users and that both address the task in an efficient and effective manner. The state space framework provides a useful tool for following the progress of users' points of view concerning the implementation, while the conceptual models cited (Markus, 1983; Joshi, 1991; Marakas & Hornik, 1996; Martinko, Henry & Zmud, 1996; Lapointe & Rivard, 2005) suggest pertinent axes for observations.

In the analysis of implementation, the analyst should remember Ford, Ford, and D'Amelio (2008) who suggest that if there is "resistance," it may be a reasonable reaction to a flawed system or even a way for change agents to explain away their own management failings. Change agents should be careful and open in their interpretations of users' actions and be willing to negotiate accommodations. The following points summarize the point of view of Ford, Ford, and D'Amelio:

- In a radical change, old ways of work and agreements will be cast aside, perhaps creating *violations of trust* or perceptions of such violations. A radical change should, therefore, be managed as a dialog that is not driven by a rigid timetable
- Change agents must develop *compelling arguments* in favor of proposed changes (legitimization) to avoid the early development of strong counterarguments that are harder to rebut once enunciated

- Change agents should call for a *concrete first step*, such as attendance at a training course, system demonstration, or site visit. If there is no call to action, users may conclude that there will be no change
- If change agents describe only favorable outcomes, they will provoke a strong reaction should a hidden difficulty actually arise. The analyst can use the state space framework

References

Brown, G. G., DeGrange, W. C., Price, W. L., & Rowe, A. A. (2017, December). Scheduling Combat Logistics Force Replenishments at Sea for the US Navy. *Naval Research Logistics (NRL)*, *64*(8), 677–693. doi: 10.1002/nav.21780

Campion, M., & Medsker, G. (1992). Job Design. In G. Salvendy (Ed.), *Handbook of Industrial Engineering* (2nd Edition ed., pp. 845–881). New York: Wiley.

Cokins, G. (2012, February). Obstacle Course for Analytics. *ORMS-Today*.

Ford, J. D., Ford, L. W., & D'Amelio, A. (2008). Resistance to Change: the Rest of the Story. *Academy of Management Review, 33*(2), 362–377.

Joshi, K. (1991, June). A Model of Users' Perspective on Change: the Case of Information Systems Technology Implementation. *MIS Quarterly*, Vol 15, Issue 2, 229–240.

Kübler-Ross, E. (1969). *On Death and Dying*. Routledge.

Lapointe, L., & Rivard, S. (2005, September). A Multilevel Model of Resistance to Information Technology Implementation. *MIS Quarterly, 29*(3), 461–491.

Lapointe, L., & Rivard, S. (2006). Getting Physicians to Accept New Information Technology: Insights from Case Studies. *Canadian Medical Association Journal (CMAJ)* Vol 174, Issue 11, 1573–1578.

Leclerc, L., & Thiboutot, S. (2003, June). A.MAZE Routes & Zones: Transportation Management System Offers Fleet Scheduling Solutions for Large Organizations. *OR/MS Today, 30*.

Marakas, G., & Hornik, S. (1996, September). Passive Resistance Misuse: Overt Support and Covert Recalcitrance in IS Implementation. *European Journal of Information Systems*, Vol 5, Issue 3, 208–220.

Markus, G. (1983, June). Power, Politics, and MIS Implementation. *Communications of the ACM*, Vol 26, Issue 6, 430–444.

Martinko, M., Henry, J., & Zmud, R. (1996). An Attributional Explanation of Individual Resistance to the Introduction of Information Technologies in the Workplace. *Behaviour & Information Technology*, Vol 5, Issue 3, 313–330.

Meister, D. (2003). Conceptual Foundations of Human Factors Measurement. CRC Press.

Price, W. L., & Rousseau, R. (1994). Modeling of Processes and Tasks; RGTI Research Group, Faculty of Business Administration, Laval University, Quebec.

Shingo, S. (1989). A Study of the Toyota Production System from an Industrial Engineering Viewpoint (Produce What Is Needed, When It's Needed). Productivity Press.

Stone, D., Deci, E., & Ryan, R. (2009). Beyond Talk: Creating Autonomous Motivation Through Self-Determination Theory. *Journal of General Management, 34*, 75–91.

Chapter 13

From BOGSAT to TurboTeam: Collaboration for National Security Teams in the Age of Analytics

F. Freeman Marvin

13.1 From BOGSAT to TurboTeam: Analytic Collaboration Comes of Age

The late Senator John McCain did not like meetings. He thought most meetings were unfocused, led to unnecessary argument instead of healthy debate, and usually failed to resolve the issues that were the reason for the meeting in the first place (Kostman, 2018). Many analysts in the national security community have had the same experiences in team meetings.

The key to better teamwork is better team meetings, and the key to better team meetings is a process facilitator or a team leader with the skills of a facilitator (Schwarz, 1994). The role of a facilitator is to ensure that the meeting has clear objectives and an agenda, and to keep the discussion on track, manage conflicts, and record action items. The goal is to prevent a BOGSAT – Bunch of Guys/Gals Sitting Around a Table. As Senator McCain observed, meetings often end with little to show for the time and money invested. Teams supported by a process facilitator stand a much better chance of achieving meaningful results. The facilitator focuses on the meeting process, so team members are able to focus on the issue or problem at hand.

However, in the past decade, traditional process facilitators have reached the limit of their ability to improve collaboration in team meetings. One reason is that the role of meetings in organizations has evolved from being mostly about information exchange and discussion to a focus on information analysis and decision-making. The current flat structure and fast pace in many organizations also require process facilitators who can take a leadership role during team meetings. Access to more data and information, new analytic software tools, and remote conferencing technology are clearly challenging the traditional "soft" skills of facilitators.

Today, teams working in national security organizations are more likely to be cross-functional, multi-disciplinary, and inter-agency than ever before. Collaborative teams work on increasingly complex issues that require problems to be identified, data organized and analyzed, decisions implemented, and results evaluated without a lot of wasted time. Meetings are now the place where issues are resolved – not simply debated, where consensus is built – not just put off to the next meeting, and where commitment to action by all participants is critical to success.

In 2005, two experienced national intelligence analysts, Dick Heuer and Randy Pherson, assembled a set of thinking aids they called Structured Analytic Techniques (SATs) to improve the problem-solving and decision-making of facilitated teams (Heuer and Pherson, 2008). SATs are mostly whiteboard exercises that help teams frame problems, brainstorm ideas, and mitigate cognitive biases. Heuer and Pherson developed a few new, innovative SATs, such as Analysis of Competing Hypotheses, or ACH (Heuer, 1999), but most of their SATs had been around for a while and had already been adopted by many national security teams. For example, SATs to organize information and frame problems included Affinity Diagrams and Mind Maps. Techniques to encourage contrarian thinking included Devil's Advocate and Red Hat analysis, and techniques for analysis, diagnosis, and prediction included Scenario Analysis, Role Playing, and SWOT. These techniques are notable in that they do not require computers to implement, and, in fact, most do not involve any calculations.

However, when national security teams are faced with complex problems or tough decisions to make during a meeting, they may be unable to reach a resolution using SATs alone. Recent experiences with a range of national security and public safety teams have shown that there are new ways to combine process facilitation, analytic techniques, and computer software tools right in the meeting room to improve the quality of team collaboration and to do it in near real-time.

Analytic Collaboration involves building on the foundation of process facilitation skills and structured analytic techniques with Simple Analytic Models (SAMs). SAMs are computer models that can be constructed right in the team meeting room when the complexity of a problem exceeds the ability of team members to keep track of all the moving parts in their heads or with a whiteboard SAT (Vennix, 1996). SAMs allow a team to change variables, test possible relationships, and predict outcomes. SAMs

TABLE 13.1: Simple Analytic Models

Task	Simple Analytic Models
Deciding among alternatives	Value Trees
Deciding among courses of action	Strategy Trees
Diagnosing the chances of a failure	Fault Trees
Diagnosing the causes of a failure	Bayesian Networks
Designing a process	Queueing Networks
Designing a portfolio	Portfolio Models

are transparent and can be explained by the team members themselves because they helped to create them. SAMs also help to communicate results to senior leadership and other stakeholders. SAMs provide a collaboration bridge to move teams from intuition to analysis and from debate to action.

Facilitators and team leaders acting as facilitators need to know how to use analytic collaboration tools to turn SATs – diagrams, charts, and matrices – into SAMs. This means that facilitators need to learn some basic modeling skills and some computer software tools, but they do not need to learn how to code. Most SAMs can be built in an Excel spreadsheet or inexpensive propriety software.

This chapter describes how six national security and public safety teams used Analytic Collaboration and SAMs to move from BOGSAT to TurboTeam (Reagan-Cirincione, 1994). The examples span three common team tasks: Deciding, Diagnosing, and Designing. The table above summarizes these examples.

The hope is that this chapter gets your mental wheels turning, prompts you to learn more about Analytic Collaboration, and leads you to discover new ways to improve your own organization's performance. As one team leader exclaimed, "Analytic collaboration enables collective action!"

13.2 Deciding

One common challenge faced by national security teams is choosing among alternatives or different courses of action. Making a decision among alternatives, such as the best weapon system to buy or where to build a facility, could involve assessing and weighing many competing factors. Choices among different courses of action may involve two or more decisions, where the structure, sequence, and timing of the decisions are important to determining the best strategy and where significant uncertainty exists. Teams facing choices of both types can improve their chances of success by moving beyond SATs to SAMs.

13.2.1 From Pros and Cons to Value Trees

The US Food and Drug Administration (FDA) evaluates and approves medical treatment devices as part of its mission to protect the health and safety of the public. The FDA approval process had come under scrutiny after problems with several approved medical devices were reported in the media. A study by the National Academy of Sciences found that, "approvals of devices were inconsistent, [and there was] a perceived lack of transparency in decision making" (Wizemann, 2010).

The FDA decided to develop a new approval process – one that took the uniqueness of each device into account, but was flexible and transparent. The agency chose a class of therapeutic devices that treat mental depression as a test case. A team of clinical psychiatrists and regulatory specialists was formed to recommend a new evaluation and approval process. The team leader wondered how they would be able to find an alternative to the traditional approval process that could optimally protect patients and still promote innovation in support of public health. Just as importantly, she wondered how to provide an easily understood and defensible rationale for each approval decision to convey to the companies submitting devices for approval and the medical and patient advocacy groups.

13.2.1.1   Ben Franklin Weighs the Pros and Cons

The FDA looked at using a variation of a structured technique used by decision makers for over 200 years.

In 1772, Benjamin Franklin offered a method of pros and cons to his friend and fellow scientist, Joseph Priestly, to help him made a tough career choice. Franklin wrote, 'I cannot for want of sufficient Premises, advise you what to determine, but if you please, I will tell you how. Divide half a Sheet of Paper by a Line into two Columns, writing over the one Pro, and over the other Con. Then during three or four Days Consideration, put down under the different Heads short Hints of the different Motives that at different Times occur to you for or against the Measure. When you have thus got them all together in one View, endeavour to estimate their respective Weights; and where you find two, one on each side, that seem equal, strike them both out: If you find a Reason pro equal to some two Reasons con, strike out the three. If you judge some two Reasons con equal to some three Reasons pro, strike out the five; and thus proceeding you will find at length where the Ballance lies. And tho' the Weight of Reasons cannot be taken with the Precision of Algebraic Quantities, yet when each is thus considered separately and comparatively, and the whole lies before you, I think I can judge better, and am less likely to make a rash Step; and in fact I have found great Advantage from this kind of Equation, in what may be called Moral or Prudential Algebra. Yours most affectionately, B. Franklin'

(Jones, 1995).

The approach considered by the FDA team resembled Franklin's technique – a type of Pro/Con checklist to answer the question, "Should the FDA approve medical device A to treat disease X?" For example, a pro might be that device A addresses a severe form of disease X that, if left untreated, could cause death. A con might be that device A is only intended to treat the most mildly affected patients.

Using the Pro/Con checklist, an evaluator assigns one or more pluses or minuses to each pro and con (Quinlivan-Hall & Renner, 1990). Evaluators then add up the pluses and minuses. One plus cancels out a minus. If the evaluator considers the plus very important, a plus might cancel out two minuses. If the evaluator is cautious, it might take two pluses to cancel out a minus. Each factor can also be weighted, and the pluses and minuses multiplied by the factor weights. Although efficient and flexible, it was unlikely this technique could produce a defensible approval process for the FDA.

13.2.1.2   Planting a Value Tree

The FDA team moved beyond its initial approach by using a simple analytic model called a value tree, which is based on multiple objective decision analysis (MODA). MODA models allow a team to include additional considerations, or criteria, beyond the attributes of current alternatives, when making choices. Some of the criteria may conflict with each other, such as quality vs. cost, setting up a tradeoff. Criteria may also include features that do not show up in any of the current options, leading to a search for better alternatives. To build a value tree, the FDA team extended the Pro/Con checklist in three ways.

First, the team reframed its problem from how to approve a particular medical device to how to define the goals and objectives of the FDA that would apply to any device. This approach to making choices is called Value Focused Thinking (VFT) (Keeney, 1992). Instead of beginning with specific alternatives, the team reversed the decision process by first developing a list of the goals and objectives that were important to the FDA using the Nominal Group Technique (NGT), a useful facilitation technique that lets each team member contribute equally to the discussion (Bens, 1997).

The team organized the criteria using a SAT, an Affinity Diagram, and turned the list into a SAM, a value tree. In a value tree, the objectives flow from the top down, creating a taxonomy of criteria and sub-criteria. The team first thought about all the criteria that should be in any new device. For example, a pro such as "the device should address severe forms of the disease that, if left untreated, could cause death," was called "Severity of Disease." A con such as, "the device does not work on severely depressed patients," was called "Patient Population."

Next, the FDA team clarified what distinguished a good medical device from a poor one by creating a measurement scale for each criterion. The measurement scale, or value function, shows how each additional increment of "goodness" for a criterion provides slightly less marginal value, or diminishing returns to scale. The team defined end points at the top and bottom of each scale to show the best and worst possible levels.

Finally, the FDA team assessed the relative priority of each criterion, accounting for both the importance of the factor and the range, or swing weight, from the bottom to the top of its measurement scale. Swing weights account for both the range variation in the measurement scale and the relative importance of the criterion. A common mistake in value tree models is to only capture the importance of the criterion and ignore the significance of the range variation (Hammond, Keeney & Raiffa, 1999).

The FDA team used a simple elicitation technique called "Allocate one hundred coins" for weighting the value tree. For example, the criterion Safety was very important, but the scale on Safety only ranged from Safe to Very Safe. Since the difference between the top and bottom of the scale was not significant, Safety received a relatively low swing weight and therefore would not have much impact on the final approval. All devices had to be safe to be considered for approval. Value trees can easily be built in Microsoft Excel (Kirkwood, 1997). There are also many affordable value tree packages such as Logical Decisions for Windows (n.d., Retrieved from https://www.logicaldecisions.com).

By adding up the weighted scores on the criteria, any medical device for treating depression could be rated using the team's value tree. Competing medical devices are assessed against the same standards. Since the evaluation criteria are carefully defined and documented, there is maximum transparency. Using value trees, the FDA team produced an approval process that could be defended to senior leadership, other stakeholders, and to the public.

13.2.2 From Roadmaps to Strategy Trees

The Tech Base Program (TBP) team at the Army's Chemical and Biological Defense Command (CBDCOM) located at Edgewood Arsenal, Maryland, was responsible for creating a strategy for funding research and development to improve the military's defense against chemical and biological weapons. Since the Army was the executive agent for acquiring all chemical and biological defense equipment for the Department of Defense, the TBP team's work had a large impact on future US defensive capabilities.

Each year, CBDCOM published a set of requirements and requested proposals for new technologies that could meet future needs. These proposals were binned into four business areas: detection, individual protection, collective protection, and decontamination. After proposals were selected one at a time for funding, a technology roadmap would be developed for each business area. The purpose of a technology roadmap was to provide an overall strategy for the projects within a business area, facilitating a more comprehensive and integrated investment program.

13.2.2.1 Highway to the Danger Zone

The team began with a literal interpretation of a roadmap – a line for each project with a start date and an end date. Roadmaps were based on a timeline, in a Gantt chart format, which did not account for any uncertainties in the projects. This meant that a technology roadmap could easily lead down the wrong path if conditions changed or some projects failed.

There were a number of other concerning issues with the approach: stakeholders' perception of a closed process, no consensus on the allocation of resources, and no way to compare the risks associated with each project. The program also faced dwindling dollars and constraints imposed by Joint Service regulations. The technology projects were found to have significant duplication and gaps. In addition, there were concerns about integrating among different business areas. Was there a better way to develop technology roadmaps?

13.2.2.2 Swinging from the Strategy Trees

The TBP team decided to take a new approach that could help create better strategies in an uncertain environment. For each business area, the team built a value tree from a set of strategic goals and objectives defined by senior leadership. It then replaced the Gantt chart scheduling technique with a strategy tree. A strategy tree is a simple analytic model that links the series of decisions that must be made, and the structure, sequence, and timing of each decision to show the potential consequences. Strategy trees have been used for many years in business to explore the expected return on business investments. This SAM is sometimes called a "multi-stage decision model under conditions of uncertainty" or a decision tree (Clemen, 1996). A strategy tree combines the organization's goals and objectives, uncertain events and conditions that may change, and the decisions the organization can make into a SAM to test alternative strategies. The TBP team took the strategy tree technique and added the ability to calculate the uncertainty in each proposed R&D effort to assess the probability of an overall successful outcome. The set of R&D projects taken together provided a realistic picture of TBP investments.

A technology roadmap using strategy trees is a flexible planning model to support strategic and long-range planning, by matching short-term and long-term goals to specific technology solutions. It is an approach that applies to a new product or process and may include using technology forecasting or technology scouting to identify suitable emerging technologies. It is a good way to help manage the risky front-end of innovation efforts. It can also help a program survive in turbulent budget environments by showing viable, but cheaper, paths to success. The team used one of the excellent software tools for making strategy trees called DPL from Syncopation (n.d., retrieved from https://www.syncopation.com).

The SAM used by the TBP team provided three critical advantages. First, the team was able to use the flexibility of the strategy tree to mirror the familiar roadmap structure – the list of decisions that must be made, the options the team has (decision nodes), the things that are uncertain (nature nodes or chance nodes), and the things the team is trying to achieve (value or utility nodes). A value tree shows the relative value of each endpoint or path outcome in the roadmap using a common set of goals and objectives.

Second, the team constructed better options for the sequence of projects. A decision is defined as an "irrevocable commitment of resources." The team identified the key decision points on the roadmap and then was able to make new strategies by simply rearranging the sequence of the decisions and assessing the new outcomes.

Finally, the team could adjust the timing of the funding decisions. The separation in time between decisions can have a big impact on the value of the outcome. The TBP team was able to slide the decision points, future events, and the path outcomes back and forth, left and right on the timeline to evaluate the impact of the timing of program decisions.

13.3 Diagnosing

A second area of team problem-solving and decision-making involves predicting the likelihood of different causes and effects in a system. Diagnostics are important in many organizations, from medical teams to maintenance crews. A team uses inferential reasoning when it confronts a problem or a symptom and is tasked to diagnose the potential causes. A team uses deductive reasoning when it begins with a given cause – a disease or component failure – and must predict the potential effects. Diagnostic SAMs can help a team do both types of reasoning.

13.3.1 From Fishbones to Fault Trees

Commercial aviation safety, both in the air and on the ground, is of critical importance to the economy. The Federal Aviation Administration (FAA) had identified a number of potential accident sequences it wished to evaluate. These accident sequences included problems during take-off such as aircraft system failures and air traffic control (ATC) issues, and emergency events such as fire onboard an aircraft, flight crew member incapacitation, and ice accretion on an aircraft in flight (Dillon-Merrill, 2015). A team of experienced air traffic controllers and aviation safety experts was formed to evaluate the likelihood of various accident sequences in order to develop preventive and mitigation measures.

13.3.1.1 Choking on Fishbones

The team began with 30 Event Sequence Diagrams (ESD) describing the sequence of events from an undesirable initiating event (cause) to its possible outcomes (effects).

This can be thought of as a deductive thinking approach. For example, a bird strike could cause an aircraft engine to fail, shatter control surfaces, or cause no damage at all. The FAA team needed a way to evaluate the reverse question: what are all the ways that an engine could fail, a bird strike being just one way?

The team first looked at using root cause analysis (RCA). RCA is a technique of problem-solving used to identify the root causes of faults or problems – an "effect-to-cause" approach (Bens, 1997). A factor is considered a "root cause" if removal from the problem-fault-sequence prevents the final undesirable outcome from recurring, whereas a "causal factor" is one that affects an event's outcome, but is not a root cause.

Of course, a problem such as an engine failure could have many possible root causes. In RCA, each potential causal factor is traced back to find the root cause, often using a SAT called a Fishbone Diagram, or Ishikawa Diagram. Fishbone Diagrams allow the team to work backward from an "end event" to find the "root causes" of a problem. The usual way that a team works is to ask the "Five Whys" (Bens, 1997). The goal of the technique is to determine the root cause of a defect or problem by repeating the question: why? Each answer forms the basis of the next causal factor. The possible causes are grouped into categories on the main branches off the spine of the Fishbone Diagram. There is an implicit timeline on the Fishbone, where causal factors to the left must occur or be in place before causal factors to the right.

The FAA team ran up against the major problem with Fishbone Diagrams – how to determine which of the potential root causes were enough to cause an event by itself or which could do so only in combination with other causal factors.

13.3.1.2 Finding Fault Trees

The FAA team moved beyond Fishbone Diagrams and developed fault trees for each end event in the ESDs. A fault tree uses logic gates to identify the various combinations of failures, or fault events, unexpected errors, and normal events involved in causing a specified undesired event to occur. A fault tree works from effect-to-cause. For example, an engine failure could be caused by a bird strike, a fuel blockage, or a mechanical defect. The fault tree is a SAM that can compute failure probabilities and the relative importance of system components.

In a fault tree, the undesired outcome is taken as the top event. For example, the undesired outcome of an aircraft operation is an engine failure. Working backward from this top event, the team might determine there are two ways this could happen: a bird strike or a maintenance problem. Here the team would place a gate. The gates in a fault tree work in different ways. In an OR gate, the output event occurs if any of the input events occur. In an AND gate, the output occurs only if all inputs occur (assuming inputs are independent). There are several other types of gates with more specific purposes. The team might identify a design improvement by requiring a backup system in case of bird strikes – this is a safety feature that would change the gate to a logical AND. When fault trees are assigned event probabilities, the SAM can calculate overall failure probabilities. The FAA team also found that some common causes can introduce dependency relationships between events in the ESDs, which would have been difficult to see in a Fishbone SAT.

Fault trees provided three important advantages for the team. First, the FAA team was able to use fault trees to re-organize the ESDs into a more logical hierarchical structure. As the team transferred the causal factors to the fault trees, the team identified the critical AND/OR relationships at each branch.

Second, the team was able to assess the relative probabilities at each branch. The team ensured that the probabilities of OR branches added to one and AND relationships were mutually exclusive. If two AND causes depended on each other, the team combined them into a single causal factor. The fault tree calculated the overall probability of each end event for comparison with other events to see where cost-effective preventive and mitigation measures should be taken.

Third, the team was able to use fault trees to identify the minimal cut sets. Cut sets are all the combination of events that could cause the system to fail. A minimal cut set is a cut set, such that if any event is removed from it, the top event will not necessarily occur, even if all remaining events in the cut set occur. Sorting a cut set by cost allows the team to focus on the "cheapest" failure points; that is, the points of failure that would be most likely to naturally fail or that a potential adversary could most easily attack. The team used a software tool called DPL Fault Tree (Syncopation, 2004) that provides a step-by-step template for building the fault tree and finding the minimum cut sets.

Using the rigorous and structured methodology of fault tree SAMs allowed the FAA team to model the various combinations of fault events that could cause system failures to occur and the relative likelihood of each.

13.3.2 From Causal Diagrams to Bayesian Networks

The Army Communications and Electronics Command (CECOM) had a deadly problem. Army troops fighting in Afghanistan and Iraq had to set up small, forward operating bases in remote locations, often in rugged, mountainous terrain. Their only source of electric power at these bases was from the diesel-fueled generators they hauled with them. But when a generator broke down, the soldiers had to radio for a generator mechanic to fly by helicopter out to the forward base, diagnose the problem and then fly back, a dangerous and costly round trip. Was there a way that an average soldier in the field could perform the diagnostic procedure on a faulty generator?

The Army formed a team of generator mechanics and logistics experts to find a solution to this problem – the Virtual Logistics Assistance Representative (VLAR) team. The VLAR team needed to gather and codify expert knowledge about diesel generator operations and apply that knowledge to troubleshooting the equipment in remote combat zones (Aebischer, 2016).

13.3.2.1 Colliders and Confounders

The primary technique for aiding soldier mechanics who are troubleshooting problems in the field has been the technical manual. These can include simple schematics, such as wiring diagrams, or very comprehensive if-then logic diagrams. For example, if the engine runs erratic, then the fuel could be dirty, and if there is dirty fuel, then there could be water in the fuel filter. However, for complex mechanical equipment, the number of possible combinations of potential causes and failures modes becomes unmanageable with a rule-based diagnostic approach.

Other diagnostic techniques, such as ACH, take advantage of probabilistic inference. The effects of a problem, or observable symptoms, are compared with hypotheses of possible causes to infer the most likely cause. But these SATs have no way to handle two common types of cause-effect situations: colliders and confounders. A collider is a factor which can be caused by two or more other factors, but which can impact those factors "in reverse." In other words, colliders have a common effect on their causal factors. For example, an engine could be caused to run erratically by dirty fuel or by a leaky fuel line. If the soldier knows the fuel is dirty, that fact has no impact on the probability that the line is leaking. However, if the soldier knows from observation that the engine is running erratically, then finding out that the fuel line is OK will increase the probability that the fuel is dirty. This is called explaining away and makes diagnosis by causal diagram complicated.

A confounder is an unknown factor that lurks in the background in a complex causal system and may impact other factors without being explicitly accounted for. For example, a technical manual's causal diagram may not mention that the altitude at which the generator is being used has an impact on its performance. Confounders are found in many situations and can make correct diagnosis of problems nearly impossible by untrained soldiers without a simple-to-use tool that could be built into an automated onboard system.

13.3.2.2 Diagnosis with Bayesian Networks

Adopting a Bayesian network model helped the team move beyond the limitations of causal diagrams and other structured techniques. Bayesian networks are SAMs based on Bayes' Rule that show how effects are inferred to be caused by various conditions or events, and how a fix, or an intervention, will impact the problem (Fenton, 2012). Judea Pearl, professor at UCLA, created the first practical Bayesian network algorithms in the 1980s, and since then they have become easy to apply to a wide range of problems. A Bayesian network looks very much like a causal diagram with nodes connected by arcs to indicate the causal direction.

Bayesian networks provided the VLAR team with three critical advantages. First, the team built the structure of the problem in a format that was familiar to subject matter experts to help them make their assessments. A Bayesian network looks like a causal diagram that establishes the visual framework – a structure – within which to analyze a cause-and-effect system.

Next, the team built a Conditional Probability Table (CPT) for each node in the model. A CPT shows the conditional probabilities of the cause-effect relationship. A true positive is the probability that the presence of a certain factor causes the effect. A true negative is the probability that the effect will not happen in the absence of the factor.

The VLAR team used a new facilitated approach to build the Bayesian network called the DSEV process (Tatman et al., 2015). DSEV stands for define, structure, elicit, and verify. The team carefully defined all the terms in the problem. This prevented confusion about what the team was actually assessing. The initial structure of the model was built in small sections of cause and effect nodes. Before moving on to the next section, the facilitator elicited from the team the confusion matrix for each node. The section of the model was exercised and verified that it behaved the way the team intended it to act. Finally, the team used the Bayesian network to identify the "colliders and confounders" in the generator system. This involves exercising the model over a variety of fault conditions.

Through 2015, the VLAR team saved the Army millions of dollars in direct labor costs and prevented many casualties by reducing requirements for helicopter and ground-convoy movements. The VLAR team used BayesiaLab by Bayesia (Bayesia.com) to create the Bayesian networks. Another tool suitable for smaller organizations is Netica by Norsys (Norsys.com).

Based on the success of the VLAR team, the Army decided to expand the use of the DSEV process and Bayesian network SAMs to build diagnostic tools for all CECOM field equipment (CECOM, 2016).

13.4 Designing

Teams today not only have to make tough choices and evaluate uncertain situations, they may have to design a new system, process, or program. Design teams can improve their designs by using various SATs, but TurboTeams take design thinking to the next level by creating and testing a prototype. A prototype helps gather feedback on the performance of a design in a risk-free environment. Prototypes speed up the process of innovation because they allow a team to understand the strengths and weaknesses of new ideas by "failing many times, quickly and cheaply, in order to reach success." The following sections describe two TurboTeams who used SAMs to improve their designs.

13.4.1 From Flowcharts to Queueing Networks

Three years before the terrorist attacks on 9/11, the US Congress funded the National Domestic Preparedness Program in anticipation of rising international terrorism. The most visible portion of the program was the training and equipping of emergency medical personnel in 120 cities across the country to respond to an attack by terrorists using biological or chemical weapons. The program manager needed to develop common emergency response templates so that when a city was attacked, medical responders from other cities could quickly move in to assist.

One of the proposed response templates was for a Neighborhood Emergency Help Center (NEHC) which could be set up in a school or other local building to provide emergency medical triage for people who thought they might be infected following a biological attack. An NEHC design team was formed with emergency managers, epidemiologists, physicians, nurses, and first responders from five different cities.

13.4.1.1 When Flows Back Up

The NEHC design team began by creating a process flowchart. A flowchart is a logic diagram, often used in software programming that shows the various paths that an item can take based on decisions made at each step in the process. The NEHC flowchart was a map of how people would arrive at the facility, get routed to different stations where they would be processed, and either sent home with medication, held for observation, or immediately rushed to an acute care center.

The flowchart represented the steps as blocks connected with arrows. However, it soon became apparent that team members had several different concepts for how patients should be routed. One process would move people quickly through the facility with minimal treatment in order to handle a large volume of patients. Another option focused on taking lifesaving actions that might prevent more deaths, but risked being overwhelmed by the large number of people expected to seek help.

The team modified the flowchart several times to try to visualize each option. But the team had no way to compare which one was better, or even to assess whether any of the designs would work at all. It was hard to know if people would trickle in throughout the day or descend on the facility all at once. The various kinds of biological agents also had very different effects on people. Some agents produced a massive number of fatalities quickly, while others created large populations of sick people who needed medical attention over a period of time in order to survive. Some data existed on how long it takes to triage and treat patients in hospital emergency rooms during flu season, but for treating most biological agents, the NEHC team would have to take a guess. Most importantly, if the number of medical personnel assigned to operate one NEHC was too big, fewer NEHC facilities could be set up during a crisis while some doctors and nurses might be underutilized. If the staff was too small, patients might be waiting in long lines for medical attention. Which approach was best?

Flowchart SATs are an excellent way to help a team begin to think through alternative designs of a system. Sometimes a flowchart can be used to show the flows of a continuous quantity, like water flowing from a faucet into a bathtub (stock and flow). A flowchart can also try to show the movement of discrete items, such as people shopping in a grocery store and waiting in line at the checkout counter. But a flowchart is only a static representation of the structure of a system and is not able to show what happens when the bathtub starts to overflow or when the checkout line extends out the door and down the street.

13.4.1.2 Jumping into Test Beds

The NEHC team chose a simple analytic model called a queueing network (Cochran, 2019) that would help it explore and validate the alternative designs. Using a laptop with the model projected onto a projection screen in the meeting room, the team was able to drag and drop icons representing various work stations onto a blank canvas and connect them with arrows, replicating each alternative design concept. The team used the available data and expert judgment to assign probability distributions to the icons to define parameters like patient arrivals, service times, number of triage and treatment stations, and other quantitative data. When the team ran its first model, they were astounded by what they saw – a long line of sick people and worried well people extended out the door and down the street!

After several adjustments, patients were flowing smoothly through the facility, but an unrealistically large number of stations and medical staff was needed. Perhaps a different treatment protocol was needed. With the SAM, the team could continue to make changes to the design, in real time, collaborating on new ideas, and testing out their ideas quickly.

A simple analytic queueing network provided four critical advantages over a flowchart. First, the waiting lines the team saw when the initial model ran are not seen in a flowchart. How long could these queues grow before people turned around and left, or worse still, start to die? If the team set up fewer NEHCs in the city and added more staff to each facility to handle peak patient loads, would the staff be underutilized during low demand hours? In addition, when people see a queue, they tend to exhibit certain behaviors. People may decide not to stand in the line at all (balk), to stand for a while, but leave and come back later or not at all (renege), or if there are several lines, to switch from line to line to stay in the shortest (jockey). All these behaviors can greatly impact the number of doctors and other resources needed.

Second, a SAM allows the use of probability distributions to represent the uncertainty in a process, such as patent arrivals. In a flowchart, it is possible to assess *throughput* capability for a station by comparing the average input rate to the output rate. For example, if the team expected an average of ten incoming patients per hour, and the station took no longer than six minutes to triage each patient, it might reasonably assume that, on average, the center could throughput ten patients per hour. However, people do not usually arrive at a grocery store or an NEHC in regular intervals, but sometimes arrive in bunches, and sometimes no-one might come in for hours. Because of this uneven arrival distribution, only a SAM will show the queues that are likely to form.

Third, the queueing network showed the utilization of resources. In any system design, the team wants to acquire and operate just enough processing capacity to keep the flow flowing. A flowchart does not show how these servicing resources are utilized. How many stations should be open during different times of the day? With the SAM, the team could change the number of stations and see the impact on the length of the queue and the waiting times of patients. Not enough capacity, and things bog down. Too much capacity, and medical staff sits idle, wasting time and money.

Finally, the SAM shows how the system represented by the static flowchart will work under different scenarios or different biological agents. The team is able to hold its assumptions constant, while varying the parameters of interest and changing the process flow. This allows a consistent analysis of the problem, letting the team compare one design with another. The team can also perform sensitivity analysis, or a what-if analysis, to see the impact of changing assumptions. This is why a queueing network simulation SAM is a great test bed for a design team. The team used a discrete event simulation tool called ExtendSim (n.d., retrieved from https://www.extendsim.com). ExtendSim is one of several excellent desktop graphical simulation software packages, some with free versions for limited modeling.

The NEHC team was able to model the different design concepts in the meeting room with team members systematically testing each idea in near-real time. The simulations showed the team the impacts of each design on staffing requirements, casualty flows, and other key variables. This rapid, facilitated "build-test-build" approach takes advantage of the collaborative thinking of a multi-disciplinary team, making maximum use of their valuable and limited time, and allows them to "try before they buy" into a particular design concept. The models were then used to predict performance, to provide a baseline to compare with live test results, and to make a final recommendation for an NEHC template to the program manager.

13.4.2 From Knapsacks to Portfolios

Following the terrorist attacks on 9/11, Congress directed the Department of Health and Human Services (HHS) to buy and maintain a national stockpile of drugs and other medical countermeasures (MCM) that could be used to protect the country from a biological weapons attack by terrorists or a naturally occurring pandemic. HHS established the Strategic National Stockpile in 2005 and planned to upgrade and add to it in 2010 using a new program called *Project Bioshield*. An inter-agency Strategy and Implementation Plan Working Group (SIPWG) was formed to develop the acquisition strategy for *Project BioShield*. Deciding what drugs to buy to upgrade the stockpile proved to be a challenge. How had the biological threat evolved over time? What new drugs were under development that might be more effective than what was already in the stockpile? Would it be better to protect additional people with more of the older, cheaper, but less-effective drugs, or invest in more effective, but less available new drugs?

13.4.2.1 Hiking through the Desert

The SIPWG laid out a traditional strategic planning framework, with a mission statement, vision, goals, and objectives. Then the team leader created a matrix using an Excel spreadsheet. Across the top of the matrix, she made column headers for each expected biological threat such as anthrax, tularemia, and the plague, and a few chemical and radiological threats. Down the left side, she listed the current MCMs in the stockpile, such as vaccines, antitoxins, and antibiotics. Some MCMs could be used to address two or more threats, while a few MCMs were specific to one threat. In the cells of the matrix, she noted information about the amounts in the current stockpile, the effectiveness of the drugs for each threat, and the cost to buy more. The team printed this matrix out on a large wallchart and posted it in the SIPWG team room.

A general purpose SAT, such as a matrix, was a good place to start. It helped the team organize the available data and make connections among the different problem elements. The SIPWG members agreed that their mission was to "prioritize near, mid, and long-term goals for acquisition of medical counter-measures," but there was no consensus on a way forward. Someone suggested that they prioritize and sort the rows in the matrix – the list of current MCMs – from highest priority to lowest. Then they could simply go down the list, calculate the cost for adding more of each MCM, and propose to acquire MCMs in priority order until the BioShield funds ran out. In setting the priorities, another member suggested using a SAT called *Weighted Ranking* to assign a "risk" weight to each of the threats according to how likely and how severe they expected the threat to be. If some MCMs still had usable quantities in the stockpile, that amount could be subtracted from the quantity needed for new acquisition.

When weighted and sorted, the SIPWG noticed that by using this "cut line" approach, several of the more broadly applicable MCMs used up all the BioShield funds and left nothing to buy the many more "targeted" drugs needed to counter specific threats – some with significant risk weight. To counter this problem, another suggestion was made to simply buy a flat percentage of each MCM until the funds ran out, somewhere in the range of 30% of each requirement. This seemed like a fair and equitable allocation of the BioShield funds.

But neither of these methods accounted for potential new MCMs that were in development at various Technology Readiness Levels (TRLs) and might not be available for several years, or for the differences in medical efficacy of each MCM. Some scientists on the SIPWG had experience applying mathematical programming to maximize the return on investment for drug development projects. They called these "knapsack" problems because the idea is to fit as many items from a list as possible into a constrained space (like a hiker's knapsack) in order to maximize value to the organization. The SIPWG leads thought that this optimization approach, while being theoretically sound, might take too long and not be transparent enough to explain to senior leadership.

13.4.2.2 Learning to Play Football

The SATs applied by the SIPWG were a good beginning in framing the problem and organizing the data. However, when they had finished prioritizing the items on the list, they had no direction for where to go next. In fairness to the team, SATs for resource allocation are simply not powerful enough to be useful.

On the other hand, optimization modeling approaches to resource allocation can take too long, require data that are not available, or are not transparent to the team, stakeholders, or decision-makers. And most disturbing, when a budget constraint is tightened or relaxed, the mix of items funded in the optimum knapsack can change illogically. Items that were funded prior to an increase in budget, may get deleted and replaced by other items that previously were not funded!

The SIPWG selected a SAM called Portfolio Decision Analysis, also called Multiobjective Portfolio Analysis with Resource Constraints (Parnell, Bresnick, Tani & Johnson, 2013), or more simply, Marginal Benefit to Cost Analysis (MBCA). The idea behind this SAM is to produce an easy to understand "Order of Buy" that the SIPWG desired, but in a logically defensible way.

Portfolios are simply bundles of things that you want to buy. A bag of groceries is a portfolio. The idea is to look at all the combinations of food that a family might want to buy, including how much of each, and pack the shopping cart with small, medium, or large amounts of each that will give the best mix of food for the budget at the checkout counter. The shopper always wants to put in the cart the items that will give the overall best value, or the most "bang for the buck." If the shopper gets to the counter and finds that he or she is a little short on funds, an item is removed from the cart that gives the least bang for the buck.

The shopper can plot all combinations of food items on a diagram of cost vs. benefit. This is called the Pareto space, bounded on the top by the Pareto frontier. The frontier curve shows the set of portfolios that give the best "bang for the buck" – the best value for any given cost. The other points in the Pareto space are also feasible "bags of groceries," but just don't give us the best value for our money. The shape of the Pareto space looks like an American football – pointy on both ends and fat in the middle.

Working in weekly meetings over several months, the SIPWG first created a portfolio model structure to solve their acquisition strategy problem. The team used one of many inexpensive software tools called *LDW Portfolio*. First, the team extended their planning matrix by taking the MCM rows and adding any additional MCMs that were known to be in development. The team then created three to five potential funding levels for each row extending out from left to right. These were realistic levels that indicated the cost and relative benefit of buying different amounts of the MCM. The lowest levels (normally the status quo or baseline) were assigned zero marginal benefit, while the highest levels under consideration were assigned 100.

A bottom level score of zero did not mean there was no benefit. In fact, for MCMs that had quantities currently in the stockpile, there may have been significant benefit, but no marginal benefit, as the baseline was the lowest level. Then the benefit assessments took into account the special issue of when the MCM would be available to buy. For example, if the MCM would not be available to enter the stockpile for three years, an appropriate discount factor was applied to its benefit score.

Second, the team reviewed and refined the list of biological and other threats and their relative risks. These became the evaluation criteria and "across criteria weights" for the overall benefit of each MCM row.

Third, the team used a decision analysis weighting exercise to develop "within criteria weights" in order to normalize the 0–100 scales on each row for each criterion, since some rows had more levels than other rows. In addition, the team accounted for the variation in medical efficacy of each MCM against the threat criterion. For example, an MCM that was very effective against the anthrax criterion would get a large relative weight, while an MCM that had no effectiveness against anthrax would get a weight of zero on that criterion.

Fourth, the team did a through reviewing the order of buy and identifying the best mix of MCMs. New or existing MCM levels that provide a large improvement over the baseline and that are available earlier are preferred in the order of buy. The team then ran two models. The simpler model examined the strategy for each MCM of buying the required amount of the next generation product when it becomes available. The simplified model also assumed that a next generation MCM should receive 100% of its benefit regardless of when it becomes available. A more complex model examined several possible acquisition strategies for each MCM, including buying more of the current MCM, waiting for the next generation MCM before buying, or combinations of these two acquisition strategies.

The SIPWG finished with a workshop attended by the team members and a broad set of homeland security officials and other stakeholders. The purpose of the workshop was to gain consensus among the national security community on the priorities for meeting medical countermeasure preparedness and response goals using the decision framework identified by the team. They examined the key tradeoffs that emerged from the first discussion to reach consensus on the implementation plan. The workshop was supported by the portfolio model which integrated the alternative MCM strategies to help the SIPWG visualize key tradeoffs over the three timeframes and to help develop the best value MCM acquisition strategy.

13.5 Lessons Learned

Analytic Collaboration combines group process facilitation, problem-structuring techniques, and simple analytic models in a powerful combination to take any team through a hard decision or problem to an informed and transparent solution. Here are ten tips for using and growing Analytic Collaboration in your own organization.

1. **Team meeting rooms should enable Analytic Collaboration**. Good meeting management is critical to Analytic Collaboration. A good meeting space to collaborate is critical to meeting management. The team collaboration room should be free from distractions such as large windows, pictures on the walls, and unused equipment. If the team doesn't have a dedicated meeting room, the facilitator should prepare an available conference room before each meeting. Move the tables and chairs into a horseshoe shape to allow team members to see each other when they are talking and to turn to face the projector screen when they are building a model. Have lots of wall space to hang flipchart paper in order to keep track of the team's progress.

2. **Focus team discussions on "beating up" the model, not each other**. Analytic Collaboration introduces a valuable new team member: SAM! The model ties together all the information and judgment of the team and makes clear where there are inconsistencies or conflicts. Team members can debate with the model without directly criticizing the other team members. The model allows multiple perspectives or "alternative facts" to be tested and encourages team learning. The model is a neutral participant in the team meeting who gives honest feedback without being judgmental.

3. **Practice good group process**. The best team meetings require more than models and meeting rooms. They require a group facilitator or other designated person to manage a good group process. Good group process involves managing the pace of the meeting and alternating between divergent thinking (NGT, brainstorming,

and open discussion) and convergent thinking (coming to conclusions, a resolution, and action.) The facilitator should create a "parking lot" for issues that may be addressed later and stay focused on the meeting goals and agenda.

4. **Collect relevant information before the meeting.** Preparing for Analytic Collaboration requires bringing data and information to the meeting. Data calls and read-aheads should be sent out to team members as a matter of routine. Be specific about what content and format are expected. Senior leaders, subject matter experts, and other stakeholders should be interviewed or sent a questionnaire when value preferences, risk tolerance, and probability assessments are needed. Bringing data and information to meetings will allow the team to build a SAM that can transform that information into insights.

5. **Set ground rules for Analytic Collaboration during meetings.** Team meetings are more productive when the organization's culture and senior leadership reinforces collaborative values. Meeting ground rules are a good way for an organization to set expectations for responsibility and accountability. Start with the four core principles of collaborative teams: 1) a common purpose, 2) a shared understanding, 3) informed choices, and 4) commitment to action. Have clear ground rules for who sets deadlines and makes the decisions.

6. **Avoid using averages and create probability distributions.** In problem-solving, averages tend to produce misleading results (Savage, 2009). Yet, without a computer, it is hard for a collaborative team to think in terms of distributions. For example, if our 08:30 weekly staff meetings take on average 30 minutes, we might assume that we will be late for a standing 09:00 project meeting half the time. However, even using normal distributions, a simple analytic model shows that after two weeks we will only have a 25% chance of making the new meetings on time and after a month only about a 6% chance! SAMs are easily able to handle these calculations. The hard part is getting your team to think in terms of distributions.

7. **Mitigate biases: cognitive, organizational, and motivational.** Use good facilitation and SATs to counteract potential individual and group biases. Common biases, such as groupthink (Janis, 1982), can be avoided using ground rules (e.g., the team leader does not state his or her opinion first) and by taking turns being the Devil's Advocate. There are many other good techniques a facilitator can use to improve the quality of model assessments and judgments provided by the team.

8. **Learn how to use simple tools well, rather than using complex tools that you don't have time to master.** Analytic Collaboration must be practiced in the meeting room to be effective. Only model clean-up and documentation should be done after the meeting. The facilitator or team leader should select and practice using the software tools before the meeting. Most SAMs can be built in Excel. If the facilitator is not comfortable typing, a computer "technographer" can handle the keyboarding. Grow the model-building skills of the team during meetings for continuous improvement of Analytic Collaboration in the organization.

9. **Build consensus, not compromise.** Analytic Collaboration often increases the conflict within a team because new ideas must be tested in real-time. Some team members may not be comfortable with conflict. Compromise is a common escape tactic used by teams facing conflict, especially teams that are short on time. Contrary to what you learned in kindergarten, you should avoid compromise when practicing analytics. It sounds counterintuitive, but compromises such as horse-trading, agreeing on the lowest common denominator, and "splitting the baby,"

tend to simply postpone inevitable bad results. Resolve large differences first, and if the team can't reach consensus, use the model to conduct sensitivity analysis to show what the results would be under different assumptions (Wilkinson, 2005).

10. **Use voting to understand team differences, but make team decisions by consensus**. SAMs help teams to converge on the best solutions or decisions. The key to convergence in a collaborative team is to poll team members frequently for their inputs to the model. However, model results usually require interpretation by the team. "Up or down" votes can cause unnecessary conflict within the team, when it is more important to understand and explore the reasons behind disagreements. Always strive to reach consensus around the final solution, decision, or recommendation.

References

Aebischer, D. (2016). CECOM Virtual Logistics Assistance Representative (VLAR): Bayesian Networks for Combat Equipment Diagnostics, Presented at INFORMS Business Analytics Conference Franz Edelman Award. Orlando, FL. 2016.

BayesiaLab. (n.d.). Retrieved from https://www.bayesialab.com

Bens, I. (1997). *Facilitating with Ease*. Sarasota, FL: Participative Dynamics.

Clemen, R. T. (1996). *Making Hard Decisions, 2nd Ed*. Pacific Grove, CA: Duxbury Press.

Cochran, J. J. (2019). ed., *INFORMS Analytics Body of Knowledge (ABOK)*. Hoboken, NJ: Wiley.

Dillon-Merrill, R. L. (2015). Quantifying Risk in Commercial Aviation with Event Sequence Diagrams and Fault Trees, Presented at 2013 Winter Meeting, American Nuclear Society, Washington, DC. 2013.

ExtendSim. (n.d.). Retrieved from https://www.extendsim.com

Fenton, N. E. (2012). *Risk Assessment and Decision Analysis with Bayesian Networks*. Boca Raton, FL: CRC Press.

Hammond, J. S., Keeney, R. L., & Raiffa, H. (1999). *Smart Choices: A Practical Guide to Making Better Decisions*. Boston, MA: Harvard Business School Press

Heuer, R. J. (1999). *Psychology of Intelligence Analysis*. Center for the Study of Intelligence.

Heuer, R. J., & Pherson, R. H. (2008). *Structured Analytic Techniques for Intelligence Analysis*, 2nd Ed. Thousand Oaks, CA: CQ Press.

Janis, I. L. (1982). *Groupthink, 2nd Ed*. Boston, MA: Houghton Mifflin.

Jones, M. D. (1995). *The Thinker's Toolkit*. New York: Three Rivers Press.

Keeney, R. L. (1992). *Value-Focused Thinking: A Path to Creative Decision Making*. Cambridge, MA: Harvard Univ. Press.

Kirkwood, C. W. (1997). *Strategic Decision Making: Multi-objective Decision Analysis with Spreadsheets*. Belmont, CA: Duxbury Press.

Kostman, J. T. (2018). The McCain Meeting Rule. Retrieved from https://www.linkedin.com/feed/update/urn:li:activity:6441999670280482816/

Logical Decisions. (n.d.). Retrieved from https://www.logicaldecisions.com

Norsys. (n.d.). Retrieved from http://www.norsys.com

Parnell, G. S., Bresnick, T. A., Tani, S. N., & Johnson, E. R. (2013). *Handbook of Decision Analysis*. Hoboken, NJ: Wiley.

Quinlivan-Hall, D., & Renner, P. (1990). *In Search of Solutions: Sixty Ways to Guide Your Problem-Solving Group*. Vancouver, BC: PFR Training Associates Limited.

Reagan-Cirincione, P. (1994). Improving the Accuracy of Group Judgment. *Organizational Behavior and Human Decision Processes*, 58(2). 58, 246–270.

Savage, S. L. (2009). *The Flaw of Averages*. Hoboken, NJ: Wiley.

Schwarz, R. M. (1994). *The Skilled Facilitator*. San Francisco, CA. Jossey-Bass Publishers.

Syncopation. (n.d.). Retrieved from https://www.syncopation.com

Tatman, J. T., et al, (2015). *Bayesian Network Elicitation Facilitator's Guide*. Vienna, VA: Innovative Decisions, Inc.

Vennix, J. A. M. (1996). *Group Model Building*. Hoboken, NJ: Wiley.

Wilkinson, M. (2005). Consensus Building. In S. Schuman (Ed.), *The IAF Handbook of Group Facilitation* (pp. 361–380). San Francisco, CA: Josey-Bass.

Wizemann, T. (2010). *National Academy of Sciences Report of Public Health Effectiveness of the FDA 510(k) Clearance Process*. Washington, DC: The National Academies Press.

Section III

Applications

Chapter 14

A Model for and Inventory of Cybersecurity Values: Metrics and Best Practices

Natalie M. Scala and Paul L. Goethals

14.1 Introduction

This chapter outlines several methodologies in operations research that are useful for cyber defense, then establishes a value model for cybersecurity metrics and best practices. The framework developed in this chapter can be customized based on organizational values and needs. An inventory of values from a survey of information technology professionals provides context, but each individual organization should use its own data and assessments to populate this model. Students can benefit from this chapter by learning the Value Focused Thinking and multi-objective decision analysis frameworks, as well as cybersecurity values. Practitioners can benefit from understanding an illustration of the model and then applying the approach in their own organizations.

14.1.1 Threat Modeling: A Defensive Mindset

Military defensive operations strive to establish strongholds or create fortifications that are impenetrable to attackers. Since the middle of the twentieth century, military operations researchers have formally sought modern analytical tools and techniques to support this effort. To reduce the likelihood of a defensive breach and increase the possibility of enemy detection, threat models are frequently employed. The threat model may incorporate both the attacker and defender perspective, consider decision-making priorities for action or response, and account for multiple layers of protection. It may also take the form of a variety of constructs, such as with network topologies, pure mathematical models, or measurement-driven analytics.

Perhaps one of the most popular tools for threat modeling is the attack tree, first documented in the literature by the cryptographer Schneier (1999). The concept, illustrated in Figure 14.1, utilizes a diagram consisting of branches to represent various attack avenues, whereby an adversary achieves its objectives through a combination of A (and), O (or), or T (terminal) nodes. By identifying the different attack scenarios and establishing a scoring framework to assess the likelihood of each adversarial objective, the risk of an attack can then be quantified. The method involves estimating the degree of cost and benefit for an attacker; in some instances, an adversary's motivations are further approximated using capability or behavior-based mathematical models.

Another method used to assess risk in systems is the attack surface, first introduced by Howard (2003) to address vulnerabilities in computer software. While several different definitions of the attack surface exist, it is generally used to describe the internal and external accesses or privileges via hardware or software; the union of system components, features, and services; and the protocols established for a given organization (Theisen et al., 2018). At the macroscopic level, an attack surface may be used to evaluate vulnerabilities or identify attack vectors across the physical architecture of an organization, including its servers, routers, and other devices connected to the network. It can also involve very detailed constructs at a minuscule level, such as with specific application susceptibilities, interfaces between email and the internet, an individual's network behavior, and data storage modules. Illustrations like Figure 14.2 are often created to give decision-makers an increased awareness of high-risk areas within their security environment. The objective is then to strengthen system security by focusing resources on minimizing the organization's attack surface.

Beyond the attack tree and attack surface, most threat models incorporate some type of metric to quantify the level of risk in a system or process for a decision-maker. For example, trust metrics are used to detect malicious bot activity in networks, aid in authenticating user credentials, or support cryptographic methods. Metrics developed to infer the intent of an individual utilize pattern matching algorithms, hypothesis testing, or statistical process control to differentiate between anomalous and normal

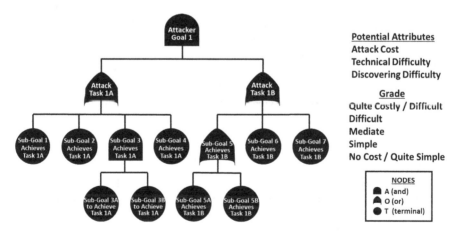

FIGURE 14.1: Attack tree architectural diagram with potential scoring framework.

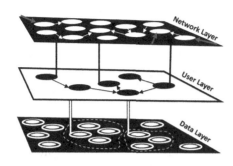

Potential Metrics
Estimated Damage
Entry / Exit Points
Untrusted Data Items
Attack Complexity
User Interaction
Report Confidence

Grade
Difficult / Costly
Mediate
No Cost / Simple

FIGURE 14.2: Attack surface construct with potential scoring framework.

behavior. To properly design a metric to aid in strengthening defense, the security environment must be fully understood. This includes the nature of the threat and also previous attempts to quantify risk in this space. The remaining portion of this chapter is afforded to designing or selecting metrics grounded in defensive principles for the cyber domain.

14.1.2   The Cybersecurity Environment

It is important to understand values and secure systems using metrics and best practices, as cybersecurity is a topic of immense interest for organizations. Much of the attention can be attributed to attacks and breaches at high-profile organizations such as Equifax, the Office of Personnel Management (OPM), JP Morgan Chase, Home Depot, Target, and Yahoo. Large companies and government entities are not the only targets; hackers may target any organization. For example, Farahani et al. (2016) reviewed organizations breached in the State of Maryland in the United States, showing that both small and large organizations can be affected. While most security studies provide an estimate of the costs associated with an attack on or a breach of various entities, the true cost is really unknown. The diminished trust between an organization and its clients, the loss of credibility in securing valuable information, and the damage to the reputation of a company are all abstract costs that are difficult to estimate in practice. For these reasons and many others, organizations are operating in an environment characterized by uncertainty and a high consequence of failure.

14.1.3   Disconnects in the Problem Space

The Science of Security (SoS) initiative (NSA, n.d.), sponsored by the United States Department of Defense, supports and indexes research related to cybersecurity and organizes the problem space into five thrusts, or "hard problems." These problems are (1) scalability and composability, (2) policy-governed secure collaboration, (3) resilience, (4) human behavior, and (5) security metrics; each problem addresses a specific area yet presents challenges. Specifically, *scalability and composability* examines combining secure components into a larger secure item, as vulnerabilities often lie in the gaps between two components; *policy-governed secure collaboration* develops methods and requirements to guarantee data protection while enabling information sharing and collaboration; *resilience* measures the ability of a system to resist an attack by unauthorized parties; *human behavior* addresses the unpredictability and complexity of human behavior which aids in the development of models that have increased accuracy while minimizing potential vulnerabilities from humans interacting, possibly unsecurely, with systems; and *security metrics* measure the security or vulnerability of a system (Nicol et al., 2015). Further reading on each problem is available in Nicol et al. (2012); Nicol et al. (2015); and Scala et al. (2019).

In general, the *security metrics* problem is especially challenging. Systems are extremely complex, and small nuances may have significant impact. The challenge is to develop security metrics, models, and best practices that are capable of predicting if or confirming that a cyber system preserves a given set of security properties. Metrics must be quantifiable, feasible, repeatable, and objective (Goethals, Farahani & Scala, 2015). A system having security properties, however, does not mean it is immune to an attack or potential breach; rather, the goal is to prevent these activities. In contrast, the analytics field considers a metric to be descriptive, defined as a variable of interest or as a unit of measurement that provides a means to objectively quantify performance (Evans, 2013). By definition, descriptive analytics use data from the past and present to understand current and historical performance; unlike predictive analytics, they do not make a forecast about future events. Thus, a disconnect in semantics exists, with the cybersecurity field attempting to utilize descriptive analytics (such as a metric on a static system) to make an assessment or prediction about future events. Such a disconnect may cause confusion between members of a project team or among client interactions. It is important to understand that the two fields view metrics with differing purpose, and a team must be clear as to how metrics will be managed and defined within the project. For this chapter, the SoS definition of metrics is used to examine both metrics and best practices that can augment security of a cyber system.

In addition, there is a lack of translation of cybersecurity research into action. Most of the work indexed by SoS addresses theory and architecture but does not consider how guidelines and best practices should be implemented, how the overall risk of a breach of the system should be measured within the hard problem environment, and how risk should be managed and quantified. Because uncertainty is not quantified, risk related predictions of what comes next, such as if the next breach can be predicted, the probability of a breach, and the risk to the system, have not been addressed. Opportunities to include risk in the hard problem environment are described in Scala et al. (2019) and include policy-related risk assessments that adapt to an evolving adversary, analyses of ranges of consequences, decision-making under uncertainty for actors' objectives and motivations, and mitigations for composability. Such analyses are especially important to the metrics and human behavior hard problems. Thus, decision analysis techniques are needed to assist in translating the current cybersecurity theory and infrastructure research into practice. Specifically, teams need to apply research into their system design and projects, to help ensure the most recent advancements are deployed into practice. This chapter identifies one such way to translate research into practice.

It is important to note that a disconnect also exists between just identifying and actually implementing cybersecurity metrics and best practices. If cybersecurity is thought of as actions taken to prevent attacks on or breaches of a system, then a success becomes the lack of an attack or breach. This is a clear reversal from a traditional risk or probability problem, where success is typically defined as the occurrence of a desired event. Furthermore, attacks can happen quite frequently, especially to high-profile organizations in both defense and corporate industry. Therefore, it may be futile to focus on preventing an attack and instead the focus should be on preventing a breach or access into a network.

To clarify, a breach and an attack are dissimilar events. Attacks are attempts by an outside entity to gain access to or information from a network. A breach is an attack that is successful, with the actor gaining access or information. Breaches have high consequences. Customer loyalty may be lost, whereby the trust between the organization and its clients must be regained (Farahani et al., 2016). Moreover, breaches do not become known events until they are discovered; some breaches may never be discovered. Although the frequency of known breaches is rising, it is difficult to fully understand the true number of breaches that are occurring or the true threat to an organization's systems.

For example, if the probability of a breach is low, a model of cybersecurity systems considers these events as low probability high consequence problems, akin to nuclear accidents, and employs models designed for these conditions. Unfortunately, these conclusions cannot be readily drawn. Due to the fact that most cybersecurity research has materialized in the computer science and network infrastructure management fields, models to assess and measure a system's current security and/or predict the probability of a breach do not widely exist. As a result, the work thus far has primarily focused on theory, infrastructure, and best practices, downplaying risk, decision analysis, and practical applications to cybersecurity systems.

A value model framework to identify the preferred metrics and best practices for organizations to implement in augmenting their cybersecurity addresses these gaps. In this model, metrics and best practices are alternatives between which a decision-maker needs to choose, specifically selecting which ones that should be implemented. Those alternatives are then evaluated against attributes or values of secure cyber systems. The model employs Value Focused Thinking (VFT) and uses value functions to perform the analysis. A survey of information technology (IT) professionals enables understanding of (1) what they value in secure cyber systems, and (2) their organizations' histories of attacks and/or breaches. These survey results are further synthesized with interviews of subject matter experts (SMEs) to develop the value model framework; consequently, preferred metrics and best practices for cybersecurity are identified for implementation in organizations. This is a framework that can be then adapted or modified to fit the needs of a specific organization. That is, practitioners can use the steps and structure outlined here with data and values from their organization to build a model for their cybersecurity needs.

14.2 Past Research

The literature proposes many candidate metrics and best practices that are appropriate for cybersecurity. Organizations need to identify which metrics and best practices are appropriate for their systems. The first step is to create a list of potential metrics and best practices for the model to evaluate. The literature can assist with creating this list, and this review includes metrics that have already been identified. The Science of Security initiative indexed 110 metrics-related papers as of October 2016, and these papers are all tied to research funded or related to the initiative. Outside of the SoS, Viduto et al. (2012) suggest 24 candidate metrics for cybersecurity, including the use of a system administrator, firewalls, patching, and separation of duties. These 24 metrics are organized into four categories: support, prevent, detect, and recover. Schilling and Werners (2016) use the IT protection catalog from the German Federal Office for Information Security in their research; that catalog contains over 1,200 safeguard alternatives. Santos, Haimes, and Lian (2007) identify four categories of metrics when protecting SCADA systems: cybersecurity cost, network vulnerability, equipment downtime, and production delay. These authors argue that key considerations in cybersecurity should include identifying threats, establishing countermeasures, recognizing the role of temporal factors in assessing risk, and modeling management decision dynamics. Clearly, the choices for metrics and best practices are extensive and not all can be implemented by a single organization. Methods to discern or identify preferred metrics, such as this model, are needed to reasonably manage security.

While needs exist, research in implementation and management of cybersecurity metrics is emerging. In their survey, Wright, Liberatore, and Nydick (2006) identify that literature on cyber countermeasures is increasing but needs exploration, as cyber is an evolving application area. In 2005, the President's Information Technology Advisory Council argued for shifting the perception that greater security is not worth the cost; Rees et al. (2011) interpret that as a need to determine a best set of countermeasures to implement, given the array of options. Ryan et al. (2012) address risk in information security by expert judgment elicitation, focusing on the frequency of system attack, number of successful attacks, need for cyber investment, and the probability of a successful attack. The authors conclude that investment in protection is rational for firms with high exposure or valuable assets. However, the authors do not present a model of or recommendation for how to make such an investment.

Cyber policy literature addresses the overall state or design of systems but does not necessarily provide detailed implementation guidance. Scala and Goethals (2016) provide a review of three policy models; each model has a limitation in terms of practicality or scope. In particular, Viduto et al. (2012) argue that policies based on attack trees and attack graphs, using shortest paths and related costs, lack practical sense and cannot support cost-effective decisions.

14.2.1 Evaluating Metrics

Some literature exists on evaluating cybersecurity metrics and best practices in order to provide implementation recommendations, but only financial approaches are taken. Cost and/or budget are primary components of the models. Furthermore, the literature in this area uses the term *countermeasures,* which Rees et al. (2011) define as tools and actions to block intrusions or mitigate damaging effects of an intrusion. Note that, generally speaking, countermeasures can be thought of as best practices; however, metrics and best practices can extend beyond intrusions. For example, the practice of anonymization, a best practice, is to remove personal information associated with data (Cormode et al., 2013). However, the papers that do rank or evaluate countermeasures, especially from a financial perspective, are relatively recent at the time of this writing. Financial based models include Rees et al. (2011), who employ a genetic algorithm and fuzzy set theory to examine the tradeoff between the cost of a countermeasure portfolio and subsequent risk, while meeting the overall budget for countermeasures. Viduto et al. (2012) create a risk assessment for countermeasures, with an optimization routine using a multiobjective Tabu Search heuristic and financial cost as a variable. Sawik (2013) proposes a mixed integer program and financial engineering measures to identify countermeasures to implement under a limited budget in order to minimize potential losses and mitigate the impact of disruptions. Yevseyeva et al. (2015) offer a resource allocation problem to identify security controls to purchase and then employ portfolio optimization and financial management techniques to identify optimal controls within a limited budget. Fielder et al. (2016) propose the cybersecurity investment challenge to balance the cost of implementation defense against the subsequent impact on the business; in their work, the authors consider game theory, combinatorial optimization, and a hybrid model with a focus on small- to medium-sized enterprises. Schilling and Werners (2016) use a combinatorial optimization model to address the optimal amount to invest in security as well as the countermeasures that should be selected for investment; their research utilizes a knowledge base of IT security as the foundation for a model, with costs of specific countermeasures considered as a second step after optimization. Schilling and Werners (2016) also provide a review of other papers that address investment selection of countermeasures, using techniques such as financial analysis, real options analysis, optimization, and fuzzy set theory. These papers present prescriptive models to identify optimal countermeasures, driven by budgetary concerns.

However, not all organizations or industries value the same components of a secure cyber system, nor have the same budget or access to countermeasure resources. A more holistic, robust, and impactful approach would be to base the selection on value to the firm, identifying metrics, best practices, and countermeasures that aim to support an organization's unique preferences and needs. A broad policy view applied to all industries and organizations is not ideal, as the adversary can quickly adapt to a one-step prescriptive methodology. Furthermore, a general policy might not be ideal for every organization. Therefore, an approach using VFT can help to identify value to the organization, so that the metrics, best practices, and countermeasures that are selected for implementation truly reflect the organization's needs and requirements for a secure cyber system. When using VFT, the analysis is driven by value instead of only budget or cost.

14.2.2 Value Focused Thinking and Multi-Objective Decision Analysis

By definition, Value Focused Thinking is a decision analysis process that is designed to stimulate meaningful development of a decision model while supporting creative thinking about the problem at hand; its purpose is to enable clear and explicit definition of the decision problem (Keeney, 2008). The decision-maker functionally brainstorms all possible objectives and alternatives. This allows for value to drive inclusion of all that the decision-maker cares about, even alternatives or objectives that may have not been originally considered. Further details on the VFT process can be found in Keeney (1992, 2008).

Multi-objective decision analysis (MODA) is a decision analysis technique for evaluating a decision under multiple objectives or criteria, and the objectives may be conflicting (Parnell, 2007; Keeney & Raiffa, 1976; Kirkwood, 1997). Related methods to MODA include multiple attribute utility theory (MAUT) and multiple criteria decision analysis (MCDA). MODA is a utility approach; objectives are defined into measures or characteristics that are important to the decision problem. A value function is then defined for each measure, with 0 representing the worst-case scenario and 100 for the best-case scenario. Alternatives are then scored by creating an additive value function across the value objectives and measures for each alternative; the highest-scoring alternative is preferred (Keeney & Raiffa, 1976).

Most MODA analyses have at least three to five measures, while very complex problems can have up to 100 measures (Scala et al., 2012). Further details on the MODA process can be found in Parnell (2007); Keeney and Raiffa (1976); and Kirkwood (1997). A step-by-step outline of the process, especially for defense applications, can be found in Dillon-Merrill et al. (2008). A review of VFT applications can be found in Parnell et al. (2013).

VFT and MODA have been used in defense and security applications, but to the best of our knowledge, not for cybersecurity metrics and best practices. Keeney (2007) presents value model principles for the United States Department of Homeland Security and terrorist organizations. He argues that value models quantify objectives and what decision-makers need to achieve; knowledge of values is critical for informed choices. Dillon et al. (2012) apply VFT for a multiattribute utility model for domestic intelligence choices post-9/11. Beauregard (2001) used VFT to assess the level of information assurance within United States Department of Defense units. Buckshaw et al. (2005) create an approach to measure risk in critical information systems and use MODA as one model within the overall risk assessment methodology; their focus is on information assurance in a hostile and malicious operating environment. Feng and Keller (2006) formulate a multi-objective decision analysis model for nuclear terrorism. Parnell et al. (2015) use multi-objective decision analysis for building cyber infrastructure as one component in a cyber investment model for the United States Air Force. Finally, Keeney and von Winterfeldt (2011) propose a value model for homeland security. The authors propose a framework, using primarily literature review and a survey, to support United States Department of Homeland Security (DHS) staff and policymakers. Their survey was distributed to managers and researchers at a DHS University Center of Excellence, eliciting objectives and consequence measures for impacts of terrorist attacks and counterterrorism decisions. The research created an inventory of strategic objectives for homeland security and demonstrated that a value model for DHS can make a substantive contribution of both relevance and legitimacy.

VFT and MODA are appropriate for evaluation of cybersecurity metrics and best practices, as the creative thinking process allows for identification of values by the decision-maker, both consciously and subconsciously. The brainstorming process is critical to the VFT process, as it assists in identifying what is truly most important (Keeney, 1992, 2008). In this problem, brainstorming considers values in light of demographics, history of cyber awareness, and current operating climate. Such an approach enables better quantitative decision-making, as opposed to just following trends or recent media reports of breaches. In order for an organization to implement appropriate cyber metrics and best practices, it must understand what it cares about, or its values, and the VFT process provides a formalized method to scale back the overwhelming number of metrics and best practices options in order to focus on the preferred ones for that organization. Experts can fully participate in the process, even without a decision analysis or risk background, as the framework is accessible and easy to update to fit the unique profile of an organization. While Viduto et al. (2012) note that security measure implementation decisions are typically made using decision-makers' personal experiences, the VFT and MODA processes allow for those experiences and related values to be quantified and evaluated in a formalized manner. This chapter provides an outline of the structure of our model and then an illustration of how it can be used.

14.3   A Cybersecurity Value Model

14.3.1   Methodology Development

Development of the cybersecurity value model follows the approach of Keeney and von Winterfeldt (2011) and creates a framework that can be customized. While those authors used literature review and a survey, the methodology outlined in this section employs the platinum standard, as defined by Parnell et al. (2013), to formulate the model hierarchy, utilizing direct interviews with SMEs (subject matter experts) and survey response data. To start, elicitation sessions were held with two SMEs in separate interviews. The first SME is a technology manager at a thinktank, with over 12 years of experience in information technology, including software development and systems integration efforts for human capital and learning management systems, library and information retrieval systems, knowledge and content management systems, dashboards, portals, and other financial/administrative systems of record. This individual served a primary role for hierarchy and value elicitation. As such, the SME participated in a series of conference calls, with the first call addressing potential measures for the value tree, followed by value functions for those measures.

With this process, a mutually exclusive and collectively exhaustive hierarchy was developed, a key tenet of a MODA analysis (Parnell et al., 2013), and the value functions were defined. These measures were then reviewed by a second SME with more than 40 years of extensive cyber and IT experience ranging from iron core memory and plugboards to advanced microelectronic technologies, e.g., nano-scale components and graphene. His varied background also includes foreign languages, technical intelligence operations, space and airborne operations, and biometrics. He has served in the military, worked in industry, and has a long record of government service. The primary role of the second SME was to provide validation for the original elicitation; as such, this individual gave feedback and insight into the measures and functions identified by our first SME. The process of alternating between the first and second SME in individual sessions for feedback was then repeated until they reached agreement. While participating in the interviews, the SMEs were asked to define attributes that are important to a secure cyber system. Those attributes became measures in the model. When implementing in any organization, a similar approach can be taken, starting with a group of core SMEs who can provide an initial assessment. Those SMEs can work iteratively, as in this example, or may have their values assessed in a group format, using a method, such as Delphi method (Helmer-Hirschberg, 1967), to gain consensus.

As a result of the iterative SME process, six measures were identified: (1) *data integrity*, (2) *end-to-end security*, (3) *cloud security*, (4) *security policies*, (5) *intrusion and threat detection*, and (6) *vulnerability mitigation. Intrusion and threat detection* were broken into two submeasures: (5a) *monitoring*, and (5b) *reaction*. Table 14.1 presents the corresponding measure definitions, and Figure 14.3 presents the value hierarchy.

The second step in model development is to create value functions. Value functions can be loosely defined as the marginality of the decision-maker's preferences, and they do not consider risk attitudes (Goodwin & Wright, 2009). For each measure, value functions were elicited from the first SME, who provided the functions in terms of a scoring rubric. Although a rubric is not a traditional approach for eliciting value functions, the method is appropriate for this model. The approach was to identify activities or actions that support performance of secure cyber systems, assigning relative preference related to the value

TABLE 14.1:   Measures and Definitions

Measure	Definition
Data Integrity	Confidence that data stored on the system is the original, uncompromised data. This includes actions to keep the chain of custody intact, with modification and authorization logs.
End-to-End Security	Ability to keep data safe in communications such as VPN, mobile devices, and software packages, so that inappropriate human behavior becomes physically impossible.
Cloud Security	Security in all cloud-based services, both in-house and third party. This includes secure communication to/from the cloud to the outside world as well as security between servers in the cloud. Software as a service in the cloud, both in-house and vendor-developed, must also remain secure.
Security Policies	Existence of robust organizational policies to promote secure behavior. This includes enforcement of policies with repercussions to prevent affiliates from ignoring policy.
Intrusion and Threat Detection	Live monitoring as well as reactive response to an intrusion. Live monitoring may include watching for anomalies using packages installed on network hardware and other means, such as anti-virus measures. Reactive actions may include immediate response and periodic network probing. These responses may be directed at ransomware, malware, etc.
Vulnerability Mitigation	Ability to remain current with updates, such as patch sets and hot fixes. This includes hardware, firmware, the operating system, and software on the operating system.

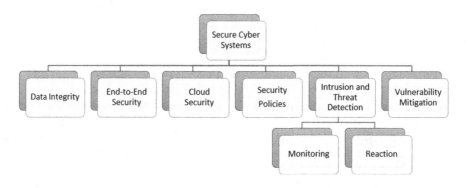

FIGURE 14.3: Value hierarchy.

of each activity. In that sense, an alternative that supported all activities for a given measure would receive a score of 100 (full value), while alternatives that employed some activities would receive a value corresponding to the sum of the values associated with each relevant activity. An alternative that did not support any activity would receive a score of 0 (no value). As a metric or best practice supports more activities or features, it has higher worth to securing the system in line with the organization's values. Following the same pattern, the first SME initially provided the value functions via the activity rubric, which were then reviewed by the second SME. The feedback process iterated

TABLE 14.2: Scoring Rubric for Value Functions

Measure	Activity	Value
Data Integrity	Existence of security and access permissions	50
	Existence of transaction or version logs	30
	Having and using an audit process	20
	None of these	0
End-to-End Security	Existence of a VPN	40
	Mobile device management	20
	Use of SSL in all communications	40
	None of these	0
Cloud Security	SSL is used for all communications	25
	Presence of third-party security audit	20
	Existence of security back-end integration	20
	Mechanism to ensure that only people who have access to system do so	20
	Key based authentication for server management	15
	None of these	0
Security Policies	Having a password policy	40
	Sensitive or confidential data policy	15
	Ability to configure software to be policy compliant	25
	Clear approach for addressing policy violations	20
	None of these	0
Monitoring	Network anomaly monitoring	20
	Tools to monitor log files	25
	Proactively scanning or crawling network	15
	Existence of anti-virus and anti-malware software	40
	None of these	0
Reaction	Existence of critical incident response plan	35
	Existence of response team in house or outsource	65
	None of these	0
Vulnerability Mitigation	Having process and timeline on which updates made	50
	Existence of mechanism to receive notification	30
	End of life policy	20
	None of these	0

between the SMEs until they reached agreement. Table 14.2 provides the scoring rubric for the value functions. Table 14.3 identifies the monotonic nature of creating a value function from the scoring rubric for the *data integrity* measure; a similar table could be constructed for all value functions. Value functions should be monotonic; as more activities of use are added, the metric or best practice continues to increase in worth. When implementing in practice, the SMEs could use an approach such as the Delphi method (Helmer-Hirschberg, 1967) to agree on characteristics that make a measure valuable. All value functions should have a best-case scenario (100 points), a worst-case scenario (0 points), and some middle-case scenarios valued between 0 and 100 points.

TABLE 14.3: Value Function for *Data Integrity*

Activities	Value
Existence of security and access permissions, existence of transaction or version logs, having and using audit process	100
Existence of security and access permissions, existence of transaction or version logs	80
Existence of security and access permissions, having and using audit process	70
Existence of transaction or version logs, having and using audit process	50
Existence of security and access permissions	50
Existence of transaction or version logs	30
Having and using audit process	20
None of these	0

Elicitation of weights for the hierarchy measures was also iterative, beginning with the first SME, who ranked the measures in order. The second SME provided a different rank order list, and the list of measures in rank order was iteratively revised until the two SMEs agreed. Such conflict is not uncommon; Edwards and von Winterfeldt (1987) and Dillon et al. (2012) note that conflict between decision-makers is expressed via different weights and priorities. For the purposes of this framework, the disagreement in priority was resolved via iterations and communication between the SMEs; however, such conflict supports the need for customization of the framework for each organization's unique values. Not all are the same, and care should be taken to ensure the organization's values are correctly implemented into the model. It is also important to include a range of decision-makers with varied experience so that all relevant values and input can be incorporated into the model.

The rank order centroid method (ROC) was used to convert the ranks to weights, by using the following equation, which sums the weights to unity: $w_k = \frac{1}{n} \sum_{i=k}^{n} \left(\frac{1}{i} \right)$, where the weight of the kth measure is computed among n total measures (Barron, 1992). ROC weights are not ad hoc (Barron and Barrett, 1996); moreover, the method has been used in other value model approaches (e.g., Scala et al., 2012; Scala & Pazour, 2016). Such an approach is appropriate for this illustration, as it is a framework which would be augmented in practice. Swing weights (Kirkwood, 1997; Parnell et al., 2013) are a more direct elicitation method to assess measure weights and can be used during model implementation. ROC can also be used in implementation if decision-makers do not reach consensus on the swing weights. Further reading on ROC can be found in Edwards and Barron (1994) and Buede (2009). Table 14.4 presents the final rank order of measures along with the corresponding ROC weights.

14.3.2 Illustration

This framework of cyber values and preferred metrics can be applied to many different industries, with organizations evaluating metrics and best practices that are relevant to them and modifying the value functions as appropriate. For illustration purposes, generally consider the supply chain sector with notional scoring of metrics using the model's measures and value functions. The significance of the cybersecurity

TABLE 14.4: Measures and Weights

Measure	Elicited Rank	ROC
End-to-End Security	1	0.370
Cloud Security	2	0.228
Data Integrity	3	0.156
Vulnerability Mitigation	4	0.109
Monitoring	5	0.073
Reaction	6	0.044
Security Policies	7	0.020

problem for the supply chain cannot be overstated. High-profile cybersecurity breaches such as with Target in late-2013, Home Depot in 2014, and OPM in 2015, are all known to have involved compromises to their supply chain in some manner (Shackleford & Douglas, 2015). Each of these breaches impacted millions of people worldwide at a tremendous cost; around the time of this writing, the first chief information security officer for the government estimated the OPM breach alone could cost more than $1 billion over the next decade (Townsend, 2017). Specific examples within the software supply chain include hackers inserting a backdoor into Cisco's free CCleaner software, fake versions of Xcode for Apple developers, and code repositories for the Python computer language infected with malicious code (Greenberg, 2017). The risk of breach for any one organization can exist in many facets of the supply chain network, such as vendor policies, transportation systems, supply sourcing, internal practices, and the manufacturing infrastructure or componentry (NIST, n.d.; Lord, 2017). Some researchers and government entities perceive cyber risks to the supply chain increasing exponentially in the future, with the expanding Internet of Things, the evolution of digitization, and the growth of cybercrime globally (Boyens, 2016; Wallace, 2016; Camillo, 2017; Simpson, 2017).

For evaluation, consider ten metrics and best practices as defined by the United States National Security Agency's Information Assurance Mission. These metrics and best practices are: (1) *application whitelisting*, (2) *control administrative privileges*, (3) *limit workstation-to-workstation communication*, (4) *use anti-virus file reputation services*, (5) *enable anti-exploitation features*, (6) *implement host intrusion prevention system rules*, (7) *set a secure baseline configuration*, (8) *use web domain name system reputation*, (9) *take advantage of software improvements*, and (10) *segregate networks and functions* (NSA, 2013). Granted, any organization implementing this framework could sample any subset of metrics from the Information Assurance list, the Science of Security index of metrics, the metrics identified in the literature review, or other emerging topics in cybersecurity. However, for this example, the Information Assurance list assists in demonstrating the use and robustness of this model.

Policy documents and support from the literature (i.e., Shackleford, 2009; Shackleford, 2010; Johnson et al., 2011; Anderson, 2011; Vengurlekar, Clouse & Kammend, 2012; Souppaya & Scarfone, 2013; Anderson, 2014; Sedgewick, Souppaya & Scarfone, 2015) are used to score each alternative (in this case, metric or best practice) against each measure using the value functions. Using these types of documents is considered to be the gold standard by Parnell et al. (2013), as SMEs are not interviewed for the scoring analysis. When scoring, each activity on the rubric for every value function is evaluated to determine if the metric or best practice being scored supports that activity. If so, points are earned. The policy documents and literature are used as evidence that the metrics or best practices do or do not have value. Table 14.5 shows the scoring process for *application whitelisting*. Each activity on the value function rubrics is listed, and the activities that are supported or addressed by *application whitelisting* are in italics. Then the score for *application whitelisting* is the sum of rubric points for all italicized items. All other metrics and best practices followed the same scoring approach.

The SMEs reviewed the gold standard scoring to provide validation for the policy-based analysis. Both SMEs provided feedback, and the scoring was revised accordingly to correct any errors or misinterpretations. Then, to develop a weighted score, s_a, for each metric or alternative a: $s_a = \sum_{i=1}^{m} w_i v_{ia}$, where w_i is the weight of measure i and v_{ia} is the value for measure i, alternative a, from the scoring rubric, with m total measures. This function should be used in all MODA implementations and is the standard equation to determine the score. The final scores, in descending order, for the ten candidate metrics or best practices are shown in Table 14.6. In an organizational setting, the literature and/or policy documents could be used for scoring. However, it is best to

TABLE 14.5: Scoring Example for *Application Whitelisting*

Measure	Activity	Value	Application Whitelisting
Data Integrity	*Security and access permissions*	*50*	70
	Transaction and version logs	30	
	Audit process	*20*	
End-to-End Security	Existence of VPN	40	20
	Mobile device management	*20*	
	SSL in all communications	40	
Cloud Security	SSL in all communications	25	75
	Presence of 3rd party security audit	*20*	
	Security back-end integration	*20*	
	Mechanism to ensure access	*20*	
	Key based authentication	*15*	
Security Policies	*Password policy*	*40*	65
	Sensitive or confidential data policy	15	
	Configure software to be policy compliant	*25*	
	Clear approach for policy violations	20	
Monitoring	Network anomaly monitoring	20	65
	Tools to monitor log files	*25*	
	Proactively scanning or crawling network	15	
	Anti-virus and anti-malware software	*40*	
Reaction	Critical incident response plan	35	0
	Response team	65	
Vulnerability Mitigation	*Process and timeline for updates*	*50*	80
	Mechanism to receive notification	*30*	
	End of life policy	20	

TABLE 14.6: Final Scores for Candidate Metrics

Metric/Best Practice	Score
Control administrative privileges	81.32
Limit workstation-to-workstation communication	78.86
Take advantage of software improvements	73.17
Use anti-virus file reputation services	71.82
Enable anti-exploitation features	71.63
Use web domain name system reputation	69.15
Implement host intrusion prevention system rules	65.19
Segregate networks and functions	63.27
Application whitelisting	50.19
Set a secure baseline configuration	30.26

use the group's data to support the scoring of each metric or best practice under consideration by the model.

These ten best practices are popular tools that can help to secure a cyber system. The top two practices, *control administrative privileges* and *limit workstation-to-workstation communication*, earned close to the same score and should be implemented if this were a real scenario. Other practices, such as *take advantage of software improvements* and *use anti-virus reputation software* also score highly. In fact, the scoring is somewhat consistent for the first eight metrics; the last two practices, *application whitelisting* and *set a secure baseline configuration* exhibit a large drop-off in score. Although important, this suggests these two may not be as relevant or important for the notional firm. Of course, this list is not a prescriptive recommendation for every supply chain firm or every organization looking to improve their cybersecurity. Experience, expertise, and other factors will drive the values considered by an organization. The model then discerns the most valuable best practices for that organization to implement.

14.3.3 Sensitivity Analysis

The sensitivity of any ranked list should be analyzed. To do so, vary the weight of one metric while normalizing and holding all others constant. Test weights between 0.05 and 0.95 to perform this analysis for every measure. The goal of the analysis is to determine if the rank order of the preferred metrics and best practices remains constant, regardless of the weight of a given measure. If so, the ranking is robust. If not, examine how sensitive the overall ranking is to changes in measure weights.

The highest-weighted measure is *end-to-end security*, with a model weight of 0.37. Testing weights for *end-to-end security* between 0.05 and 0.95, while not changing the weights for any other measures, other than to normalize so that the weights add to up 1.0, determines that *control administrative privileges* remains the highest-rated practice, unless the weight for *end-to-end security* falls below about 0.1. The second highest-scoring practice, *limit workstation-to-workstation communication*, remains as such unless the weight for *end-to-end security* falls below about 0.15. Because the current weight is 0.37, a moderate shift in importance of this measure would have to occur before the top two practices are no longer the same. Figure 14.4 depicts this analysis, graphing the tested weight for *end-to-end security* on the *x*-axis and the corresponding

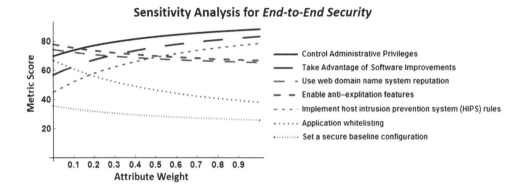

FIGURE 14.4: Sensitivity analysis for *end-to-end security*.

score on the *y*-axis. The graph depicts that once the weight for *end-to-end security* falls below about 0.1, *enable anti-exploitation features* becomes the top preferred metric or best practice for implementation in our illustration. Please note that, for purposes of readability, three metrics are removed from Figure 14.4. Those metrics are *limit workstation-to-workstation communication, use anti-virus reputation services,* and *segregate networks and functions.* Each of those metrics would not be preferred and score with a low rank when the weight for *end-to-end security* is examined for sensitivity.

Now consider *data integrity.* For this measure, *control administrative privileges* remains the top preferred practice with *limit workstation-to-workstation communication* remaining in second, regardless of the weight assigned to *data integrity* in the model. Therefore, this measure is robust with respect to its model weight. Figure 14.5 depicts the corresponding sensitivity analysis graph. Like Figure 14.4, the three low-ranking metrics are removed from Figure 14.5 to improve readability.

This analysis is similarly done for every other measure in the value model and should be done for a model that is being implemented. *Reaction* and *security policies* are robust, as the top two practices do not change, regardless of measure weight. For *cloud security* and *vulnerability mitigation,* a moderate to large shift in weight is needed before the top metric changes. Specifically, the weight for *cloud security* has to be greater than

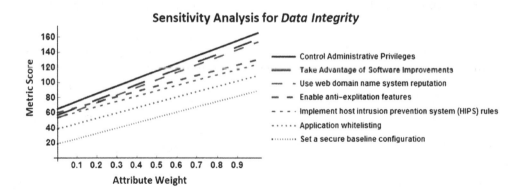

FIGURE 14.5: Sensitivity analysis for *data integrity.*

about 0.6 for *control administrative privileges* to no longer be the highest-scored metric, and the weight for the measure must be greater than about 0.42 for *limit workstation-to-workstation communication* to no longer be the second highest-rated metric. The current weight for *cloud security* is 0.228. For *monitoring, control administrative privileges* is no longer the top-rated metric once the measure weight is greater than about 0.17; the current weight is 0.073. Thus, a shift in weight is sensitive for this measure. When a measure is sensitive, care should be paid that SMEs correctly identified relative importance. This example provides empirical evidence that the model is robust and can be used as a framework for candidate metrics and best practices. This means that even with some slight to moderate error or shifting in preference or weights, essentially the same metrics and best practices would be recommended for implementation, to align with value. Robustness in this case is good, as ROC was used for illustration, and actual elicited weights may be different. Weights elicited from SMEs are assessments based on experience and other factors; allowing some room for change without affecting the final solution can increase the overall confidence a decision-maker has in implementing the results of the model. For this example, these illustrative results create a framework to identify preferred metrics and best practices for organizations to implement in order to augment cyber systems as well as some general guidance to supply chain firms regarding specific top metrics and best practices to consider.

14.4 Survey

Illustration of this model with notional data for the supply chain sector does not imply the same values and metrics are appropriate for all organizations or economic sectors. To examine potential differences, consider a national survey of cyber professionals in order to have a broad and representative range of responses. Values are not consistent across industries or organizations, and a large sample should highlight potential differences. Questions examined organization demographics such as industry and size of firm as well as the respondent's experience working in the cybersecurity profession. Respondents also assessed the strength of their organizations' cyber systems and identified if the organization's systems have been breached or attacked. If the respondent indicated a breach, a follow-up question inquired the effects of the breach and any use of a mitigation plan. If the respondent indicated an attack, a question followed for frequency. Finally, the respondent identified valued attributes or measures of a secure cyber system along with corresponding weights of those measures.

The survey was IRB reviewed and distributed via IT security professional listservs, social media, and authors' personal networks. Potential respondents were provided a link to the online survey via Qualtrics, and the survey was open for six weeks during summer 2016. In total, 98 responses were collected, with 79 usable, as some respondents did not proceed past the informed consent or basic demographics questions. Table 14.7 identifies the percentage of respondents by industry. Please note that a majority of respondents who identified with academia were professionals from university and college technology services and not necessarily professors.

Regarding the security of their organization's cyber systems, about one-half of the respondents generally felt their systems were secure. However, respondents in small firms gave the strongest assessment of security, with 85% identifying their system as secure or very secure. Large and mid-size organizations were more cautious, with only 47% and 41% identifying as secure or very secure, respectively.

TABLE 14.7:   Survey Respondents by Industry

Industry	Percentage of Respondents
Academia	73.40%
Computers and technology	5.10%
Engineering	5.10%
Banking	2.50%
Government	2.50%
Transportation and distribution	2.50%
Aerospace	1.30%
Energy Utilities	1.30%
Entertainment and media	1.30%
Healthcare	1.30%
Military and defense	1.30%
Other or none specified	1.30%
Retail and wholesale sales	1.30%

In total, about 52% of respondents reported their organization had been breached, with the breakdown by sector shown in Table 14.8. Although the small sample sizes cannot extend to general industry trends, note that attacks and breaches can happen to any organization. In the survey, of those who reported a breach, 71% were large organizations, 17% were mid-size organizations, and 12% were small organizations. This is compared to the total responses, whereby 57% of the large organizations reported a breach, 47% of mid-size organizations reported a breach, and only 38% of small organizations reported a breach. In total, 65% of organizations reported an attack without a breach. Moreover, it is quite possible that survey respondents that did not report their system breached could be unaware that one may have occurred.

Finally, the survey asked respondents to identify measures that are valued in a cyber system. This was an open response question, where respondents were not prompted with any specific measures or suggestions. As a result, the survey received a wide variety of responses, with some more relevant than others. For example, a measure of

TABLE 14.8:   Breaches by Industry

Industry	Number of Breaches
Academia	33
Engineering	2
Aerospace	1
Computers and technology	1
Entertainment and media	1
Government	1
Military and defense	1
Transportation and distribution	1

TABLE 14.9: Valued Measures of a Secure Cyber System, by Industry Sector

Category	Measure (semi-colon separating each response)	Industry
Access Controls	Administrative controls; control of information assets	Academia
	Authenticated access; controlled access to personalized data	Transportation and Distribution
Automation	AI driven (Artificial Intelligence)	Academia
	Automated	Academia, Banking
Awareness	Awareness; employee awareness; informed user base; phishing awareness; security awareness	Academia
	User awareness	Other
Consistency	Consistent	Academia
Monitoring	Continuous monitoring	Academia, Computer and Network Security
	Event monitoring; flow monitoring; scanning	Academia
	Network monitoring	Aerospace
	Reduced false positive rate	Other
Human Behavior	Behavior based; communication with community; people; skilled staff or team	Academia
Other	Identity Management; non-invasive; Peoplesoft; technology	Academia
	Tooling	Banking
	Dependence	Engineering
	Contextual data	Other
Prevention	IT security governance; procedure/policy/authority document linkages; process; encryption; IDS (intrusion detection system); intrusion prevention; IPS (internet provider security); prevention; strong two-factor authentication	Academia
	Risk posture	Computer and Network Security
	Risk assessment; identification of vulnerabilities	Energy Utilities
	Comprehensive security program	Technology
	Secure authentication and certificates	Other
	IAST (interactive application security testing) and RASP (runtime application self-protection)	Engineering
	Encryption	Healthcare, Technology
	Validated inputs	Transportation and Distribution

(Continued)

TABLE 14.9 (CONTINUED): Valued Measures of a Secure Cyber System, by Industry Sector

Category	Measure (semi-colon separating each response)	Industry
Resilience/Design	Firewall; design; multiple defensive layers; NGFW; red forest	Academia
	Red team	Banking
	Multi-factor authentication	Computer and Network Security
	Integrity; most of security features in AWS; pentest (penetration testing)	Engineering
	Resilience	Engineering, Transportation and Distribution
	Secure data APIs (application programming interfaces)	Transportation and Distribution
Robustness	Robustness in the face of attack	Academia
Threat Detection	Event management; ability to separate network traffic; defense in depth; early threat detection	Academia
	Detection	Academia, Banking
	Threat detection	Academia, Aerospace, Computer and Network Security, Energy Utilities, Engineering, Government
	Threat modeling	Banking
	Unique tools that recognize threats	Energy Utilities
	Threat detection	Retail and Wholesale Sales
Training	Training	Academia, Aerospace
	User training	Academia, Banking, Technology
	Developer training	Banking
	Administrator training	Other
Usability/ Transparency	Transparent to users	Academia
	Alerts/notifications	Banking
	Usability	Military and Defense

"continuous monitoring" is a best practice, which through metrics, can determine the degree to which system monitoring is being achieved. On the other hand, "PeopleSoft" is a human capital management system and not necessarily a security practice that can be measured. Table 14.9 presents the final inventory of responses; a summary of these responses also appears in Black et al. (2018).

For simplicity, like responses or measures are grouped into themes or categories with the industry from which the measures were elicited. What an organization or IT

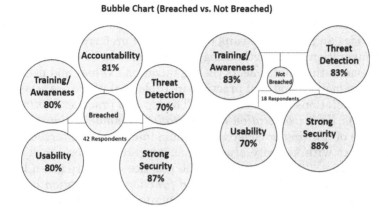

Bubble Chart (Breached vs. Not Breached)

FIGURE 14.6: Survey responses (breached versus non-breached).

professional values may vary by industry or other factors, such as degree of experience with cyber or a history of breaches. For example, companies or organizations that were breached may value different measures or may provide different relative importance to like measures. To highlight differences in value based on an individual's history of cyber breaches, consider the major themes within the elicited measures from the survey along with the corresponding 0–100 weight for the measure provided by the survey respondents. Figure 14.6 provides an illustration of the respondent themes when the weights are averaged across all measures; the responses are further separated into those who reported a breach and those who did not report a breach. Note that a portion of the survey respondents did not answer as to if their organization was breached, and the measures or values provided by those respondents were removed for this illustration. The figure shows that breached organizations were consistent in valuing threat detection, training/ awareness, and strong security. Organizations that were breached, regardless of industry sector, also valued accountability, which is not seen in the non-breached organizations.

Students and practitioners can use the results of the survey to understand trends in cybersecurity. Becoming aware of what IT professionals consider and care about can lead to better decision-making and reporting that supports the organization's needs. It is important to develop and build models that truly address the problem at hand. Specifically, models should be verified (no math or logic errors) and valid (truly answering the question posed by the decision-maker). The survey results should start a discussion within a practitioner's organization: are the same or different measures valued? What is relevant for this organization?

14.5 Conclusions

Within the academic literature, most research in cybersecurity focuses on theory, infrastructure management, and best practices associated with computer science and network infrastructure. To translate cybersecurity research into practice, however, decision modeling and risk assessment are needed. This chapter outlines several of the operations research techniques used in modeling cybersecurity threats, then proposes a cybersecurity value model framework, supported by data and interviews with subject matter experts, that offers measures and values, scores metrics on their contribution to value, and provides an illustrative rank ordered list of preferred metrics and best practices for implementation. This methodology can be followed by organizations to build a model for implementation that reflects the organization's values and needs.

With the list of metrics and best practices in Table 14.6, there are several suggestive insights or interpretations as to their rank. The top two practices, *control administrative privileges* and *limit workstation-to-workstation communication*, are most closely tied to the theme of limiting or constraining human-computer interaction. This seems to align in context with security studies that suggest most breaches are caused by human error (Verizon, 2017). In contrast, the practices that are given less priority in the listing, such as *segregating networks and functions*, *application whitelisting*, and *setting a secure baseline configuration*, are commonly performed only by IT administrators rather than the common workplace employee. Between these extremes, there are technological routines such as anti-virus and anti-exploitation schemes, as well as software improvements, which are services that may be provided and administered external to an organization's network infrastructure. In summary, the ranking validates a spectrum of influence, putting greater emphasis on those practices that are less likely to be controlled in terms of organizational cybersecurity.

While the model presented in this chapter is illustrated using a notional supply chain case study, it can be customized to assess the performance of cyber systems for any organization. Aside from broadening the survey to investigate different industry perspectives or expanding the value criteria, there are several logical extensions to this model. This work can be further extended to examine various ranking techniques, along with the rank-order centroid method, to compare scoring tendencies and potential biases for the metrics and best practices. Additional value function elicitation techniques may also be used to refine the scoring and provide alternative perspectives for prioritizing criteria. Finally, scoring based upon the contribution to value of a practice may be adapted to consider principles of cost or damage in the context of organizational risk.

As technology develops, the measures, metrics, and best practices used in this framework may no longer be current or robust. However, the methodology and approach will hold, allowing for organizations to adapt and update their security practices to keep up with the times. Exercises for students include researching current measures, metrics, and best practices in order to become well-versed in current trends. Other exercises include notional scoring for a specific firm or industry. Practitioners can take the framework and directly apply it to their organization, using their firm's data and values, to customize and increase security based on the organization's needs. Successful models in practice directly address the problem at hand and are understandable and relatable by decision-makers. This framework, which is an example of using metrics grounded in defense principles, allows for growth and customization, hence translating research into practice with direct impact.

Acknowledgements

The authors would like to thank Rachael Artes, Jasmin Farahani, Rachel Fredman, and Emil Manuel for their help with data collection and scoring. The authors would also like to thank the two SMEs for their time and invaluable insights. This research was partially supported by a Towson University College of Business and Economics

Summer Research Grant; the views expressed in this paper are those of the authors and do not represent the official policy or position of the United States Military Academy, the United States Army, or the United States Department of Defense.

References

Anderson, D. (2011). Increase security posture with application whitelisting. https://cdn. selinc.com/assets/Literature/Publications/Technical%20Papers/ 6477_IncreaseSecurity_ DA_20110201_Web.pdf?v=20150812-081845. Accessed 08.02.18.

Anderson, D. (2014). Protect critical infrastructure computer systems with whitelisting. https:// www.sans.org/reading-room/whitepapers/ICS/protect-critical-infrastructure-systems-whitelisting-35312. Accessed 07.15.18.

Barron, F. H. (1992). Selecting a best multiattribute alternative with partial information about attribute weights. *Acta Psychologica*, 80(1–3), 91–103.

Barron, F. H., & Barrett, B. E. (1996). Decision quality using ranked attribute weights. *Management Science*, 42(11), 1515–1523. doi:10.1287/mnsc.42.11.1515. Accessed 07.15.18.

Beauregard, J. E. (2001). *Modeling information assurance*. Air Force Institute of Technology.

Black, L., Scala, N. M., Goethals, P. L., and Howard II, J. P. (2018). Values and trends in cybersecurity. Proceedings of the 2018 Industrial and Systems Engineering Research Conference. http://tinyurl.com/ValuesTrends. Accessed 02.19.2019

Boyens, J. (2016). Integrating cybersecurity into supply chain risk management. https://www. rsaconference.com/writable/presentations/file_upload/grc-w03_integrating_cybersecurity_ into_supply_chain_risk_management.pdf. Accessed 08.02.18.

Buckshaw, D. L., Parnell, G. S., Unkenholz, W. L., Parks, D. L., Wallner, J. M., & Saydjari, O. S. (2005). Mission oriented risk and design analysis of critical information systems. *Military Operations Research*, 10(2), 19–38.

Buede, D. M. (2009). *The engineering design of systems: Models and methods* (2nd ed). Hoboken, NJ: John Wiley & Sons.

Camillo, M. (2017). Cyber risk and the changing role of insurance. *Journal of Cyber Policy*, 2(1), 53–63.

Cormode, G., Procopiuc, C. M., Shen, E., Srivastava, D., & Yu, T. (2013). Empirical privacy and empirical utility of anonymized data. Proceedings of the IEEE 29th International Conference on Data Engineering Workshops, 77–82.

Dillon, R. L., Lester, G., John, R. S., & Tinsley, C. H. (2012). Differentiating conflicts in beliefs versus value tradeoffs in the domestic intelligence policy debate. *Risk Analysis*, 32(4), 713–728.

Dillon-Merrill, R. L., Parnell, G. S., Buckshaw, D. L., Hensley, W. R., & Caswell, D. J. (2008). Avoiding common pitfalls in decision support frameworks for department of defense analyses. *Military Operations Research*, 13(2), 19–31.

Edwards, W., & Barron, F. H. (1994). SMARTS and SMARTER: Improved simple methods for multiattribute utility measurement. *Organizational Behavior and Human Decision Processes*, 60(3), 306–325.

Edwards, W., & Winterfeldt, D. (1987). Public values in risk debates. *Risk Analysis*, 7(2), 141–158.

Evans, J. R. (2013). *Business analytics: Methods, models, and decisions*. Upper Saddle River, NJ: Pearson Education.

Farahani, J., Scala, N. M., Goethals, P. L., & Tagert, A. (2016). Best practices in cybersecurity: Processes and metrics. *Baltimore Business Review: A Maryland Journal*, 28–32.

Feng, T., & Keller, L. R. (2006). A multiple-objective decision analysis for terrorism protection: Potassium iodide distribution in nuclear incidents. *Decision Analysis*, 3(2), 76–93.

Fielder, A., Panaousis, E., Malacaria, P., Hankin, C., & Smeraldi, F. (2016). Decision support approaches for cyber security investment. *Decision Support Systems*, 86, 13–23.

Goethals, P. L., Farahani, J., & Scala, N. M. (2015). Metrics for tracking and evaluating cyber-security posture. Presentation at the Army Operations Research Symposium, Aberdeen Proving Ground, Maryland.

Goodwin, P., & Wright, G. (2009). *Decision analysis for management judgment* (4th ed). Chichester, West Sussex, UK: Wiley.

Greenberg, A. (2017, September 18). Software has a serious supply-chain security problem. *Wired.* https://www.wired.com/story/ccleaner-malware-supply-chain-software-security/. Accessed 08.02.18.

Helmer-Hirschberg, O. (1967). Analysis of the future: The Delphi method. https://www.rand.org/pubs/papers/P3558.html. Accessed 02.19.2019

Howard, M. (2003). Fending off future attacks by reducing attack surface. *Microsoft Developer Network (MSDN) Magazine.*

Johnson, A., Dempsey, K., Ross, R., Gupta, S., & Bailey, D. (2011). Guide for security-focused configuration management of information systems. http://nvlpubs.nist.gov/nistpubs/ Legacy/SP/nistspecialpublication800-128.pdf. Accessed 08.02.18.

Keeney, R. L. (1992). *Value-focused thinking: A path to creative decision making.* Cambridge, MA: Harvard Univ. Press.

Keeney, R. L. (2007). Modeling values for anti-terrorism analysis. *Risk Analysis*, 27(3), 585–596.

Keeney, R. L. (2008). Applying value-focused thinking. *Military Operations Research*, 13(2), 7–17.

Keeney, R. L., & Raiffa, H. (1976). *Decisions with multiple objectives: Preferences and value tradeoffs.* New York: Wiley.

Keeney, R. L., & von Winterfeldt, D. (2011). A value model for evaluating homeland security decisions. *Risk Analysis*, 31(9), 1470–1487.

Kirkwood, C. W. (1997). *Strategic decision making: Multiobjective decision analysis with spreadsheets.* Belmont: Duxbury Press.

Lord, N. (2017, July 27). Supply chain cybersecurity: Experts on how to mitigate third party risk. https://digitalguardian.com/blog/supply-chain-cybersecurity. Accessed 08.03.18.

National Institute of Standards and Technology. (n.d.). Best practices in cyber supply chain risk management: FireEye supply chain risk management. https://csrc.nist.gov/CSRC/media/Projects/Supply-Chain-Risk-Management/documents/case_studies/USRP_NIST_FireEye_081415.pdf. Accessed 08.02.18.

National Security Agency. (n.d.). Science of Security. https://www.nsa.gov/What-We-Do/Research/Science-of-Security. Accessed 12.05.19.

National Security Agency Information Assurance Directorate. (2013). IAD's Top 10 Information Assurance Mitigation Strategies. https://www.sans.org/security-resources/IAD_top_10_info_assurance_mitigations.pdf. Accessed 11.01.17.

Nicol, D., Sanders, W., Scherlis, W., & Williams, L. (2012). Science of Security hard problems: A lablet perspective. Science of Security and Privacy Virtual Organization: https://cps-vo.org/node/6394. Accessed 01.01.18.

Nicol, D. M., Scherlis, W. L., Katz, J., Scherlis, W. L., Dumitras, T., Williams, L. M., & Singh, M. P. (2015). Science of Security lablets: Progress on hard problems. Science of Security and Privacy Virtual Organization: http://cps-vo.org/node/21590. Accessed 07.15.18.

Parnell, G. (2007). Value-focused thinking using multiple objective decision analysis. In A. G. Loerch & L. B. Rainey (Eds.), *Methods for conducting military operational analysis: Best practices in use throughout the Department of Defense.* Alexandria, VA: MORS, 619–656.

Parnell, G. S., Bresnick, T. A., Tani, S. N., & Johnson, E. R. (2013). *Handbook of decision analysis.* Hoboken, NJ: Wiley.

Parnell, G. S., Butler, R. E., Wichmann, S. J., Tedeschi, M., & Merritt, D. (2015). Air force cyberspace investment analysis. *Decision Analysis*, 12(2), 81–95.

Parnell, G. S., Hughes, D. W., Burk, R. C., Driscoll, P. J., Kucik, P. D., Morales, B. L., & Nunn, L. R. (2013). Invited review - Survey of value-focused thinking: Applications, research developments and areas for future research: Survey of value-focused thinking. *Journal of Multi-Criteria Decision Analysis*, 20(1–2), 49–60.

Rees, L. P., Deane, J. K., Rakes, T. R., & Baker, W. H. (2011). Decision support for cybersecurity risk planning. *Decision Support Systems*, 51(3), 493–505.

Ryan, J. J. C. H., Mazzuchi, T. A., Ryan, D. J., Lopez de la Cruz, J., & Cooke, R. (2012). Quantifying information security risks using expert judgment elicitation. *Computers & Operations Research*, 39(4), 774–784.

Santos, J. R., Haimes, Y. Y., & Lian, C. (2007). A framework for linking cybersecurity metrics to the modeling of macroeconomic interdependencies: Linking cybersecurity metrics to the modeling of macroeconomic interdependencies. *Risk Analysis*, 27(5), 1283–1297.

Sawik, T. (2013). Selection of optimal countermeasure portfolio in IT security planning. *Decision Support Systems*, 55(1), 156–164.

Scala, N. M., & Goethals, P. L. (2016). A review of and agenda for cybersecurity policy models. Proceedings of the 2016 Industrial and Systems Engineering Research Conference. http://www.tinyurl.com/PolicyModels. Accessed 02.19.2019

Scala, N. M., Kutzner, R., Buede, D., Ciminera, C., & Bridges, A. (2012). Multi-objective decision analysis for workforce planning: A case study. Proceedings of the 2012 Industrial and Systems Engineering Research Conference.

Scala, N. M., & Pazour, J. A. (2016). A value model for asset tracking technology to support naval sea-based resupply. *Engineering Management Journal*, 28(2), 120–130.

Scala, N. M., Reilly, A., Goethals, P. L., & Cukier, M. (2019). Perspectives: Risk and the five hard problems of cybersecurity. *Risk Analysis: An International Journal*, 39(10), 2119–2126.

Schilling, A., & Werners, B. (2016). Optimal selection of IT security safeguards from an existing knowledge base. *European Journal of Operational Research*, 248(1), 318–327.

Schneier, B. (1999). Attack trees: Modeling actual threats. Social and Affective Neuroscience Society (SANS) Conference on Network Security, New Orleans, Louisiana, October 6.

Sedgewick, A., Souppaya, M., & Scarfone, K. (2015). Guide to application whitelisting. http://nvlpubs.nist.gov/nistpubs/SpecialPublications/NIST.SP.800-167.pdf. Accessed 07.15.18.

Shackleford, D. (2009). Application whitelisting: Enhancing host security. https://www.sans.org/reading-room/whitepapers/analyst/application-whitelisting-enhancing-host-security-34820. Accessed 07.15.18.

Shackleford, D. (2010). Keys to the kingdom: Monitoring privileged user actions for security and compliance. https://www.sans.org/reading-room/whitepapers/analyst/keys-kingdom-monitoring-privileged-user-actions-security-compliance-34890. Accessed 07.15.18.

Shackleford, D., & Douglas, J. (2015, September 10). Combatting cyber risks in the supply chain. https://www.sans.org/webcasts/combatting-cyber-risks-supply-chain-100657. Accessed 08.02.18.

Simpson, D. (2017). Cybersecurity risk reduction. https://apps.fcc.gov/edocs_public/attachmatch/DOC-343096A1.pdf. Accessed 08.02.18.

Souppaya, M., & Scarfone, K. (2013). Guidelines for managing the security of mobile devices in the enterprise. http://nvlpubs.nist.gov/nistpubs/SpecialPublications/NIST.SP.800-124r1.pdf. Accessed 07.15.18.

Theisen, C., Munaiah, N., Al-Zyoud, M., Carver, J., Meneely, A., & Williams, L. (2018) Attack surface definitions: A systematic literature review. *Information and Software Technology*, 104, 94–103.

Townsend, C. (2017). OPM breach costs could exceed $1 billion. https://www.symantec.com/connect/blogs/opm-breach-costs-could-exceed-1-billion. Accessed 08.02.18.

Vengurlekar, N., Clouse, B., & Kammend, R. (2012). Network isolation in private database clouds. http://www.oracle.com/technetwork/database/database-cloud/ntwk-isolation-pvt-db-cloud-1587225.pdf. Accessed 08.02.18.

Verizon Enterprise. (2017). Data breach investigations report. http://www.verizonenterprise.com/verizon-insights-lab/dbir/2017/?utm_source=pr&utm_medium=pr&utm_campaign=dbir2016. Accessed 07.15.18.

Viduto, V., Maple, C., Huang, W., & López-Peréz, D. (2012). A novel risk assessment and optimisation model for a multi-objective network security countermeasure selection problem. *Decision Support Systems*, 53(3), 599–610.

Wallace, M. (2016). Mitigating cyber risk in IT supply chains. *The Global Business Law Review*, 6(1–2), 1–55.

Wright, P. D., Liberatore, M. J., & Nydick, R. L. (2006). A survey of operations research models and applications in homeland security. *Interfaces*, 36(6), 514–529.

Yevseyeva, I., Basto-Fernandes, V., Emmerich, M., & van Moorsel, A. (2015). Selecting optimal subset of security controls. *Procedia Computer Science*, 64, 1035–1042.

Chapter 15

Applying Information Theory to Validate Commanders' Critical Information Requirements

Mark A.C. Timms, David R. Mandel and Jonathan D. Nelson

15.1 Introduction

The primary aim of this chapter is to introduce a novel approach to strengthen contemporary intelligence community practices for establishing intelligence collection priorities based on expected information value. We propose the integration of quantitative measures of information utility that have been discussed in the literature on information theory (Lindley, 1956; Nelson, 2005; Crupi & Tentori, 2014) as a method for optimizing intelligence collection planning. We argue that enhancing the effectiveness through which command information requirements are established can improve consequent intelligence collection priorities. We contrast this approach with the structured analytic technique (SAT) approach that is currently described as a method for prioritizing information requirements in intelligence collection. Specifically, we proceed with a review of the Indicators Validator™ (IV) SAT (Heuer & Pherson, 2008, 2014) for establishing information value, illustrating how it works, and where it falls short as an analytic method. Next, we introduce a quantitative information-theoretic measure of information utility called *information gain* (Lindley, 1956). We illustrate the contrast between these approaches using a practical example featuring a hypothetical North Atlantic Treaty Organization (NATO) dilemma. This analysis shows how information gain overcomes many limitations of the IV technique, along with how it might be applied to modern NATO operational practice.

15.1.1   Background: The NATO Intelligence Community Dilemma

Intelligence organizations iteratively explore new ways to assess information value. NATO intelligence professionals inform complex, high-consequence, operational decisions on a routine basis. First established in 1949, NATO's stated purpose is to "guarantee the freedom and security of its members through political and military means" (NATO, 2017a). Where a NATO force has been deployed to monitor another government's adherence to ceasefire agreements, the success of its mandate could become entirely dependent on its ability to accurately interpret indicators of imminent aggression. Under these circumstances, failing to act when required can be just as damaging as taking action when none is warranted. Intuitively, when charged with such a fragile task, a NATO commander would want to position his forces in such a way that allow the initiation of swift, deliberate, and decisive intervention (if and when required), without adopting a force posture that inadvertently encourages or re-ignites existing tensions between hostile states.

The NATO intelligence community (IC) enhances command understanding of complex environments through the delivery of predictive assessments founded in the deliberate analysis of threat event indicators and warning signs. Whether the analysis is intended to provide context or early warning, or to identify opportunities, it is fundamentally about improving decision-making under conditions of uncertainty (CFINTCOM, 2016). Once a mission or political mandate is defined, intelligence professionals are often left to identify which questions, if answered, can most efficiently improve stakeholder decision-making in the context of that mission. We suggest that the consistent, coherent, and precise evaluation of information usefulness during the earliest stages of operational planning is vital to ensuring sound intelligence collection planning, although some literature suggests that members of the operational community often only pay lip service to this aim (US Government, 2013).

15.1.2   Establishing Command Information Priorities

In order to prioritize organizational resourcing, decision-makers issue a series of information requirements (IR) to subordinate units that subsequently drive intelligence collection efforts. Some of those IR (i.e., questions), when answered, may compel stakeholders to take action, cease action, or have some form of immediate impact on organizational posture. These are often called Commander's Critical Information Requirements (CCIR) (Commander of the Canadian Army, 2013, p. 3-2). Although commanders issue planning guidance, individual planners (or groups) often develop CCIR through subjective, multi-stage planning processes that vary across organizations. CCIR can drive the assignment of collection resources to fill information gaps (Chief of Defence Staff, 2002), and they are often presented to non-expert decision-makers for approval without having been formally evaluated to ensure that their answers will actually reduce uncertainty in command decision-making in some appreciable manner (US Government, 2013).

Figure 15.1 depicts the hierarchical relationship between Commanders' Critical Information Requirements (CCIR) and related information priorities developed through consequent processes, including Priority Intelligence Requirements (PIR) and Friendly Force Information Requirements (FFIR), each of which drive the defining of Essential Enemy Information (EEI), Essential Elements of Friendly Information (EEFI), and more (adapted from US Government, 2013). Numerous agencies may define CCIR (and

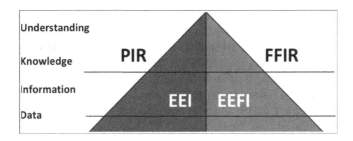

FIGURE 15.1: Hierarchy of information requirements.

other consequent IRs) through subjective group (or individual) brainstorming. The success of such exercises will likely depend on the abilities of the individual planner (or planning group), varying in terms of the diversity, size, and breadth of experience (among other items) of the members involved. At the very least, planners must be capable of shaping proposed CCIR from their translation of such intent, through formulating plans to meet those requirements.

15.2 The SAT Approach

The IV technique that we consider later in this chapter is an example of the SAT approach adopted over several decades within many NATO countries. The development of SATs started out as a rather idiosyncratic, path-dependent endeavor spearheaded by US intelligence tradecraft mavericks, such as Richards Heuer Jr and Jack Davis, who took it upon themselves while working at the Central Intelligence Agency (CIA) to develop back-of-the-napkin techniques to aid intelligence analysts (Mandel & Tetlock, 2018). The impetus for developing such methods was, in large part, the belief that analysts were prone to cognitive biases that SATs would help effectively overcome. In this view, SATs could be used to structure the otherwise unbridled intuitions of analysts and to tame their purported wanton subjectivity. The effort to develop SATs and require that analysts be trained to use them received intermittent commitment within the US IC until pivotal geopolitical events – namely, the 9/11 terrorist attacks against the US by Al Qaeda and the faulty, invasion-prompting intelligence estimate that Saddam Hussein was developing weapons of mass destruction in Iraq – triggered congressionally mandated institutional reforms that required the use of SATs by the US IC (Artner, Girven & Bruce, 2016; Chang et al., 2018; Coulthart, 2017; Marchio, 2014). The CIA's tradecraft manual originally included 12 SATs (US Government, 2009), but the list of SATs has burgeoned, now including several dozen such techniques (Heuer & Pherson, 2008, 2014).

The body of scientific research on SATs (and analytic tradecraft, more generally) remains scant (Chang et al., 2018; Dhami et al., 2015; Mandel, in press; Pool, 2010). Unfortunately, the IC has tended to view SATs as benign if not beneficial. While SAT proponents will often admit that SATs "aren't perfect," they are usually quick to add that they are "better than nothing." However, recent evidence indicates that the latter supposition may be false. Mandel, Karvetski, and Dhami (2018) studied the effects of training analysts in the use of one particular SAT, the Analysis of Competing Hypotheses (Heuer, 1999; Heuer & Pherson, 2014). Remarkably, analysts who used that SAT to assess the probability of alternative hypotheses were significantly *less* coherent and also less accurate in their judgments than analysts who were not instructed to use any SAT (also see Dhami, Belton & Mandel, in press). Such findings should not be unexpected given that SATs, more generally, are subject to two important conceptual shortcomings (Chang et al., 2018; Mandel & Tetlock, 2018). First, they neglect the fact that most cognitive biases are bipolar (e.g., calibrated confidence is offset by underconfidence *or* overconfidence) and they fail to assess the types of biases analysts are in fact prone to before intervening. Second, SATs neglect the cost of noisy judgments that follow from techniques that, though supposedly objective, in fact invite a range of implementation-related decisions that are left to analysts' discretion. Thus, SATs may do more to redirect subjectivity from substantive assessment to resolving methodological vagueness.

Few SATs focus explicitly on evaluating information utility. Those that do are geared towards establishing the predictive value of threat event indicators in the context of impending hypothetical threat events; namely, the Indicators and IV SATs (Heuer & Pherson, 2008, 2014). Indicators are defined as: "observable phenomena that can be periodically reviewed to help track events, spot emerging trends, and warn of anticipated changes" (Heuer & Pherson, 2008, 2014, p. 149). The Indicators SAT encourages analysts to leverage their personal experience in concert with easily accessible information in the development of a detailed indicators list. This list reflects a "pre-established set of observable or potentially observable actions, conditions, facts, or events whose simultaneous occurrence would strongly argue that a phenomenon is present, or at least highly likely to occur" (Heuer & Pherson, 2008, 2014, p. 149). Heuer and Pherson's Indicators SAT encourages analysts to build a list of indicators presumed to be associated with hypothetical threat events.

For instance, if you were wondering whether it was going to rain, you might reflexively consider checking the ambient barometric pressure, listening for thunder, or perhaps looking for lightning. Heuer and Pherson's (2014) Indicators SAT suggests that NATO intelligence professionals perform similar exercises when attempting to predict events of operational interest. Hypothetical examples include (but are not limited to): whether a political official will be re-elected before cease-fire agreements are signed, the volume of refugees that might move down a series of different corridors in the aftermath of a natural disaster, or whether a small military force might suddenly annex a sovereign bordering state. Whereas the Indicators SAT focuses on indicator definition, its companion IV SAT aims to assist analysts in establishing the predictive value of indicators in the context of a given threat event scenario. In the next section, we review the IV SAT and analyze its performance. Later, to illustrate differences between the IV SAT and more formal models of information utility, we introduce a relevant example inspired by NATO's Enhanced Forward Presence (EFP) forces stationed in Eastern Europe.

15.2.1   The Indicators Validator™ SAT

Introduced by Heuer and Pherson in 2008, the IV SAT focuses on establishing the predictive value of an indicator based on how exclusively it indicates a focal hypothesis or threat scenario among a set of scenarios. According to Heuer and Pherson, indicators are "observable phenomena that can be periodically reviewed to help track events, spot emerging trends, and warn of anticipated changes" (2014, p. 149).

To use the IV SAT, analysts must first identify a list of mutually exclusive and collectively exhaustive threat scenarios (sometimes called hypotheses) to be predicted. These can be multialternative (Event A: Person X is elected, Event B: Person Y is elected, Event C: Person Z is elected) or binary (yes/no, happened/did not happen). Each scenario is accompanied by a list of primary indicators that analysts believe would be likely to be present if that scenario were to occur (or if the hypothesis were true). Indicators that are generated for a particular scenario are said to be "at home" for that scenario and are not at home for the alternative scenarios. That is, any given indicator can only be at home in one scenario for the IV SAT to work as intended. However, any given scenario may have multiple indicators that are at home in it.

After assigning indicators to their home scenarios, the analyst must judge whether each indicator is to be rated as *likely* or *highly likely* in its home scenario, as this will affect the consequent information value scoring procedure (see Figure 15.2). At this stage, the analyst cannot select other probability values (e.g., *very unlikely*) to represent the indicator. In other words, an indicator that is "at home" must be judged to be either *likely* or *very likely*, given that scenario. Next, the likelihood of each indicator given each of the alternative scenarios is assessed. For example, if an indicator is deemed to be *highly likely* given the home scenario, then numerical values would be assigned to the indicator in the alternative

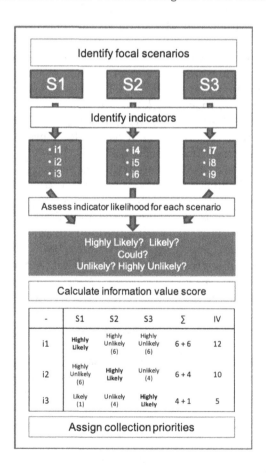

FIGURE 15.2: The Indicators Validator™ technique.

scenarios as a function of how divergent they are from the original at home rating using the following coding scheme: *highly likely* = 0 (i.e., no divergence), *likely* = 1, *could* = 2, *unlikely* = 4 *highly unlikely* = 6 (i.e., maximum divergence; Heuer & Pherson, 2014, p. 159). The IV SAT makes an adjustment for whether the indicator is *highly likely* or only *likely* in the home scenario; if it is judged to be *likely*, a similar rating scheme is applied but the "distance scores" are smaller in magnitude (Heuer & Pherson, 2014). Finally, the analyst would sum the distance scores assigned to the alternative scenarios. The greater the summed distance score, the more useful a given indicator is deemed to be.

15.2.2   Analysis of the IV SAT

The IV SAT is a simple method for scoring the usefulness of different indicators. The ordering of the steps in the technique is easy to follow and the application of the IV SAT does not require mathematical sophistication. Nevertheless, in spite of its ease of application, the IV SAT has important limitations that could lead collectors astray, and misinform analysts and intelligence consumers.

One problem with the technique is the arbitrariness of matching indicators to home scenarios. For an indicator to be at home it must be judged either *likely* or *very unlikely* given the scenario. However, it may be judged equally likely under other scenarios, in which case the indicator might just as well have been at home in those scenarios. This aspect of the technique highlights another sense in which it is arbitrary – namely, it disallows negative hypothesis tests in which one searches for indicators that may be (*very*) *unlikely* if the scenario were true. Detecting the presence of such low probability events can be highly informative, yet the IV SAT precludes such a focus in prioritization of information requirements. One remarkable implication of this constraint is that the complement of a good indicator (i.e., one that has low probability in the home scenario but high probability in all of the other scenarios) would be precluded from being considered. This is not only arbitrary; it is incoherent.

Another limitation of the IV SAT is that it does not require or even prompt analysts to consider the prior probability of scenarios or indicators. It does not do so either in terms of collection of objective, relative frequency data that could be used to establish base-rate estimates or in terms of subjective estimates of these relative frequencies, which could also be useful. This critical base-rate information is accounted for in virtually all information-theoretic models (Lindley, 1956; Nelson, 2005; Wu et al., 2017), and also in Bayesian approaches to belief revision (Navarrete & Mandel, 2016). Collection resources are not unlimited (Folker, 2000), necessitating careful evaluation and IR prioritization. We suggest that it would be useful to first establish how likely a commander is to correctly predict an outcome based on what is already known about event base rates, and then to evaluate the utility of information in the context of how much it improves event predictability over reliance on prior probabilities alone. If that which is already known about a given threat event can enable a commander to confidently predict its occurrence, collection assets might more efficiently be directed towards answering questions that would measurably reduce uncertainty associated with other events.

Consider an indicator that would almost certainly signal the occurrence of Scenario 1 (Attack) – such as an intercepted Russian correspondence directing military forces to cross their border at a given time – where the actual probability of an attack occurring was assessed to be extremely low. The IV SAT would award a high information value score to that indicator, plausibly resulting in it becoming a priority collection item, despite the fact that its likelihood of appearance is significantly less likely than competing indicators in other scenarios, because it is at home in an improbable scenario.

The IV technique is the only information evaluation SAT featured in open-source intelligence analytic tradecraft manuals for both Canadian military (CFINTCOM, 2016) and American (US Government, 2013) intelligence agencies. Unfortunately, for the reasons noted, it lacks a sound, logical foundation. However, its vague quantification of individual probability judgments using linguistic probabilities on an ordinal scale, and the procedures used for calculating a final information value score for an indicator, can lead analysts to believe that they are following a valid, even objective, method.

The IV SAT assigns an indicator value as a function of how strongly its presence predicts a single threat scenario. The final information value score focuses only on the extent to which the indicator can discriminate between the scenario in which it is at home and alternatives in the same set. This could dramatically reduce efficiency in collection planning. In our example, for simplicity, we limit the number of threat scenarios to three. But in many cases, there will be multiple threat scenarios of interest that are thematically similar but ultimately distinct. In such cases, a low information value score for an indicator could be deceiving. Furthermore, consider an indicator that is strongly associated with the occurrence of all but one scenario. IV would give such an indicator a low information value score, despite the fact that it could reliably help a NATO commander predict the event in which it is not present.

With this in mind, we present an alternative approach to evaluating and prioritizing command information requirements, using information gain (Lindley, 1956; Nelson, 2005). In many ways, the structure of information-theoretic measures compels increased analytic reasoning, as the various inputs of the information gain formula may require the conduct of research, or the deliberate assignment of a numeric probability to a threat event scenario or its co-occurrence frequency with indicators. Importantly, information gain requires some input values for the probability of each scenario. The need to include estimates about each scenario's probability can encourage analysts to reduce the uncertainty associated with an entire problem, rather than pursuing information associated with events that may already be relatively easy to predict.

15.3   Information Gain: A Principled Approach to Evaluating Indicator Usefulness

In this section, we describe a quantitative information-theoretic measure of information utility called *information gain* that measures the average reduction in uncertainty achieved by using a specific indicator or cue (Lindley, 1956; Nelson, 2005). Information gain is an example of an information utility function, a mathematical formula designed to compute a quantitative estimate of utility for a piece of information. Superficially, information utility functions, like information gain, require similar inputs to those required when using the IV SAT. The basic principle remains: first, define information gaps (i.e., what one wants to know); next, identify what questions (and answers) might help fill them. The expected information value of a question is ultimately defined as the expected value of the not-yet-obtained answer, although the value of specific answer could also be calculated (Nelson, 2005, 2008). In contrast to the IV SAT, information gain has been effectively used in a variety of domains, such as automatic face recognition systems (Imaoka & Okajima, 2004), image registration (Chen, Arora &

Varshney, 2003), predicting human queries (Crupi et al., 2018), philosophy of science (Crupi & Tentori, 2014), and modeling neurons in visual (Ruderman, 1994; Ullman, Vidal-Naquet & Sali, 2002) and auditory perception (Lewicki, 2002).

15.3.1 Computing Information Gain

Information gain quantifies the utility of a given indicator as a function of how effectively its presence or absence reduces uncertainty about a hypothetical event of interest (Nelson, 2005). Lindley (1956), Box and Hill (1967), and Fedorov (1972) quantified this idea explicitly, using Shannon's (1948) entropy to measure the uncertainty in the outcome of a specific event. We measure information with base 2 logarithms (bits). Other bases could also be used. If the natural logarithm is used, the unit is *nats*.

For the purposes of defining information gain, let $Q = \{q_1, q_2, \ldots q_m\}$ represent a query (in mathematical terms, a random variable), in this case, the option of querying the value of a particular threat event indicator or command information requirement. Let each q_j represent one of the m possible answers to the question Q. Let $H = \{h_1, h_2 \ldots h_n\}$ represent the unknown hypothesis (or category or threat scenario) one is trying to predict. Finally, let each of the n possible h_i represent a specific hypothesis in the set of possibilities (i.e., a list of mutually exclusive and exhaustive threat event scenarios). Equation 1 shows the information gain calculation:

$$I(H,Q) = \left[\sum_{i=1}^{n} {}^*P(h_i) * \log_2 \frac{1}{P(h_i)}\right] - \left[\sum_{q=1}^{m} P(q_j) * \sum_{i=1}^{n} {}^*P(h_i \mid q_j) * \log_2 \frac{1}{P(h_i \mid q_j)}\right] \quad (15.1)$$

Information gain for a given indicator is equal to the initial entropy minus the entropy that is expected (on average) to be remaining after the indicator's state (e.g., present or absent) is observed. In other words, information gain measures the change in Shannon entropy from before (i.e., base-rate scenario uncertainty) to after consideration of the indicator's state (Nelson, 2005). Information gain can be used with indicators having two or more possible states. If information gain were used to prioritize collection, then the indicator with greatest expected reduction in uncertainty across the whole set of threat scenarios would be rated as the top priority for subsequent collection activities. Information gain is also known as the mutual information (Cover & Thomas, 2012) between the hypotheses of interest H and the indicator Q.

15.4 Applying the IV Sat and Information Gain to the NATO Example

Currently, there are numerous battalion-sized (300–1,300 soldiers) military units from contributing NATO member nations occupying a defense and deterrence posture in several countries along the Russian border (NATO, 2017b). Hypothetically, if one of these units intercepted correspondence that Russia intended to conduct a large-scale training event in the near future, this might trigger the formation of an incident-based planning group, where available staff officers would convene and think through new information with a view to presenting their commander with options for implementation. Imagine that a planning group is convened. The group must generate a prioritized list of information requirements associated with the Russian exercise, with a view to helping their commander determine their force posture during the exercise, whether reinforcements will be required, and more.

Using the IV SAT, the group would first flesh out a list of mutually exclusive and collectively exhaustive hypotheses, each of which might compel their commander to take or delay a specific action (kept relatively simple here, see Table 15.1). Scenario 1, from the infantry: the Russians are staging for an attack. The infantry planner also proposes her top indicator for this scenario: live ammunition. The idea is that if the Russians were staging for an attack, they would most certainly be carrying live ammunition. Scenario 2, from the logistician: the Russians intend to carry out a training exercise, sincerely aiming to improve the quality and professionalism of their forces through the practice of large military maneuvers. The logistician highlights that soldiers feed differently under combat conditions than they do in training. Largescale training events are likely to implicate the use of a non-tactical field feeding kitchen system for soldiers participating in training. Finally, the public affairs officer proposes another possibility, Scenario 3, namely that the Russians are actually posturing, conducting a show of force to NATO, to communicate that the multinational posture has not impacted their resolve. The public affairs officer further suggests that, if this scenario were to occur, the Russians would communicate their message through deliberate media events, such as press conferences.

Next, planners debate indicator/scenario co-occurrence frequencies. In Table 15.1, I1: Live Ammunition is at home in S1: Attack. The planners judge that Russian soldiers are *highly likely* to be carrying live ammunition when they stage before a combat event. I1 is agreed to be *highly unlikely* to be present in the other two scenarios, which earns it an IV information value score of 12, thus moving it to top priority for collection assets, followed closely by I2 (IV score: 10), with I3 well behind its companions (IV score: 5). Thus, the IV SAT prioritizes the indicators as follows: I1 > I2 > I3. Because these numbers are generated on an ordinal scale, differences in information value score do not directly reveal proportional increases (i.e., the fact that I3 = 5, and I2 = 10 does not mean that I3 is half as valuable as I2).

Next, we consider how information gain might be applied in this scenario. In Table 15.2, we have included additional planner estimates of event base rates for each of the threat event scenarios, where the probability (P) of S1, a deliberate military attack, is considered low (1%); the other scenarios are deemed much more likely to occur, with the probability of an exercise (S2) at a 33% chance, and the probability of posturing (S3) at a 66% chance. Table 15.2 illustrates that the information value scores applied to the same indicators using information gain produce the opposite prioritization as the IV

TABLE 15.1: IV Matrix for the Scenario. Estimates for "At Home" Indicators Are Bolded

Indicators	S1: Attack	S2: Exercise	S3: Posturing	Information Value Score	Collection Priority
i1: Live Ammunition	**Highly Likely**	Highly Unlikely (6)	Highly Unlikely (6)	12	1
i2: Field Kitchen	Highly Unlikely (6)	**Highly Likely**	Unlikely (4)	10	2
i3: Media Events	Likely (1)	Unlikely (4)	**Highly Likely**	5	3

TABLE 15.2: Information Gain Assessment for the Scenario

Indicators	S1: Attack	S2: Exercise	S3: Posturing	Information Value Score	Collection Priority
Prior Probabilities	**0.01**	**0.33**	**0.66**	–	–
i1: Live Ammunition	0.9	0.1	0.1	0.0249	3
i2: Field Kitchen	0.1	0.9	0.75	0.0408	2
i3: Media Events	0.75	0.3	0.9	0.2721	1

SAT. That is, using information gain, the expected values (in bits) of the indicators are: I1 = 0.0249, I2 = 0.0408, and I3 = 0.2722. This results in a collection priority assignment of I3 > I2 > I1. Clearly, the choice of method used to evaluate information usefulness can have dramatic consequences, including the full reversal of recommended collection priorities.

15.5 Discussion

As demonstrated in our example scenario in Tables 15.1 and 15.2, both the IV SAT and information gain can be used to facilitate prioritization of IR in support of information collection and decision-making. The IV SAT generates a rank-ordered list of indicator usefulness. Information gain provides a continuous ratio measure of the expected information value of alternative indicators. Although both the IV SAT and information gain require analysts to assess probabilities, the IV SAT imposes arbitrary constraints on what probabilities may be applied. Indeed, for the focal ("home") hypothesis, only two possibilities are permitted: likely or highly likely. In contrast, information gain does not impose arbitrary rules on probability assignment. Moreover, unlike the IV SAT, information theoretic measures such as information gain do not impose coarseness on the probabilities assigned. Recent research has shown that imposing coarseness on more granular probability judgments reduces accuracy across a wide range of conditions (Friedman, 2019). Although analysts may initially balk at the idea of providing more granular judgments, Barnes (2016) found that they rapidly adjust to making granular assessments and are willing to debate about differences that would otherwise have been obscured. The use of information-theoretic measures is therefore in step with recent recommendations to use numeric probabilities in intelligence production (Barnes, 2016; Dhami et al., 2015; Friedman, 2019; Irwin & Mandel, 2019).

Strikingly, our hypothetical NATO example shows that the IV SAT can deliver information collection priorities in total opposition to those generated by information gain. Given that information gain (in contrast to the IV SAT) has proven itself robust and useful over many decades in diverse contexts, this striking discrepancy should be disconcerting to operational communities that rely on the IV SAT or something like it. Why do the IV SAT and information gain contradict each other in this example? A key reason is that information gain accounts for the base rate of each threat scenario when computing information value. By comparison, the IV SAT is fully insensitive to these base rates. In this sense, the IV SAT formalizes, and perhaps even reinforces, base-rate neglect (Bar-Hillel, 1980; Kahneman & Tversky, 1973), whereas information-theoretic measures like information gain should mitigate this form of bias.

Information-theoretic metrics such as information gain could be tested through future research, completed in collaboration with military intelligence professionals, or anyone seeking to strengthen the manner in which they establish and validate command information requirements, on and off the battlefield. Information gain, based on expected reduction in Shannon entropy across the set of all possible hypotheses, is a widely used method. However, expected reduction in other kinds of entropy measures or qualitatively different information utility functions could also be used. The differences between information gain and related information-theoretic approaches are in many cases not dramatic (Nelson, 2005), in contrast with the differences between information gain and the IV SAT. For conciseness and because of its robustness and wide use in many domains, we have based the numeric example in this chapter on information gain. Nelson (2005, Appendix A) provides example numeric calculations, and Crupi et al. (2018) show how many different entropy and information measures from mathematics and physics can be articulated within a unified formal framework.

In future work, it would be worthwhile to go beyond our toy example and to consider actual relevant scenarios using these methods, considering also which of many possible methods are most appropriate. It would also be useful to consider the implications of including extrainformational factors, such as the cost of acquiring specific pieces of information, and potentially asymmetric costs of different kinds of mistakes, such as false-positive versus falsenegative errors, when evaluating the expected usefulness of possible indicators. A further qualification is that in our scenario, we consider the usefulness of each indicator individually. Depending on the dependency structure in the domain, if more than one indicator can potentially be queried, it may be necessary to evaluate the information value of possible sequences (decision trees) of indicator evaluation. The information-theoretic approaches generalize naturally to situations with known dependencies among indicators (see Nelson, Meder & Jones, 2018).

Since access to quantitative, frequentist data for intelligence analysis is often lacking (Spielman, 2016), future research might also explore optimal methods for compiling, organizing, and structuring such intelligence data. Additionally, since there is currently considerable apprehension in the intelligence community to using quantitative methods that require at least a rudimentary understanding of statistics and probability, intelligence management would have to better train analysts in these respects. Even brief training in probabilistic reasoning using natural frequency formats has been shown to improve analysts' logical coherence and accuracy (Mandel, 2015). Intelligence organizations might do well to de-emphasize some training that is of questionable value, such as current SAT tradecraft training (Chang et al., 2018; Mandel, in press). Rather than introducing new SATs that may have no more than face validity, it would be useful – as in the case of evaluating the expected usefulness of potential information sources – to consider whether other disciplines have robust techniques that could be adopted in defense and security contexts. As noted elsewhere (Dhami et al., 2015; Mandel, in press; Mandel & Tetlock, 2018; Pool, 2010), in support of that objective, the defense and security community should exploit the interdisciplinary decision sciences.

Author's Note: Correspondence concerning this chapter should be sent to David Mandel at David.Mandel@drdc-rddc.gc.ca. We thank the two anonymous reviewers and the editors for their feedback on earlier drafts of this chapter. This work was funded by Canadian Safety and Security Program projects CSSP-2016-TI-2224 and CSSP-2018-TI-2394 under the scientific direction of David Mandel. The work described in this chapter also contributed to NATO Systems Analysis and Studies Panel Research Task Group on Assessment and Communication of Uncertainty in Intelligence to Support Decision Making (SAS-114).

References

Artner, S., Girven, R. S., & Bruce, J. B. (2016). *Assessing the Value of Structured Analytic Techniques in the U.S. Intelligence Community*. Santa Monica, CA: RAND Corporation. https://www.rand.org/pubs/research_reports/RR1408.html.

Bar-Hillel, M. (1980). The base-rate fallacy in probability judgments. *Acta Psychologica, 44*, 211–233. doi:10.1016/0001-6918(80)90046-3

Barnes, A. (2016). Making intelligence analysis more intelligent: Using numeric probabilities. *Intelligence and National Security, 31*, 327–344.

Box, G., & Hill, W. (1967). Discrimination among mechanistic models. *Technometrics, 9*, 57–71.

Canadian Forces Intelligence Command. (2016). *Analytic Writing Guide* (Version 3.0). Ottawa, ON: Department of National Defence

Chang, W., Berdini, E., Mandel, D. R., & Tetlock, P. E. (2018). Restructuring our thinking about structured techniques in intelligence analysis. *Intelligence and National Security, 33*(3), 337–356.

Chen, H. M., Arora, M. K., & Varshney, P. K. (2003). Mutual information-based image registration for remote sensing data. *International Journal of Remote Sensing, 24*(18), 3701–3706.

Chief of Defence Staff. (2002). *CF Operational Planning Process*. B-GJ-005-500/FP-000. Ottawa, ON: Department of National Defence.

Commander Canadian Army. (2013). *Intelligence, Surveillance, Target Acquisition, and Reconnaissance (ISTAR) Volume 1 – The Enduring Doctrine*. B-GL-352-001/FP-001. Ottawa, ON: Department of National Defence.

Coulthart, S. J. (2017). An evidence-based evaluation of 12 core structured analytic techniques. *International Journal of Intelligence and CounterIntelligence, 30*, 368–391. doi:10.1080/08850607.2016.1230706

Cover, T. M., & Thomas, J. A. (2012). *Elements of Information Theory*. John Wiley & Sons.

Crupi, V., & Tentori, K. (2014). State of the field: Measuring information and confirmation. *Studies in History and Philosophy of Science Part A, 47*, 81–90.

Crupi, V., Nelson, J. D., Meder, B. Cevolani, G., & Tentori, K. (2018). Generalized information theory meets human cognition: Introducing a unified framework to model uncertainty and information search. *Cognitive Science, 42*, 1410–1456.

Dhami, M. K., Belton, I. K., & Mandel, D. R. (in press). The "analysis of competing hypotheses" in intelligence analysis. *Applied Cognitive Psychology*. doi:10.1002/acp.3550

Dhami, M. K., Mandel, D. R., Mellers, B. A., & Tetlock, P. E. (2015). Improving intelligence analysis with decision science. *Perspectives on Psychological Science, 106*(6), 753–757.

Fedorov, V. V. (1972). *Theory of Optimal Experiments*. New York: Academic Press.

Folker Jr, R. D. (2000). *Intelligence Analysis in Theater Joint Intelligence Centers: An Experiment in Applying Structured Methods*. Washington, DC: Center for Strategic Intelligence Research.

Friedman, J. (2019). *War and Chance: Assessing Uncertainty in International Politics*. New York: Oxford University Press.

Heuer, R. J. Jr., & Pherson, R. H. (2008). *Structured Analytic Techniques for Intelligence Analysis*. Washington, DC: CQ Press. Developed by Pherson Associates, LLC.

Heuer, R. J. Jr., & Pherson, R. H. (2014). *Structured Analytic Techniques for Intelligence Analysis*. Washington, DC: CQ Press. Developed by Pherson Associates, LLC.

Heuer, R. J. Jr. (1999). *Psychology of Intelligence Analysis*. Washington, DC: Center for the Study of Intelligence.

Imaoka, H., & Okajima, K. (2004). An algorithm for the detection of faces on the basis of Gabor features and information maximization. *Neural Computation, 16,* 1163–1191.

Irwin, D., & Mandel, D. R. (2019). Improving information evaluation or intelligence production. *Intelligence and National Security, 34*(4), 503–525.

Kahneman, D., & Tversky, A. (1973). On the psychology of prediction. *Psychological Review, 80,* 237–251.

Lewicki, M. S. (2002). Efficient coding of natural sounds. *Nature Neuroscience, 5*(4), 356–363.

Lindley, D. V. (1956). On a measure of the information provided by an experiment. *Annals of Mathematical Statistics, 27,* 986–1005.

Mandel, D. R. (2015). Instruction in information structuring improves Bayesian judgment in intelligence analysts. *Frontiers in Psychology, 6,* 387. doi:10.3389/fpsyg.2015.00387

Mandel, D. R. (in press). Can decision science improve intelligence analysis? In S. Coulthart, M. Landon-Murray, & D. Van Puyvelde (Eds.), *Researching National Security Intelligence: Multidisciplinary Approaches*. Washington, DC: Georgetown University Press.

Mandel, D. R., & Tetlock, P. E. (2018). Correcting judgment correctives in national security intelligence. *Frontiers in Psychology, 9,* 2640.doi:10.3389/fpsyg.2018.0264.0

Mandel, D. R., Karvetski, C. W., & Dhami, M. K. (2018). Boosting the accuracy of intelligence analysts' judgment accuracy: What works, what fails? *Judgment and Decision Making, 13*(6), 607–621.

Marchio, J. (2014). Analytic tradecraft and the intelligence community: Enduring value, intermittent emphasis. *Intelligence and National Security, 29*(2), 159–183.

Navarrete, G., & Mandel, D. R. (Eds.) (2016). *Improving Bayesian Reasoning: What Works and Why?* Lausanne, Switzerland: Frontiers Media. doi:10.3389/978-2-88919-745-3

Nelson, J. D. (2005). Finding useful questions: On Bayesian diagnosticity, probability, impact, and information gain. *Psychological Review, 112*(4), 979–999.

Nelson, J. D. (2008). Towards a rational theory of human information acquisition. In M. Oaksford., & N. Chater (Eds.), *The Probabilistic Mind: Prospects for Rational Models of Cognition* (pp. 143–163). Oxford, UK: Oxford University Press.

Nelson, J. D., Meder, B., & Jones, M. (2018, December 17). *Towards a Theory of Heuristic and Optimal Planning for Sequential Information Search*. doi:10.31234/osf.io/bxdf4

North Atlantic Treaty Organization. (2017a). Website homepage. Last accessed on 22 September, at URL: http://www.nato.int/nato-welcome/index.html

North Atlantic Treaty Organization. (2017b). Boosting NATO's presence in the east and southeast. Last accessed on 24 August 2018, at URL: https://www.nato.int/cps/en/natohq/topics_136388.htm

Pool, R. (2010). *Field Evaluation in the Intelligence and Counterintelligence Context: Workshop Summary*. Washington, DC: National Academies Press.

Ruderman, D. L. (1994). Designing receptive fields for highest fidelity. *Network: Computation in Neural Systems, 5,* 147–155.

Shannon, C. E. (1948). A mathematical theory of communication. *The Bell System Technical Journal, 27,* 379–423.

Spielmann, K. (2016). I got algorithm: Can there be a Nate Silver in intelligence? *International Journal of Intelligence and CounterIntelligence, 29*(3), 525–544.

Ullman, S., Vidal-Naquet, M., & Sali, E. (2002). Visual features of intermediate complexity and their use in classification. *Nature Neuroscience, 5,* 682–687.

US Government. (2009). *Structured Analytic Techniques for Improving Intelligence Analysis*. Washington, DC: Center for the Study of Intelligence.

US Government. (2013). Insights and best practices paper: Commander's Critical Information Requirements. Deployable Training Division (DTD), Deputy Director Joint Staff J7, Joint Training. Last assessed on 23 Oct 18 from URL: http://www.dtic.mil/doctrine/fp/fp_ccirs.pdf

Wu, C. M., Meder, B., Filimon, F., & Nelson, J. D. (2017). Asking better questions: How presentation formats influence information search. *Journal of Experimental Psychology: Learning, Memory, and Cognition, 43*, 1274–1297

Chapter 16

Modeling and Inferencing of Activity Profiles of Terrorist Groups

Vasanthan Raghavan

16.1 Introduction

Activity profiles of many terrorist groups with different ideological (e.g., political, religious, and socio-economic) affiliations and across distant geographies are now available via open-source databases such as the ITERATE database (ITERATE, 2006), the Global Terrorism Database (GTD) (LaFree & Dugan, 2007), the RAND Database on Worldwide Terrorism Incidents (RDWTI) (RDWTI, 2009), etc. These databases capture a good representative subset of all attacks made by different terrorist groups and are thus useful in studying broader terrorism trends. In spite of the availability of this data, extensive quantitative studies such as modeling and inferencing are still lacking in many aspects. In particular, broader questions of interest that could be answered with a quantitative analysis include the best model fits that could be possible for terrorist activity and what these models intuitively mean in terms of terrorist behavior and dynamics, inferencing problems such as early detection of spurts and downfalls (see brief definitions of some of the terms introduced in this chapter in Table 16.1) in activity

TABLE 16.1: Brief Description of Certain Terms Used in This Chapter

Spurt/Downfall	Sudden increase/decrease in the activity of a terrorist group.
Intentions	Ideological attributes (e.g., political, religious, linguistic, socio-economic, cultural, etc.) that contribute towards the grievance profile of a certain terrorist group.
Capabilities	Assets (e.g., manpower, material, skill-sets) that allow a group to terrorize and realize its *Intentions*.
Tactics	Specific attack attributes (e.g., attacks that capture a higher degree of resilience or coordination) that are used by the group to achieve short- and long-term objectives.
Resilience	Ability of the group to sustain terrorist activity over a number of days.
Coordination	Ability of the group to launch multiple attacks over a given time-period.
Changepoint detection	A statistical theory of detection of abrupt changes in an observation's density by trading off false alarms of non-existent changes with quick detection of true changes.
Majorization theory	A theory of comparing two positive vectors (with the same sum) so as to decide whether one of them is more well-spread than the other (e.g., income inequality).
Parametric	Typically associated with a model or parameter(s).
Non-parametric	Typically oblivious of models and not explicitly learning model parameters.

and what they mean in terms of changes in organizational dynamics, etc. While some of these aspects have been addressed via qualitative methods in the wider terrorism literature earlier, a quantitative approach could provide additional approaches/methodologies/techniques for uses in the field and render human biases less important.

The scope of this chapter is to first present in Section 16.2 different quantitative modeling approaches for terrorist group activity: interrupted time-series techniques, self-exciting hurdle models (SEHM), and hidden Markov models (HMM). In Section 16.3, we then perform a comparative study of these three distinct approaches from a number of viewpoints such as model efficacy, ease in parameter learning, etc. In particular, we consider aspects such as the explanatory power behind a model (the ability of the model to explain terrorism observations) which is studied via a widely used metric such as the Akaike Information Criterion and the predictive power of a model (the ability of the model to predict future observations given past data) which is studied via the Symmetric Mean Absolute Percentage Error metric. From these studies, we show that a simple two parameter hurdle-based geometric model with a two-state HMM framework effectively captures the activity of a number of terrorist groups. The intuition behind this model is that the group remains oblivious of its past activity and continues to attack with the same type of *Tactics* (see Table 16.1) as before, as long as its short-term objective is not met, provided a certain group resistance/hurdle has been overcome. The theme of how the current activity of a group is influenced by past history unifies both the SEHM and HMM frameworks, providing further insights into the mechanistic aspects of terrorism. In addition, the HMM approach offers a number of computational and robustness advantages in model learning making it attractive for terrorism analysis and monitoring.

Changes in the organizational dynamics of terrorist groups, changing socio-economic and political contexts at either the local, regional, or global levels, counterterrorism activity, etc., lead to either spurts or downfalls in their activity profiles. It is of utmost importance to detect such changes, associate these changes to specific macroscopic changes in group dynamics, and track these dynamics over time since a majority of counterterrorism activity is often geared to these objectives. Of particular interest in this chapter are practical approaches to monitor and detect changes in resilience and coordination (see Table 16.1) in a terrorist group. Resilience and coordination are quantified in this work in terms of the group's ability to sustain multiple attacks over the long- and the short-term, respectively.

Towards the goal of spurt detection, in Section 16.4, we start with a parametric (see Table 16.1) approach that exploits the HMM structure by learning the model parameters and then estimating the most probable state sequence using the Viterbi algorithm. We show that this approach allows one to not only detect *non-persistent* changes in terrorist group dynamics, but also to identify key geopolitical undercurrents that lead to sudden spurts/downfalls in a group's activity. Despite its good performance, model-based parametric approaches in general suffer from fundamental difficulties associated with model learning, a difficult task for on-the-field deployment of inferencing algorithms.

To overcome these difficulties, we start with a non-parametric (see Table 16.1) solution motivated by the theory of changepoint detection (see Table 16.1). This approach is a stopping-time based on the Exponentially Weighted Moving Average (EWMA) filter that tracks the quantities of interest. This non-parametric approach does not require knowledge of the underlying model and is hence useful in settings where model learning is difficult or where model uncertainty is high. However, as a price for this capability, only *persistent* changes (changes that last for a sufficiently long duration) can be detected with this approach. To remedy the poor performance of the EWMA solution, we then propose a non-parametric solution, motivated by ideas from majorization theory (see Table 16.1), that monitors and tracks quick changes in resilience and coordination. We show that this non-parametric solution can lead to good performance.

16.2 Modeling of Activity Profiles

16.2.1 What Are Activity Profiles?

The observations capturing the dynamics of a terrorist group are of a multivariate nature. Specifically, these observations are made of categorical, ordinal, and interval variables.[1] Examples of these variables include interval variables such as time of attacks, categorical variables such as location of attacks, type(s) of ammunition used, (apparent) sub-group of the group involved, ordinal variables such as intensity and impact of the attacks, etc. While probabilistic models can be generated for the entire set of observations that make terrorism dynamics, many observations have missing data for a number of these attributes. Thus, it is often simplistic to consider only the

---

[1] A categorical variable corresponds to mutually exclusive, but not ordered, categories. An ordinal variable is one where the order matters, but not the difference between values. An interval variable is one where the difference between two values is meaningful.

time-series capturing the number of attacks instigated, orchestrated, or organized by the group over a time period of interest. Such a time-series is denoted as the *activity profile* of a terrorist group.

The activity profile of a terrorist group can be modeled as a discrete-time stochastic process. Let the first and last time-window of interest be indexed as 1 and $\mathcal{N}$, respectively. Typical time-windows include a day, a week, or a 15-day ($\delta = 1$, 7, or 15 days) observation period. Let M_i denote the number of terrorism incidents over the ith time-window Δ_i, $i = 1, \cdots, \mathcal{N}$ where $\Delta_i = \{(i-1)\delta + 1, \cdots, i\delta\}$. Note that M_i can take values from the set $\{0, 1, 2, \cdots\}$ with $M_i = 0$ corresponding to no terrorist activity over Δ_i. On the other hand, there could be multiple terrorism incidents over this period reflecting a high level of coordination between various sub-groups of the group. Let $\mathcal{H}_i$ denote the history of the group's activity until Δ_i. That is, $\mathcal{H}_i = \{M_1, \cdots, M_i\}$, $i = 1, 2, \cdots, \mathcal{N}$ with $\mathcal{H}_0 = \varnothing$. The point process model is complete if the conditional probability of current activity given the past history of the group, $\mathsf{P}(M_i = r | \mathcal{H}_{i-1})$, is specified as a function of $\mathcal{H}_{i-1}$ for all $i = 1, \cdots, \mathcal{N}$ and $r = 0, 1, 2, \cdots$.

A typical example of a terrorism activity profile for the *Fuerzas Armadas Revolucionarias de Colombia* (FARC) terrorist group from Colombia is provided in Figure 16.1. This plot corresponds to attacks by FARC binned over disjoint 15-day time-windows and as recorded in the RDWTI database (RDWTI, 2009) for the 1998 to 2006 time period. Note that there are 219 time-windows in this dataset. Also, the indication "1998" on the X axis of Figure 16.1 corresponds to January 1, 1998. In addition to the attacks, important geopolitical events associated with major and minor spurts in the

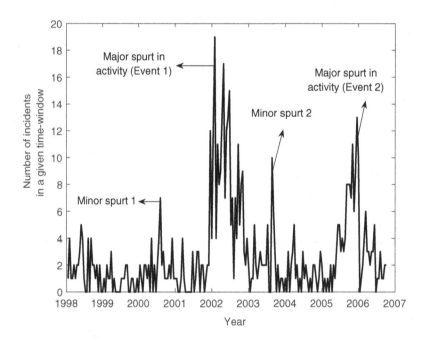

FIGURE 16.1: Activity profile corresponding to the number of attacks over disjoint 15-day time-windows from 1998 to 2006 along with key geopolitical events for FARC, a terrorist group from Colombia.

activity profile are also indicated. The geopolitical events behind the major spurts of FARC are now described.

- *Event 1:* With Colombia becoming the leading cultivator (United Nations Office on Drugs and Crime, 2010, Figure 21) of coca by 1997, the Colombian government under President Andrés Pastrana and the US government under President William J. "Bill" Clinton instituted legislation, popularly known as *Plan Colombia*, under which, among other tasks, US aid for counter-narcotics efforts increased significantly. In the ensuing Presidential elections in 2001–2002, Álvaro Uribe won the popular mandate on an anti-FARC platform. The impact of *Plan Colombia* and the impending election of President Álvaro Uribe are the primary reasons behind Event 1, a spurt in the activity profile of FARC in the late-2001–2002 time period
- *Event 2:* With FARC adapting itself to the new equilibrium established by the institution of *Plan Colombia* by 2003–2004, we observe a return to "normalcy" in terms of the activity profile of FARC. However, with elections to Colombia's Congress and a re-election bid by President Uribe in the mid- to late-2005–2006 period ensured that FARC, which had remained relatively passive in this period, attacked with renewed vigor. This is associated with Event 2, another spurt in the activity profile of FARC. See Beittel (2010); Cragin and Daly (2004) (and references therein) for a more detailed description of these events and FARC activities

Alternate to the number of incidents over a time period, since an activity profile captures a temporal point process, it could be equivalently represented by counting the time to the next incident (over an appropriate time-scale). Different from terrorism activity, Figure 16.2 illustrates the time (in minutes) for the next social media posting or inter-tweet duration by a highly active user on Twitter. This activity profile serves as an additional example of a completely different application where some of the methods illustrated in this chapter would be useful. The readers are referred to Raghavan et al. (2014) for more background and details on this setup.

Returning to terrorism, an important attribute of terrorism activity profiles is the associated *sparsity* of data. In particular, Figure 16.1 corresponds to one of the most active terrorist groups in the world with 604 incidents over a nine-year period for an average of ≈ 1.29 incidents per week. The fact that significant amounts of resources need to be invested by the terrorist group for every new attack acts as a natural dampener towards more attacks. Sparsity of data motivates the need for parsimonious models with few parameters and motivated by few hypotheses capturing the mechanism of terrorism dynamics.

Further, as with any data collection exercise, terrorism activity profiles can suffer from impairments such as missing or mislabeled data, temporal or attributional ambiguity of attacks, transcribing errors, etc. Thus, the robustness of models to such impairments is also critical. With this background, we now study modeling of terrorism activity profiles.

16.2.2   Time-Series Analysis Techniques

Time-series analysis techniques have been studied for a long time (see, e.g., Brockwell & Davis, 2010; Shumway & Stoffer, 2017; Peña, Tiao & Tsay, 2000; Hyndman & Athanasopoulos, 2013 and references therein) and these techniques have been used in terrorism analysis for a number of years. In particular, different versions of

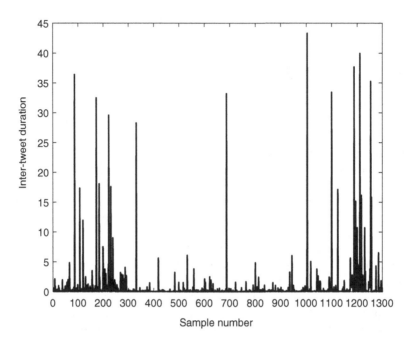

FIGURE 16.2: Time (in minutes) to next social media posting of a highly active user on **Twitter**.

interrupted time-series analyses[2] have been used to study whether certain strategic policy interventions lead to statistically significant reductions in certain types of attacks and/or if different types of attacks act as substitutes for (or complements of) one another (Brophy-Baermann & Conybeare, 1994; Cauley & Im, 1988; Enders, Sandler & Cauley, 1990; Landes, 1978; Enders & Sandler, 1993, 2000, 2002). In particular, the main focus of works such as Brophy-Baermann and Conybeare (1994), Cauley and Im (1988), Enders, Sandler, and Cauley (1990), and Landes (1978) is the study of the efficacy of interventions such as strengthening airport security barriers, fortification of US embassies/missions abroad, adoption in the US of stringent anti-terrorism laws, adoption of international conventions on hijackings, retaliatory bombings, etc.

In an example of the time-series approach (Enders & Sandler, 1993), a simple first-order threshold vector auto-regression (TAR) model studying the impact of a certain policy intervention on two types of attacks (denoted by the time-series $\{M_{1,i}\}$ and $\{M_{2,i}\}$, respectively) is given as:

$$M_{1,i} = a_1 M_{1,i-1} + b_1 M_{2,i-1} + c_1 \mathsf{p}_i + \varepsilon_{1,i}, \tag{16.1}$$

$$M_{2,i} = a_2 M_{2,i-1} + b_2 M_{1,i-1} + c_2 \mathsf{p}_i + \varepsilon_{2,i} \tag{16.2}$$

In (16.1) and (16.2), $M_{1,i-1}$ and $M_{2,i-1}$ stand for the delayed/lagged versions of the two types of attacks. The above model considers dependence of $\{M_{1,i}\}$ and $\{M_{2,i}\}$ only

---

[2] An interrupted time-series differs from a classical time-series in that the data is known (or conjectured) to be affected by interventions or controlled external influences.

on their first-order lags and also assumes that they are cross-correlated with appropriately chosen model coefficients (a_j, b_j, and c_j, $j = 1, 2$) capturing the interdependence between them. The policy intervention is captured by the binary indicator variable p_i, which is 1 before a certain time i when the policy was in effect, and is 0 after this point in time when the policy was no longer in effect. The other components in (16.1) and (16.2) such as $\varepsilon_{1,i}$ and $\varepsilon_{2,i}$ correspond to the modeling of random effects.

The main conclusion from the TAR modeling approach is that certain policy interventions result in an unanticipated increase in certain types of substitution attacks. For example, installation of metal detectors, barbed wire fences around US consulates, and airport security barriers that render certain types of attacks (such as skyjackings) more costly for the terrorist group tend to result in the substitution of these attacks with other types of attacks that are less costly for the group (such as other types of hostage events not protected by metal detectors). Another example of this substitution effect is the rise in assassinations of protected persons as a consequence of increased security barriers at US missions abroad, even as kidnappings and hostage events decrease.

The time-series approach has also been used to study both the short- and long-run spurts in world terrorist activity over the period from 1970 to 1999 (Enders & Sandler, 1993). In a specific example of this approach (Enders & Sandler, 2000), a classical time-series model[3] for trans-national terrorism incidents, M_i, is fitted as

$$M_i = T_i + C_i + S_i \qquad (16.3)$$

where T_i, C_i, and S_i denote the linear or non-linear trend component, cyclical or seasonal component, and the stationary or random component, respectively. The trend component corresponds to either a historical growth or decay of terrorism, while the cyclical component corresponds to "wave-like" behavior of terrorism with high activity and occasional calm. Using the ITERATE dataset, Enders and Sandler identify a rough four-and-a-half-year cycle in terrorism events corresponding to increased terrorist activity (perhaps of a substitution or complementary nature) in response to certain policy interventions that then results in depletion of terrorist group resources leading to a subsequent phase of low activity. In particular, the US embassy hostagetaking in Iran in 1979 and the post-Cold War period of 1991 are explored as potential switching points in the TAR framework contributing to a structural change in the nature of trans-national terrorist activity. The fitted models with the TAR framework are provided in Enders and Sandler (2000, Table 4, p. 327) and these models show that both the hostage-taking and post-Cold War events have a tremendous impact on the number of casualties in terrorism incidents.

Following the same theme as above, in Enders and Sandler (2002), a model that switches from one auto-regressive process to another with the switches corresponding to key geopolitical events is studied. Some of the geopolitical events considered here include the Iran hostage-taking event, embassy barrier strengthening, the start of the use of metal detectors at entry points to secure locations, retaliatory raids in Libya, etc. The fitted models are presented in Enders and Sandler (2002, Equations 6 and 12). The non-linear trend and seasonality components in the time-series model explain the

---

[3] Note that the classical time-series approach fits a trend, seasonality, and a stationary random error component to the data (Brockwell & Davis, 2010). Typically, there are no constraints (binary or otherwise, Poisson or otherwise) on these three components, or on the data.

observed boom and bust cycles in terrorist activity. Further, since terrorist activity are cyclical and irregular, trends in terrorism cannot be accurately predicted.

Other examples of similar approaches include the use of Cox proportional hazards or zero-inflated Poisson models (Dugan, LaFree & Piquero, 2005; LaFree, Morris & Dugan, 2010) for the short- and long-run behavior of worldwide terrorist activity. The hazard model is more appropriate to capture the rate of certain rare events. In this sense, this approach is of the same flavor (but not quite the same) as the TAR model. The zero-inflated Poisson model is appropriate in settings such as terrorism where most days have no attacks till the group changes gears and becomes *Active*. The hazard function of an event random variable T at time t is defined as

$$\mathsf{Haz}(t) = \lim_{\Delta t \to 0} \frac{\mathsf{P}(t \le T < t + \Delta t | T \ge t)}{\Delta t}. \tag{16.4}$$

Note that $\mathsf{Haz}(t)$ can be viewed as the probability of occurrence of the event in an infinitesimally small time period, given that the event has not occurred till then. Thus, it measures the instantaneous rate of occurrence of the event. Mathematically, it is defined as the ratio of the density function and the complementary distribution (or commonly denoted as the survival) function.

In Dugan, LaFree, and Piquero (2005), the hazard function $\mathsf{Haz}(t)$ of the time to the next hijacking (denoted as T) is modeled as a function of a few underlying control variables (policies or interventions, purpose behind hijackings, and their context):

$$\log(\mathsf{Haz}(t)) = \log(\mathsf{Haz}_0) + \beta_1 \cdot \mathsf{Policies} + \beta_2 \cdot \mathsf{Major\,Purpose} + \beta_3 \cdot \mathsf{Context}. \tag{16.5}$$

In the above equation, Haz_0 is a baseline hazard function, and $\{\beta_i\}$ are optimized from a certain model class. Further, while the terms in (16.5) appear to be categorical variables, they are captured with specific numerical possibilities in this work; see Dugan, LaFree, and Piquero (2005, Table 1, p. 1048) for details.

16.2.3 Self-Exciting Hurdle Models

A common theme that ties all the above works is a *contagion* theoretic viewpoint (Midlarsky, 1978; Midlarsky, Crenshaw & Yoshida, 1980) that the current activity of the terrorist group is *explicitly* dependent on the past history of the group, which accounts for clustering effects in the activity profile. A theoretical formalism for this viewpoint is provided with the use of self-exciting hurdle models (Cox & Isham, 1980; Hawkes, 1971). These models have found utility in diverse applications such as seismic activity and gang warfare/insurgency modeling. For example, a major tremor in a seismic activity profile is typically followed by a number of aftershocks (Ogata, 1988, 1998), which can be viewed as a *self-excitation*. Similarly, an action by a first gang can result in reactions and counteractions by a second (of possibly many) gang(s), leading to an activity profile that reflects a *self-excitatory* behavior (Mohler et al., 2011; Lewis et al., 2011).

In its simplest form, the SEHM approach consists of: i) a hurdle component and ii) a self-exciting component. The hurdle component creates data sparsity by ensuring a pre-specified density of zero events. In the context of terrorism, the intuitive meaning of the hurdle component is to endow a certain resistance in the group that needs to be overcome to launch a new series of attacks. The self-exciting component induces clustering of attacks after the hurdle has been overcome. The SEHM formalism is described by the likelihood equation:

$$P(M_i = r \mid \mathcal{H}_{i-1}) = \begin{cases} e^{-(B_i + SE_i(\mathcal{H}_{i-1}))}, & r = 0 \\ \dfrac{r^{-s}}{\zeta(s)} \cdot \left(1 - e^{-(B_i + SE_i(\mathcal{H}_{i-1}))}\right), & r \geq 1 \end{cases} \qquad (16.6)$$

where B_i is an underlying baseline process, and $SE_i(\cdot)$ is the self-exciting component given as

$$SE_i(\mathcal{H}_{i-1}) = \sum_{j:j<i,M_j>0} a_j g(i-j) \qquad (16.7)$$

for an appropriate choice of decay function $g(\cdot)$ and influence parameters $\{a_j\}$.

A typical choice for $g(\cdot)$ in (16.7) is a shifted negative binomial density function (see Porter & White, 2012, Equation 5, p. 113 for how to set up this density function) and the parameters a_j are chosen to capture the impact of previous events on current activity. On the other hand, the s parameter of the zeta distribution in (16.6) is chosen appropriately from the set $(1, \infty)$, and $\zeta(s) = \sum_{n=1}^{\infty} n^{-s}$ is the Riemann-zeta function used to normalize a probability measure. While a constant s parameter leads to the simplest modeling framework, s can in general be driven by another self-exciting process trading-off efficacy with model complexity. With this background, the HMM framework discussed next provides an alternate generative mechanism for terrorist activity.

16.2.4   Hidden Markov Models

Hidden Markov processes build on Markov processes that endow memory in a stochastic process, with the extension being that the observations in a hidden Markov process depend on an underlying state process that evolves in a Markovian[4] manner, but remains hidden or unobservable (Rabiner, 1989; Ephraim & Merhav, 2002). Hidden Markov processes have been widely explored as generative models for engineering applications such as speech processing, pattern recognition, dynamic systems, etc. (Rabiner, 1989; Ephraim & Merhav, 2002). Due to this application-driven background, hidden Markov models are mature and numerous stable software packages (e.g., HMM Toolbox in **MATLAB** (MATLAB, n.d.), HMM (R, n.d.) or depmixS4 (Visser & Speekenbrink, n.d.), Packages in **R,** etc.) are available for parameter learning, inferencing, prediction, etc.

In the context of terrorism dynamics, the hidden Markov approach provides an alternate evolutionary mechanism to explain clustering of attacks (Cragin & Daly, 2004; Raghavan, Galstyan & Tartakovsky, 2013). In this approach, an increase (or decrease) in attack intensity can be naturally attributed to certain changes in the group's internal states that reflect the dynamics of its evolution, rather than the fact that the group has already carried out attacks on either the previous day, or week, or month. The HMM framework is motivated by two hypotheses. The first hypothesis assumes that

---

[4] The term "Markovian" means that the current set of observations depends only on a finite set of the more recent past of observations, but not the infinite past set of observations. In the HMM framework, the current state typically depends only on the most immediate past state.

M_i depends only on certain hidden states S_i such as *Intentions, Tactics*, or *Capabilities* of the group in the sense that M_i is conditionally independent of $\mathcal{H}_{i-1}$ and $S_1, \cdots, S_{i-1}$ given S_i. Such an evolutionary model for terrorism dynamics was first hypothesized in Cragin & Daly (2004).

It is clear that the dynamics of terrorism are well-understood if the underlying states $\{S_i\}$ are known. However, in reality, S_i cannot be observed directly and we can only make indirect inferences about it by observing $\{M_j, j = 1, \cdots, i\}$. To simplify modeling, we make a second hypothesis that S_i can be described by a time-homogenous one-step Markovian evolution with a d-state model to capture the dynamics of the group over time. That is, $S_i \in \{0, 1, \cdots, d-1\}$ with each distinct value corresponding to a different level in the underlying attribute of the group. See the discussion later for good choices of d. A pictorial illustration of the assumptions leading to the HMM framework is presented in Figure 16.3.

Using these two hypotheses, the temporal point process model can be simplified as

$$P(M_i = r | \mathcal{H}_{i-1}) = \sum_{j=0}^{d-1} \sum_{k=0}^{d-1} P\left(M_i = r, S_i = j, S_{i-1} = k | \mathcal{H}_{i-1}\right) \qquad (16.8)$$

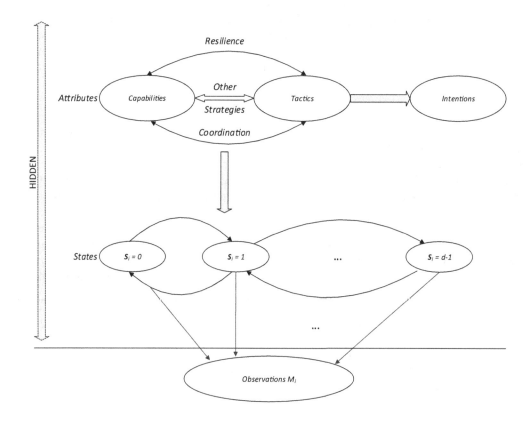

FIGURE 16.3: A model capturing the dynamics of a terrorist group with connections between the underlying attributes, states and observations.

$$= \sum_{j=0}^{d-1} \sum_{k=0}^{d-1} \mathsf{P}\Big(M_i = r \Big| S_i = j, S_{i-1} = k, \mathcal{H}_{i-1} \Big) \cdot \mathsf{P}(S_i = j, S_{i-1} = k | \mathcal{H}_{i-1}) \qquad (16.9)$$

$$= \sum_{j=0}^{d-1} \sum_{k=0}^{d-1} \mathsf{P}(M_i = r | S_i = j) \cdot \mathsf{P}\Big(S_i = j, S_{i-1} = k \Big). \qquad (16.10)$$

Note that (16.9) follows from writing the joint conditional density function $P(A,B|C)$ as $P(A,B|C) = P(A|B,C) \cdot P(B|C)$. On the other hand, (16.10) follows from the conditional independence of M_i and $\{\mathcal{H}_{i-1}, S_{i-1}\}$ given S_i and the assumption of non-causality between the states and observations. From (16.10), we observe that the likelihood function $\mathsf{P}(M_i = r | \mathcal{H}_{i-1})$ is expressed by knowing the transition probabilities and the conditional observation density function in each state.

In terms of modeling, there is a trade-off between a careful reproduction of the group's behavioral attributes (larger d is better for this goal) versus estimating more model parameters (smaller d is better for this goal). Note that the model parameters in the hidden Markov framework include the transition probability matrix parameters and the observation density parameters. The transition probability matrix is a $d{\times}d$ matrix with the (j, k)-th entry denoting the transition probability (or the likelihood of state transition(s)) $\mathsf{P}(S_i = j | S_{i-1} = k)$. Since $\sum_{j=0}^{d-1} \mathsf{P}(S_i = j | S_{i-1} = k) = 1$, each column has only $d{-}1$ independent parameters leading to $d \cdot (d-1)$ parameters for the transition probability matrix. With ℓ being the number of observation density parameters in each state, we have $d \cdot \ell$ parameters for the observations and thus $d(d-1+\ell)$ parameters in all to capture the description in (16.10). The trade-off between model complexity and data representation can be simplistically resolved by focusing on mature terrorist groups (where the *Intentions* and *Tactics* attributes remain stable) and by considering a $d = 2$ setting. This trade-off corresponds to a binary quantization of the group's *Capabilities* into *Active* and *Inactive* states. Such an approach is pursued in Raghavan, Galstyan, and Tartakovsky (2013) and is shown to lead to good performance in modeling terrorism data, which is measured via comparison with the Akaike Information Criterion (AIC) metric (Akaike, 1974) (see discussion on AIC later).

16.2.5 Fitting the Correct Observation Model in the HMM Framework

For the HMM framework, multiple models that have support on the non-negative integers can be considered for $\{M_i\}$. Some of the popular models include one-parameter models such as Poisson and geometric, and two-parameter models such as Pòlya, (non-self-exciting) hurdle-based Zipf, and hurdle-based geometric. These models are described as follows:

$$\mathsf{P}\Big(M_i = r \Big| S_i = j \Big)\Big|_{\text{Poisson}} = \frac{\big(\lambda_j\big)^r e^{-\lambda_j}}{r!} \qquad (16.11)$$

$$\mathsf{P}\Big(M_i = r \Big| S_i = j \Big)\Big|_{\text{Geometric}} = \big(1 - \gamma_j\big) \cdot \big(\gamma_j\big)^r \qquad (16.12)$$

$$\mathsf{P}\left(M_i = r \middle| S_i = j\right)\bigg|_{\text{Polya}} = \frac{\Gamma\left(r + r_j\right)}{\Gamma\left(r + 1\right) \cdot \Gamma\left(r_j\right)} \times \left(1 - y_j\right)^{r_j} \left(y_j\right)^r \tag{16.13}$$

$$\mathsf{P}\left(M_i = r \middle| S_i = j\right)\bigg|_{\text{Hurdle-based Zipf}} = \begin{cases} 1 - \gamma_j & \text{if } r = 0 \\ \dfrac{\gamma_j \cdot r^{-y_j}}{\zeta(y_j)} & \text{if } r \geq 1 \end{cases} \tag{16.14}$$

$$\mathsf{P}\left(M_i = r \middle| S_i = j\right)\bigg|_{\text{Hurdle-based Geometric}} = \begin{cases} 1 - \gamma_j & \text{if } r = 0 \\ \gamma_j\left(1 - \mu_j\right) \cdot \left(\mu_j\right)^{r-1} & \text{if } r \geq 1. \end{cases} \tag{16.15}$$

In the above five models, the associated parameters in the jth state are λ_j, γ_j, $\{r_j, y_j\}$, $\{\gamma_j, y_j\}$, and $\{\gamma_j, \mu_j\}$, respectively. In terms of the complementary cumulative distribution functions, it can be seen that models based on Zipf distribution can accommodate heavy tails (see Raghavan, Galstyan & Tartakovsky, 2013, Supplementary A for an explanation), whereas the other four models have exponentially vanishing tails. This distinction is important in understanding model fits across disparate terrorist groups.

Out of these five models, the *hurdle-based geometric* model has an interesting generative mechanism explanation for terrorism. To understand the intuition behind this model, note that a geometric density captures *memorylessness* in a stochastic process. The intuition behind the hurdle-based geometric model is that the terrorist group remains *oblivious* of its past activity and continues to attack with the same *Tactics* as before (with a mean number of attacks per day in the state $\{S_i = j\}$ of $\dfrac{1}{1 - \mu_j}$), as long as its objective is met. This *Tactic* is followed by the group provided that a certain group resistance/hurdle (captured by the parameter γ_j) has been overcome. The special case where there is no group resistance to this aforementioned strategy is obtained by setting $\mu_j = \gamma_j$, resulting in a geometric observation density. This special case corresponds to a purely memoryless group with no sophistication in attacks other than a *Tactic* of repeated attacking with no distinction in strategies. Such a group could become entirely predictable from a counterterrorism perspective.

For the five models (as described in (16.11)–(16.15)), we study the goodness-of-fit captured by the AIC for the observational data from FARC. The AIC is defined as

$$\mathsf{AIC}\big|_{\text{model}} \triangleq 2k_{\text{model}} - 2\log\left(\mathsf{P}\left(\{M_i\} \middle| \text{model}\right)\right) \tag{16.16}$$

where a model with k_{model} parameters is used for fitting $\{M_i\}$. Note that the AIC score captures the negative of the log-likelihood and thus a model with a smaller AIC score is better than a model with a larger AIC score.

In Table 16.2, we present the frequency counts of the number of days with $\ell\left(\ell = 0, 1, \cdots\right)$ attacks per day for the FARC data. Using the Viterbi algorithm (Rabiner, 1989) for binary state classification with the FARC data, it is observed that FARC stays in the *Inactive* state for 2,657 days and in the *Active* state for 630 days. Also, presented in the same table are the (rounded-off) expected number of days with ℓ attacks for the five models with parameters fitted using the maximum likelihood principle as well as the AIC for these model-fits. From Table 16.2, it is seen that in

TABLE 16.2: Frequency Counts of Observed Number of Attacks per Day for FARC Data with Different Model Fits

No. of attacks (*Inactive*)	Obs.	Poisson	Geometric	Pòlya	Hurdle-Based Zipf	Hurdle-Based Geometric
0	2420	2421	2430	2421	2420	2421
1	227	225	207	225	229	226
2	9	11	18	11	7	10
3	1	0	2	0	1	0
4	0	0	0	0	0	0
>4	0	0	0	0	0	0
AIC		1690.3	1696.7	1692.3	1962.6	1691.9
Parameter		0.09	0.09	$\hat{r}_0 = 24.48$	$\hat{\gamma}_0 = 0.09$	$\hat{\mu}_0 = 0.04$
Estimate				$\hat{y}_0 = 0.004$	$\hat{y}_0 = 5.10$	$\hat{\gamma}_0 = 0.09$

No. of attacks (*Active*)	Obs.	Poisson	Geometric	Pòlya	Hurdle-Based Zipf	Hurdle-Based Geometric
0	384	359	404	389	384	384
1	174	202	144	160	189	171
2	46	57	52	56	31	52
3	19	11	19	17	11	16
4	4	1	7	6	5	5
>4	3	0	4	2	10	2
AIC		1313.9	1291.7	1288.9	1308.1	1287.1
Parameter		0.57	0.36	$\hat{r}_1 = 1.48$	$\hat{\gamma}_1 = 0.39$	$\hat{\mu}_1 = 0.31$
Estimate				$\hat{y}_1 = 0.28$	$\hat{y}_1 = 2.61$	$\hat{\gamma}_1 = 0.39$

the *Inactive* state, all the models result in comparable fits. Specifically, the hurdle-based geometric model differs from the observed frequency counts in only one day and results in the second-lowest AIC value. On the other hand, in the *Active* state, the hurdle-based geometric model produces the best fit with only the Pòlya model resulting in a comparable fit. In the *Active* state, the one-parameter models (such as geometric and Poisson) poorly estimate either the tail or the days of no activity. In fact, the heavy tail afforded by a Zipf-based model may not always be necessary. In contrast, the Indonesia/Timor-Leste data studied in Porter and White (2012) exhibit several extreme values (e.g., days with 36, 11, and 10 attacks) and the authors observe that a self-exciting hurdle-based Zipf model captures the heavy tails well. The FARC dataset used here shows a maximum of seven attacks per day. Even simple non-self-exciting models are enough to capture these datasets sufficiently accurately. This study suggests the following:

- If parsimony of the model is of critical importance, the geometric model serves as the best one-parameter model with the Poisson model either underestimating or overestimating the number of days with no activity in the *Active* state

- If parsimony is not a critical issue and the data does not have (or has very few) extreme values, the hurdle-based geometric model serves as the best or a near-best model in either state
- However, if the data has several extreme values (as seen in Porter & White, 2012), the self-exciting hurdle model offers the best model-fit, albeit at the expense of learning several model parameters. The heavy-tailed zeta distribution is also explored in Clauset, Young, and Gleditsch (2007) for modeling extremal terrorist events

---

16.3   Comparison across Models

We now present a comparative study of the three models introduced in Section 16.2 from different viewpoints.

16.3.1   Qualitative Comparisons

While the TAR, SEHM, and HMM frameworks assume that the current observation or activity in a terrorist group is dependent on the past history of the group, each model differs precisely in how this dependence is realized.

- In the TAR framework, the current observation is explicitly dependent on the past set of related observations (such as complementary or substitution attacks) along with (possibly) the impact from other independent variables corresponding to certain geopolitical events/interventions
- In the SEHM framework, the probability of a future attack is enhanced by the history of the group according to the formula:

$$\frac{\mathsf{P}\left(M_i > 0 \middle| \mathcal{H}_{i-1}\right)\Big|_{\text{SEHM}}}{\mathsf{P}\left(M_i > 0 \middle| \mathcal{H}_{i-1}\right)\Big|_{\text{Non-SEHM}}} = 1 + \frac{e^{-B_i}}{1 - e^{-B_i}} \cdot \left(1 - e^{-SE_i(\mathcal{H}_{i-1})}\right) \geq 1. \tag{16.17}$$

- The HMM framework combines both these facets by introducing a hidden state sequence (which can capture facets such as switching to different *Tactics*). The state sequence depends explicitly on its most immediate past (one-step Markovian structure), whereas the probability of an attack is enhanced based on the state realization

The TAR and HMM frameworks are similar from the viewpoint of regime-switching as these features are modeled explicitly. However, the mechanism of regime-switching is different in the two cases: the former assumes a change in the auto-regressive process with an independent control variable, whereas the latter assumes a state transition in a Markov model. The SEHM framework also incorporates a switch between states (induced by the self-exciting component), but this switch is more of an implicit feature of the model rather than an explicit driving function.

More importantly, while the TAR model could be used to study the behavior of individual terrorist groups, historically it has been used to study global terrorism trends rather than trends constrained to a specific region or a specific group. Similarly, the

Indonesia/Timor-Leste dataset considered by the SEHM framework in Porter and White (2012) is a collation of *all* attacks in Indonesia and Timor-Leste from diverse groups with significantly different activity profiles (e.g., Islamist, independence-seeking, socialist, etc.) such as *Dar-ul-Islam, Gerakan Aceh Merdeka, Jemaah Islamiyah*, etc. On the other hand, the FARC dataset considered in the HMM framework studied by Raghavan, Galstyan, and Tartakovsky (2013) is exclusively the actions of the many sub-groups of FARC with a similar ideological affinity. While this comparison may appear to be imprecise, the main distinctions are that the TAR model has not been applied to individual groups, while the SEHM and HMM have been. On the other hand, there are some issues with the dataset used to study the efficacy of the SEHM.

These subtle (yet important) differences lead to distinctive abilities for each framework in terms of the explanatory power (of past attacks) and the predictive power (of future attacks). These aspects are studied in Sections 16.3.2 and 16.3.3, and the power of each framework is illustrated with the FARC dataset studied in Raghavan, Galstyan, and Tartakovsky (2013) and the Indonesia/Timor-Leste dataset studied in Porter and White (2012).

16.3.2 Explanatory Power

To understand the explanatory power of the HMM framework, we let T_k, $k = 1, 2, \cdots$ denote the time to the kth day of terrorist activity (with T_0 set to $T_0 = 0$) and define $\Delta T_k \triangleq T_k - T_{k-1}$ to denote the time to the subsequent day of activity (inter-arrival duration of attack days). With a two-state HMM over $\{M_i\}$, we are interested in the ability of the HMM framework to explain $\{\Delta T_k\}$. For this, model parameters (denoted by the simplistic notation λ_{HMM}) are learned with the classical Baum-Welch algorithm (Rabiner, 1989) to locally maximize the log-likelihood function of the inter-arrival durations, $\mathsf{P}\left(\Delta T_1^n \middle| \lambda_{\mathsf{HMM}}\right)$. Note that we perform model learning with $\{\Delta T_k\}$ instead of $\{M_i\}$ since model learning with the latter data often mirrors the randomness in it, instead of exposing the macroscopic features of the terrorist group (Raghavan & Tartakovsky, 2018). Technical details on the iterative update equations for parameter learning with the Baum-Welch algorithm builds on well-established optimization strategies (Bilmes, 1998; Rabiner, 1989) and can be found in Raghavan, Galstyan, and Tartakovsky (2013, Supplementary A).

For the SEHM framework, the different baseline and self-exciting models considered in Porter and White (2012) are used to model $\left\{\Delta T_1^n\right\}$. The **fmincon** function[5] in **MATLAB** is used to learn model parameters that maximize the likelihood function, $\mathsf{P}\left(\Delta T_1^n \middle| \lambda_{\mathsf{SEHM}}\right)$ (see Porter & White, 2012, Equation 8). A class described by eight parameters in the SEHM framework of Section 16.1.2.3 is studied in Porter and White (2012). It is shown that a four-parameter model, where one parameter is used for the trend component and three parameters are used for the negative binomial self-exciting component, optimizes an AIC metric for terrorism data from Indonesia/Timor-Leste over the period from 1994 to 2007.

For the explanatory power, we focus on the HMM's and SEHM's ability to explain the times to the subsequent day of activity $\left\{\Delta T_1^n\right\}$. This is captured by the AIC for the two models, defined as,

---

[5] See https://www.mathworks.com/help/optim/ug/fmincon.html.

TABLE 16.3: Comparison between AIC Scores with the SEHM and HMM Frameworks for the FARC and Indonesia/Timor-Leste Datasets

	FARC			Indonesia/ Timor-Leste	
n	SEHM	HMM	n	SEHM	HMM
100	671.68	671.06	100	723.78	729.47
200	1117.40	1112.07	165	1091.78	1116.92
300	1521.93	1521.36	200	1283.08	1305.27
400	2127.55	2121.81	250	1589.43	1615.87
450	2333.88	2327.02	300	2018.92	2041.35

$$\text{AIC}(n)\big|_{\text{HMM}} \triangleq 2k_{\text{HMM}} - 2\log\left(\mathsf{P}\left(\Delta T_1^n \big| \lambda_{\text{HMM}}\right)\right) \tag{16.18}$$

$$\text{AIC}(n)\big|_{\text{SEHM}} \triangleq 2k_{\text{SEHM}} - 2\log\left(\mathsf{P}\left(\Delta T_1^n \big| \lambda_{\text{SEHM}}\right)\right). \tag{16.19}$$

Table 16.3 shows the AIC score comparison between the optimal HMM and the optimal four parameter SEHM for the two datasets. The HMM and SEHM are fitted for the whole dataset and their fits are studied with n events where n could be a subset of the whole dataset. The precise choice of n may be application-specific (see discussion later). AIC is a measure of the model fit over the data used, whether it is a subset of the dataset or the whole dataset. From the study in Table 16.3, we see that in terms of explanatory power, both the HMM and SEHM frameworks perform reasonably well, with neither framework clearly outperforming the other. The HMM framework is better for the FARC dataset, whereas the SEHM is seen to be better for the Indonesia/Timor-Leste dataset. If HMM scores better than SEHM (as is consistently seen for FARC and *vice versa* for Indonesia/Timor-Leste), then it suggests that the HMM is a better framework to capture any temporal subset of the data than the SEHM framework for FARC. An explanation for this observation is that the Indonesia/Timor-Leste dataset has a heavier tail than the FARC dataset, which is better captured with the SEHM framework.

16.3.3 Predictive Power

For the predictive power, we focus on each approach's ability to predict ΔT_{n+1} given $\left\{\Delta T_1^n\right\}$. For this, model parameters are learned with $\left\{\Delta T_1^n\right\}$ as training data and a conditional mean estimator of the form $\tilde{\Delta} T_{n+1} = \mathsf{E}[\Delta T_{n+1} | \Delta T_1^n]$ is used for prediction. For the HMM framework, it can be checked that

$$\tilde{\Delta} T_{n+1}\big|_{\text{HMM}} = \sum_{i=0}^{1} \beta_i \mathsf{E}\left[\Delta T_{n+1} \big| S_{n+1} = i\right], \tag{16.20}$$

where $a_n(j) \triangleq \mathsf{P}\left(\Delta T_1^n, S_n = j\right)$ is updated via the forward procedure (Rabiner, 1989) and

$$\beta_i = \frac{\sum_j a_n(j)\mathsf{P}\left(S_{n+1} = i \middle| S_n = j\right)}{\sum_j a_n(j)}. \tag{16.21}$$

For the SEHM framework, from (16.6), we have

$$\left.\tilde{\Delta T}_{n+1}\right|_{\mathsf{SEHM}} = \frac{1}{1 - e^{-\left(B_n + SE_n\left(H_{n-1}\right)\right)}}. \tag{16.22}$$

For the sake of comparison, we also use a sample mean estimator as a baseline:

$$\left.\tilde{\Delta T}_{n+1}\right|_{\mathsf{Baseline}} = \frac{1}{n}\sum_{i=1}^{n}\Delta T_i. \tag{16.23}$$

To compare the three prediction algorithms, we use the Symmetric Mean Absolute Percentage Error (SMAPE) score, defined as,

$$\mathsf{SMAPE}(n) \triangleq \frac{1}{n}\sum_{i=1}^{n}\left|\frac{\Delta T_i - \tilde{\Delta T}_i}{\Delta T_i + \tilde{\Delta T}_i}\right| \tag{16.24}$$

as the metric. Note that the SMAPE score captures the relative error in prediction and is a number between 0% and 100% with a smaller value indicating a better prediction algorithm. An estimation error of the form $\left|\Delta T_i - \tilde{\Delta T}_i\right|$ only captures the absolute deviation between the true quantity and estimated quantity. On the other hand, a metric such as in (16.24) normalizes this absolute deviation by the weight of the true quantity itself and is thus a better reflection of the predictive power of the algorithm.

The SMAPE scores of the time to the next day of activity for the three estimators (HMM, SEHM, and baseline) are plotted as a function of the training period for model learning in Figure 16.4 for the FARC dataset and in Figure 16.5 for the Indonesia/Timor-Leste dataset. Table 16.4 also shows the SMAPE comparison between the two frameworks for the two datasets. It can be seen from these results that for both the datasets, the HMM framework results in a better prediction than the SEHM and the baseline frameworks, provided the training period is long to ensure accurate model learning for the HMM. Further, for the Indonesia/Timor-Leste dataset, even the baseline sample mean estimator outperforms the SEHM estimator for large n. Note that it is reasonable to expect some crossovers between the performance of HMM and SEHM at some n values, depending on how well a specific model has been learned. However, a sample mean estimator is not a sophisticated model. To observe that this estimator outperforms a more sophisticated model such as SEHM shows that either the SEHM is unwarranted, or it is a poor fit for predictive applications with this specific dataset.

In practice, the training period n that is used is dictated by when we would like to make predictions about future terrorist attacks given a certain past history of the group. In general, the longer the training period, the better (or at least more stable) the predictive power of a good model should be (and *vice versa*). For example, with the FARC dataset, the predictive power appears to stabilize at around $n = 150$ and for Indonesia/Timor-Leste, the predictive power seems to be getting better even at $n = 200$ observation points. From these observations, it appears that a choice of $n = 100$–150 observation points would be good for prediction in a practical setting (see Figures 16.4 and 16.5 for illustration).

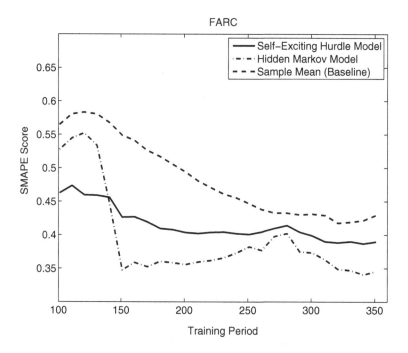

FIGURE 16.4: SMAPE scores for the three models with the FARC dataset.

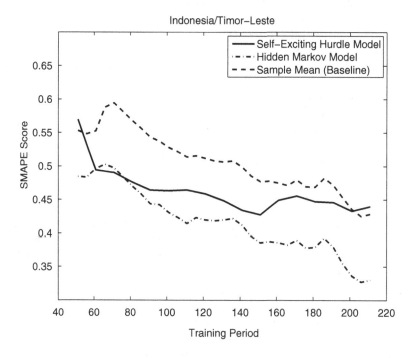

FIGURE 16.5: SMAPE scores for the three models with the Indonesia/Timor-Leste dataset.

TABLE 16.4: Comparison between SMAPE Scores with the SEHM and HMM Frameworks for the FARC and Indonesia/Timor-Leste Datasets

	FARC			Indonesia/ Timor-Leste	
n	**SEHM**	**HMM**	**n**	**SEHM**	**HMM**
100	46.27%	52.78%	100	46.33%	43.32%
150	42.95%	35.75%	125	45.47%	41.89%
200	40.40%	35.61%	150	42.84%	38.75%
250	40.09%	38.14%	175	45.23%	38.00%
300	39.92%	37.35%	200	43.46%	33.99%

16.4 Inferencing of Activity Profiles

The main inferencing goal with terrorism data is the early detection of abrupt spurts and downfalls in the activity profile. This is a problem of significant bearing in counterinsurgency operations as well as policy framing to mitigate terrorism and terrorist groups.

16.4.1 Parametric Approach: Viterbi Algorithm

The simplest approach to leverage the underlying HMM structure is to develop a parametric scheme to classify the hidden states (independent of whether they capture *Capabilities* or *Tactics*) via the use of the Viterbi algorithm (Rabiner, 1989) with the converged model parameter estimates from the Baum-Welch algorithm on $\{M_i\}$. A notable disadvantage of this approach is that inferencing on the group's *Capabilities* on a daily basis could lead to a performance mirroring the potential rapid fluctuations in the observations. This is particularly disadvantageous in making global policy decisions based on local inferencing of group dynamics. See more discussion in Raghavan and Tartakovsky (2018).

To overcome this difficulty, we propose inferencing over a $\delta > 1$ day time-window. For this, we decompose the time period of interest into disjoint time-windows, Δ_n, $n = 1, 2, \cdots, \mathcal{N}$, where $\Delta_n = \{(n-1)\delta+1, \cdots, n\delta\}$. The appropriate choice of δ is determined by the group dynamics and the timelines for inferencing decisions with typical choices being seven or 15 days corresponding to a weekly or a fortnightly decision process (Porter & White, 2012; Raghavan, Galstyan & Tartakovsky, 2013; Raghavan & Tartakovsky, 2018). We then assume that the hidden state remains fixed over Δ_n:

$$S_i\big|_{i \in \Delta_n} = s_n, s_n \in \{0, 1\}, \tag{16.25}$$

and our goal is to infer s_n with the aid of an appropriate set of observations corresponding to Δ_n.

To aid in inferencing, we associate a spurt in activity with either a change in the resilience of the group or a change in the level of coordination in the group (Sageman, 2004; Cragin & Daly, 2004; Santos, 2011; Lindberg, 2010; Blomberg, Gaibulloev &

Sandler, 2011; Bakker, Raab & Milward, 2012) (or perhaps both features). Resilience and coordination are quantified in terms of the group's ability to sustain multiple attacks over a long and a short term, respectively. To capture these attributes, we focus on a set of *attack metrics* that capture the resilience and coordination in the group: i) X_n, the number of days of terrorist activity, and ii) Y_n, the total number of attacks, both within the Δ_n time-window,

$$X_n = \sum_{i \in \Delta_n} \mathbb{1}\left(\{M_i > 0\}\right) \tag{16.26}$$

$$Y_n = \sum_{i \in \Delta_n} M_i, \quad n = 1, 2, \cdots, \tag{16.27}$$

where $\mathbb{1}(\cdot)$ denotes the indicator function of the set under consideration. Note that Y_n/δ is the average number of attacks per day and thus Y_n is a reflection of the intensity of attacks launched by the group. In general, X_n is more indicative of resilience in the group, whereas Y_n captures the level of coordination better.

With the hurdle-based geometric model in (16.15), it can be seen (Raghavan, Galstyan & Tartakovsky, 2013) that

$$\mathsf{P}\left(X_n = k, Y_n = r \middle| \mathbf{S}_i\big|_{i \in \Delta_n} = j\right) = \binom{\delta}{k}\binom{r-1}{r-k} \cdot \left(1 - \gamma_j\right)^{\delta - \kappa}\left(\gamma_j\right)^k \cdot \left(1 - \mu_j\right)^k\left(\mu_j\right)^{r-k}, \quad r \geq k. \tag{16.28}$$

Note that γ_j and μ_j are the model parameters, whereas k and r denote the values of number of days of activity and number of attacks, respectively. Model parameters learned with the Baum-Welch algorithm with $\{(X_n, Y_n)\}$ as observations are then used *retrospectively* (or non-causally) with the Viterbi algorithm for state classification. The output of the Viterbi algorithm is a state estimate for the period of interest

$$\left\{\mathbf{S}_i = \hat{s}_n \in \{0,1\} \text{ for all } i \in \Delta_n \text{ and } n = 1, \cdots, K\right\}. \tag{16.29}$$

A state estimate of 1 indicates that the group is *Active* over the period of interest, whereas an estimate of 0 indicates that the group is *Inactive*. Transition between states indicates spurt/downfall in the activity.

The approach described as above is applied to the FARC dataset with a $\delta = 15$ day time-window and the results are illustrated in Figure 16.6. As can be seen from this study, the state classification approach detects even *small* and *non-persistent* changes. However, this performance comes at the cost of model learning (which implicitly assumes model stationarity) and retrospective state classification (that renders it almost impractical from an applications standpoint). In particular, model learning can take a significant/certain amount of latency that can render it unusable in a delay-constrained decision-making setting.

16.4.2 Non-parametric Approach: EWMA Algorithm

While Section 16.4.1 considered a parametric approach exploiting the HMM structure, this approach of Section 16.4.1 also requires a reasonable knowledge of the

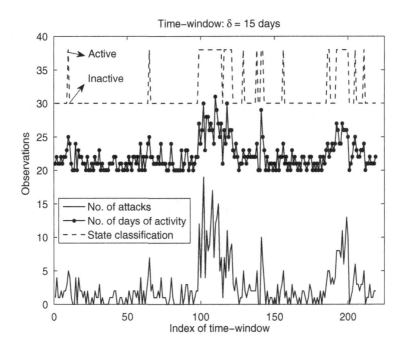

FIGURE 16.6: State classification with the hurdle-based geometric model for the observation sequence $\{(X_n, Y_n)\}$ of the FARC dataset.

underlying parameter estimates for state classification. Acquiring such knowledge leads to a latency in inferencing. From a terrorism analysis perspective, terrorist activity is "rare," even for some of the most active terrorist groups. For example, the FARC dataset of Figure 16.1 corresponds to an average of approximately 1.29 incidents per week. As a crude illustration, learning a four-parameter model with 100 observation points (on average) per parameter leads to a model learning latency of $\frac{4 \times 100}{1.29} \approx 310$ weeks (or ≈ 6 years).

A closely related and more challenging problem in applications is the fact that most models capture some underlying attribute of the group dynamics, which in itself can change dramatically over a long time period (such as that incurred in model learning). This fact renders assumptions of model stability over such periods questionable. The use of the proposed parametric approaches in the previous sections over a long time-horizon (with time-varying parameter estimates) opens up an array of issues on the stability of inferencing decisions in the short time-horizon. A problem that makes the approach of Section 16.4.1 unattractive as an *online* approach is its use in a *retrospective* state classification manner after model learning.

With the difficulties using the parametric approach, we note that the spurt detection problem is similar in spirit and objective to the changepoint problem of detecting sudden/abrupt changes in the statistical nature of observations. The theory of changepoint detection has matured significantly and different detection procedures such as the Shewhart chart, Cumulative Sum (CUSUM) chart, Exponentially Weighted Moving Average (EWMA) control chart, Shiryaev procedure, Shiryaev-Roberts (SR) procedure, generalized SR procedure, etc., have been developed over the last few decades (see books by Shiryaev, 1978; Basseville & Nikiforov, 1993; Poor & Hadjiliadis, 2008 for details). Fundamentally speaking, a changepoint procedure is equivalent to an update equation for the test statistic based on the likelihood ratio of the observations. This test statistic is tested against a threshold (which is chosen to meet appropriate false alarm constraints) to lead to a change decision. Developing the structure of the update equation as well as setting the threshold require knowledge of the pre-change and post-change densities/parameters.

Motivated by the rich literature of changepoint detection (Shiryaev, 1978; Basseville & Nikiforov, 1993; Poor & Hadjiliadis, 2008), we now propose a non-parametric spurt detection approach based on the EWMA algorithm. The EWMA algorithm was first introduced by Roberts (1959) for (continuously) tracking and detecting a change in the mean of a sequence of observations. Here, the test-statistic R_n is a first-order auto-regressive version of the observation process Z_n to be tracked with smoothing effected by an appropriately chosen parameter λ as given by:

$$R_n = \left(1 - \lambda\right) R_{n-1} + \lambda Z_n, \quad n \geq 1 \tag{16.30}$$

and $R_0 = 0$. The test-statistic is tested continuously against a threshold A_γ and change is declared at the first instance (this first instance is denoted as τ_{EWMA}) the test-statistic exceeds the threshold:

$$\tau_{\text{EWMA}} = \inf\left\{n \geq 1 : R_n \geq A_\gamma\right\}. \tag{16.31}$$

A_γ is chosen to ensure that the average run length (ARL) to false alarm is at least γ. The average run length to false alarm is the average number of observations taken to declare a change when there is none. In general, we would like to minimize this metric along with a constraint on minimizing the average number of observations in declaring a change where there is actually one (the mean detection delay). Small values of the smoothing parameter λ usually work best in changepoint detection (Srivastava & Wu, 1997; Polunchenko, Sokolov & Tartakovsky, 2012) as they smooth big changes and enhance small changes.

The EWMA framework can be applied to spurt detection in the activity profile of a terrorist group by repeatedly applying (16.31) with X_n and Y_n as observations. Two parameters $\left\{\lambda_1, \lambda_2\right\} \in \left[0,1\right]$ are chosen and used to update the variables $R_{\{1,n\}}$ and $R_{\{2,n\}}$ as follows:

$$R_{1,n} = \left(1 - \lambda_1\right) R_{1,n-1} + \lambda_1 X_n \tag{16.32}$$

$$R_{2,n} = \left(1 - \lambda_2\right) R_{2,n-1} + \lambda_2 X_n \tag{16.33}$$

for $n > 1$ with $R_{1,0} = 0 = R_{2,0}$. As far as we understand, the state-of-the-art in EWMA design is such that smoothing parameter design is still open, even for simple models such as Gaussian and exponential densities (Srivastava & Wu, 1997; Polunchenko, Sokolov & Tartakovsky, 2012). There is no prior work that considers the specific observation densities that we study in this work. Short of this, the best choices of smoothing parameters λ_1 and λ_2 for changepoint detection are obtained experimentally/numerically (see discussion that follows next).

We propose three stopping-times for declaring change: one based on $R_{\{1,n\}}$, another based on $R_{\{2,n\}}$, and the third on a weighted combination (with weights α and $\sqrt{1-\alpha^2}$ for some appropriate $\alpha \in [0, 1]$) of the two test-statistics:

$$\tau_1 = \inf\{n \geq 1 : R_{1,n} \geq A_1\} \tag{16.34}$$

$$\tau_2 = \inf\{n \geq 1 : R_{2,n} \geq A_2\} \tag{16.35}$$

$$\tau_{\text{weighted}} = \inf\{n \geq 1 : \alpha R_{1,n} + \sqrt{1-\alpha^2} R_{2,n} \geq A\}, \tag{16.36}$$

where the thresholds A_1, A_2, and A are chosen to meet the corresponding ARL constraints. While design of the threshold requires further work, experimental studies suggest that a threshold of the form

$$\{A_1, A_2, A\} = \mathcal{O}(\log(\gamma)) \tag{16.37}$$

ensures that $\{\text{ARL}(\tau_1), \text{ARL}(\tau_2), \text{ARL}(\tau_{\text{weighted}})\} = \mathcal{O}(\gamma)$. In other words, from the theory of changepoint detection, a good choice of threshold that trades off false alarms with quick detection should be such that it is $\mathcal{O}(\log(\gamma))$ (Polunchenko, Sokolov & Tartakovsky, 2012). Since τ_{weighted} combines the information contained in both $\{X_n\}$ and $\{Y_n\}$, it should empirically be a better test than both τ_1 and τ_2. Nevertheless, all the three tests could be of potential utility depending on the nature of the terrorist group.

We now study the performance of the proposed EWMA approach with the FARC data. In the first study, presented in Figure 16.7, we study how the tracking properties of $R_{2,n}$ (to changes in Y_n) behave with different values of the smoothing parameter λ_2. The true behavior of Y_n is also provided as a benchmark to compare the behaviors of $R_{2,n}$ with different λ_2. In general, small values of λ_2 smoothen big changes in Y_n considerably, whereas large values of λ_2 ensure that the evolution of $R_{2,n}$ is vulnerable to even small changes in Y_n. As can be seen from the plot, a value of $\lambda_2 = 0.10$ trades off the phenomena of smoothing out big changes and enhancing small changes, and thus renders the test-statistic $R_{2,n}$ effective from a changepoint detection perspective. The precise choice of λ_2 would be application-specific and needs to be chosen in terms of the above tradeoffs. One specific criterion for the choice of λ_2 could be to measure the overshoots/enhancements at every instance of a small change enhancement and also to measure the smoothing/undershoots at every instance of a big change being smoothed and find the specific choice that optimizes a weighted combination of these metrics.

In Figure 16.8, we study the performance of the three EWMA tests proposed here. We plot the test-statistics: $R_{1,n}$ with $\lambda_1 = 0.05$ for τ_1, $R_{2,n}$ with $\lambda_2 = 0.10$ for τ_2, and $\alpha R_{1,n} + \sqrt{1-\alpha^2} R_{2,n}$ with $\alpha = 0.25$ for τ_{weighted}. The threshold is designed as $\{A_1, A_2, A\} = 3\log(\gamma)$ for $\gamma = 10$. From Figure 16.8, we see that an appropriate weighted combination of the metric that captures resilience and the level of coordination in the group performs better than either test statistic taken separately (with the same threshold for all the three tests). For this, note that the weighted metric is larger than either of the individual test statistics, which allows easier detectability with the same threshold. That is, the weighted metric crosses the threshold faster than either $R_{1,n}$ or $R_{2,n}$, reducing the detection delay. While with the FARC data, the weighted sum performs only

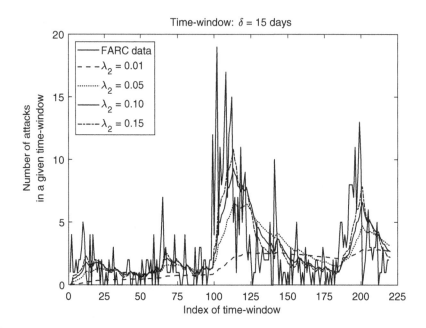

FIGURE 16.7: Tracking properties of $R_{2,n}$ in response to changes in Y_n as a function of the smoothing parameter.

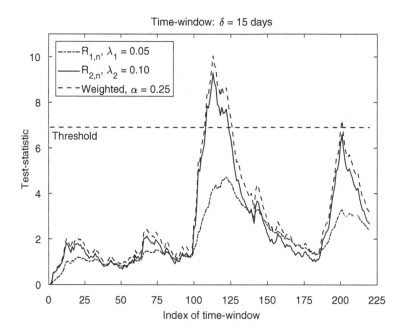

FIGURE 16.8: Performance of the three EWMA stopping-times in spurt detection for the FARC dataset.

marginally better than $R_{2,n}$, in general, we expect τ_{weighted} to significantly improve the performance over either τ_1 or τ_2. But more importantly, the EWMA approach detects only *persistent* changes or changes that last for a sufficiently long duration so that the changepoint detection approach can work accurately. In other words, the major spurts in FARC activity are detected, while the approach *cannot* detect the minor spurts. This is because the approach does not incorporate the statistical information of $\{X_n\}$ or $\{Y_n\}$, but only tracks them.

16.4.3 Non-parametric Approach: Majorization Theory

We now consider an alternate non-parametric approach for spurt detection motivated by majorization theory. To illustrate this approach, consider two extreme scenarios: i) a group conducting δ attacks on a specific day over a δ-day time-window and no other attacks in this period, and ii) a group conducting one attack on each day of the δ-day period. If $\delta > 1$, the former setting correlates well with a group having a high degree of coordination, whereas the latter setting would be more amenable with the belief that the group has a high degree of resilience. Rephrasing the above, a metric that measures the degree of "well-spreadedness" of attacks (or its lack thereof) over an appropriately chosen time-window can be used as an indicator of high resilience (or coordination). On this note, majorization theory provides a theoretical framework to compare two vectors on the basis of their "well-spreadedness" (Marshall & Olkin, 1979). For example, notions such as Gini index, income inequality, etc., can be studied via majorization theory. This theory provides a formal framework for comparing whether two vector quantities with a fixed sum can be arranged in such a way that one of them is more well-spread out or more unambiguously random/"bursty" than the other. See Marshall & Olkin, 1979; Raghavan and Tartakovsky, 2018 for details on how else majorization theory can be used.

The main problem with majorization theory is that it provides a partial ordering to compare two vectors with a fixed sum. In other words, if (and only if) two vectors are comparable, then we could make a statement whether one of them is more well-spread out than the other (or *vice versa*). But there is no guarantee that any two arbitrarily chosen vectors of a fixed sum are comparable. To overcome this fundamental difficulty, we take a recourse to the theory of catalytic majorization which is applicable over a larger space of vectors (than majorization theory) and allows a comparison across two vectors from this bigger space. In short, if and when two vectors are comparable in a catalytic majorization relationship, there exist certain functionals that *bijectively*[6] capture this relationship and with which these vectors can be far more easily compared since they are scalar quantities that can be ordered definitively. More theoretical details on these aspects can be found in Raghavan and Tartakovsky (2018).

We apply the theoretical framework of catalytic majorization to detect changes in resilience and coordination. Let $\underline{M} = [M_1, \cdots, M_\delta]$ capture the distribution of frequency of attacks over a certain time-window. That is, M_i denotes the fraction of attacks on the ith day with $\sum_{i=1}^{\delta} M_i = 1$, provided that there is at least one attack over the time period of interest. We call $\underline{M}$ the *attack frequency vector* and note that by definition $\underline{M} \in \mathcal{P}(\delta)$,

---

[6] Note that bijective means one-to-one and onto. In other words, there is a unique relationship between the vector under consideration and the values taken by the functionals as evaluated with the vector under consideration.

where $\mathcal{P}(\partial)$ denotes the space of probability vectors (each component of the vector is non-negative and the sum of the components is one).

To study resilience and coordination, we consider the following three functionals in comparing two different attack frequency vectors:

- Number of attacks over the time-window (denoted as Y_n, where we borrow the notation from our earlier studies)
- Normalized power mean of attack frequencies for some $\alpha > 1$, defined as

$$\text{NPM}\left(\underline{\boldsymbol{M}}\big|_{\Delta_n}, \alpha\right) \triangleq \frac{\left(\sum_i \mathsf{M}_i^\alpha\right)^{1/\alpha}}{\sum_i \mathbb{1}\left(\{\mathsf{M}_i > 0\}\right)}. \tag{16.38}$$

- Shannon entropy of the attack frequency vector, defined as

$$\text{SE}\left(\underline{\boldsymbol{M}}\big|_{\Delta_n}\right) \triangleq -\sum_i \mathsf{M}_i \log\left(\mathsf{M}_i\right). \tag{16.39}$$

A vector that corresponds to a large Y_n and is more spread-out (indicating a higher resilience in the group) results in a larger value for $\text{SE}\left(\underline{\boldsymbol{M}}\big|_{\Delta_n}\right)$. On the other hand, a vector that corresponds to a large Y_n and is less spread-out (indicating a high coordination in the group) results in a larger value for $\text{NPM}\left(\underline{\boldsymbol{M}}\big|_{\{\Delta n\}}, \alpha\right)$. Finally, a small value for Y_n suggests that the group is in an *Inactive* state.

We now propose a simplistic birth-death process model to track changes in resilience and coordination. For this, we define two functions $\tilde{X}_n$ and $\tilde{Y}_n$ that compare the Shannon entropy and the normalized power mean over Δ_n with the corresponding running sample means as follows:

$$\tilde{X}_n = \frac{\text{SE}\left(\underline{\boldsymbol{M}}\big|_{\Delta_n}\right)}{\frac{1}{\Delta}\sum_{i=1}^{\Delta} \text{SE}\left(\underline{\boldsymbol{M}}\big|_{\Delta_{n-i}}\right)} \tag{16.40}$$

$$\tilde{Y}_n = \frac{\text{NPM}\left(\underline{\boldsymbol{M}}\big|_{\Delta_n}, \alpha\right)}{\frac{1}{\Delta}\sum_{i=1}^{\Delta} \text{NPM}\left(\underline{\boldsymbol{M}}\big|_{\Delta_{n-i}}, \alpha\right)}. \tag{16.41}$$

In (16.40) and (16.41), an appropriate choice of Δ is made. We then update two functions (resilience and coordination functions) that capture the two facets of interest (for $n \geq 1$), $R(n)$ and $C(n)$, as follows:

$$R(n) = R(n-1) + \tau_{\mathcal{R}}, \quad R(0) = 0, \tag{16.42}$$

$$C(n) = C(n-1) + \tau_{\mathcal{C}}, \quad C(0) = 0, \tag{16.43}$$

where

$$\tau_{\mathcal{R}} = \mathbb{1}\left(\tilde{X}_n > \bar{\gamma}_{\mathcal{R}}, Y_n > \tau\right) - \mathbb{1}\left(\tilde{X}_n < \underline{\gamma}_{\mathcal{R}}, Y_n > \tau\right) - p_{\mathcal{R}} \cdot \mathbb{1}\left(Y_n \leq \tau\right) \tag{16.44}$$

$$\tau_C = \mathbb{1}\left(\tilde{Y}_n > \bar{\gamma}_C, Y_n > \tau\right) - \mathbb{1}\left(\tilde{Y}_n < \underline{\gamma}_C, Y_n > \tau\right) - p_C \cdot \mathbb{1}\left(Y_n \le \tau\right) \qquad (16.45)$$

and $0 < p_R < 1$ and $0 < p_C < 1$ are appropriately chosen *Inactive* state penalties. The thresholds $\bar{\gamma}_R$ and $\underline{\gamma}_R$ are chosen so as to correspond to a unit increase and a unit decrease in the resilience of the group in the *Active* state, whereas the threshold τ is chosen to determine an *Inactive* state which is penalized with a decrease of resilience corresponding to the parameter p_R. Similarly, for coordination, thresholds $\bar{\gamma}_C$ and $\underline{\gamma}_C$, and the penalty parameter p_C are chosen.

To restate the above, τ_R (and τ_C) take four possible values: 1, –1, 0, and p_R (or p_C), depending on whether the group is resilient/coordinating, non-resilient/non-coordinating, neither resilient nor coordinating, and *Inactive*, respectively. With this background, the proposed approach detects changes in resilience and coordination (and allows these changes to be categorized) without suffering from explicit model learning delays. Thus, the proposed approach is of tremendous advantage in practice.

We now illustrate these advantages by considering state classification with the FARC dataset. For this, we use the following parameters in our study: $\delta = 15$ days, $\Delta = 5$, $\alpha = 2.5$, $\tau = 4$, $p_R = 0.2$, $p_C = 0$, $\bar{\gamma}_R = \bar{\gamma}_C = 0.6770$, and $\underline{\gamma}_R = \underline{\gamma}_C = 0.4513$. Figure 16.9 plots the two statistics, $R(n)$ and $C(n)$, against the backdrop of $Y(n)$. It can be seen that $R(n)$ decreases initially before starting to rise in early 2002 (coinciding with *Plan Colombia*) and again in 2006 coinciding with the re-election period. Also, observe Figure 16.1 for the time axis of events. On the other hand, $C(n)$ shows only minor spurts over the same period, indicating that FARC was a more resilient group than a group coordinating multiple attacks. Note the trends for $R(n)$ and $C(n)$ in Figure 16.9, where $R(n)$

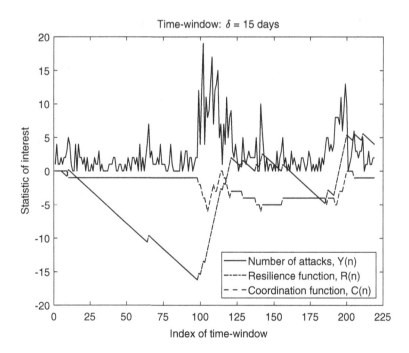

FIGURE 16.9: Resilience and level of coordination functions for the FARC dataset.

shows a far more dramatic variation than $C(n)$ and coinciding with the time periods as described above.

At this point, it is important to note that the design of parameters in the test of this section needs further exploration. Short of this, what we have illustrated here is a proof-of-concept that the proposed approach can work well if the parameters are designed appropriately.

16.4.4 Qualitative Comparisons Between Different Approaches

While Section 16.4.3 has explored one flavor of majorization theory-based approaches with a simple thresholding function to capture resilience and coordination metrics, more detailed approaches can be constructed and the readers are referred to Raghavan and Tartakovsky (2018) for some of these approaches. The parametric approach based on HMM and the majorization theory-based approaches are different and complementary solutions to the same problem of spurt detection and tracking. In general, these approaches are similar performance-wise; see Raghavan and Tartakovsky, 2018 for some comparisons. The major striking difference lies in the implementation pitfalls associated with the parametric approach where the model has to be learned over a training period, and a retrospective (due to the nature of the Viterbi algorithm) state classification has to be performed over a testing period. The majorization theory-based approach avoids these issues and is well suited for practical deployments. However, the tuning or design of parameters in the proposed algorithms need more exploration/ study. Table 16.5 broadly compares the pros and cons of the different approaches that can be pursued for spurt detection.

TABLE 16.5: Comparison between Different Spurt Detection Approaches

Method	Pros	Cons
Changepoint detection theory	1. Optimal/best performance given an observation model	1. Ignores connection between states and observations 2. Design of test parameters (e.g., thresholds) 3. Model learning
Section 16.4.1: Parametric approach using Viterbi algorithm	1. Good (comparable to optimal) performance given a state and observation model	1. Hypothesizes state and observation model for data 2. Retrospective state classification 3. Model learning
Section 16.4.2: Non-parametric approach using EWMA algorithm	1. Reasonable/acceptable performance with big changes 2. Instantaneously useable	1. Design of test parameters 2. Can detect only persistent changes
Section 16.4.3: Non-parametric approach using majorization theory	1. Good performance and allows the tracking of attributes over time 2. Instantaneously useable	1. Design of test parameters

16.5   Concluding Remarks

The focus of this chapter has been on developing good models for time-series data associated with terrorism activity. This problem is of importance in understanding broader trends in terms of temporal evolution of terrorism, switches due to certain geopolitical events and their ramifications on the nature and scope of terrorism, etc. In this context, this chapter summarized prior works based on classical time-series analysis techniques and self-exciting hurdle models for terrorism activity. Alternate to these modeling approaches, this chapter explored an HMM framework to model the activity profile of terrorist groups. Key to this development is the hypothesis that the current activity of the group can be captured completely by certain states/attributes of the group, instead of the entire past history of the group. In the simplest example of the proposed framework, the group's activity is captured by a $d = 2$ state HMM with the states reflecting a low state of activity (*Inactive*) and a high state of activity (*Active*), respectively.

This chapter then focused on detecting sudden spurts in the activity profiles of terrorist groups. This is an important inferencing task in combating terrorist activity. Most work in this area is parametric in nature (Enders & Sandler, 1993, 2000; Porter & White, 2012; Raghavan, Galstyan & Tartakovsky, 2013; Lewis et al., 2011), which renders their real-life application computationally difficult. In particular, parametric approaches to spurt detection often rely on past behavior for prediction, but terrorists' behavior changes quickly enough to make some of this analysis questionable. To overcome this fundamental difficulty, we proposed a novel statistical non-parametric approach based on majorization theory to detect sudden and abrupt changes in the *Capabilities* of the group. Leveraging the notion of catalytic majorization, we developed a simple approach to increment/decrement an appropriate statistic that captures resilience and coordination. Performance comparison shows that in general, non-parametric approaches typically have a better performance than the HMM-based parametric approach (Raghavan & Tartakovsky, 2018).

16.5.1   How to Apply the Developed Techniques in Practice?

The ideas exposed in this chapter can be used with minor modifications in many practical inferencing applications on time-series data where the observations reflect an evolution across a number of hidden states and attributes (such as those described in Figure 16.3). For example, spurt detection is a problem of importance across many other fields: social network analysis (Raghavan et al., 2014), credit-card fraud/financial data monitoring, criminal network activity monitoring (Mohler et al., 2011), geological/seismic analysis (Ogata, 1988), etc. Applications in the social network setting could include activity monitoring for the purposes of efficient network resource allocation (e.g., Google FluTrends, insider threats as illustrated by Azaria et al., 2014, etc.), user-specific information dissemination (e.g., advertisements), user classification, rapid detection of anomalous behavior such as bot or compromised/spamming accounts, identification of macroscopic behavioral trends in online users for social trend analysis, etc.

The first step in applying the proposed ideas is to verify whether the data under consideration follows the general principles as expounded in Figure 16.3. The next step is to fit an appropriate model class for the data. These model classes could be of the classical time-series type with trend, seasonality, and stationarity components fitted either on the data itself, or on the data after an appropriate set of transformations, or on lagged and differenced versions of the data (innovations). Other approaches could be pursued with the SEHM or HMM classes discussed in great detail in Section 16.2. From within these model classes, appropriate observation densities such as those in (16.11)–(16.15) can be fitted. Following this, the learned model can be used to perform an appropriate validation task such as explanation/prediction of historical/future data with the AIC/SMAPE metrics used for model efficacy evaluation.

The learned model can then be leveraged for an appropriate inferencing task (e.g., spurt detection) via either a parametric or a non-parametric approach as discussed in Section 16.4. As summarized earlier, even though non-parametric approaches can be application-dependent and often capture/track primarily the signatures behind the spurts, their low-complexity and ease of design, comparable performance efficacies, and robustness can make them advantageous over more carefully designed parametric approaches. All this said, it is important to note that this chapter only provided a specific methodological approach for a specific data type and is not meant to be a universal prescription for all data types. Thus, while the techniques herein can be adapted to other applications, some caution is essential in understanding the properties of the data, and the nature and scope of the algorithms discussed here.

16.5.2   Additional Reading Material

More information on classical time-series approaches for terrorism modeling can be found in Brophy-Baermann and Conybeare (1994), Cauley and Im (1988), Enders, Sandler and Cauley (1990), Landes (1978), Enders and Sandler (1993, 2000, 2002), Dugan, LaFree, and Piquero (2005), LaFree et al. (2010). Self-exciting hurdle models are considered in Hawkes (1971), Mohler et al. (2011), Porter and White (2012). Hidden Markov models for terrorism and social networks are studied in Raghavan, Galstyan, and Tartakovsky (2013) and Raghavan et al. (2014). The parametric approach for spurt detection based on HMMs as well as the EWMA algorithm are studied in Raghavan, Galstyan, and Tartakovsky (2013). The majorization theory-based approach is pursued in Raghavan and Tartakovsky (2018) and theoretical foundations of majorization theory can be found in Marshall and Olkin (1979). Also, see references within these works for further exploration.

References

Akaike, H. (1974, Dec.). A new look at the statistical model identification. *IEEE Transactions on Automatic Control*, *19*(6), 716–723.

Azaria, A., Richardson, A., Kraus, S., & Subrahmanian, V. S. (2014, June). Behavioral analysis of insider threat: A survey and bootstrapped prediction in imbalanced data. *IEEE Transactions on Computational Social Systems*, *1*(2), 135–155.

Bakker, R. M., Raab, J., & Milward, H. B. (2012, Winter). A preliminary theory of dark network resilience. *Journal of Policy Analysis and Management*, *31*(1), 33–62.

Basseville, M., & Nikiforov, I. V. (1993). *Detection of Abrupt Changes: Theory and Applications*. Prentice Hall, Englewood Cliffs, NJ.

Beittel, J. S. (2010, Apr.). Colombia: Issues for the Congress. *Congressional Research Service Report for the U.S. Congress*. (Available: [Online]. http://www.fas.org/sgp/crs/row/RL32250.pdf)

Bilmes, J. A. (1998, Apr.). *A gentle tutorial of the EM algorithm and its application to parameter estimation for Gaussian mixture and hidden Markov models* (Tech. Rep.). International Computer Science Institute, Berkeley, CA.

Blomberg, B. S., Gaibulloev, K., & Sandler, T. (2011, Dec.). Terrorist group survival: Ideology, tactics, and base of operations. *Public Choice, 149*(3–4), 441–463.

Brockwell, P. J., & Davis, R. A. (2010). *Introduction to Time Series and Forecasting* (2nd ed.). Springer.

Brophy-Baermann, B., & Conybeare, J. A. C. (1994, Feb.). Retaliating against terrorism: Rational expectations and the optimality of rules versus discretion. *American Journal of Political Science, 38*(1), 196–210.

Cauley, J., & Im, E. I. (1988, May). Intervention policy analysis of skyjackings and other terrorist incidents. *The American Economic Review, 78*(2), 27–31.

Clauset, A., Young, M., & Gleditsch, K. S. (2007). On the frequency of severe terrorist events. *Journal of Conflict Resolution, 51*(1), 58–87.

Cox, D. R., & Isham, V. (1980). *Point Processes* (1st ed.). Chapman and Hall.

Cragin, K., & Daly, S. A. (2004). *The Dynamic Terrorist Threat: An Assessment of Group Motivations and Capabilities in a Changing World.* RAND Corporation, Santa Monica, CA.

Dugan, L., LaFree, G., & Piquero, A. (2005, Nov.). Testing a rational choice model of airline hijackings. *Criminology, 43*(4), 1031–1065.

Enders, W., & Sandler, T. (1993, Dec.). The effectiveness of antiterrorism policies: A vector autoregression-intervention analysis. *The American Political Science Review, 87*(4), 829–844.

Enders, W., & Sandler, T. (2000, June). Is transnational terrorism becoming more threatening? A time-series investigation. *Journal of Conflict Resolution, 44*(3), 307–332.

Enders, W., & Sandler, T. (2002, June). Patterns of transnational terrorism, 1970–1999: Alternative time-series estimates. *International Studies Quarterly, 46*(2), 145–165.

Enders, W., Sandler, T., & Cauley, J. (1990). U.N. conventions, technology and retaliation in the fight against terrorism: An econometric evaluation. *Terrorism and Political Violence, 2*(1), 83–105.

Ephraim, Y., & Merhav, N. (2002, June). Hidden Markov processes. *IEEE Transactions on Information Theory, 48*(6), 1518–1569.

Hawkes, A. G. (1971, Apr.). Spectra of some self-exciting and mutually exciting point processes. *Biometrika, 58*(1), 83–90.

Hyndman, R. J., & Athanasopoulos, G. (2013). *Forecasting: Principles and Practice* (1st ed.). Otexts.

ITERATE. (2006). International Terrorism: Attributes of Terrorist Events. (Available: [Online]. http://www.icpsr.umich.edu/icpsrweb/ICPSR/studies/07947)

LaFree, G., & Dugan, L. (2007, June). Introducing the Global Terrorism Database. *Terrorism and Political Violence, 19*(2), 181–204.

LaFree, G., Morris, N. A., & Dugan, L. (2010, July). Cross-national patterns of terrorism, comparing trajectories for total, attributed and fatal attacks, 1970–2006. *British Journal of Criminology, 50*(4), 622–649.

Landes, W. M. (1978, Apr.). An economic study of U.S. aircraft hijackings, 1961–1976. *Journal of Law and Economics, 21*(1), 1–31.

Lewis, E., Mohler, G. O., Brantingham, P. J., & Bertozzi, A. (2011). Self-exciting point process models of civilian deaths in Iraq. *Security Journal.* doi:10.1057/sj.2011.21

Lindberg, M. (2010, May). *Factors Contributing to the Strength and Resilience of Terrorist Groups.* Grupo de Estudios Estrategicos (GEES) Publication.

Marshall, A. W., & Olkin, I. (1979). *Inequalities: Theory of Majorization and its Applications.* Academic Press, NY.

MATLAB. (n.d.). HMM Toolbox for MATLAB. (Available: [Online]. https://www.mathworks.com/help/stats/hidden-markov-models-hmm.html)

Midlarsky, M. I. (1978, Sept.). Analyzing diffusion and contagion effects: The urban disorders of the 1960s. *The American Political Science Review, 72*(3), 996–1008.

Midlarsky, M. I., Crenshaw, M., & Yoshida, F. (1980). Why violence spreads: The contagion of international terrorism. *International Studies Quarterly, 24*(2), 262–298.

Mohler, G. O., Short, M. B., Brantingham, P. J., Schoenberg, F. P., & Tita, G. E. (2011, Mar.). Self-exciting point process modeling of crime. *Journal of the American Statistical Association, 106*(493), 100–108.

Ogata, Y. (1988, Mar.). Statistical models for earthquake occurrences and residual analysis for point processes. *Journal of the American Statistical Association, 83*(401), 9–27.

Ogata, Y. (1998, Jun.). Space-time point process models for earthquake occurrences. *Annals of the Institute of Statistical Mathematics, 50*(2), 379–402.

Peña, D., Tiao, G. C., & Tsay, R. S. (2000). *A Course in Time Series Analysis* (1st ed.). Wiley-Interscience.

Polunchenko, A. S., Sokolov, G., & Tartakovsky, A. G. (2012). Optimization and efficiency analysis of the EWMA procedure for detecting changes in the exponential distribution. *Proceedings of Quality and Productivity Research Conference*, Long Beach, CA.

Poor, H. V., & Hadjiliadis, O. (2008). *Quickest Detection.* Cambridge University Press.

Porter, M. D., & White, G. (2012, Mar.). Self-exciting hurdle models for terrorist activity. *Annals of Applied Statistics, 6*(1), 106–124.

R. (n.d.). HMM Package for R. (Available: [Online]. https://cran.r-project.org/web/packages/HMM/HMM.pdf)

Rabiner, L. R. (1989, Feb.). A tutorial on hidden Markov models and selected applications in speech recognition. *Proceedings of the IEEE, 77*(2), 257–286.

Raghavan, V., Galstyan, A., & Tartakovsky, A. G. (2013, Dec.). Hidden Markov models for the activity profile of terrorist groups. *Annals of Applied Statistics, 7*(4), 2402–2430.

Raghavan, V., Steeg, G. V., Galstyan, A., & Tartakovsky, A. G. (2014, Apr.). Modeling temporal activity patterns in dynamic social networks. *IEEE Transactions on Computational Social Systems, 1*(1), 89–107.

Raghavan, V., & Tartakovsky, A. G. (2018, Sept.). Tracking changes in resilience and level of coordination in terrorist networks. *IEEE Transactions on Computational Social Systems, 5*(3), 639–659.

RDWTI. (2009). RAND Database of Worldwide Terrorism Incidents. (Available: [Online]. http://www.rand.org/nsrd/projects/terrorism-incidents.html)

Roberts, S. W. (1959, Aug.). Control chart tests based on geometric moving averages. *Technometrics, 1*(3), 239–250.

Sageman, M. (2004). *Understanding Terror Networks.* University of Pennsylvania Press, Philadelphia, PA.

Santos, D. N. (2011, May). What constitutes terrorist network resiliency? *Small Wars Journal, 7*(5), 1–8.

Shiryaev, A. N. (1978). *Optimal Stopping Rules.* Springer-Verlag, New York, NY.

Shumway, R. H., & Stoffer, D. S. (2017). *Time Series Analysis and Its Applications: With R Examples* (4th ed.). Springer.

Srivastava, M. S., & Wu, Y. H. (1997). Evaluation of optimum weights and average run lengths in EWMA control schemes. *Communications in Statistics - Theory and Methods, 26*(5), 1253–1267.

United Nations Office on Drugs and Crime. (2010, May). *World Drug Report (Tech. Rep.).*

Visser, I., & Speekenbrink, M. (n.d.). depmixS4: An R Package for Hidden Markov Models. (Available: [Online]. https://cran.r-project.org/web/packages/depmixS4/vignettes/depmixS4.pdf)

Chapter 17

Expert COSYSMO Systems Engineering Cost Model and Risk Advisor

Raymond Madachy and Ricardo Valerdi

17.1 Parametric Cost Modeling for Systems Engineering

This chapter describes a method for systems engineering cost estimation and risk management support. The cost comprises the systems engineering labor for the development of complex military systems. The risks to manage are inherent in the program management of the system development. The method is organized around best practices for parametric cost modeling.

Parametric cost modeling uses cost estimating relationships (CERs) as mathematical algorithms relating cost factors. The parametric models use numeric inputs for explanatory variables reflecting system characteristics to compute cost. A parametric cost model is defined for a specific aspect of a project, product, or process. The models are developed using regression analysis and other methods.

A cost estimating relationship between parameters provides a logical and predictable correlation between the physical or functional characteristics of a system and the resultant cost. A parametric CER may account for factors such as cost quantity relationships, inflation, staff skills, schedules etc. They also provide decomposition of generic cost elements for work breakdown structures (WBS). Parametric models in general are fast, easy to use, and helpful early in a program. They serve as valuable tools for systems engineers and project managers to estimate engineering effort or perform tradeoffs between cost, schedule, and functionality.

The Constructive Systems Engineering Cost Model (COSYSMO) is a parametric model to help you reason about the cost and schedule implications of systems engineering decisions you may need to make. Expert COSYSMO is a knowledge-based tool implementation available online[1] as are additional COSYSMO resources.[2] It can also be chosen as a selectable model within the integrated COCOMO Suite tool (Madachy, 2009).

COSYSMO is called "constructive," meaning the user knows why the model produces a resulting estimate and becomes better informed for making decisions. A primary objective of the model is to be constructive, in which the job of cost and schedule estimation helps people better understand the nature of the project being estimated. This was the reason for the name of the original COnstructive COst MOdel (COCOMO) for users to understand the complexities of development lifecycle processes through the medium of a cost model (Boehm, 1981). A user of constructive cost models can tell why they give the estimates they do.

COSYSMO is in the public domain, with the formulas open and available for use. It is more valuable this way when all stakeholders can understand and discuss the model estimates with common definitions of the inputs and outputs.

Before users begin to work with COSYSMO, there should be an awareness of the inherent assumptions embedded in the model. The first is that the function of systems engineering explicitly exists in an organization. In some organizations, systems engineering exists as a formal role while in others it is combined with the activities done by hardware or software engineers and even program management. The second assumption is that the organization develops large complex systems typified by and including virtually all military systems.

17.1.1 Government Usage

COSYSMO is also useful to government organizations that acquire systems from contractors; the model can be used to: (1) evaluate estimates provided in proposals, (2) manage existing systems' engineering efforts, or (3) benchmark systems' engineering performance across organizations.

Experience shows that organizations that calibrate COSYSMO with their own historical data consistently obtain more accurate cost estimates and develop a deeper understanding of the model parameters. The organization must be able to obtain historical data from their own completed programs to calibrate the model.

17.2 Systems Engineering Risk Management Approach

Systems engineering risk management aims to reduce potential risks to an acceptable level before they occur, throughout the life of the product or project. Risk management is thus a continuous, forward-looking process that can be considered both a project management and a systems engineering process (SEBoK Authors, 2017).

---

[1] http://csse.usc.edu/tools/ExpertCOSYSMO.php.

[2] http://cosysmo.mit.edu.

Project risk analysis is a primary activity within risk management. It has been repeatedly demonstrated to be critical for successful systems engineering projects (Keeney & von Winterfeld, 1991; Chittister & Haimes, 1994; Pate-Cornell, 1996; Browning & Eppinger, 2002; and Kujawski & Miller, 2007).

Two principal aspects of risk analysis are quantifying the uncertainty of cost estimates provided by cost models (statistics-based) and identifying specific project risk items (knowledge-based). In this chapter we focus on the second aspect of risk analysis by demonstrating how to identify and mitigate risks through the use of cost parameters in COSYSMO, which is also used to estimate systems engineering effort for a project (Valerdi, 2008a).

A knowledge-based approach solves complex problems within a specific substantive domain, which this method applies to systems engineering risk. Knowledge-based systems solve problems that can't be solved by individuals without domain specific knowledge. Expert systems and decision support systems are common types of knowledge-based systems. An expert system that is rule-based operates on a collection of rules that a human expert would follow in dealing with a problem.

Our approach uses an expert-defined rule set to flag systems engineering risks for further analysis and mitigation and provide associated risk control advice. It includes the ability to update the rule base, and the opportunity to integrate it into a more comprehensive risk management framework. A rigorous knowledge engineering process was used to codify the rule set with many experts over several years.

Our Expert COSYSMO tool (Madachy & Valerdi, 2008) leverages parametric cost model factors, which are themselves based on knowledge of the structure and function of a systems engineering project.

17.2.1   Cost Estimation and Risk Management Alliance

Cost estimation and risk management are strongly connected since cost estimates are used to evaluate risk and perform risk tradeoffs; risk methods such as Monte Carlo simulation can be applied to cost models; and the likelihood of meeting cost targets depends on effective risk management strategies (Madachy, 1997). The same cost inputs can also be used to assess risk using sensitivity analysis or Monte Carlo simulation such as the COSYSMO-R approach (Valerdi & Gaffney, 2007), but the approach described here uses cost model parameters to infer specific risk situations.

Approaches for identifying systems engineering risks are usually separate from cost estimation. However, risk management practice can be improved by leveraging on existing knowledge and expertise during cost estimation activities through the use of cost factors to detect patterns of project risk. This alliance allows us to identify and rank sources of project risk for mitigation plans in conjunction with cost estimation.

With this approach, cost estimation and risk management can be jointly performed by automatically inferring risks from the cost model inputs. It saves effort by not requiring separate analyses and resources.

17.3   Methodology

The method uses Expert COSYSMO, which is an extension of the standard COSYSMO cost model. It uses factors to analyze patterns of cost driver ratings submitted for a cost estimate (Madachy & Valerdi, 2010). It automatically identifies project risks in conjunction with cost estimation similar to Expert COCOMO for software development cost (Madachy, 1997). It uses size and cost driver inputs for both the cost estimating relationship and risk heuristics to produce an integrated management view of the estimate and risk analysis. This consolidated view provides management with related information to meet the project cost budget by addressing specific risks.

Risk situations are characterized by combinations of cost driver values indicating increased effort with a potential for more uncertainties. Expert COSYSMO currently covers over 600 fine grained risk conditions using the 15 inputs of the cost model.

In practice, risks must be identified as specific instances to be manageable. Our method identifies individual risks that an experienced systems engineering manager might recognize but often fail to take into account. It also helps calibrate and rank collections of risks, a process which many managers would not do otherwise. This information is then used to develop and execute project risk management plans. With these risks, mitigation plans can be created based on the relative risk severities and provided advice.

17.3.1   Constructive Systems Engineering Cost Model

The cost factors used in the method are represented as inputs in the COSYSMO effort equation. COSYSMO predicts the systems engineering effort for a project per (17.1). The factors are encapsulated in the term $\prod_{i=1}^{n} EM_i$.

$$\text{Effort} = A * \text{Size}^E * \prod_{i=1}^{n} EM_i \qquad (17.1)$$

where
  Effort   is measured in Person-Months
  Size   is a weighted sum of requirements, interfaces, algorithms, and scenarios
  A   is a calibration constant derived from project data
  E   represents a diseconomy of scale
  EM_i is the effort multiplier for the i_{th} cost driver.

The geometric product ΠEM_i results in an overall effort adjustment factor (*EAF*) to the nominal effort. The overall effort is then decomposed by phase and activities into constituent portions.

The cost drivers are summarized in Table 17.1. On a given project, each is rated on a scale from low to high. Full definitions of the cost drivers are in Valerdi (2008a) and are available online at Madachy (2017).

17.3.2   Knowledge Engineering

Knowledge engineering is the construction of a knowledge base used in an expert system. Expert COSYSMO has been developed as a collaborative effort between industry, government, and academia and has benefited from numerous focused workshops, surveys, and follow-up interviews conducted with several dozen systems engineering practitioners from major organizations. The underlying expert system structure is relatively simple, but is populated with rules generated with an extensive process for elicitation and refinement from the systems engineering experts.

A functionally complete knowledge base requires adequate coverage of domain, appropriate structure, and accurate knowledge (Adelman & Riedel, 1997). The domain

TABLE 17.1:   COSYSMO Cost Drivers

Product	Requirements Understanding
	Level of Service Requirements
	Architecture Understanding
	Technology Risk
Process	Documentation to Match Life Cycle Needs
	Tool Support
	Process Capability
	Multisite Coordination
People	Stakeholder Team Cohesion
	Personnel Experience/Continuity
	Personnel/Team Capability
Platform	Number and Diversity of Installations/Platforms
	Migration Complexity
	Number of Recursive Levels in the Design

of systems engineering is covered in this approach as the COSYSMO model covers a full range of project phenomena as reflected in the cost drivers in Table 17.1.

The risk taxonomy described later from these cost drivers shows the granularity of knowledge constructs is useful and appropriate for managing projects. Iterating with the highly experienced practitioners also ensured that the collective knowledge base structure was appropriate as the groups assessed its suitability, soundness, and extensibility.

Elicitation of knowledge came from systems engineering domain experts through a succession of six focused workshops beginning in 2006, structured surveys, and user and student feedback. We iteratively refined risk identification rules, risk item quantification, and later automated mitigation advice for users to help develop their project-specific mitigation plans.

In the first workshop pass, we began identifying possible risk conditions from the full superset of all interactions from the factors in Table 17.1. An initial survey was used to identify and quantify risks at a coarse level using over 20 experts. An N-squared matrix of cost drivers in the COSYSMO model yielded pairwise risk conditions for further refinement.

The workshops and ongoing Delphi surveys (Dalkey, Brown & Cochran, 1969) included participants from defense contractors, FFRDCs,[3] government agencies, DoD services and academia. The workshops were sometimes held in conjunction with the International Forum on COCOMO and Systems/Software Cost Modeling (Valerdi, 2008b; Madachy & Valerdi, 2009).

The follow-on workshops were organized into three parts:

1. All participants were provided with full definitions of the COSYSMO cost drivers to ensure reliability across participants.
2. Participants were led through a Delphi survey exercise where they had the opportunity to identify the subset of combinations of COSYSMO parameters that would result in the highest risks on projects.
3. Participants were asked to recommend mitigation strategies for a subset of risk scenarios.

---

[3] Federally Funded Research and Development Center.

To ensure measurement reliability, experts underwent a half-day tutorial on COSYSMO and its cost driver definitions. Moreover, the survey was discussed in a group while individual respondents provided their inputs, similar to the Wideband Delphi technique (Dalkey, Brown & Cochran, 1969). We iterated the initial survey to reach consensus on the risk conditions and classified them into high, medium, and low risks per the systems engineering experts (Madachy & Valerdi, 2010).

We created more granular risk quantification rules as the initial risk assessment scheme was elaborated through iterations with the experts. In subsequent workshops the risks were decomposed into more detailed conditions requiring no further inputs from the user. The initial coarse risk conditions at the individual cost driver level were elaborated into more than 600 fine-grained rules when decomposed at the cost driver rating levels (i.e., ranging from very low to very high for each driver).

Next we addressed risk mitigation by making the outputs more actionable, so explicit risk management steps can be undertaken by users. With more expert involvement we documented risk mitigation advice for each risk condition, and provide that automated guidance to users to help develop their own mitigation actions.

The opportunity trees from Madachy (2008) were used as initial checklists for representing the risk mitigation knowledge and making associations with relevant mitigation actions. The expert workshop participants elaborated the network of mitigation advice and scored the opportunities for specific risks. Later workshops were used to refine for mitigation strategies from the base rules implemented from previous workshops and Delphi surveys.

17.3.3   Systems Engineering Risk Taxonomy

Various types of risks are exhibited on systems engineering projects for different aspects of the work. They can also interact with each other. The taxonomic constructs in COSYSMO provide a holistic framework to address the different risk types and their interactions.

A primary distinction is technical risk versus programmatic, or non-technical risk. Technical risk is defined as a measure of probability and severity of an adverse event inherent in the development of a system that does not meet its intended functions. This aspect is represented in COSYSMO through product and platform cost factors. Non-technical risk involves programmatic aspects in the development process of systems (Chittister & Haimes, 1994). COSYSMO covers these risk types in process and people factors.

A significant difference between technical and non-technical risk is that they require different expertise in order to assess and mitigate. The expertise needed to identify technical risk associated with determining the validity of a system and its verification is cognitively different from the expertise required to identify the management risk associated with cost, schedule, or the spending rate of a project. Knowledge and expertise are required in the application domain to adequately identify and measure the technical uncertainties and associated impacts on the system. Moreover, the methodologies employed in the identification, quantification, measurement, evaluation, and management of technical and non-technical risks are inherently different.

Despite these differences, there are important interactions between the risk types. Technical risks may manifest themselves as non-technical risks when system requirements may be too complex to implement with the budgeted cost and schedule for a project. A holistic approach that considers the interactions between categories of risks is the best strategy for avoiding potential clashes between assumptions (Boehm & Port, 1999). The knowledge-based risk analysis approach in this method builds on this philosophy, and unifies the risk types embedded in the parametric cost model.

Table 17.1 showed the engineering cost drivers in COSYSMO are categorized as Product, Process, People, or Platform factors. These natural groupings illustrate the most critical aspects for systems engineering decision makers when developing a cost estimate. They are also consistent with previous cost model categories including the Constructive Cost Model (COCOMO II) (Boehm, 1981; Boehm et al., 2000).

The factor clustering provides a natural hierarchical relationship that leads to the formation of a tree. The tree connects program risks of four top-level categories, 14 underlying cost drivers, and their combinations in the ruleset. The full taxonomy representing cost and risk factors is in Figure 17.1. The acronyms used are for the input cost factors in Figure 17.1, per the legend at the bottom.

Product factors account for variation in the effort required for systems engineering due to characteristics of the product under development. A product that is complex, has high performance requirements, or uses unproven technologies will require more effort to complete. Moreover, poor understanding of system requirements or system architecture may also result in additional effort.

Process factors account for influences of process maturity, such as the Capability Maturity Model Integrated (CMMI) (CMMI Product Team, 2010), and the integration of tools to support the systems engineering process. Customer-imposed processes such as extensive documentation requirements may impact process-related costs, for example.

People factors have one of the strongest influences in determining the amount of effort required for systems engineering. The personnel factors are for rating the development team's capability and experience, not individuals. These ratings are most likely to change during the course of a project reflecting the gaining of experience or the rotation of people onto and off the project. In systems engineering, the definition of a project team may expand beyond corporate affiliations and could include end users, subcontractors, and integration partners. Therefore, the ability of all stakeholders to work together is also an important risk consideration in systems engineering.

Platform factors refer to the diversity of possible versions or installations of a system. A case where there are multiple versions to be deployed, documented, provide training for, and maintained will require much more systems engineering effort compared to a single version implementation. The effect of legacy systems may also be important in cases where the system being developed needs to be migrated into operation while the constituent system remains in operation.

17.3.4 Knowledge-Based Risk Assessment

17.3.4.1 Expert-Defined Risk Conditions

The cost drivers represented in the COSYSMO effort multiplier term ΠEM_i are used for expressing the risk rules. Predetermined combinations of driver ratings provide red flags of possible risks as the project progresses along its lifecycle. These scenarios are predetermined and configured into the model as a set of rules that can automate and improve the risk analysis process. The knowledge base now covers the systems engineering-specific risk conditions in Figure 17.1 that are further discretized into over 600 finer risk conditions. The fine grain risk conditions for risk levels are described next.

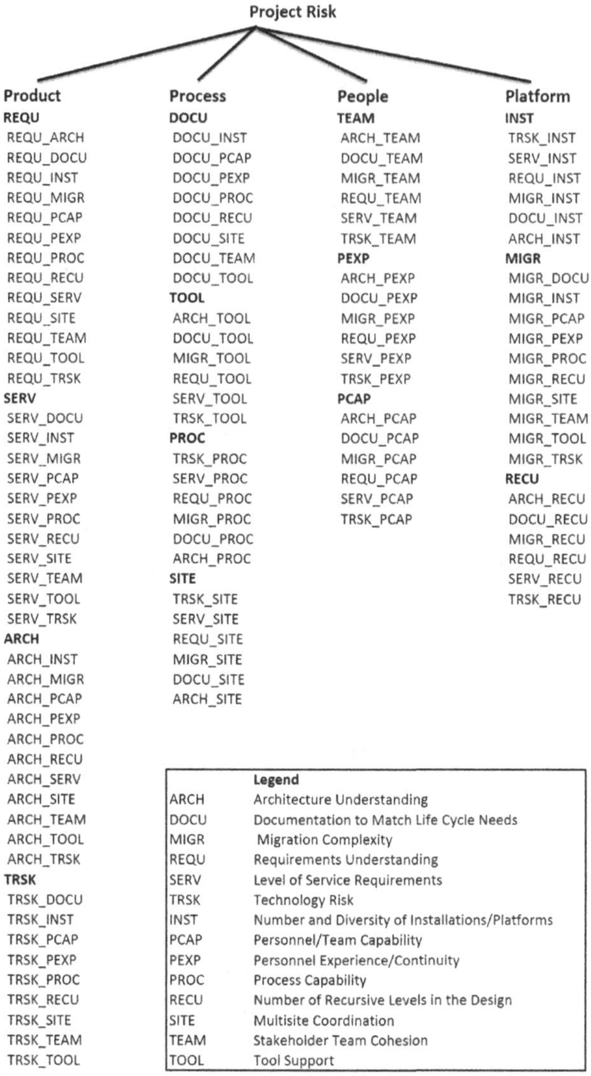

FIGURE 17.1: Risk taxonomy.

17.3.4.2 Risk Quantification

The risk taxonomy is used as a basis for risk quantification algorithms. Risk impact, or risk exposure, is traditionally defined as the probability of loss multiplied by the cost of the loss (SEBoK Authors, 2017; Boehm, 1989). The quantitative risk weighting scheme accounts for the nonlinearity of the assigned risk levels and effort multipliers, per Equation (17.2). It sums all category risks (CR) in each category, and across all categories (C). The *Effort Multiplier Product* is the $\prod_{i=1}^{n} EM_i$ term in the effort model. Each risk item may have 2 through n interacting cost factors in its effort multiplier computation. The risk levels were determined iteratively by the systems engineering process experts.

$$\text{Project Risk} = \sum^{C} \sum^{CR} \text{Risk Level} * \text{Effort Multiplier Product} \qquad (17.2)$$

where
    Risk Level = 1 Moderate
                 2 High
                 4 Very High

$$\text{Effort Multiplier Product} = EM_1 * EM_2 \ldots * EM_n$$

The risk level corresponds to the nonlinear relative probability of the risk occurring, and the effort multiplier product represents the cost consequence of the risk. The product includes effort multipliers for the cost factors involved in the risk. Each of the risk categories includes rules that include relevant cost factors. For example, process risks would include conditions that involve the factors for *Process Capability*, *Multi-site Coordination*, and *Tool Support*.

Figure 17.2 shows how the risk conditions are decomposed into finer grained conditions. The figure shows how the continuous risk iso-contours are discretized into cost driver interaction tables that the rules operate on. In the figure the term *Attribute* refers to a generalized cost factor from Table 17.1.

Figure 17.3 shows a portion of the risk network between risk categories, risks as combinations of cost drivers, and (highly abbreviated) potential mitigation actions. Risk exposure (RE) calculations are also shown centered in the risk item column, aggregated within the categories, and summed for potential risk reduction (PRR) of the mitigation actions.

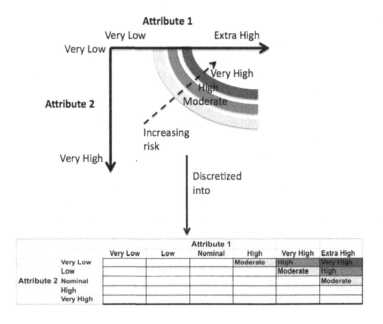

FIGURE 17.2: Example assignment of risk probability levels.

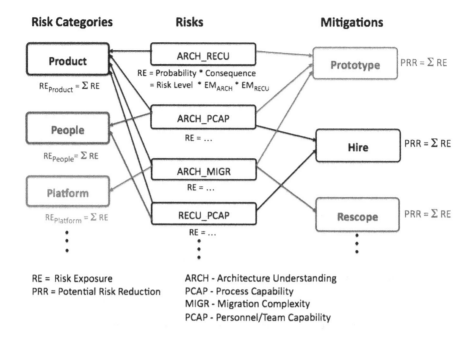

FIGURE 17.3: Risk network portion.

17.4 Examples

17.4.1 Risk Assessment

An example risk assessment of cost factor interactions between *Architecture Understanding* and *Level of Service Requirements* will demonstrate the method. The factors are defined in Tables 17.2 and 17.3 with their respective rating scales and effort multipliers at each rating.

The expert-determined probabilities for risks associated with coupling between these factors at the different rating conditions are shown in Figure 17.4. Figure 17.5 shows the analogous risk consequences using their effort multipliers.

Suppose for example that *Architecture Understanding* is rated "Very Low" and the *Level of Service Requirements* is "Very High." This indicates a potential risk in the project given that systems with high service requirements are more difficult to implement especially when the architecture is not well understood. The associated risk is calculated per below, where the *Risk Consequence* is the product of effort multipliers per Figure 17.5:

$$\text{Risk} = \text{Risk Probability} * \text{Risk Consequence}$$

$$= 4 * 1.64 * 1.85$$

$$= 12.1 \text{ Risk Exposure Points.}$$

The *Risk Exposure Points* is a dimensionless quantity. As such we quantify the risks on a scale relative to the maximum risk exposures for overall project and the constituent product, process, platform, and people risks.

TABLE 17.2:   Architecture Understanding Rating

Very Low	Low	Nominal	High	Very High
Poor understanding of architecture and COTS, unprecedented system	Minimal understanding; of architecture and COTS, many unfamilar areas	Reasonable understanding of architecture and COTS, some unfamiliar areas	Strong understanding of architecture and COTS, few unfamiliar areas	Full understanding of architecture, familiar system and COTS
>6 level WBS	5–6 level WBS	3–4 level WBS	2 level WBS	

Architecture Understanding: the relative difficulty of determining and managing the system architecture in terms of platforms, standards, components (COTS/GOTS/NDI/new), connectors (protocols), and constraints. This includes tasks like systems analysis, tradeoff analysis, modeling, simulation, case studies, etc.

TABLE 17.3:   Level of Service Requirements Rating

Very Low	Low	Nominal	High	Very High
Difficulty				
Simple; single dominant KPP	Low, some coupling among KPPs	Moderately complex, coupled KPPs	Difficult, coupled KPPs	complex, tightly coupled KPPs
Criticality				
Slight inconvenience	Easily recoverable losses	Some loss	High financial loss	Risk to human life

Level of Service Requirements: the difficulty and criticality of satisfying the ensemble of level of service requirements, such as security, safety, response time, interoperability, maintainability, Key Performance Parameters (KPPs), the "ilities", etc.

		Architecture Understanding				
		Very Low	Low	Nominal	High	Very High
	Very Low					
Level of Service	Low					
Requirements	Nominal	1				
	High	2	1			
	Very High	4	2	1		

FIGURE 17.4: Example risk probabilities (weights).

		Architecture Understanding				
		Very Low	Low	Nominal	High	Very High
	Very Low					
Level of Service	Low					
Requirements	Nominal	1.64 * 1.0				
	High	1.64 * 1.36	1.28 * 1.36			
	Very High	1.64 * 1.85	1.28 * 1.85	1.0 * 1.85		

FIGURE 17.5: Example risk consequences (effort multiplier products).

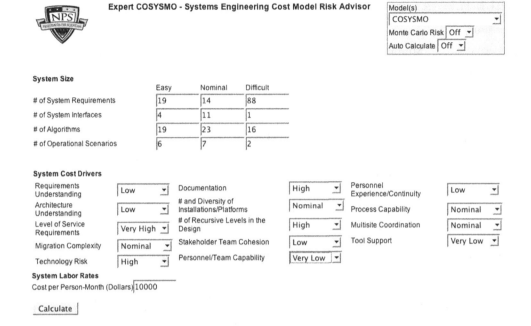

FIGURE 17.6: Expert COSYSMO inputs.

17.4.2   Example Project Estimate and Risk Outputs

Expert COSYSMO model inputs for an illustrative project are shown in Figure 17.6. Figures 17.7, 17.8, and 17.9 show the cost estimation and risk assessment outputs. The outputs include an overall risk assessment and category risks in Figure 17.8. Note that reuse risk is an added category to the base taxonomy. The risk scale in Figure 17.8 was calibrated to historical data using actual risk exposure points. The ranges were also determined with the experts assessing data from their own organizations. Figure 17.9 shows a prioritized ranking of the individual risks with mitigation guidance as a starting point for management consideration.

17.5   Conclusions and Future Work

The COSYSMO cost factors served well as a core set of abstractions for project risk assessment. They adequately cover the domain for systems engineering. The completeness of the attribute set for cost estimation was vital for generating a critical mass of rules from them. Common inputs between the expert system and cost model also ensured unambiguous mapping from project data to the rule set.

Practitioners can use this additional information extracted from cost models to improve risk planning early in the lifecycle. Users can also update the rule base for

Results
Systems Engineering
Effort =5118.9 Person-months
Schedule = 25.1 Months
Cost = $51188861

Total Size =1023 Equivalent Nominal Requirements

Acquisition Effort Distribution (Person-Months)

Phase / Activity	Conceptualize	Develop	Operational Test and Evaluation	Transition to Operation
Acquisition and Supply	100.3	182.7	46.6	28.7
Technical Management	191.4	330.7	217.6	130.5
System Design	522.1	614.3	261.1	138.2
Product Realization	99.8	230.3	245.7	192.0
Product Evaluation	285.6	428.5	634.7	238.0

FIGURE 17.7: Expert COSYSMO cost and schedule outputs.

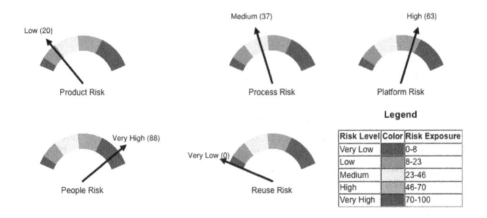

FIGURE 17.8: Expert COSYSMO risk outputs (1/2).

local conditions and have the opportunity to integrate it into a more comprehensive risk management framework.

The method also supports common process and measurement frameworks such as the Capability Maturity Model Integration (CMMI) (CMMI Product Team, 2010) and Practical Software Measurement (PSM) (Rhodes, Valerdi & Roedler, 2009), or ISO-9000 (ISO 9000:2000, 2000) both as a standalone process tool and a provider of essential data for risk metrics indicators.

Risk Mitigation Guidance

The risk mitigation guidance below shows alternatives for consideration in specific project environments. Risks are ordered by highest risk exposure.

Risk Exposure Points	Description	Alternatives
12.9	*Requirements Understanding* = Very Low and *Technology Risk* = Very High	Get customers involved, early prototypes, do trade studies, prioritize requirements
11.0	*Requirements Understanding* = Very Low and *Personnel/Team Capability* = Very Low	Facilitate workshops, get training, hire experts, set up IPTs, subcontract, get customers involved, prioritize requirements
10.3	*Technology Risk* = Very High and *Personnel/Team Capability* = Very Low	Restructure people and/or augment with high capabiity people in specific technical risk areas and prototyping expertise
5.3	*Technology Risk* = Very High and *# and Diversity of Installations/Platforms* = Very High	Early identification of potential installations, upfront effort including prototyping to cover each installation
4.5	*Requirements Understanding* = Very Low and *Personnel Experience/Continuity* = Low	Provide incentives to stay, subcontract, hire experts/consultants, get customers involved, prioritize requirements
4.3	*Requirements Understanding* = Very Low and *Tool Support* = Low	Hire tool vendors as consultants, use groupware tools to increase requirements understanding
4.2	*Technology Risk* = Very High and *Personnel Experience/Continuity* = Low	Restructure people and/or augment with more experienced people in specific technical risk areas and prototyping expertise, training
4.0	*Technology Risk* = Very High and *Tool Support* = Low	Invest in tools
1.9	*Requirements Understanding* = Very Low and *Stakeholder Team Cohesion* = Nominal	Report to a watchdog, setup key stakeholder IPTs, facilitated workshops, get customers involved, prioritize requirements
1.9	*Requirements Understanding* = Very Low and *Multisite Coordination* = Nominal	Prototyping/simulations, pay attention to interfaces across sites, outreach team, outreach to all sites, prioritize builds
1.9	*Requirements Understanding* = Very Low and *Level of Service Requirements* = Nominal	Get customer involvement early, do trade studies, prioritize requirements

FIGURE 17.9: Expert COSYSMO risk outputs (2/2).

The Expert COSYSMO methodology characterizes project risk with 14 systems engineering cost drivers. However, there may be other risks which may not be captured by the COSYSMO cost drivers in the general cost model. For example, inefficient training or lack of training (assumed in the cost model to be separate from a project) may create serious risks that could cause a program to fail.

The risk assessment is best used early in the lifecycle to help plan mitigation activities. The nature of the risks provided are static and not phase-specific risks. The risk exposures can change throughout the lifecycle, and are ideally being reduced with good risk management practices. The knowledge-based approach also assumes that certain combinations of drivers always yield the prescribed risks, which may not be true on every project with its unique circumstances.

Despite these limitations, integrating technical and non-technical risks through the use of a cost estimation model is a critical step towards more sophisticated risk analysis. Risk assessment matrices can also be generated (Kujawski & Miller, 2007) to integrate multiple risk management viewpoints into a more complete risk management framework.

As evidence of its utility, Expert COSYSMO is continuously used across a broad variety of organizations and research environments. We receive continuous feedback from users and periodically make enhancements. A parametric schedule formula calibrated to industrial data has thus been added, and all calibrations are subject to ongoing dataset refreshes. It can be further improved as an estimation expert assistant to detect COSYSMO input anomalies by flagging inconsistent inputs for the user.

Empirical systems engineering cost and risk data from ongoing projects will be incorporated in the future. For example, hundreds of program records in the Defense Automated Cost Information Management System (DACIMS) (United States Office of the Secretary of Defense, Defense Cost and Resource Center, 2012) are being assessed to enhance and refine the technique. Domain experts from industry, government, and academia will continue to provide feedback and clarification.

Acknowledgments

This work depended on dozens of expert participants across several years. The primary supporting companies and government institutions providing systems engineering practitioners include: the Aerospace Corporation, BAE Systems, Boeing, Lockheed Martin, Northrop Grumman, Raytheon, the Software Engineering Institute, Master Systems and others. Cost estimation tool companies also supported this work including Price Systems, Softstar Systems, and Galorath Inc. We also thank the University of Southern California Center for Systems and Software Engineering and Massachusetts Institute of Technology for hosting the workshops and providing student support.

References

Adelman, L., & Riedel, S. (1997). *Handbook for evaluating knowledge-based systems*. Boston, MA: Kluwer Academic Publishers.

Boehm, B. (1981). *Software engineering economics*. Prentice Hall.

Boehm, B. (1989). *Software risk management*. Los Alamitos, CA: IEEE Computer Science Press.

Boehm, B., Abts, C., Brown, A., Chulani, S., Clark, B., Horowitz, E., ... Steece, B. (2000). *Software cost estimation with COCOMO II*. Prentice Hall.

Boehm, B., & Port, D. (1999). Escaping the software tar pit: Model clashes and how to avoid them. *ACM Software Engineering Notes, 24*(1), 36–48.

Browning, T. R., & Eppinger, S. D. (2002). Modeling impacts of process architecture on cost and schedule risk in product development. *IEEE Transactions on Engineering Management, 49*(4), 428–442.

Chittister, C., & Haimes, Y. (1994). Assessment and management of software technical risk. *IEEE Transactions on Systems, Man, and Cybernetics, 24*(2), 187–202.

CMMI Product Team. (2010). *CMMI for development, version 1.3* (Tech. Rep. No. CMU/SEI-2010-TR-033). Pittsburgh, PA: Software Engineering Institute, Carnegie Mellon University. Retrieved from http:// resources.sei.cmu.edu/library/asset-view.cfm?AssetID=9661

Dalkey, N., Brown, B., & Cochran, S. (1969, November). *The Delphi method, III: Use of self ratings to improve group estimates* (Tech. Rep. No. RM- 6115-PR). RAND Corporation.

International Organization for Standardization. *International standards for quality management* (Vol. 2000; Standard). (2000). Geneva, CH.

Keeney, R., & von Winterfeld, D. (1991). Eliciting probabilities from experts in complex technical problems. *IEEE Transactions on Engineering Management, 38*(3), 191–201.

Kujawski, E., & Miller, G. A. (2007). Quantitative risk-based analysis for military counterterrorism system. *Systems Engineering, 10*(4), 273–289.

Madachy, R. (1997). Heuristic risk assessment using cost factors. *IEEE Software, 14*(3), 51–59.

Madachy, R. (2008). *Software process dynamics.* Hoboken, NJ: John Wiley and Sons.

Madachy, R. (2009). *Integrated cocomo suite tool for education.* Cambridge, MA. (24th International Forum on COCOMO and Systems/Software Cost Modeling, Cambridge, MA)

Madachy, R. (2017). *Systems engineering cost estimation workbook.* Retrieved from https://github.com/madachy/Systems-Engineering-Cost-Estimation-Workbook

Madachy, R., & Valerdi, R. (2008). *Knowledge-based systems engineering risk assessment* (Tech. Rep. No. USC-CSSE-2008-818). University of Southern California Center for System and Software Engineering.

Madachy, R., & Valerdi, R. (2009). Expert cosysmo workshop. In *24th international forum on cocomo and systems/software cost modeling.* Cambridge, MA.

Madachy, R., & Valerdi, R. (2010). Automating systems engineering risk assessment. In *Proceedings of the 8th annual conference on systems engineering research*, Hoboken, NJ.

Pate-Cornell, M. E. (1996). Uncertainty in risk analysis: Six levels of treatment. *Reliability Engineering and System Safety, 54*, 95–111.

Rhodes, D., Valerdi, R., & Roedler, G. (2009). Systems engineering leading indicators: Assessing effectiveness of programmatic and technical performance. *Systems Engineering, 12*(1), 21–35.

SEBoK Authors. (2017). *Risk management in guide to the Systems Engineering Body of Knowledge (SEBoK), version 1.8.* Retrieved from http://sebokwiki.org/wiki/Risk_Management

United States Office of the Secretary of Defense, Defense Cost and Resource Center. (2012). *CSDR overview and policy.* Retrieved from http:// dcarc.cape.osd.mil/CSDR/CSDROverview.aspx

Valerdi, R. (2008a). *The constructive systems engineering cost model* (cosysmo): *Quantifying the costs of systems engineering effort in complex systems.* Saarbrucken, Germany: VDM Verlag.

Valerdi, R. (2008b). Cosysmo 2.0 workshop. In *23rd international forum on cocomo and systems/software cost modeling.* Cambridge, MA.

Valerdi, R., & Gaffney, J. (2007). Reducing risk and uncertainty in cosysmo size and cost drivers: Some techniques for enhancing accuracy. In *5th conference on systems engineering research.* Hoboken, NJ.

Section IV

Perspectives

Chapter 18

How Data Science Happens

James P. Howard, II

18.1 Introduction

Data science has changed how we use information (van der Aalst, 2016). Growing from roots spread across data mining, statistics, computer science, and other fields (Provost & Fawcett, 2013a), data science embraces analytic approaches applied to data generated across nearly every field. The result is a nearly unprecedented growth in the demand for data science skills among job seekers (Debortoli, Müller & vom Brocke, 2014) as every business attempts to find the truths hidden in their data. This has also expanded into government, both military and civilian sectors, as social leaders push to use data available to them (Bertot & Choi, 2013). There are many potential applications of big data to the government, regulatory, military, and law enforcement (Kim, Trimi & Chung, 2014) and the military views big data as disruptive technology Symon & Tarapore, 2015).

Despite the rush to adopt data science, there is little to tell non-practitioners and students what data science is. Data science is complicated by the lack of a single definition (Provost & Fawcett, 2013b, pp. 14–17). Everything from big data (Sagiroglu & Sinanc, 2013) and data engineering (Sharma et al., 2015) to the Internet of Things (IoT) (De Francisci Morales et al., 2016) and blockchain (Dinh & Thai, 2018) get lumped into data science.

At its core, data science allows practitioners to find deep patterns within data and turn the data into actionable information (Kaisler et al., 2013). The actionable information may be advisory, an indication of a general trend, or may be specific instructions to take. Using data visualization, advanced geospatial mapping, data mining, and predictive analytics, organizations can use data science to support their objectives and field operations.

Data science, however, also embraces the application space to generate good analysis (Jagadish, 2015). In addition to being multidisciplinary, good data science teams are also interdisciplinary, bridging the data analysis to the application space through domain expertise. To support a law enforcement mission, domain experts from criminology, sociology, and other social science fields are necessary parts of the data science team (Lettieri et al., 2018). To manage logistics for a defense organization, a data science team needs supply chain experts and industrial engineers (Waller & Fawcett, 2013). These experts support a big data analysis by providing background, testable theory, and mission-specific guidance for data scientists coming from other fields. Without domain-specific knowledge, spurious patterns can be found in an analysis that leads analysts to unjustifiable conclusions.

In science fiction, governments use predictive analytics to identify "pre-crime" and stop crime before it happens (see, *inter alia*, Dick, 1956) or intelligent computers to fight wars (Badham, 1983). In reality, the techniques used by data scientists supporting agencies mirror the techniques those agencies already use, just ramped up in scale. At the threshold between reality and science fiction is using predictive analytics. Predictive analytics can benefit agencies by providing greater insight into their data.

However, despite the promise of big data and analytics to create intelligent machines that can solve problems, in any field, expectations must be managed. Large amounts of data are required to create statistically valid analyses. That volume of data may not be available when analyzing rare events. If analysis can be completed, it may not necessarily be valid. The discovered patterns may just reflect the biases and intuition of the data collection methods (Hajian, Bonchi & Castillo, 2016). Finally, getting to do analysis is itself a hurdle as data are often collected in forms unsuitable for analysis. In these cases, the data scientist supporting the analysis will spend more time in data preparation, cleaning the data and making it ready for use (Heer & Kandel, 2012).

This chapter will first outline the distinctions among the different ways agencies can use data science today. Second, this chapter will explain the process of how predictive analytics can be used to make guesses about the future. Third, this chapter will explain how the process is both similar to and different from existing techniques. Fourth, this chapter will analyze the legal requirements and how and when such predictive analytics can support agency activities. Finally, the chapter explains how final interpretation lies with the individual in the field. The objective of this chapter is to provide students and non-practitioners with guidance on how to use data science methods to solve problems.

18.2 Data Collection and Management

There are numerous data types that we use in data science. There are many ways to define data. One form, familiar from statistics courses, focuses on the content of the data (Stevens, 1946): count data including integers, categorical data assigned to bins, ratio measures, or numerical values. However, data science takes us to a different way of thinking about statistics. Instead of inferring on unknown populations from samples, we start looking at samples of data that are too large to look at holistically.

Much of data science revolves around "big data." Big data means different things in different contexts, and there are a number of different definitions (Morabito, 2015, pp. 23–45). Some argue that big data is defined by the number of records in the dataset or by bytes. Others define big data as data that requires certain tools or skills to analyze due to its volume and type. One strong definition defines big data based on properties and aspects of its use: "Big Data is the Information asset characterized by such a High Volume, Velocity and Variety to require specific Technology and Analytical Methods for its transformation into Value" (Mauro, Greco & Grimaldi, 2016). This definition captures the ideas that link big data to data science, including how it is applied to the organization. While most organizations do not produce that much data, we can use many of the same skills and techniques to analyze the data we have available to us.

"Microdata" are data that represent a single observation of something (Samarati, 2001). The speed of a vehicle at one point in time as it goes down the highway is microdata. And so is the age of the driver. This is different from the way you are used to seeing the data. News reports, briefings, and other presentations tend to focus on summary statistics, rather than the individual data elements. Summary statistics include percentages, means, and standard deviations. Microdata are the individual observations that compose those summary statistics and are a greater focus of data science than means and percentages.

When we have a lot of data, we can use statistics to make inferences about the underlying patterns. That is, we cannot establish a pattern from a single instance. A lot of data might be the speed of every vehicle traveling a certain section of the highway during an entire year. Or the ages of everyone in the organization are also a lot of data. These types of data can also be used to make inferences and predictions and collecting this information allows us to make predictions on manpower and staffing needs (Symon & Tarapore, 2015). While it is possible to use a single variable, such as speed or age, to learn something, statistics necessarily requires more information to make conclusions.

Both fingerprint (Peralta et al., 2017) and facial recognition (Samarati, 2001; Gandomi & Haider, 2015; Soyata et al., 2012) systems benefit from the big data explosion. Faster matching against existing databases is an immediate benefit as computers have grown faster, but access to more data has also improved the usability of these tools. Centralized and shared databases of fingerprints and photographs have allowed agencies to engage national and even international searches for individuals. Further, as these resources are shared, agencies can reduce costs by eliminating duplication and gain access to data they may not otherwise have acquired.

Examples like fingerprints and facial recognition provide a simple application with a single observation: a found fingerprint or a picture of someone's face. When observations of multiple variables are linked together, we can learn more information. For a vehicle's speed, we might also want information on the time of day, day of the week, weather conditions, and some sort of traffic volume metric. For age information, we might also link gender, biometric data, and other forms of data. At the very least, we might collect the date of observation, allowing us to find the age on another day. From this sort of information, we can detect patterns and perhaps empirically confirm a hypothesis. If we are luckier, we may find something we did not expect at all.

To use data in a data science analysis, the analyst must get the data to a structured format. This is often called "data cleansing" (Hernández & Stolfo, 1998). Data cleansing is the broad swath of data manipulations necessary to get into a big data analysis. This can range from date normalization, ensuring that all dates have the same format, through to extracting information from the free-form text in unstructured data.

Agencies collect information as a routine part of operations. Forms are filled out, loggers record network activity, tickets and citations are issued, and reports are written. Some of this information is collected automatically in a structured format. For instance, packages are scanned with a location, time, and data, as part of a tracking system. Those results are continuously added to a database in a structured format. However, other data, such as the contents of reports, are unstructured (Baars & Kemper, 2008). This continues to be true even if the reports are digitized.

The critical distinction between structured and unstructured data is what the nature of the data is (Park & Song, 2011). If the data is free-form, the data is unstructured. If the data are constrained, the data is structured. This can be compared to an exam. Multiple-choice and true/false questions are structured whereas short answer and essay responses are unstructured, regardless of how well they are written. If it can be put into a spreadsheet in an orderly way, it is probably structured data.

Data cleansing is not an exact science and is the first part of data preparation. The second part is about creating data aggregates. Data aggregates are new data elements created from within an existing dataset. For example, the number of packages moving through a processing center is an aggregate. We can use this sort of information to find examples of unusual spikes or delays in the supply chain. IoT devices may only report aggregates, like the average number of observations per second, rather than reporting each individual observation. In this case, the microdata is the number of observations per second, rather than "true" microdata. In many ways, the terminology is circumstance-specific.

Finally, there are outside datasets available that can be integrated with locally produced datasets to improve analysis. The World Bank provides economic aggregates that can be used to support analysis. The Global Database of Events, Language, and Tone (GDELT) provides publicly available data on news events and governments worldwide that can be used for regional analysis (Parrish et al., 2018). Analysts have combined unrelated datasets for decades but the massive merging of datasets for inference and prediction is the hallmark of data science.

While the mathematics of data cleansing and data aggregation are relatively simple, identifying what to cleanse and how to aggregate is not. Data aggregates should be justified based on an operating theory that explains the potential relevance of the data to the question. For instance, it is not immediately clear there is a reason to connect the number of soldiers in a unit with who won the World Series. Less outlandishly, there is a reason to potentially connect unemployment data for a nation to an area seeing an increase in unrest (Pervaiz, Saleem & Sajjad, 2012). Once data is cleansed and aggregated, it is possible to start analyzing to search for underlying patterns. In the next two sections, we will explore a variety of methods and show they can be used to reveal hidden information.

18.3 Data Exploration

Once data has been cleansed and prepared, we want to move into a data exploration phase. The data exploration phase allows us to examine the data and see what patterns are obvious (Tukey, 1977). In addition, a data scientist will want to have some idea of the data structure before moving too deeply into the analysis, because the structure of the data will dictate the analytic constraints. Later, in Section 18.4, we will use the scientific method to find meaning from the data. Using hypothesis testing, predictive modeling gives the analyst the ability to propose a relationship in the data and determine, within some degree of error, whether or not the relationship actually exists.

Using exploratory data analysis (EDA), we can see what information and insights are in the data, distinct from the yes and no responses from hypothesis testing. There are several types of exploration ranging from visualization, which allows us to examine the data via charts and graphs, to cluster analysis, which groups observations from the dataset based on how similar they are to each other. This section will discuss these methods and others as part of data exploration.

Visualization is often the first step in a data exploration program (Larson & Chang, 2016). At best, a table of numbers is inconvenient to review. At worst, there is more data, both by observation count and variables in each observation, than we can meaningfully understand by looking at the numbers alone. Data visualization is a suite of techniques, beyond the ubiquitous bar charts and pie charts, we can use to better understand our data.

Modern techniques include a standardized language for data plotting, called the grammar of graphics. The grammar of graphics creates a standard approach for defining the elements of a data graph (Wilkinson, 2012). On the back end, standardized presentations of data allow the viewer to understand the meaning and intent faster. Newer tools can create animations to show the change in data over time, for instance, though animation may not be an effective data analysis tool (Robertson et al., 2008). Interactive charts routinely grace the front page of major newspaper websites.

There are a number of plotting tools within Excel. Other data visualization tools are available in statistical software like R (Wickham, 2016) and some general-purpose scripting languages, like Python (Milovanovi, 2013). Many of these tools are freely available and create publication-quality graphics. Both R and Python have vibrant user communities creating a new visualization and analytic tools that address new problems and old problems in novel ways.

Common histograms are excellent for showing distinctions provided the data has already been binned by class (Freedman, Pisani, & Purves, 2007, pp. 31–56). However, our most powerful general-purpose visualization is the box plot (McGill, Tukey, & Larsen, 1978), also known as the box-and-whiskers plot, which shows several key facts of a dataset. While some packages produce slightly different versions of box plots, the basic box plot shows the mean of a dataset, the minimum, the maximum, and how the middle 50% straddle the mean. If two box plots representing different classes are presented together, any distinctions are immediately revealed.

Heat maps provide a different type of visualization (Bojko, 2009). At its core, a heat map is a two-dimensional form of a histogram. A heat map works by creating a two-dimensional matrix across two classes of the dataset, for instance, gender and age group, and provides a statistical measure for each matrix entry. In other words, a heat map is a visual representation of a cross-tabulation. Typically, color intensity is used to demonstrate values. These can be stylized such as using color intensity on a map of the world to show the relative values of some measure across counties, commonly seen in reports and the news.

Explicit classes can exist in data. For instance, we can categorize people by gender, race, age group, or where they went to college. These classes are explicit if there is variable within the dataset that provides information about the class. Box plots, histograms, heat maps, and similar are excellent for identifying distinctions when there are a small number of classes in the data and one or two variables of interest in the data. Our real-world data often exceeds these numbers. There can be multiple classes of data and dozens or more distinct variables within a dataset. Further, the distinctions between classes may not be limited to one parameter but may instead be spread among multiple parameters simultaneously.

Cluster analysis, part of a family of techniques called unsupervised learning, captures relations within the data by discovering latent classes within the data (Jain, 2010). Latent classes will exist within the data if there is a distinct grouping within the data, but the explicit class information is not available within it. An easily understood example comes from education and income. We know from numerous studies that more education generally leads to higher household income. However, if we do not have data on education attained, there is likely a latent grouping within the data by income. Cluster analysis can reveal these groupings.

Mathematically speaking, cluster analysis finds the closest observations to each other by measuring the distance among all data points, across n dimensions, where n is the number of variables in the dataset. Then, the clusters are found by finding the middle points necessary to minimize the total distance. Each middle point represents the prototypical element of the cluster, and its associated data points will lie around the midpoint. The process is complicated to visualize because it takes place across n dimensions.

However, the results of cluster analysis are not perfect. First, the results can be difficult to interpret. While the clusters may exist, it is probably not evident what the clusters represent. While the example given for income and education may be clustering based on income level, there is no reason that some other group affinity has been discovered by the process. Second, most algorithms require specifying the number of clusters to search for. Limiting the number of clusters to 3–4 can decrease the number of computations, but it may not necessarily reflect the real groupings within the data, leading to groups merged which may be distinct. Over-specifying the number of the clusters can have the opposite effect, drawing distinctions where none exist.

Beyond these elementary data processing techniques are several approaches to extract meaning loosely connected data. One example is geographic information system (GIS) (Star & Estes, 1990). GIS can take other data, however it was acquired or of whatever type, and link it to whatever geographic information is available. During our data explorations, natural clusters may emerge, especially when there are fine controls on what data is presented. For instance, the locations of terrorist events can be better understood by overlaying them on a map. In one sense, GIS is the digital equivalent of a map on the wall covered with pins.

The more detailed location is specified is generally better. But, for rarer events monitored at the national or global level, it may be sufficient only to acknowledge what jurisdiction an activity happened in. More complicated might be when proximity is not necessarily defined by the straight-line distance. One example might be a subway line. Two stations may only be a short drive apart but at many stops distant from each other. Understanding distance in the context of the relevant geography helps frame the understanding. For instance, consider a string of security violations at transportation facilities. If there is a sudden spike in violations in one area, it may be indicative of penetration testing as a prelude to a larger attack.

Whatever the geographic level of detail, policymakers and commanders will be interested in knowing if an area has become a hotbed of activity. It may not be possible to predict exactly where something will happen next, but it may be possible to document organized activity or just show where opportunistic activity is occurring. In that case, traditional sources and methods can be informally approached for support and patrolling increased in the area.

A different analytic framework for extracting meaning from data is called social network analysis (SNA) (Scott, 1988). This method is not related to social media, like Facebook and Twitter. However, social media can provide a framework for understanding social network analysis. SNA shows connections between individual observations in a population. It can be used to show who knows whom, who may be influencing whom, and who is within a certain number of connections, called hops, of someone in particular. Further, like a geographic mapping of events, the individual units of analysis may be connected to other data elements of interest. SNA reveals these patterns visually providing strong clues about where we should apply formal analytic techniques.

All of these exploratory methods can advance an analysis. GIS mapping and SNA may illuminate a relationship that was not previously available but suddenly becomes obvious. But, EDA does not, by itself, provide evidence of conclusions. There is little, if any, statistical validity in EDA. However, that does not diminish the value EDA provides to informing the modeling process and communicating results. By highlighting potential relationships, the analyst can use EDA to prepare hypotheses for testing. At the most elementary, EDA can lead to new variables, generated from the relationship data, that can be used by more advanced modeling methods.

18.4 Making Predictions

The techniques described so far allow one to examine historical data. While historical evaluation is useful to provide understanding of what has already happened, the real power of data science is the power to use historical data to make actionable predictions. This process is called predictive analytics and uses techniques from supervised learning and machine learning to produce statistical and semi-statistical models (Shmueli & Koppius, 2011). Predictive analytics provides many different approaches for making predictions, but all of them result in the basic output of answering a question. From elementary logistical regression models (DeMaris, 1992) to random forests (Breiman, 2001) to the latest deep learning networks (Goodfellow, Bengio, & Courville, 2016), all of these models take the input and put it into a bin.

These methods have become common in many fields, such as retail. Amazon.com, the large online retailer, uses its vast store of information about prior purchases by all its customers to provide the suggestions in its "Related to items you've viewed ..." list. This task is sometimes known as classification because it places inputs into classes based on how similar each is to an example provided previously (Kwak & Choi, 2002). Classes may be as simple as true or false for some property we are interested in. For instance, a simple true or false could ask a system to identify whether or not an identification card is likely valid or not. Or it could be a more complicated multiclass classification scheme that bins inputs into new categories, based, again, on similarity to previous examples. From an image, what kind of vehicle is detected is a multiclass problem (Wang & Gao, 2005). However, similarity may be defined in a number of ways and this is where data science occasionally becomes an art.

Throughout this chapter we have discussed discovering and capturing relationships within the data. Predictive analytics is where we apply the results of this information-gathering by creating statistical models that reflect the data. In practice, the process is similar to the process for data analysis presented in a research methods course, but there are many different algorithms we can chose from, as opposed to the standard linear model. These models come with names like logistic regression, naïve Bayes (Lewis, 1998), boosted decision trees (Roe, Yang & Zhu, 2006), random forest, and artificial neural networks, to name a few, and each has very different internal operations. However, despite any differences, they all function in the same basic way, accepting an input of observations and producing an output of a prediction. Fortunately, most of these model algorithms are widely available in numerous statistical packages and programming languages.

Data, called training data, is presented to the modeling algorithm that has already been class-labeled. This may be because we already know the classifications or it was hand-labeled into its distinctive classes. This sample data also includes all of the variables that will define our model. These are those elements we will have access to later, to make predictions on, but will include both source data and derived data, based on the outcomes of the exploratory analysis. Regardless of which model is used, the modeling tool will "study" the data to create a prediction engine. Depending on the amount of data and which modeling algorithm is used, the process may take from a few minutes to several hours.

The result should be a tool that can make predictions, though its usefulness is still in question. We can evaluate its validity through a process called cross-validation (Shao, 1993). Like the rest of data science, there are many options, though only a couple are commonly used. The first applies the model to test data taken from the training data. To be most effective, the test should be excluded from training. Then model accuracy can be measured by seeing how many of the test examples are correctly classified. The second option applies the model back over all of the original training data. In this option, our goal is to again ask how many are correctly identified during the test phase. A third method, called multifold cross-validation divides a training dataset into k subsets and uses all but one for training and tests on the remaining. This process is repeated for each subset, with the results of each training averaged. However, this is computationally expensive as it requires each model to be calculated k times.

During the evaluation process, regardless of which cross-validation option is used, models that are two-class classifiers use two key metrics to score the results (Baeza-Yates & Ribeiro-Neto, 1999, p. 75). The first is precision, which is the identified class that are identified correctly. Mathematically, this is,

$$\text{precision} = \frac{\text{true positives}}{\text{true positives} + \text{false positives}}.$$

The second is recall, which measures the false negative rate. Mathematically, this is,

$$\text{recall} = \frac{\text{true positives}}{\text{true positives} + \text{false negatives}}.$$

The metrics together tell the story of how good a model is. More complex metrics are available for multiclass models, but at their core, they evaluate the same error rates. Typically, many different models are used with the same data and the predictive outcomes for each scored to determine the model's effectiveness. Multiclass generalizations of these measures are also available.

By comparing the results of different models, we can determine which provides the best fit. Assuming a model with a good fit has been produced, it can be used to make predictions about data upon request. However, the most interesting possibility is to make the model and its results available through the organization's information technology infrastructure. In that case, the model can provide real-time predictions of events as they happen and make the results available to field operations (Howard & Beaumont, 2015).

18.5   In the Field

Automated analysis and predictions can provide a great advantage to decision-makers. However, in practice, the insights and predictions of data science only provide guesses. These can be used to supplement the analyst's work, but they cannot replace the sound judgment of a human. The methods themselves only provide information that can be used to point an individual in the right direction. The decisions of a predictive analytics system alone should not be used for an automated targeting or automated action system.

There are several reasons for this relating to statistical validity. The underlying predictive models are based on statistics and therefore statistical validity applies. The first of these is internal validity. Internal validity refers to the causal connection between the explanatory variables and the outcome variable (Brewer & Crano, 2000). For these sorts of predictive models, internal validity is how well the prediction is explained by the data we have given the model. In one sense, we can measure this statistically. But in a deeper sense, a subject matter expert can reasonably say the explanation makes sense. For instance, there is little reason to believe that the weather in Los Angeles would have any bearing on a baseball game in Boston. On the other hand, the weather in Boston would reasonably affect a game in Boston.

The second major reason is external validity, which describes how well a prediction based on the training data will hold up against members of the larger population (Mitchell & Jolley, 2012, p. 56). Like internal validity, we can estimate this statistically, but a subject matter expert can use their expertise to advance the analysis. An automated prediction algorithm will be inherently biased by the data it is trained on (Hajian, Bonchi & Castillo, 2016). That is, any underlying bias in the training data will be replicated in the model. For instance, if drivers of red cars are more likely to receive a ticket for speeding, and this was reflected in the training data, then a predictive model is more likely to flag red cars for speeding. That bias can be reduced or eliminated using statistical techniques and any good model will account for that. However, the model cannot account for cases that are insufficiently similar to those that it has already seen. This means that generally, a behavioral model that only used men as the training set is unlikely to be applicable to an instance where the subject was a woman.

Once these validity concerns are alleviated, an analyst must make sure that the application of the model makes sense. This means ensuring the results align with objectives and align with expectations. It does little good and erodes public goodwill to hear the words, "the computer says ..." when following up on a model-generated action (Wihlborg, Larsson & Hedström, 2016). At the same time, the model may be producing unexpected results because the model has identified a pattern previously unknown. Knowing how to use the results of a predictive model fall squarely within the range of human judgment. Ultimately, the robots will not replace humans but will work hand-in-hand with them.

18.6   Summary

This chapter has outlined the fundamentals of data science to give students or non-practitioners the vocabulary and knowledge to understand basic analysis. This chapter provides for where data may come from and how data can be extracted from other sources for analysis. The chapter has also provided for using EDA to gain insight. Finally, the chapter has explained how to create actionable predictions and put those prediction-making engines into the field. While the data can provide a great deal of insight and suggestions, it is ultimately up to the person in the field to determine how to interpret and use the information effectively (Lipsky & Hill, 1993).

Despite the warning, data use will grow and become more powerful as time goes on. Tools considered state of the art ten years ago are considered quaint by today's standards. But the increase in power, driven largely by the widespread availability of specialized hardware, has opened many new applications, from self-driving vehicles to advanced image recognition. As these applications expand, so will the amount of data available via resource sharing. Further, specialized methods invented to solve a problem unique to one sector will prove their value in others. A data scientist will have to keep abreast of new methods since options for application are not always obvious.

While new methods and datasets are developed, there is likely always a place for the elementary statistical methods on smaller datasets. Sometimes big data is not available or distinctions are clear enough that statistical learning methods are sufficient to make predictions. The analyst will have to make complex and only partially informed decisions when building data science models, but the good news is the practitioner will have a rich set of options available.

References

Baars, H. & Kemper, H.-G. (2008). Management support with structured and unstructured data: an integrated business intelligence framework. *Information Systems Management*, *25*(2), 132–148.

Badham, J. (1983). *Wargames*. Hollywood: United Artists.

Baeza-Yates, R. A. & Ribeiro-Neto, B. (1999). *Modern information retrieval*. Boston: Addison-Wesley Longman Publishing Co., Inc.

Bertot, J. C. & Choi, H. (2013). Big data and e-government: issues, policies, and recommendations. In *Proceedings of the 14th annual international conference on digital government research* (pp. 1–10). dg.o '13. Québec, Canada: ACM.

Bojko, A. A. (2009). Informative or misleading? Heatmaps deconstructed. In *International conference on human-computer interaction* (pp. 30–39). Springer.

Breiman, L. (2001). Random forests. *Machine Learning*, *45*(1), 5–32.

Brewer, M. B. & Crano, W. D. (2000). Research design and issues of validity. In H. T. Reis & C. M. Judd (Eds.), *Handbook of research methods in social and personality psychology* (pp. 3–16).

De Francisci Morales, G., Bifet, A., Khan, L., Gama, J., & Fan, W. (2016). IoT big data stream mining. In *Proceedings of the 22nd ACM SIGKDD international conference on knowledge discovery and data mining* (pp. 2119–2120). ACM.

Debortoli, S., Müller, O., & vom Brocke, J. (2014, October). Comparing business intelligence and big data skills. *Business & Information Systems Engineering*, *6*(5), 289–300.

DeMaris, A. (1992). Logit modeling: practical applications. Sage.

Dick, P. K. (1956, January). The minority report. *Fantastic Universe*, 4–36.

Dinh, T. N. & Thai, M. T. (2018). Ai and blockchain: a disruptive integration. *Computer*, *51*(9), 48–53.

Freedman, D., Pisani, R., & Purves, R. (2007). *Statistics* (4th ed.). W. W. Norton & Company.

Gandomi, A. & Haider, M. (2015). Beyond the hype: big data concepts, methods, and analytics. *International Journal of Information Management*, *35*(2), 137–144.

Goodfellow, I., Bengio, Y., & Courville, A. (2016). *Deep learning*. Cambridge, Massachusetts: MIT press.

Hajian, S., Bonchi, F., & Castillo, C. (2016). Algorithmic bias: from discrimination discovery to fairness-aware data mining. In *Proceedings of the 22nd ACM SIGKDD international conference on knowledge discovery and data mining* (pp. 2125–2126). ACM.

Heer, J. & Kandel, S. (2012). Interactive analysis of big data. *XRDS: Crossroads*, *19*(1), 50–54.

Hernández, M. A. & Stolfo, S. J. (1998). Real-world data is dirty: data cleansing and the merge/purge problem. *Data Mining and Knowledge Discovery*, *2*(1), 9–37.

Howard, J. & Beaumont, S. (2015). Analysis as a service. In *Digital leaders* Richards, J. (Ed.) (pp. 20–21). London: BCS, the Chartered Institute of IT.

Jagadish, H. (2015). Big data and science: myths and reality. *Big Data Research*, *2*(2), 49–52. Visions on Big Data.

Jain, A. K. (2010). Data clustering: 50 years beyond k-means. *Pattern Recognition Letters*, *31* (8), 651–666.

Kaisler, S., Armour, F., Espinosa, J. A., & Money, W. (2013). Big data: issues and challenges moving forward. In *2013 46th Hawaii international conference on system sciences* (pp. 995–1004). IEEE.

Kim, G.-H., Trimi, S., & Chung, J.-H. (2014). Big-data applications in the government sector. *Communications of the ACM*, *57*(3), 78–85.

Kwak, N. & Choi, C.-H. (2002). Input feature selection for classification problems. *IEEE Transactions on Neural Networks*, *13*(1), 143–159.

Larson, D. & Chang, V. (2016). A review and future direction of agile, business intelligence, analytics and data science. *International Journal of Information Management*, *36*(5), 700–710.

Lettieri, N., Altamura, A., Giugno, R., Guarino, A., Malandrino, D., Pul- virenti, A., ... Zaccagnino, R. (2018). Ex machina: analytical platforms, law and the challenges of computational legal science. *Future Internet*, *10*(5), 37.

Lewis, D. D. (1998). Naive (bayes) at forty: the independence assumption in information retrieval. In *European conference on machine learning* (pp. 4–15). Springer.

Lipsky, M. & Hill, M. (1993). Street-level bureaucracy: an introduction. In *The policy process: a reader* Hill, M. (Ed.) (pp. 381–385). New York: Routledge.

Mauro, A. D., Greco, M., & Grimaldi, M. (2016). A formal definition of big data based on its essential features. *Library Review*, *65*(3), 122–135.

McGill, R., Tukey, J. W., & Larsen, W. A. (1978). Variations of box plots. *The American Statistician*, *32*(1), 12–16.

Milovanovi, I. (2013). *Python data visualization cookbook*. Birmingham: Packt Publishing Ltd.

Mitchell, M. L. & Jolley, J. M. (2012). *Research design explained*. Cengage Learning.

Morabito, V. (2015). *Big data and analytics*. Cham, Switzerland: Springer.

Park, B.-K. & Song, I.-Y. (2011). Toward total business intelligence incorporating structured and unstructured data. In *Proceedings of the 2nd international workshop on business intelligence and the web* (pp. 1219). ACM.

Parrish, N. H., Buczak, A. L., Zook, J. T., Howard, J., Ellison, B. J., & Baugher, B. D. (2018). Crystal cube: multidisciplinary approach to disruptive events prediction. In J. I. Kantola, S. Nazir, & T. Barath (Eds.), *Advances in human factors, business management and society* (pp. 571–581). Cham, Switzerland: Springer International Publishing.

Peralta, D., Garca, S., Benitez, J. M., & Herrera, F. (2017). Minutiae-based fingerprint matching decomposition: methodology for big data frameworks. *Information Sciences*, *408*, 198–212.

Pervaiz, H., Saleem, M. Z., & Sajjad, M. (2012). Relationship of unemployment with social unrest and psychological distress: an empirical study for juveniles. *African Journal of Business Management*, *6*(7), 2557–2564.

Provost, F. & Fawcett, T. (2013a). Data science and its relationship to big data and data-driven decision making. *Big Data*, *1*(1), 51–59. PMID: 27447038.

Provost, F. & Fawcett, T. (2013b). Data science for business: what you need to know about data mining and data-analytic thinking. Sebastopol, California: O'Reilly Media, Inc.

Robertson, G., Fernandez, R., Fisher, D., Lee, B., & Stasko, J. (2008). Effectiveness of animation in trend visualization. *IEEE Transactions on Visualization and Computer Graphics*, *14*(6), 1325–1332.

Roe, B. P., Yang, H.-J., & Zhu, J. (2006). Boosted decision trees, a powerful event classifier. In L. Lyons & M. K. Ünel (Eds.), *Statistical problems in particle physics, astrophysics and cosmology* (pp. 139–142). World Scientific.

Sagiroglu, S. & Sinanc, D. (2013). Big data: a review. In *2013 international conference on collaboration technologies and systems* (pp. 42–47). IEEE.

Samarati, P. (2001). Protecting respondents identities in microdata release. *IEEE Transactions on Knowledge and Data Engineering, 13*(6), 1010–1027.

Scott, J. (1988). Social network analysis. *Sociology, 22*(1), 109–127.

Shao, J. (1993). Linear model selection by cross-validation. *Journal of the American Statistical Association, 88*(422), 486–494.

Sharma, S., Tim, U. S., Gadia, S., Wong, J., Shandilya, R., & Peddoju, S. K. (2015). Classification and comparison of NoSQl big data models. *International Journal of Big Data Intelligence, 2*(3), 201–221.

Shmueli, G. & Koppius, O. R. (2011). Predictive analytics in information systems research. *MIS Quarterly*, 553–572.

Soyata, T., Muraleedharan, R., Langdon, J., Funai, C., Ames, S., Kwon, M., & Heinzelman, W. (2012). Combat: mobile-cloud-based compute/communications infrastructure for battlefield applications. In E. J. Kelmelis (Ed.), *Proceedings of SPIE* (Vol. 8403). Modeling and Simulation for Defense Systems and Applications VII.

Star, J. & Estes, J. E. (1990). *Geographic information systems: an introduction.* Englewood Cliffs, New Jersey: Prentice Hall.

Stevens, S. S. (1946). On the theory of scales of measurement. *Science, 103*(2684), 677–680.

Symon, P. B. & Tarapore, A. (2015). Defense intelligence analysis in the age of big data. *Joint Force Quarterly, 79*(4).

Tukey, J. W. (1977). *Exploratory data analysis.* Reading, Massachusetts: Addison-Wesley Publishing Company, 4–11.

van der Aalst, W. (2016). Process mining: data science in action. Berlin: Springer.

Waller, M. A. & Fawcett, S. E. (2013). Data science, predictive analytics, and big data: a revolution that will transform supply chain design and management. *Journal of Business Logistics, 34*(2), 77–84.

Wang, J.-H. & Gao, Y. (2005). Multi-sensor data fusion for land vehicle attitude estimation using a fuzzy expert system. *Data Science Journal, 4*, 127–139.

Wickham, H. (2016). *ggplot2: elegant graphics for data analysis.* Cham, Switzerland: Springer.

Wihlborg, E., Larsson, H., & Hedström, K. (2016). "The computer says no!"-a case study on automated decision-making in public authorities. In *2016 49th Hawaii international conference on system sciences* (pp. 2903–2912). IEEE.

Wilkinson, L. (2012). The grammar of graphics. In J. E. Gentle, W. K. Hrdle, & Y. Mori (Eds.), *Handbook of computational statistics* (pp. 375–414). Springer.

Chapter 19

Modernizing Military Operations Research Education to Form the Foundation for US Military Force Modernization

Chris Arney, Michael Yankovich and Krista Watts

19.1 Introduction

The global security environment is shifting at a fast pace, creating changes in the character of warfare (Dubik, 2018) and producing challenges for military operations research (OR) analysts. The US military has new operational missions on battlefields and in environments that did not exist before this century (TRADOC G-2, 2017). The world is full of diverse and integrated systems, people, organizations, governments, and networks (internet and internet of things) with varied behaviors that produce complex situations (Kott, Swami & West, 2016). These challenges demand more data analysis, more quantitative reasoning, more innovative modeling and simulation, and more technical and intellectual decision-making to build versatility and enhance operational capabilities. These are the data-rich environments of the modern military OR analyst. Modern OR plays an important role in changing the military to become more versatile and capable of confronting emerging complexities.

In today's world, it is not unusual to have security threats that involve political uprisings, market crashes, and economic crises. Moreover, the types of missions and operation evolve faster than the military can modify its formal doctrine. Consequently, OR analysts, at all levels, are increasingly less likely able to rely on specific, detailed, routine, or algorithmic processes or techniques. OR as an academic discipline inherent in the military profession is changing to enable confrontation of modern problems and issues with new forms of complexity modeling and interdisciplinary problem-solving that are outlined in the Joint Planning Publication 5-0 (Department of Defense, 2017), one of the foundational documents for the roles of military OR. In this era of insurgencies, hybrid wars, ill-defined enemies, inter-tribal friction, and great power competition, OR analysts play important roles in both current operations – the modeling component of modern operational design – and future planning – such as the planning mission of the US Army's newly formed Futures Command. This command will need OR analysts to develop and deliver concepts and force designs, model new force structure, and integrate new technology.

Half of the OR analysts in the US Army earned their undergraduate degree at the United States Military Academy (USMA) at West Point; although not all are Operations Research majors, every cadet takes at least some core math courses and Systems Engineering is the most common 3-course Core Engineering Sequence. USMA has recently engaged in efforts to improve and modernize its OR program by seeking input from the field army, the Army OR community, and officers on its faculty who recently participated in operational deployments as analysts. That feedback has empowered a new design in the USMA OR program to help the Army deal with modern warfare needs. The OR program (curricula and courses) has changed to keep up with this paradigm-shifting modernization of the military and future warfare and is therefore an example of the modern OR education needed for future military OR analysts. In the coming years, OR-educated scholar-warriors will need to analyze global situations and complex events that necessitate military operations (MacGregor, 1997). This chapter presents the modern military challenges that affect military OR and its future, the nature of complexity modeling that is an important element in modern OR, the role of the military academies in OR education, an example case study of modern military OR education system in USMA's OR-major, information about other OR undergraduate programs and graduate programs that serve military OR analysts, and a description of the Army's OR education system.

19.2 Background

Today's US military forces face the stiff challenges of informational and kinetic dynamics and complexity in conducting military operations to defend their fellow citizens (TRADOC, 2014). The only way to meet those missions is to be versatile and adaptive. By using the tenets of modern military operations enhanced by modern methods of OR, information science, cyber science, and complexity science, military units can be prepared for these challenging global situations and events that necessitate military operations while building versatility to adjust their operations to the dynamic needs of the situation (Blacker, 2017).

The Structure of Scientific Revolutions (Kuhn, 1962) explained how and why the advancement of human knowledge and culture are not orderly processes. Rather, advancements result from interactions of people and the ideas they generate as they search for new paradigms that better describe reality (Cohen, 1987). For today's military, these new paradigms contain frameworks that consist of a combination of military principles, laws of warfare, cultural values, doctrinal knowledge, intelligent systems, and practiced processes. When assembled and integrated, these elements enable the military to be prepared for the modern mission challenges and be versatile to prepare for future changes (Wald, 2017). Modern military units are in a period where they are adjusting to new types of warfare, new doctrines, new technologies, and constantly changing missions. This creates unprecedented complexity in the analysis of military operations. The result is the creation of new and more robust military operational paradigms to overcome complexity (West 2016). This complexity comes from entity relationships, multiple scales, data size and quality, and multiple and dynamic perspectives. The modern information world is often nonlinear, nonreductive (not dividable), erratic, dynamic, nondeterministic (stochastic), chaotic, multiscaled, and complex. (Lauder 2009) The role of the military OR analyst is to work within these new systems and concepts to collect data, build models, and perform analysis that enables decision-makers to formalize effective operational concepts, doctrine, assessments, and plans.

The US military has constantly experienced these shifting paradigms as forces, units, and individual service members try to create ways to cope with the evolving security environment of turmoil, confrontation, and anxiety. The two new warfighting domains of cyber and space add immensely to the complexity of full-spectrum, multi-domain operations (Palazzo, 2017). Complicating matters is the deluge of information now available to the military as a result of micro sensors, computing power, inter-connectivity in communications, and data-generation mechanisms. While the military leadership is engaged in paradigm shifts with respect to the way it makes decisions, designs force structure, allocates resources, and executes missions, the OR analyst must shift paradigms by learning new modeling and analysis techniques to cope with and overcome these complexities of modern conflicts (Singer 2009; Snider & Watkins 2002).

For over two centuries, reductive quantitative modeling produced steady progress in the physical sciences. The reductive approach to problem-solving tends to confront a problem in a simplified form, and then iteratively adds more components or additional layers of while always ensuring each part is solvable. However, reductivity is not the only way to solve a problem. The modern world and its problems contain many forms of complexity: different and possibly contradictory views by multiple stakeholders; corrupted, uncertain, or missing data (often in hidden and dark networks); multiple and varied ideological and cultural constraints; rapidly changing military, political, and economic constraints; numerous possible intervention points; considerable uncertainty and ambiguity in assumptions; and inherent randomness or fog of war (Beyerchen, 1992; Clausewitz, 1832; Simpson, 2012). Nonreductive modeling takes a more holistic approach to building a highly integrated and connected model by minimizing assumptions. This produces an interdisciplinary model that is not separated into reductive parts. In nonreductive modeling the essence of the process is connecting the elements to understand the problem in its entirety. This nonreductive approach is used for complex problems where the simpler reductive methodology is inadequate. These complexities and the new nonreductive modeling methods characterize the essence of future military operations and form the paradigm-shifting operational domain of the modern military OR analyst (Weaver, 1948).

19.3   Modernization of the US Military

The US military is a highly structured, hierarchical institution that operates Information Age data-collection systems and complicated Industrial Age technological systems and weapons. (Barno and Bensahel, 2018). The US government often gives its military missions that are unique and dynamic. Examples of contemporaneous mission declaratives include: modernize Afghanistan, develop African security and cooperation, keep peace in the Middle East, cure Ebola in Africa, deter North Korea and other aggressive nations, end terrorism, protect US cyber systems and infrastructure from attack, and stop piracy along the African coasts (Department of Defense, 2013). To prepare for success in these missions as well as the mission to support the large-scale efforts in the major power competition with China and Russia, the US military through the National Military Strategy as part of the National Defense Strategy is being revamped to be more agile and versatile and modernized from private to general, from pistol to drone, from submarine to satellite, from basic training to the war college, and from admission to graduation at the military's service academies. The current version of the US military, even with more money, advanced technology, improved logistical facilities, and costly material resource improvements, cannot accomplish these modern complex missions without a larger, more competent, and more knowledgeable OR and technical branch workforce with more mental dexterity, critical thinking ability, and depth in the diversity of skills. Because of these challenges, OR education's new mantra at the service academies and other military schools that educate OR analysts is to embrace the complexity of the modern Information Age world and learn how to support the new military missions (Alberts, 2003). The military's OR education, like many other modern technical education programs, needs a paradigm shift to focus on the complex tasks of the OR analyst (Scales, 2017).

All these elements have complex components, which is to say that the traditional OR techniques, which tend to over-simplify problems and attempt to fit them into uniform representations, are no longer adequate for modern military OR analysts. Enhancing and modernizing the service academy OR education, the OR graduate programs attended by service analysts, and the services' OR qualification courses can jump-start the educational shift that can better enable future military analytics success.

Full-spectrum operations, information warfare, hybrid warfare, hyperwar, and cyber warfare in today's information environment require military forces to incorporate network-centric warfare, mission command, and multi-domain doctrine (Allen and Husain, 2017). The US military cannot rely only on traditional branch or service operations or its new technological developments such as artificial intelligence (AI), unmanned autonomous vehicles (UAVs), and robotic warriors. OR must play a role in analyzing operational process by enhancing jointness and cooperation and implementing the new systems, evaluating and analyzing technological and operational alternatives, and synergizing people with technologies. This can be accomplished by enhanced information and data science methodologies, appropriate complexity modeling, high-powered computing and problem-solving algorithms, and effective use of media to communicate and collaborate (Leung, 2017). Improving military information processing by embedding the OR analyst in the design of the Observe, Orient, Decide, Act (OODA) loop, using OR methods to build data-to-decision systems, and incorporating interdisciplinary modeling and gray zone thinking into OR problem-solving will play significant roles in modern force capabilities. Ultimately, the OR analyst holds the intellectual component of the prevent, shape, win mantra of the US military (Blackwell, 2010) This new futures-based orientation for the Army is manifested in the Army's newest command, the Army Futures Command.

19.4   US Army Futures Command

Effective July 2018, United States Army Futures Command (AFC) was established with the intent of being fully operating by July 2019 (Department of the Army, 2018). The Command's headquarters element are in Austin, TX, with operating agencies remaining in many diverse locations around the country. AFC will lead the Army's modernization enterprise by integrating the future operational environment, threats, and technologies to develop new concepts, requirements, opportunities, and force designs. AFC will ultimately build the concepts, requirements, and designs of combat and information systems. The Army plans to develop AFC into an organization that looks to unify the modernization effort with a combination of research innovation and practical military needs. In addition to the new headquarters, the AFC will include many of the current Army analytic and research agencies such as Army Capabilities Integration Center (ARCIC), the Capability Development and Integration Directorates (CDIDs) with their battle labs, TRADOC Analysis Center (TRAC), Research, Development, and Engineering Command (RDECOM) with its numerous laboratories such as Army Research Labs and Army Research Office, Army Materiel Systems Analysis Activity (AMSAA), Center for Army Analysis (CAA), and many other agencies. An integrating structure for the organization is in the form of Cross Functional Teams led by military officers and filled with experts from the military and research community. The first eight of those are: long range precision fires; next generation combat vehicle; future vertical lift; command, control, communications, and intelligence; assured position, navigation, and timing; air and missile defense; soldier lethality; and synthetic training environment. There will also be a Data Analytics Team of 20 scientists, military and civilian.

Much of the military's emphasis on future science and technology comes in the form of teams working on the foundations of OR, systems engineering, and high-performance computing (deSolla Price, 1986). Several of the AFC sub-agencies are OR focused and OR will play an important role in the success of this command. Using innovative research and development, realism, and practicality, along with the Army's new concept of a streamlined acquisition process, AFC hopes to produce new modern doctrine and new modern systems that can be rapidly developed and fielded. A form of sustainability is desired as well, where whatever is developed in the future will have the ability to be expanded and updated as technology improves.

19.5   Modeling in OR

Modeling, the iterative process of developing and refining abridged representations of real-world phenomena, helps analysts think more critically about problems because models help people process and organize information, make sense of both data overload and data scarcity, and focus effort on the important elements of the problem or issue

(Breece, 2017; Silver, 2012). Modeling improves the OR analyst's ability to explain problems, design and formulate strategies for solving problems, and make recommendations and decisions about which strategy or strategies to pursue and which resources to allocate toward solving problems. Through sensitivity analysis of models and their results, OR analysts can often predict the range of possible outcomes and the likelihood of outcome occurrence. Inquiry, the deliberate, thoughtful examination of a problem and the assessment of potential solution strategies and implementation systems, enables people to develop deeper knowledge of the problem scope and solution space as well as consider more innovative modes of thought. Effective modeling and inquiry skills are essential operational requirements of the modern military OR analyst.

Qualitative modeling and complexity science emerged in the twentieth century to confront the more complex phenomena in the social, behavioral, and military sciences (Ducote, 2010). The modern conception of complexity science is to use nonreductive models and interdisciplinary analysis to dovetail the quantitative and the qualitative elements of problems. Through several major paradigm shifts of society during the late twentieth century, significant information-based, military-relevant, scientific disciplines and interdisciplines developed, including operations research, decision science, information science, computer science, data science, network science, and cyber science (Buchler et al., 2016). Teams of OR practitioners and cyber experts began to incorporate complexity modeling and interdisciplinary analysis as a core feature of their scientific-processing and problem-solving methods (van Noorden, 2015) This new modernized paradigm of complexity science, with its inherent foundation of nonreductive modeling, has changed the principles of OR analysis. Some of these new OR modeling skills are:

- ability to model messy, unstructured problems with noisy data and conflicting stakeholders
- ability to use the techniques that include problem statement summaries, sensitivity analysis, and story-boarding
- Using soft skills such as problem-structuring, qualitative methodology, developing work plans, and validation of results
- ability to integrate military perspectives with other perspectives to form a whole of government approach with OR as the glue to bind and connect military branches, different services, and various government agencies

At the heart of OR modeling is measurement – finding metrics (scaled measures) that accurately represent the appropriate elements and objectives of the problem. Often the most challenging elements in modeling and OR are determining and calculating the proper metrics (Hubbard & Samuelson, 2009). Developing appropriate, often human-based, metrics that can be used to both measure elements in the problem or model and compare those measurable outcomes between alternatives are powerful concepts in modeling. Good metrics can be used to measure success or failure within a model or determine success of a mission or the validity of a plan (Silver, 2012).

Metrics must be practical in the sense that they are computable and have data that are available for the calculations. For example, metrics in health care could include life expectancy, infant mortality, maternal mortality, and deaths due to communicable diseases. Metrics for automated military satellite interpretation could include percentage of targets recognized, percentage of military vehicles identified, percentage of determination of country of origin for vehicles and tanks, and percentage of antennae located.

Complex models must balance the dynamic with the static, the micro-variables with macro-variables, the discrete with the continuous, and the big-data with the small-data phenomena. Examples of applications being modeled by military OR analysts are designing structure of cyber security systems; overcoming networked system failures; preventing tipping points and surprises; and optimizing land and maritime transportation, port logistics, supply chain networks, and manufacturing control systems. Operations research analysts must be able to model, problem-solve, and communicate results and recommendations verbally and in writing. To be of value to military leaders, they must be able to:

- model and solve complex problems
- design structures for data to include collecting useful, relevant data "cheaply" – both in terms of time and money
- develop algorithms to effectively address new and emerging problems
- make and provide supporting evidence for policy recommendations
- present data visually in a manner that communicates information clearly and efficiently to decision-makers
- make complex problems understandable to decision-makers with little or no knowledge of the underlying science

19.6   Modern Military OR

The OR communities are not strangers to adaptation and change. To deal with the latest elements of change, an adaptive, highly creative, approach to operations based on complexity modeling and operational design becomes the principal force multiplier for many military operations (Department of Defense, 2017; Air Land Sea Applications Center, 2019). Complexity models, as elements within an OODA loop, enable military forces to change to a more desired situation through the ability to learn and adjust faster than the adversary. Operational design is the military version of complexity modeling. The decisive step to achieve this capability involves developing and assessing smart, knowledgeable, interdisciplinary, and capable analysts, systems, and forces (Banach, 2009). The effectual way to prepare OR analysts to deal with change is to develop their ability to be agile in building adaptive models in evolving settings. The military specifically uses OR at the strategic, operational, and tactical levels, and OR applications cover the gamut of military activities including national policy analysis, resource allocation, force composition and modernization, logistics, human resources, battle planning, and maintenance and replenishment.

Smart, defined as the ability to learn, and knowledgeable, defined as having access to and the ability to process relevant information, are advantages for any kind of system whether physical, informational, social, biological, or hybrid (Department of the Army, 2010). OR analysts will help to create smart and knowledgeable models that are descriptive, diagnostic, predictive, and prescriptive, as needed. These intelligent systems can apply their knowledge and skills to design, plan, execute, and sustain effective military operations. Modern OR analysts contribute to the military's ability to optimize force size and talents, communicate over long distances, achieve dynamic coordination of operations, manipulate and exploit enemy weaknesses, and effectively employ kinetic, cyber, informational, and psychological warfare tactics. Many of these systems are based on multi-layered, networked systems that give leaders and service members the situational awareness to succeed by combining and integrating appropriate joint, coalition, and specialized forces. When all this happens, OR analysts become force multipliers, making their contributions greater than the sum of the parts and creating mission success.

Modern wars are fought between the intellects and actions of the combatants through confrontations of each side's ability to array personnel, to have the right materiel in the right place at the right time, and to operate appropriate systems and weapons (Biddle, 2005). These equations form a warfare calculus that holds true for nations, non-state actors, terrorists, and insurgent groups (Gray, 2005). Effective designs by OR analysts of complex systems-of-systems enable the US military to deal with determined adversaries and adverse environmental factors. Understanding and explaining the complex nature of modern systems-of-systems to the decision-makers are important for the OR analysts who develop, test, and validate network-centric and multi-domain warfare doctrine. The agility of a force is forged by OR analysts through connecting and integrating systems of communication and networks of data sensors, validating intelligence and surveillance systems, designing and monitoring effective supply chains, identifying the value of targets, and informing decision-makers. The results are shared situational awareness, increased speed of command and control, higher tempo of operations, greater lethality of weapons and systems, increased survivability, and synchronization of action. These OR-supported capabilities create the ability to sustain operations while maintaining rapid operational speed. Having the ability to make decisions and act faster than the enemy through rapid OODA loop execution are the fundamentals that produce favorable conditions for US forces (Coram, 2002).

The US modern military needs to instill the essence of OR analysis in all its leaders. OR thinking helps to integrate disciplines and concepts into the interdisciplinary problem-solving skills needed for an officer. Military success rests with the leaders' ability to think and adapt faster than the enemy and to operate effectively in a complex environment of uncertainty, ambiguity, and unfamiliar circumstances. OR modeling and inquiry build the framework for the nation's security community to anticipate and react to those changes. The military needs its OR communities to shift their educational paradigms to stay ahead of changing defense issues and needs. Such an effort requires high-quality OR analysts and OR-savvy modern leaders. Education of the force and the intellectual development of analysts and technical leaders are primary instruments of modern military power. A military force cannot be successful without smart people who can make wise decisions, solve challenging problems, know the modern language and techniques of military science, and build sustaining systems and infrastructure. Smart, dedicated, and hardworking OR analysts can help to make a military strong and capable. In today's world, digital generals have become as important to national security as military generals (Andriola, 2017). The modern military needs people who are both digitally and technically savvy and strong leaders (e.g., the military's modern OR analysts and the newly recruited and trained cyber analysts) (Negroponte, 1996).

19.7   Military Academies' Roles in OR

History shows that the US military academies have contributed to the success of the US military and the improvement of the world in the past (Ambrose, 1966). During the 19th century, USMA and USNA served as national models for undergraduate technical education in America. During the 20th century, academy graduates in the Army, Navy, Marines, Air Force, and Coast Guard managed complicated military systems and massive military forces to fight large-scale wars. The service academies must adapt to provide intellectual leadership for the US military in the 21st century and beyond. The service academies serve as de facto national universities and have been the exemplars of America's intellectual strength and set the tone for undergraduate education in the US (Betros, 2012). However, much more is needed in the complex, information-based world of today. The academies need to deliver not only capable junior combat leaders for the military, as they did in the infrastructure-development era (18th–19th century) and cold war era (20th century), but also capable leaders in the technical branches – OR, intelligence, cyber, logistics, and communications fields. This role is relatively new for USMA and the Army since the Army had very few technical branches until recently. Today's academy graduates must be able to grow in intellect, wisdom, thinking, and complex problem-solving capabilities to produce the future senior leaders for the nation. It is up to the service academies to prepare future graduates to develop intellectual talents to make the most of their opportunities. The academic programs can contribute curricula, pedagogy, and standards to the entire nation's education community to build and enhance the nation's undergraduate educational system.

Because the environment where military officers perform and serve is demanding and sometimes dangerous, the goals of the service academies are extensive. The reality of the world as a complex mix of humans, machines, bio-systems, geographic formations, cities, ports, ships, planes, roads, buildings, nations, businesses, and much more, produces a substantial intellectual requirement for academy graduates and all future officers. The problems that future officers will face are unlike any that can be studied in a classroom, yet they contain elements from the problems they often study in all their courses. This is where OR can make its contribution to all academy students, not just those studying OR. The requirement of this intellectual capacity is why the academies' programs are broad and deep in courses, subjects, ideas, and perspectives that include embedded OR concepts. The military profession requires broadly and deeply educated leaders who have a scholarly foundation in complexity science and a strong dedication to life-long learning. Academy students encounter OR and modeling as they study mathematics, science, technologies, devices, people, and machines. USMA introduces OR modeling and complexity in its required first-year mathematics course. The academies are evolving to cope with, take advantage of, and embrace the complexities and real problems and issues of the modern world.

OR is inextricably linked to the direction and management of large systems for people, machines, materials, and money in government, industry, business, and defense. Because of the increased demand for OR analysis within the Army, the OR specialty (Functional Area 49 or FA49) continues to enjoy steady growth, which require the West Point OR major to grow and modernize.

19.8   Case Study: Designing the OR Curriculum, Courses, and Projects at West Point

The new changes to the OR program at USMA can be viewed as an example for modernizing OR education. As such, the program must meet the requirements to produce modern military OR majors that are able to:

- demonstrate competence in modeling physical, informational, and social phenomena
- identify and articulating assumptions, metrics, and constraints
- apply appropriate solution techniques
- interpret results within the appropriate context
- understand the role of OR in interdisciplinary problem-solving and technological development

They achieve proficiency in OR breadth, depth, and versatility by:

- understanding and applying probabilistic and statistical models and methods;
- modeling complex phenomena
- understanding and applying simulation methods
- understanding and applying optimization methods
- communicating effectively – orally and in writing
- using technology to model, visualize, and solve complex problems

As modern problem-solvers, they develop productive habits of mind including:

- creativity
- intellectual curiosity
- an experimental disposition
- critical thought and reasoning
- a commitment to life-long learning

In addition, undergraduate OR study at the Academy prepares a cadet for future graduate-level study in OR, since many will go on to graduate school eventually. OR officers must be able to make complex problems understandable to the decision-makers they will advise. This requires an ability to present multivariate data in a manner easy to visualize, an ability to design data structures and collect data cheaply – both in terms of time and financial cost – and an ability to develop algorithms to answer questions when no readily available tool exists. OR analysts must be familiar with optimization methods, applied statistics, and data science and simulation techniques. Underlying all of this is relative comfort with computing and programming. USMA's curriculum is designed to give its OR graduates both the necessary tools as well as the ability to apply them to complex, multidimensional problems.

USMA regularly receives feedback from a variety of clients on its OR program. Officers deploy overseas or to stateside agencies to provide analytical support to commanders. Cadets participate in short-duration summer internships everywhere from the Army Research Labs to NYC investment firms. OR analysts of the future must be comfortable working with data and networks and using them to inform decisions (Coronges, Barabasi & Vispignani, 2016) To do these tasks, they must have solid programming skills that are transferable across software. The stakeholders of the West Point OR major are the operational Army, Army OR agencies, strategic leaders in the national OR professional community, and top-tier higher education institutions with OR programs or related fields. The base of knowledge expected of successful modern military OR analysts and graduate school applicants formed the basis of the program's student learning outcomes. The primary objectives of the program are for cadets to perform modern complexity-based modeling, incorporate real-world data into their models, and make evidence-based recommendations for optimal decision-making that will improve unit or system performance or manage resources effectively.

Degrees other than OR may also be useful for undergraduate education for military analysts. For instance, USMA recently started an Applied Statistics and Data Science degree. This degree was proposed in response to indications of need from a variety of sectors, including the Army's OR community. The Student Learning Outcomes for this degree differ from those for the OR degree. While the outcomes for the Operations Research major focus on solving complex problems and providing solutions to problems that are generalizable, the outcomes for Applied Statistics majors focus more on providing contextual solutions and are much more focused on the management and processing of data and using statistical modeling to quantify uncertainty in the data. It is anticipated that the need for these skills in the military and the Army OR community will continue to grow, leaving even more unmet demand. The Army OR branch has recently modified the branch vision to state that they "serve the Army as organic experts in data science, data analytics, data visualization, and other big data specialties" (Functional Area 49 website, 2019).

These majors are designed to meet the requirements for the Army's OR needs. The research projects and integrative experience give cadets hands-on experience tackling diverse, often ill-defined problems. The practical knowledge gained from these experiences will translate directly into the critical thinking and problem-solving expertise expected of successful Army officers.

Each OR major completes an integrative experience (IE), which consists of a research sequence focused on applying OR skills learned to an interdisciplinary project. This research project integrates the principles, concepts, and models explored in previous courses and gives cadets the opportunity to apply them to a real-life system. Cadets work under the supervision of a faculty member to address a large-scale scenario, providing them an integrative experience that increases the depth of their education. During the different assessments conducted throughout the course, cadets will apply the knowledge gained through their OR coursework and subject matter experts to fully develop their understanding of the topic and appropriately respond to the challenges presented. While these evaluations take on different forms, they always present problems that are best solved from multiple viewpoints using an interdisciplinary method. Research projects are vetted by the OR or Systems Engineering Program Director. The IE sequence is taught by senior faculty, and the effectiveness of the sequence in achieving the overarching goal is assessed by the senior faculty.

Each OR major takes several complementary support courses which provide breadth to the major by providing alternative perspectives to the discipline. The OR major requires understanding of technology, programming, and the interaction between information technology (IT) systems and IT-enabled decision-making. Other foundational complementary support courses include Systems Simulation and Stochastic Models. This modern-OR-focused curricular re-design resulted in the following elements:

- more skills in data analysis, more skills in technical math, and more programming and coding
- more computational mathematics through a newly developed Algorithmic Programming course
- more research opportunities to fulfill the integrative experience rather than just systems design projects
- more elective options, particularly in computer science and economics
- organized elective options into threads (supply, transportation, project management, production analysis, finance, data science)

19.9 Pedagogy for OR Courses

West Point's goal is for OR cadets to be confronted with meaningful problems that are challenging to solve. Though faculty publish a suggested solution, the purpose for doing so is not to show students the correct answer or to indicate that there is a preferred solution. Rather, the faculty publish suggested solutions to demonstrate how one could approach the problem-solving process for the assigned problem and fully communicate the results of the analysis. Homework problems are designed to confront problems that cadets might see when they become officers. The course is designed for cadets to face academic challenges that they have not faced before.

The process is started with cadets during their introductory mathematical modeling course that they are required to take during the first semester of their freshman year. This course emphasizes applied mathematics through modeling. Students develop effective strategies to solve complex and often ill-defined problems. The course exercises a wide array of mathematical concepts while nurturing creativity, critical thinking, and learning through activities performed in disciplinary and interdisciplinary settings. The course introduces calculus using continuous and discrete mathematics while analyzing dynamic change in applied problems. Students employ a variety of technological tools to enhance the ability to visualize concepts, to explore ideas through experimentation and iteration, to complete complex and time-consuming computations, and to develop numerical, graphical, and analytical solutions that enhance understanding.

Here are two examples of prompts used in the Introductory Modeling course for the first-year (freshmen) cadets:

- After years of analysis, discussion, and debate, the National Defense Authorization Act for Fiscal Year 2016 created the Blended Retirement System (BRS). The name reflects its inclusion of elements from the legacy retirement system, namely a defined pension, which will complement Thrift Savings Plans (TSP) and matching government contributions. All service members (SMs) joining the force after January 1, 2018 will be automatically enrolled in the BRS; however, current service members with less than 12 years of service (as of January 1, 2018) had the choice to opt-in to the BRS or remain under the legacy retirement system. The purpose of this assignment is to make a recommendation to your commander about whether he or she should choose the legacy retirement system or the BRS.

- As an officer supporting the battalion staff in a forward location in Afghanistan and recently finished analyzing some patrol debriefs from 2nd platoon, you discuss some of the details with the S-2, and both of you believe that the debriefs contain some information as to the whereabouts of a possible High Value Target (HVT) operating in your battalion's area of operations (AO). In an effort to gather more intelligence, you provide the platoon leader with some information requirements (IR) that you want answered. The platoon leader returns with some information which you present to the battalion commander (BC) and the rest of the battalion in a storyboard during that night's battle update brief (BUB). Due to the operational tempo of the battalion, the BC does not think he can afford to shift the focus away from the Alpha Company AO, even to catch a possible HVT. He directs Charlie Company to establish check points with their partners from the Afghan National Army (ANA) immediately along both roads leading out of the town to inspect all vehicles leaving. This will allow Charlie Company to positively identify the HVT and allow the ANA to detain him before he can leave the AO. The BC does not believe he will be able to allocate the forces required to go into the town and search for the HVT for at least another seven days. Your boss believes that the information provided from the storyboard could be used to develop a probability model which could assist the battalion when the time comes to go in and search the town. The model could also be used to convince the BC to act now or risk losing the opportunity to capture the HVT. Your task is to develop a model that predicts the location of the HVT in Charlie Company's AO each day over the course of the next two weeks. Advise your BC on the impacts of not pursuing the HVT immediately. In addition, provide the BC with possible neighborhoods to focus on if the search were to begin on day seven for the HVT.

As is shown in the first mathematics course in their required core program, cadets gain experience in modeling modern complex problems. In their calculus courses, cadets model epidemics and investigate the military operations in such situations; in the basic probability course, they have simulated repair costs for legacy systems during deployment. Cadets are required to analyze these problems in the absence of complete data, making assumptions and modeling decisions along the way.

In the OR program courses, the problem-solving is more rigorous. The Linear Optimization course required of all OR majors emphasizes the applications of optimal solutions to linear algebraic systems using the simplex method of linear programming. This includes an in-depth development of the simplex method, the theory of duality, an analysis of the dual problem, integer programming, sensitivity analysis, and the revised simplex procedure. Cadets also study additional computational techniques that are applicable to specific mathematical models such as the trans-shipment problem, assignment problem, and network problems. The course emphasizes the use of computer software to solve problems that illustrate useful applications. This course is designed to present the underlying principles and methods used in solving linear optimization problems. These models apply to a wide range of applications and develop the ability to formulate and solve a broad range of such problems. Advanced topics, such as sensitivity analysis and interior point methods for solving very large problems, are topics that interest the cadet OR students. The field of mathematical programming includes formulating and solving mathematical problems as well as interpreting and explaining results. Cadets develop the ability to use and modify the standard methods, to explain and justify solutions, and to provide clear, coherent write-ups of solutions.

The Linear Optimization course contains homework problems. In each case, they include an application problem that requires modeling the problem as a linear program, coding the model using the OPL Interactive Development Environment for CPLEX, running the model, interpreting the results, and writing up a solution using a scientific document preparation software like LaTex. The problems are based on situations that the instructors, or some of their colleagues, have faced on the job whether in the civilian or military sector. There is also a course project assigned, which involves a more complex application. Students work together in teams of two. The project requires students to clarify a vaguely stated problem, locate relevant resources and data, and create a complete, professionally written, solution to the problem.

Examples of some of the problems assigned in the course include:

- Find a least distance flight plan for a helicopter crew that was required to deliver mailbags to a set of Forward Operating Bases (FOBs) in Afghanistan as well as satisfy a list of Air Mission Requests (AMRs), including the transportation of a Congressional Delegation. Aerial imagery of the Area of Operations is shown in Figure 19.1

  This project required the cadets to locate the relevant data (helicopter load capacity, range, cruising speed as well as distances between FOBs) then to create a mixed integer program and solve it using CPLEX. This problem was inspired by real work done by a former faculty member who was an aviation company commander deployed to the theater. A fully optimal solution may not be available and therefore students must determine which AMRs cannot be filled. Additional requirements such as supporting congressional delegations may be added. Students must consider factors such as payload capacity, seating capacity, passenger and cargo weight, cruising speed, range, and weather when developing their solutions. Students must submit written solutions as well as give oral presentations to their supervisor.

FIGURE 19.1: Aerial imagery of Area of Operations to include FOB locations.

- Find an optimal route for travel by a vehicle between locations while satisfying a set of constraints on the route. Some of these constraints included required stops in several geographically dispersed locations. Other constraints included the requirement to carry a certain number of passengers and bags between various pairs of locations while considering vehicle capacity limitations, fuel tank capacity, and required sleeping time
- Formulate a linear program that minimizes the transportation costs of shipping trucks or fuel from production centers to distribution centers to regional service centers while considering various shipping options along the routes between centers, costs of production at various production facilities, storage capacity constraints at distribution centers, and demand at regional service centers
- Determine a fielding plan for the Army's new troop transport. The Army has finalized design specifications for two new models of multi-purpose troop transport: a basic model and a heavily armored version. The transports will be manufactured by private contractors and then will be sent to newly constructed test facilities where they will be extensively tested before being shipped to four different deployment locations. Cadets are provided the locations plus their demand for each model. They must decide where to construct test facilities, which contractor bids should be accepted, and how many vehicles of each type should be ordered from each contractor. They are provided the submitted bids for construction of the troop transports, the maximum number of each model that each contractor can produce within the time constraints, the per-unit cost per contractor to provide the transports, the capital cost incurred for each test facility including fixed construction and location-dependent staffing costs, the total capital budget for test facility construction, operating cost structure for each test facility, and shipping costs – both contractor to test center and test center to designated deployment center. Cadets' overall objective is to minimize the operating costs of meeting the demands at the deployment locations with tested and certified transports constructed by the selected contractors. The solution presented must be feasible with respect to contractor capacity, test facility capacity, and capital budget for construction of the test

Many cadet theses and projects have direct military applications. For instance, one student did his thesis on catastrophe mitigation. Specifically, the cadet looked at safeguarding the West Point food supply chain, but his work is transferable to many other areas. This project was inspired by work he had done at a national lab while on a summer internship. This cadet analyzed the supply chain supporting the food services provided by the West Point Mess. He considered a broad range of disruptions such as power outages, natural disasters, and intentional attacks that could affect these services and analyzed the impacts that they could have. He worked directly with the West Point director of food services and analyzed the disruption occurrences using historical data, considering both impact and probability of occurrence. He then developed a set of possible mitigation packages that would increase the robustness of mess hall services by reducing the severity of the disruptions. He used a mixed integer linear program following a capital budgeting model to compute an optimal choice of mitigation packages for the mess hall, subject to a constraint on the total amount of funds available for mitigation.

Another example of cadet thesis work is a structural analysis of the Army Mission Essential Task List (METL) network. A METL crosswalk is a primary analysis tool to help a unit plan training. This student believed that improper execution of a METL crosswalk is not a function of training, but rather an unreasonable expectation to solve a difficult problem by hand. He modeled tasks associated with a METL crosswalk as a task network and automated the crosswalk to identify key collective tasks a unit should train on to achieve battle focus. Additionally, his technique allows commanders to develop a better understanding of task relationships through data visualizations. He then applied his method to an infantry company METL.

19.10   USMA's OR Program Requirements

The academic requirements of West Point's OR program were determined by considering the modern needs of the OR profession. This program will be effective with the graduating class of 2022. Interactions with stakeholders, including Army analytics and research agencies and research sponsors, identified the new trends needed in the program. Faculty who have recently served as OR officers regularly discuss changes to the program. Recently, these assessment mechanisms identified the need for additional data analysis and programming skills. The OR academic major is a 41-course program with 17 courses in the major. These courses are divided into a required sequence of 11 foundational courses, a two-course research sequence, and four electives. Three of the four electives are chosen from the same thread. The six possible threads can be mapped to INFORMS Operations Research subdivisions. Additionally, cadets may choose to pursue an honors diploma, which requires an additional course from their selected thread as well as GPA requirements. In addition to several required mathematics and science courses, the cadets studying OR take 11 required courses. The overall breadth of topics covered by the required courses are adjusted periodically to adapt to the changes in technology, OR needs, or the Army's mission.

The required courses are:

- Theory and Practice of Military IT Systems
- Linear Algebra
- Applied Statistics
- Mathematical Statistics
- Nonlinear Optimization
- Linear Optimization
- Mathematical Computation
- Theory and Applications of Data Science
- Systems Simulation
- Decision Analysis
- Probabilistic Models

Cadets are also required to take a two-course research sequence. Typically, this is either an individual Mathematics thesis with a topic that is sufficiently interdisciplinary or a Systems Engineering capstone as part of a four-person team. Cadets may request to conduct research in another subject (for example, mechanical engineering or computer science).

Cadets are also required to take four electives, with at least three being from the same one of the following threads:

- Operations Management
- Data Science
- Networks, Computing, and IT
- Simulation
- Finance

With the exception of Theory and Practice of Military IT Systems, required courses come from either the Department of Mathematical Sciences or the Department of Systems Engineering. Elective courses come from a variety of departments in addition to Mathematics and Systems Engineering, to include Computer Science, Behavioral Sciences, and Social Sciences (Economics). See Figure 19.2 for how the typical program is sequenced.

An additional opportunity for many USMA OR majors comes in the summer when they can take a short (often just three weeks) internship. Some of the recent ones were:

- Research Associate at the Defense Manpower Data Center, Monterey, CA
- Determining Sample Size Requirements for Ballistic Testing, Army Research Labs, Aberdeen Proving Grounds, MD
- Spread of Opinions Across Complex Social Networks, Army Research Labs, Aberdeen Proving Grounds, MD
- Estimation of Energy Expenditure During Dismounted Soldier Tasks, Army Research Labs, Aberdeen Proving Grounds, MD
- Counterinsurgency Combat Modeling, Army Research Office, NC
- Social Network Analysis of Terrorist Organizations, Special Operations Command CENTCOM, MacDill AFB, FL

Fourth Class		Third Class		Second Class		First Class	
1st Term	**2nd Term**	**1st Term**	**2nd Term**	**1st Term**	**2nd Term**	**1st Term**	**2nd Term**
MA103 Math Modeling / Intro to Calc	MA104 Calculus I	MA205 Calculus II	MA206 Prob/Stats	MA376 Applied Statistics	MA476 Mathematical Statistics	MA381 Nonlinear Optimization	MA481 Linear Optimization
CH101 Chem	PH205 Physics	Science Depth	EV203 Phys Geog	MA486 Mathematical Computation	MA477 Advanced Data Science	Research I	Research II
PL100 Psych	CY105 Info Tech I	MA371 Linear Algebra	CY305 T & P Mil IT Sys	EM481 Systems Simulation	SE385 Decision Analysis	OR Elective 1	Elective Thread 2
HI105 History	HI108 History	SS201 Econ	SS202 Pol Sci	HI302 Mil Art	SE388 Stochastic Modeling	Elective Thread 1	Elective Thread 3
EN101 English Comp	EN102 Literature	LXXXX For Lang	LXXXX For Lang	PL300 Leadership	SS307 Int'l Rel	MX400 Officership	LW403 Law
			PY201 Philosophy				

FIGURE 19.2: Sample 8-Term academic plan for operations research majors. The thread electives are chosen from one of the five threads: Operations Management, Data Science, Networks, Computing and IT, Simulation, or Finance.

19.11 Other OR Undergraduate Programs

Undergraduate degrees in OR are uncommon; according to The College Board, only 19 institutions in the United States offer undergraduate degrees in Operations Research, including USMA, United States Naval Academy (USNA), and United State Air Force Academy (USAFA) (College Board Access, 2019). Although not identical, the OR major at USMA is similar in construct to programs at both USNA and USAFA. USAFA is more focused on applications, especially economics, and less on the mathematical underpinnings of OR techniques and has more interdisciplinary breadth. USAFA shows its modernization and breadth considerations by indicating the program has

> *captured the essence of the field by establishing a truly interdisciplinary major. The OR program is jointly administered by the Departments of Computer Science, Economics and Geosciences, Management and Mathematical Sciences. In addition to the basic set of OR courses, the required major's courses will include courses from each of the four departments.*
>
>   (Operations Research Program, 2019)

USNA offers more flexibility with more electives than the USMA program. USNA modernized by including courses in simulation models and dynamic and stochastic models (Operations Analysis Major, 2019). The strongest similarity between USMA and other OR programs are the shared objectives to develop mathematical models with uncertainty and incorporate real-world data into these models in order to make optimal decisions that improve performance or manage resources effectively.

19.12 Graduate OR Programs

Many OR analysts in the Army obtain a graduate degree, and some of the graduate programs the analysts attend are modernizing in similar ways to USMA. As indicated through correspondence with the program director in 2018, over 80% of the 523 Army officers designated as OR analysts have at least one graduate degree, and their graduate degrees are primarily in OR, Statistics, Network Science, or Mathematics. It is very typical for the Army OR analysts, except for the USMA OR and Mathematics majors, to have an undergraduate degree in science or business and a graduate degree in OR or Statistics. Some of these graduate programs are in the process of modernizing courses, but not all of them. Highly successful graduate programs of interest for the military are at the Naval Postgraduate School (NPS) and Air Force Institute of Technology (AFIT). The NPS program develops students "to identify relevant information, formulate decision criteria, and select alternatives" (Department of Operations Analysis, 2019). The program includes courses with modern military OR topics, such as Network Flows and Graphs, Joint Combat Modeling, Joint Campaign Analysis, Strategy and War, and Wargaming Applications. AFIT describes their program in modernization and complexity modeling terms:

> *Operations Research aims to provide rational bases for decision making by seeking to understand and structure complex situations and to use this understanding to predict system behavior and improve system performance. Much*

of this work is done using analytical and numerical techniques to develop and manipulate mathematical and computer models of organizational and operational systems composed of people, machines, and procedures.

(Department of Operational Sciences, 2019)

19.13 Army OR Programs

The Army also educates OR analysts in its own education program through the Army's Operations Research Systems Analysis Military Operations Course (ORSA MAC). Also all Army OR analysts are required to complete the Functional Area 49 Qualification (FA49Q) Course prior to promotion to Lieutenant Colonel. ORSA MAC is a 14-week, skills-based program that emphasizes fundamental ORSA skills in a military setting. This program also includes foundational topics in Probability, Calculus, Statistics, Data Analysis and Programming (Excel-based), and Math Programming. In an Application Phase, students study Linear Statistical Models, Simulation, Decision Analysis, Data Analysis and Programming (Visual Basic for Applications and R programming), and Cost Analysis. A new Data Science course adds complexity modeling and data analysis to the program. ORSA MAC is equivalent to an 18-graduate credit program. USMA is discussing its modernization with the leaders of the ORSA MAC program. The FA49 qualifying course is just six weeks. These students use the program's interdisciplinary and multidisciplinary elements to gain comfort with ambiguity and complexity. The program learning objectives, which contain some of the modernization elements, are:

- understand senior leader's perspectives in order to better communicate analysis to senior leaders
- understand the roles of OR analysts across the breadth of OR positions within the Army and DoD
- learn to lead or work as part of a team to plan, prepare, execute, and defend analysis to solve a problem
- learn to plan, create, explain, and defend technical knowledge and findings through written and oral communication and evaluate others' work

These objectives are supported through nine modules: Strategic Thinking, Analysis Planning Process, Operational Force Analysis, Generating Force Analysis, Service/Department Analysis, DoD Community Analysis, Capstone Project, Professional Development, and Data Science. This program is equivalent to six graduate credits. Professional information for Army OR analysts can be found through these links associated with Army professional oversight and professional societies:

Army Human Resources for the Operations Research Career Field https://www.hrc.army.mil/content/FA49%20-%20Operations%20Research%20and%20Systems%20Analysis%20ORSA

Institute for Operations Research and the Management Sciences https://www.informs.org

Operations Research: The Science of Better https://www.scienceofbetter.org

Military Operations Research Society https://www.mors.org/

19.14   Conclusion

To be successful in future warfare in the Information Age, the US military will use information, data, networks, and computational power and knowledge to enhance its systems' and units' performance (Leonhard, 1998). To facilitate these performance goals, OR analysts will accomplish their highly valuable analytic roles in all domains of warfare (air, land, sea, space, and cyber) and in all spectrums of operations. The modern military OR analysts will become the informational, data, and network experts who enhance the performance of the decision-making systems – both human and machine.

The future capability of the US military depends on innovative analysis, versatile abilities, and clear decision-making. Many military units will use inter-branch teams of OR analysts, domain experts (especially cyber), service members, and intelligent machines to perform their functions on the battlefields. Modern military OR analysts will need to shed the cognitive constraints and outmoded simplicity of Industrial Age thinking and learn to apply complexity modeling and skilled analytics to Information Age problems. The military OR community can achieve this by modernizing and enhancing the OR education programs at the service academies and in the services' OR education schools. Military OR education programs can follow West Point's lead to embrace complexity and inspire students to engage in deeper and more rigorous modeling and problem-solving challenges. Highly capable OR analysts of the future will develop and deliver empowering concepts and force designs, model new force structures and high-technology systems, integrate wide-ranging domain and service capabilities, and solve many of the military's most challenging problems. As experts of military-relevant analysis, modern military OR analysts will strive to improve in all facets of their profession. Central to that focus is their education in the skills that produce the intellectual preparation for future conflicts. The doctrines, technologies, systems, and equipment on the battlefield will evolve, but the most powerful weapon needed to win future battles will continue to be the knowledge and skills the service members contribute to leading and decision-making at all levels of forces and in all types of systems.

References

Air Land Sea Applications Center. (2019). Retrieved from https://www.alsa.mil/Portals/9/Documents/roadshow.pdf?ver=2016-12-02-095351-280

Alberts, D. S. & Czerwinski, T. J., Editors. (1997). *Complexity, global politics and national security*. Washington, DC: National Defense University.

Alberts, D. S., Garstka, J. J. & Stein, F. P. (1999). *Network centric warfare*, 2nd Ed. Washington, DC: C4ISR Cooperative Research Program.

Alberts, D. S. (2003). *Information age transformation*. Washington, DC: C4ISR Cooperative Research Program.

Allen, J. R. & Husain, A. (2017). On hyperwar. *Proceedings, U.S. Naval Institute, 143*. Retrieved from https://www.usni.org/magazines/proceedings/2017-07/hyperwar.

Ambrose, S. (1966). *Duty, honor, country: A history of West Point*. Baltimore, MD: Johns Hopkins Press.

Andriola, S. (2017, June). Digital generals are as important as military general (because the US is losing the digital war). *Forbes*. Retrieved from https://www.forbes.com/sites/steveandriole/2017/06/20/digital-generals-are-as-important-as-military-generals-because-the-us-is-losing-the-digital-war/#31ba23de6f25

Banach, S. J. (2009). Educating by design: Preparing leaders for a complex world. *Military Review, 89*(2).

Barno, D. W. & Bensahel, N. (2018, June 18). War in the fourth industrial revolution. Retrieved from https://warontherocks.com/2018/06/war-in-the-fourth-industrial-revolution/

Betros, L. (2012). *Carved from granite: West Point since 1902*. College Station, TX: Texas A&M University Press.

Beyerchen, A. D. (1992). Clausewitz, nonlinearity and the unpredictability of war. *International Security 17*(3), 59–90.

Biddle, S. (2005). *Military power: Explaining victory and defeat in modern battle*. Princeton, NJ: Princeton University Press.

Blacker, N. (2017). Winning the cyberspace long game applying collaboration and education to deepen the US bench. *Cyber Defense Review 2*(2), 21–32.

Blackwell, J. (2010). *The cognitive domain of war*. New York, NY: Praeger.

Breece, J. (2017, April). A new mantra for decision making: #use all the data. *Where Next Magazine*. Retrieved from https://www.esri.com/about/newsroom/publications/wherenext/new-mantra-decision-making-useallthedata/

Buchler, N., Fitzhugh, S. M., Marusich, L. R., Ungvarsky, D. M., Lebiere, C. & Gonzalez, C. (2016). Mission command in the age of network enabled operations: Social network analysis of information sharing and situation awareness. *Frontiers in Psychology 7*(937). Retrieved from https://www.ncbi.nlm.nih.gov/pmc/articles/PMC4916213/

Clausewitz, C. (1832). *On War*. Edited and translated by Howard, M. & Peter, P. Princeton, NJ: Princeton University Press.

Cohen, I. B. (1987). Scientific revolutions, revolutions in science, and a probabilistic revolution 1800–1930. In *Probabilistic revolution: Vol.1 ideas in history*. Cambridge MA: MIT Press.

College Board Access. (2019). Retrieved from https://bigfuture.collegeboard.org/college-search?major=484_Operations%20Research

Coram, R. (2002). *Boyd: The fighter pilot who changed the art of war*. Boston, MA: Little Brown.

Coronges, K., Barabasi, A. L. & Vespignani, A. (2016). *Future directions of network science: A workshop report on the emerging science of networks*. Arlington, VA: Virginia Tech Applied Research Corporation.

Department of Defense. (2013). *Joint publication 3-12 (R): Cyberspace operations*. Washington, DC.

Department of Defense. (2017). *Joint publication 5-0: Joint planning*. Washington, DC.

Department of Operational Sciences. (2019). *Air force institute of technology course catalog*. Retrieved from https://www.afit.edu/ENS/programs.cfm?p=12&a=pd

Department of Operations Analysis. (2019). *Naval postgraduate school academic catalog*. Retrieved from https://my.nps.edu/web/or/or_curric

Department of the Army. (2010). *Hybrid threats: Training* circular 7-100. Washington, DC.

Department of the Army. (2018, June 4). *General orders 2018-10. Establishment of the army futures command*. Washington, DC.

deSolla Price, D. J. (1986). *Little science, big science... and beyond*. New York, NY: Columbia University Press.

Dubik, J. M. (2018). No guarantees when it comes to war. *Association of the US Army*. Retrieved from https://www.ausa.org/articles/no-guarantees-when-it-comes-war

Ducote, B. M. (2010). *Challenging the application of PMESII-PT in a complex environment*. Fort Leavenworth, KS: School of Advanced Military Studies, United States Army Command and General Staff College.

Functional Area 49. (2019). Retrieved from http://www.fa49.army.mil/

Gray, C. S. (2005). *Another bloody century: Future warfare*. London, England: Weidenfeld & Nicolson.

Hubbard, D. & Samuelson, D. A. (2009). Modeling without measurements: How the decision analysis culture's lack off empiricism reduces its effectiveness. *ORMS Today 36*(5), 26–31.

Kott, A., Swami, A. & West, B. J. (2019). The internet of battle things. *IEEE Computer, 49*(12) 70–75.

Kuhn, T. (1962). *The structure of scientific revolutions*. Chicago, IL: University of Chicago Press.

Lauder, M. (2009). Systemic operational design: Freeing operational planning from the shackles of linearity. *Canadian Military Journal, 9*(4), 41–49.

Leonhard, R. (1998). *The principles of war for the information age.* Novato, CA: Presidio Press.

Leung, H. (2017). *Information fusion and data science.* Berlin, Germany: Springer.

MacGregor, D. A. (1997). *Breaking the phalanx: A new design for landpower in the 21st century.* Westport, CT: Praeger Publishers.

Negroponte, N. (1996). *Being digital.* New York, NY: First Vintage Books.

van Noorden, R. (2015). Interdisciplinary research by the numbers. *Nature 525*(7569), 306–307.

Operations Analysis Major. (2019). *United States Naval Academy academic program.* Retrieved from https://www.usna.edu/Academics/Majors-and-Courses/Majors/Operations-Research.php

Operations Research Program. (2019). *United States Air Force Academy academic programs.* Retrieved from https://www.academyadmissions.com/the-experience/academics/majors/operations-research-major/

Palazzo, A. (2017, November 14). Multi-domain battle: Meeting the cultural challenge. Retrieved from https://thestrategybridge.org/the-bridge/2017/11/14/multi-domain-battle-meeting-the-cultural-challenge.

Scales, R. (2017, August 16). Ike's lament: In search of a revolution in military education. Retrieved from https://warontherocks.com/2017/08/ikes-lament-in-search-of-a-revolution-in-military-education/

Silver, N. (2012). *The signal and the noise: The art and science of prediction.* New York, NY: Penguin Books.

Simpson, E. (2012). *War from the ground up: Twenty-first century combat as politics.* Oxford, England: Oxford University Press.

Singer, P. (2009, July 7). Tactical generals: Leaders, technology, and the perils. Retrieved from https://www.brookings.edu/articles/tactical-generals-leaders-technology-and-the-perils/

Snider, D. & Watkins, G. (2002). *The future of the army profession.* New York, NY: McGraw-Hill.

TRADOC. (2014). *Win in a complex world.* TRADOC Pamphlet 525-3-1. Fort Eustis, VA: TRADOC.

TRADOC G-2. (2017). *The operational environment and the changing character of future warfare.* Retrieved from https://community.apan.org/wg/tradoc-g2/mad-scientist/m/visualizing-multi-domain-battle-2030-2050/200203

Wald, C. F. (2017, June 13). Military omnipresence: A unifying concept for America's 21st-century fighting edge: The Pentagon should converge its technological and doctrinal efforts towards a perpetual, networked presence that enables operations and awareness anywhere in the world. Retrieved from https://www.defenseone.com/ideas/2017/06/military-omnipresence-unifying-concept-americas-21st-century-fighting-edge/138640/

Weaver, W. (1948). Science and complexity. *American Scientist, 36*(4), 536–544.

West, B. J. (2016). *Simplifying complexity: Life is uncertain, unfair and unequal.* Sharjah, UAE: Bentham Books

Chapter 20

Strategic Analytics and the Future of Military Operations Research

Greg H. Parlier

20.1 Introduction: The Future of Military Operations Research?

As the Battle of Britain loomed early in World War II, in a converted English manor along Suffolk's southeastern coast overlooking the North Sea, a new profession was being invented. The phrase to define these new multi-disciplinary teams supporting Air Chief Marshall Hugh Dowding and his Royal Air Force Fighter Command with the integration and use of newly invented radar was, of course, "Operational Research" (OR) (Her Majesty's Stationery Office, 1963). The fate of Western Civilization was indeed at stake in those early war years. Prime Minister Winston Churchill would later describe this period after the fall of France as Britain's "Darkest Hour" (Fisher, 2005).

It has been nearly 80 years since combat scientists first emerged at Bawdsey Research Station to integrate disruptive technologies into combat operations and strategic planning, thus representing the "marriage" between the art of warfare and the application of the scientific method. Following the war, OR continually evolved by capitalizing on advances in mathematical sciences, computational capabilities, information technologies, economics, and decision sciences. Both technological and theoretical advances enabled OR to expand its unique contributions to other government functions and a diverse array of commercial applications including manufacturing and industrial operations, transportation, telecommunications, energy, and, more recently, agriculture and healthcare.

Contributions from the *practice* of Operations Research subsequently encouraged OR as an academic *discipline* formally taught at the undergraduate level in our military academies and the graduate level in our major research universities. Since its inception, and despite some inevitable philosophical and pedagogical battles along the journey, the profession of Operations Research – both the practice and discipline – has flourished. Nonetheless, rather than investigating past technical accomplishments, significant as they surely are, this essay instead orients on the future *practice* of our Military Operations Research (MOR) profession: what do we need to be doing? Where *should* we be going?

Our national security concerns today extend well beyond an exclusive military focus to now include geopolitical (Huntington, 1996), geo-economic (Blackwill & Harris, 2016), and socio-demographic (Kotlikoff & Burns, 2012) considerations, as well as energy security, population migration, environmental shifts, and various interacting dynamics among all these challenges. These broader influences on national security, including regional competition, global tensions, *and* especially domestic conditions, suggest the scope of our professional discipline must be addressed. This expanded view of national security indicates the term *"military"* OR has become too restrictive.

To address these growing, overlapping challenges and respond to the question "What *can* the OR community do?," a recent project undertook a comprehensive assessment from which emerged a new approach for transforming large-scale, complex, interdependent organizations that comprise an extended enterprise. Characterized as "Strategic Analytics," critical components include (1) OR-based decision support systems; (2) a transformational approach to strategic planning guided by "engines for innovation"; and (3) creative management policies to enable successful organizational change using "analytical architectures" in public sector organizations, government institutions, and our national defense establishment.

We need to coordinate, integrate, and focus our analytical horsepower on the pressing national and international security issues we increasingly face. We must assess the trajectory of our professional discipline by examining the adequacy of our current capabilities, capacity, organization, and contributions to address the defense challenges of our time (Parlier, 2015). Recognizing these challenges, along with evolving trends we can discern, what would a forward-looking appraisal suggest? What could a more expansive, re-invigorated *national security* OR community do?

20.2   Geopolitical Conditions: The Global Landscape

By assessing contemporary affairs and the evolving US national security predicament, a portrayal of future challenges emerges. Recent global events and current domestic trends, described in this section and the next, help us to frame persisting problems within longer perspectives enabling contemporary challenges to come into better focus through the long lens of history. We can then visualize expanded roles for our MOR profession and better anticipate opportunities to apply Strategic Analytics.

Following the sudden collapse of the Soviet Union nearly three decades ago, the rapid secession of 15 states from the former Soviet Empire in 1991 constituted the greatest peaceful revolution in modern history. A third wave of global democratization then followed during the 1990s, including an amazing series of "color revolutions" culminating in the Arab Spring. But now a reversal is occurring among international forces of integration and disintegration, away from supranational institutions, cosmopolitan globalism, and the international liberal order with its utopian promise of "universal empire" offering salvations of peace and prosperity (Hazony, 2018). Strong countervailing forces of realism and nationalism are returning as the world re-enters an era of regional hegemony and great power competition (Mearsheimer, 2018). The continual clash between forces of order and disorder is now shifting in ominous ways (Brands & Edel, 2019).

The NATO defense alliance of European and North American democracies deters external threats through the collective security agreement mandated in Chapter V of the NATO Charter: an attack against one is an attack against all. NATO rapidly expanded eastward after 1991, initially into Central Europe and then Eastern Europe, including the Balkans and Baltics. In response, Russia "annexed" Crimea and continues to threaten eastern Ukraine. Although a declining economic power, Russia has become a revisionist state refusing to join the international liberal order (Stent, 2019).

In the Middle East, humanitarian crises and barbaric acts shocked our senses as an Islamic caliphate gained hold in Syria, Iraq, and Lebanon. After losing most of the territory they had rapidly gained across the Levant and Maghreb, ISIL remnants are now contained in small pockets within Syria. The Syrian Civil War, ablaze now for seven long and tragic years, has resulted in genocide with nearly half a million deaths, and mass refugee flows overwhelming Europe. In Gaza, the only certainty is another ceasefire … that will soon be broken again. Meanwhile Iran continues its evasive march toward nuclear weapons and missile delivery capabilities, exacerbating ancient disputes between Sunni and Shia Muslims which predate the Protestant Reformation by several centuries.

Within Central Asia, the "forever war" persists across the "graveyard of empires" (Lieven, 2018). Ancient rivalries persist between the Pashtun and Punjab, between Afghanistan, Pakistan, and India, and among a dozen insurgent groups. In contrast, across Iran on its northwestern border, Kurdistan remains a self-governed, but unrecognized "nation" within Iraq, Turkey, and Iran. These are persisting consequences of post-colonial state boundaries arbitrarily imposed by victorious powers after both World Wars without consideration for the tribal communities and cultures within.

In the Far East, we wonder if North Korea will implode before it explodes. Their *songbun* social classification system deprives food and medical aid to the vast majority of their population, condemning them to malnourishment and starvation. While their nuclear missile test flights over Japan have recently abated, most of the 1.2 million strong North Korean Army remains poised along the DMZ with long-range conventional artillery capable of obliterating Seoul (Kazianis, 2017). As diplomacy and international economic sanctions impose a containment strategy, we wonder if our missile defense systems will keep our Allies, and us, safe (Slayton, 2013).

Historically rooted animosities among major Asian cultures make future conflict seem inevitable (Ayson & Bell, 2014). Japan recently reinterpreted its constitutional "self-defense" clause in response to an assertive and belligerent China which confronts its neighbors, challenging international norms, fomenting tension and dangerous arms races. Across the Taiwan Straits, 1,600 Chinese missiles remain pointed toward Taiwan (Auslin, 2017). China is not only pursuing naval dominance in the East and South China Seas, but also extending their maritime influence into the Indian Ocean and around the Arabian Peninsula with new deepwater ports in Djibouti and Pakistan to control access to the Red Sea and Persian Gulf. Simultaneously, their massive "Belt and Road" infrastructure project is designed to extend access and influence across all of Asia. China is also investing heavily in quantum computing, directed energy, anti-satellite, and hypersonic technologies. The history of international relations warns that great power clashes often arise from friction between two powers, one dominant and the other growing. US–Chinese relations may be at an inflection point, presenting a dilemma between increasing economic growth and global stability or another tragic "Thucydides Trap" (Kaplan, 2014).

In Africa, operating from northeast Nigeria, the insurgent Islamist organization Boko Haram continues its mass abductions, especially of schoolgirls, forcing over two million refugees to flee. In the past two decades, the terrorist group has killed tens of thousands including young female Red Cross workers. The largest outbreak of Ebola, a highly infectious and acutely lethal virus, eventually caused over 11,000 deaths of hemorrhagic fever. The virus quickly spread from Equatorial Guinea to nine different nations, including Europe and the US. Since then, three more mass outbreaks have occurred in the Democratic Republic of Congo.

In our own hemisphere, Latin American drug gangs, corrupt governance, weak institutions, and extreme poverty threaten our neighbors. Venezuela, once the wealthiest nation in South America with the world's largest proven oil reserves, is now a failed state. Thanks to the Chavez-Maduro socialist revolution, the misery of Marxism is widespread: hyperinflation, disease, crime and corruption with rising infant mortality, malnutrition, and starvation causing millions to flee. During the past 15 years, 2.5 million people have been murdered, 35,000 in just Mexico alone in 2018. Now, with over 400 murders per day, Latin America is by far the deadliest region on Earth. At home, these consequences overwhelm us with refugee "caravans" causing border security challenges and the humanitarian drama of children pouring across our own southern boundary.

We are also confronted with ever-evolving forms of geo-economic warfare, including intellectual property theft, trade wars, and currency manipulation, all occurring when global debt is higher than ever. Now at $250 *trillion*, it is three times what it was just two decades ago. Interactions between resource depletion, climate change, and especially cross-border water flows, further stress international tensions.

The promise of a New World Order vanished less than a decade after the First Gulf War. Indeed, the world seems to be unravelling as we are confronted with major challenges to humanity. Are we entering another "Dark Hour," where the only alternative to *Pax Americana* is global disorder (Stephens, 2014)? One need not be a "declinist" to concur with former Secretary of State Madeleine Albright's recent observation: "The world is a mess" (Albright, 2014). So, what can *we* do?

20.3 Domestic Conditions: The State of Our Union

While a broader international security perspective must certainly address the current and foreseeable geopolitical environment, we are also afflicted with troubling domestic issues: disconcerting socio-demographic patterns and education challenges;

national infrastructure decay in energy, transportation, and civil works; sagging personal savings and national investment levels; and increasingly intractable social entitlement policies. Policy-makers and defense officials are increasingly addressing the relationship and interactions between foreign and domestic policy and their impact on national security (Sullivan, 2019). A conspicuous example is the federal budget process and deficit spending. When asked what the greatest threat was to national security, then-Chairman of the Joint Chiefs of Staff Admiral Mike Mullen responded: "The national debt."

We are running near *trillion* dollar annual budget deficits, national debt now exceeds $22 trillion, and our current "fiscal gap," which looks beyond current debt to unfunded future obligations for Medicare and Social Security, exceeds an incomprehensible $100 trillion. The *interest* alone on this debt is now expected to total $7 trillion over the next decade. Debt interest payments are crowding out public investment and national security, and are projected to exceed the defense budget in less than five years. Traditionally, the purpose of federal debt was to support long-lifespan investments in public goods since borrowing spread costs among both current and future taxpayers, both of whom benefit. Today, however, we are increasingly borrowing to fund individual consumption-driven personal benefits, not public goods (DeMuth, 2014).

This rapidly deteriorating predicament, which has been on legislative auto-pilot for several decades now, is becoming both addictive and profligate (White, 2014). We are mortgaging not only our future but our posterity's as well, imposing impossible burdens on future generations of Americans. Realistic projections of these converging trends illuminate enormous fiscal risk to the Nation – a potentially *catastrophic gap* between future expectations and unfolding realities. Our current and projected fiscal condition is dismal. Amazingly, as Harvard historian Niall Ferguson explains, the greatest risk to Western democratic economic security seems increasingly to be our self-imposed precarious financial conditions and increasingly dysfunctional governments. He refers to "the great degeneration" in Western political and social institutions (Ferguson, 2013). Sadly, our public discourse, increasingly split along class lines, has been distorted by identity politics and poisoned by political partisanship (Huntington, 2005; Hatch, 2018).

Following several decades of decline in violent crime, a resurgence is now occurring fueled by gang violence in our cities. A heart-wrenching epidemic of illicit drug abuse, consisting of multiple sub-epidemics of heroin, cocaine, prescription opioids, synthetic opioids (fentanyl), and methamphetamines, is alienating and destroying a wide cross-section of American society (McKay, 2018; Carney, 2019). We are also afflicted with another, less recognized epidemic, but one due to *prescribed* drug use. The past three decades have seen an explosion in psychotropic medications with one in five Americans now taking psychiatric drugs and nearly five million on social security disability insurance for various mental disorders (Whitaker, 2010).

Our defense industrial base has been diminished. While we increasingly resort to automation, unmanned systems, and AI, our military fleets – aircraft, ships, combat vehicles, satellites – remain vulnerable to debilitating cyberattacks due to poor design, software flaws, and stolen technology (GAO, 2018). At the same time, our shrinking industrial base is increasingly dependent upon foreign nations for key technologies and especially strategic minerals. This is now having a significant impact on our supply chains ranging from specialized ball-bearings to aircraft carrier propellers for the US Navy, and from the Patriot air and missile defense system to the F-35 Joint Strike Fighter.

Now, following the diamond anniversary of OR, we are emerging from the longest sustained war in American history. Facing post-war pressures during a period of increasing global tension and turmoil, our Nation confronts enormous challenges over these next few decades. At a crucial period when the "state of military OR" appears to be at a crossroads, we must harness and apply the full power of analysis – OR, data analytics, management innovation – across our national security establishment in these turbulent, uncertain times (Parlier, 2014). Given the challenges we now face and the opportunities we can discern, is the current trajectory of our unique community aligned with what really *needs* to be done? What *can* our military operations research community do?

20.4   Bridging the Ingenuity Gap: Strategic Analytics

We are beginning to capitalize on the exploding Big Data environment and the surging Analytics movement to expand reach and application, and to promote the potential contributions of our unique, multi-disciplinary profession to these pressing challenges of our time. We must better link academic advancements to real world problems to bridge sound theory, engineering design, management concepts, and effective practice. We especially need to focus and apply the power of analysis to improve decision-making for large-scale, complex systems and organizations. We should also develop the capacity for rapid experimentation, without sacrificing the scientific method, in order to generate insight and to rapidly climb steep learning curves. Significant potential for dramatic improvement exists.

Management advances often lag technology advances, yet they are essential for improved policy responses needed to achieve breakthrough performances, overcome bureaucratic paralysis and political gridlock, and resolve international security dilemmas. A fundamental question is whether our capacity for *social ingenuity* will be adequate for solving persisting problems, averting catastrophic failures, and managing seemingly intractable 21st century issues. Social ingenuity involves solving problems, developing rules for governance, creating flexible institutions for transitional challenges, and better understanding recent socio-cultural, demographic, economic, and geopolitical developments (Homer-Dixon, 2000). We must be smart and agile enough to bridge our growing social ingenuity gap while continuing to promote technological ingenuity.

Ideally, management options should be developed within a strategic, ends-ways-means framework focusing on the ultimate purpose for which an organization exists (Vaill, 1982). Descriptive analytics are used to segment problems, systematically diagnose structural disorders, and identify enabling remedies and potential "catalysts for innovation" ("means"). Next, integration challenges are addressed using prescriptive analytics to attain objectives that characterize organizational goals, for example: efficiency, resilience, and effectiveness (desired "ends"). The design and evaluation phase incorporates predictive analytics to develop an "analytical architecture" ("ways") providing a comprehensive roadmap for organizational transformation.

Strategic Analytics also encourages shaping innovative policies and strategies around new concepts and technologies. As we look to the future, what are some of the analytical concepts and multi-disciplinary capabilities that could further empower OR across a broader context of national security? The next few paragraphs examine foundational building blocks: decision support systems, engineering systems, and transformational strategic planning.

20.4.1   Information Technologies

During most of our professional existence, OR has been "cursed" with both data challenges and computational power. But that is clearly changing in this new era of Big Data. Indeed, data has become ubiquitous; the challenge now is to somehow *make sense of it all*. So these twin banes of our past, data and computational power, rather than hindering our future, are more likely to offer opportunities. We already have the link between Big Data and Analytics, the extensive use of data, statistics, and quantitative algorithms for descriptive (explanatory), predictive (forecasting), and prescriptive (optimization) modeling and analyses. Through sensor technology, RFID, ERP systems, and the internet, IT has expanded to capture, track, monitor, and make visible data in near-real time across disparate, dislocated entities comprising the entire enterprise, thereby providing total asset visibility.

But we have yet to fully integrate *analytical architecture* into our enterprise system challenges. Complementary decision support systems have not yet been developed which could capitalize on all this (overwhelming) enterprise data and, using analytically based methods, dramatically improve performance for defense enterprise systems.

If uncertainty is viewed as the complement of knowledge, then for a fixed demand the three quantities shown in Figure 20.1 (inventory, capacity, and knowledge) are substitutes in the following sense: if more of one is available, then less of one or both of the others is necessary for the same level of system performance needed to meet that demand. This trade-off suggests a fundamental truth: if the amount and timeliness of useful data and good information for actionable decisions improves (i.e., increased knowledge or "what we know"), then with the same capacity ("what we can do") as before, it now becomes possible to improve system performance with fewer resources ("what we have").

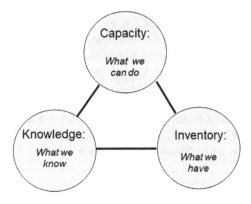

Capacity, Inventory, and Knowledge

Substitutable Ingredients of System Performance

FIGURE 20.1: Substitutable capabilities for enterprise system performance.

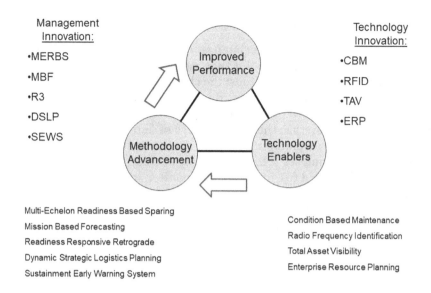

Management
Innovation:

•MERBS

•MBF

•R3

•DSLP

•SEWS

Technology
Innovation:

•CBM

•RFID

•TAV

•ERP

Multi-Echelon Readiness Based Sparing

Mission Based Forecasting

Readiness Responsive Retrograde

Dynamic Strategic Logistics Planning

Sustainment Early Warning System

Condition Based Maintenance

Radio Frequency Identification

Total Asset Visibility

Enterprise Resource Planning

FIGURE 20.2: Differentiating technology from management innovation.

20.4.2   Decision Support Systems

Empirical studies have consistently shown that "IT solutions" alone, even when implemented with the best information systems tools, have *not* produced the desired or expected results without accompanying business process changes. Although so-called "IT solutions" have ubiquitous appeal and enormous investment levels, without using the analytical, integrative power of OR to focus business process re-engineering on desired outcomes, this obsession with IT results in growing complexity, exceeding the interpretive capacities of organizations responsible for developing and using the IT solutions. The effects of this information and cognitive overload has been termed an "ingenuity gap" (Homer-Dixon, 2006). A complementary relationship between decision support systems (DSS) and management information systems (MIS) – both symbiotic and synergistic – is needed (Figure 20.2).

Ultimately, it is this *management innovation* approach that will enable senior leaders and managers to generate knowledge and better decisions from the growing amounts of information and improved situational awareness made available by advances in information systems technologies. Clearly, there is a powerful link between commercial organizations with strong analytical orientations and market-leading performance. The goal should be effective integration of Analytics into management policies for organizational decision-making. Hence, for large-scale complex organizations, the greatest return on investment is derived from incorporating relevant analytical tools (OR) *with* the appropriate IT. Acknowledging these needs and developing the capacity to address them represent first steps toward Strategic Analytics.

20.4.3   The Internet of Things

Early in World War II, one of the emerging technologies that provided the impetus for creating and rapidly developing OR was Radio Detection and Ranging, now simply known by its acronym "radar." Today, the Internet of Things (IoT) offers another disruptive opportunity for OR to integrate new technologies into enterprise systems. The internet, which evolved from mainframe computers, to connecting personal computers, then mobile devices, is now linking digital with physical worlds using interconnected networks.

Defined as networks of devices, objects, and people, IoT reflects the convergence of multiple technologies, including: real time analytics, machine learning, sensors, and embedded systems including wireless networks, micro-control systems, and automation. This next wave of the IT revolution is integrating human with machine intelligence by connecting digital and physical worlds to improve performance through greater automation and sensor-based analytics across consumer, commercial, industrial, and infrastructure applications (Gershenfeld & Vasseur, 2014). Examples include global logistics and supply chains, autonomous vehicles and aircraft, "smart" buildings and cities, and healthcare.

IoT is also enabling a variety of prognostic early warning systems which capitalize on predictive analytics to anticipate change, then pre-empt system degradation or failure through proactive management interventions in large-scale enterprise systems. Two such IoT applications for defense enterprise systems are described further in the Strategic Analytics application section: "connecting" Condition-Based Maintenance to military supply chains for a Sustainment Early Warning System; and the Enlisted Early Warning System to support the Army recruiting enterprise for the All-Volunteer Force.

20.4.4 Engineering Systems

Traditional systems engineering optimizes performance based upon a set of requirements or design specifications generated from an assumed operating environment. Yet, experience and history reveal that systems and their uses change over time, often in unanticipated ways. Furthermore, many of our systems seem increasingly fragile and vulnerable. They are subject to catastrophic failure due to the debilitating effects of age and decay, "tight-coupling" in complex enterprise systems, or human error, whether negligent or deliberate (Perrow, 1984). This awareness and recognition is now leading to new engineering design and operations management concepts.

This emerging field for large-scale, socio-technical systems has been referred to as "Engineering Systems," and is defined as: "A class of systems characterized by a high degree of technical complexity, social intricacy, and elaborate processes, aimed at fulfilling important functions in society" (MIT ESD, 2008). Solutions to socio-technical problems require a mix of technical innovation, organizational strategies, and carefully designed policies. A holistic perspective is emphasized, recognizing that a high degree of interdependency among elements requires decomposition to isolate and identify interactions, and integration to understand their effects across the enterprise. Network and matrix methods are used for analyzing and visualizing system structure, and system dynamics and simulation modeling methods are used to understand, measure, and predict how these complex systems behave over time.

Engineering Systems represents the next epoch of scientific innovation beyond inventions and complex systems, and can exist at the project, enterprise, or societal levels. Long-term, life cycle "ilities" are analyzed: quality, reliability, adaptability, sustainability, interoperability, agility, scalability, maintainability, and especially flexibility. This new and evolving approach constitutes a paradigmatic shift in systems design by moving from the traditional focus on meeting fixed specifications toward the active management of uncertainty in the implementation of socio-technical systems (deWeck, Roos & Magee, 2011; Giachetti, 2010). Just as nanotechnology is increasing our understanding of very small-scale structures, the discipline of Engineering Systems is expanding our macroscopic understanding of very large-scale enterprise systems defined by their technical, managerial, and social complexity.

20.4.5 Dynamic Strategic Planning

Most system design methods generate a precise, "optimized" solution based on a set of very specific conditions, assumptions, and forecasts. While these conditions and assumptions may be appropriate in the short-term for tactical operations, a practical limitation of these techniques is that they are rarely valid over longer planning horizons as strategic designs for technological systems (de Neufville, 1990).

In contrast, Dynamic Strategic Planning (DSP) instead presumes forecasts to be inherently inaccurate ("the forecast is always wrong") and therefore incorporates flexibility as part of the design process. Originally developed, refined, and applied at MIT by the Engineering Systems Division led by Professor Richard de Neufville and his colleagues, this method incorporates and extends earlier best practices including systems optimization and decision analysis. It has evolved by adapting "options analysis," now commonly associated with financial investment planning. DSP allows for the optimal policy, which cannot be preordained at the beginning of the undertaking, to reveal itself over time while incorporating risk management: a set of "if-then-else" decision options evolving as various conditions unfold that, even when anticipated, cannot be predicted with certainty.

The goal of DSP, to paraphrase one of its creators, is to optimize expected performance by building flexibility into the project to enable adaptability to changing circumstances that inevitably prevail. Rather than optimizing to specific conditions, flexibility accommodates inevitable change over time by adapting to a range of future possibilities. Though perhaps easier to engineer and manage, traditional optimal designs can quickly degenerate toward instability when such conditions no longer exist. In contrast, DSP accommodates and adapts to change as it occurs, rather than catastrophically failing because the underlying assumptions for which the system was originally optimized are no longer valid.

DSP incorporates flexibility into the system design to accommodate and respond to changes as they occur in the real world, even though we cannot predict exactly when or how they may occur. This built-in flexibility actually creates additional value for the system, and in many cases this additional value can be quantified (de Neufville and Scholtes, 2011).

In essence, DSP provides a multiperiod decision support system that encourages and assists in identifying, clarifying, and quantifying risk during the system design effort. As implementation evolves, and subsequent events occur, a mechanism is needed to routinely update the current solution. This "optimal" solution will inevitably change over time due to an inability to perfectly forecast future conditions or the consequences of past decisions that do not always reveal the results expected. And, as well, opportunities provided by adaptation and innovation will materialize, offering improved solutions requiring new decisions. This DSP capacity for adaptation enables a resilient enterprise system that can adjust gracefully as needed, rather than fail catastrophically.

20.4.6   Engines for Innovation (EfI)

Innovation is typically accompanied by the disruptive consequences resulting from the synergistic effects of multiple inventions converging in time. Hence, the phrase "creative destruction" originated by the esteemed Austrian economist Joseph Schumpeter (McCraw, 2007). How, then, can innovation be better understood and then accelerated in a controlled way to minimize the debilitating effects of disruption? A virtual test bed is needed to provide a synthetic, non-intrusive environment for experimentation and evaluation of innovative ideas and concepts. This synthetic environment, or microworld, guides and accelerates transformational change along cost-effective paths while providing the "analytical glue" to integrate and focus what otherwise would be disparate initiatives and fragmented research efforts.

While institutional adaptation requires a culture of innovation, inertia remains a powerful force within bureaucratic organizations. Consequently, sources to enable and encourage innovation must exist for the culture to embrace. An EfI provides such a source by building a capacity for low-risk, low-cost experimentation using a synthetic environment where analytically rigorous cost-benefit analyses can be performed to differentiate between desirable objectives and attainable (affordable) ones that can actually be implemented.

The organizational construct for an EfI consists of three components which comprise core competencies:

(1) an R&D model and supporting framework to function as a generator, magnet, conduit, clearinghouse, and database for "good ideas";
(2) a modeling, simulation, and analysis component which contains a rigorous analytical capacity to evaluate and assess the improved performance, contributions and associated costs that promising "good ideas" might have;
(3) an organizational implementation component which then enables the transition of promising concepts into existing organizations, agencies, and companies by providing training, education, technical support, and risk reduction/mitigation methods to reduce organizational risk during transformational phases.

The purpose of this deliberative, cyclical discovery process is to sustain continuous improvement through experimentation, prototyping, field testing, and rigorous analysis. The EfI provides low-risk, low-cost, high-velocity experimentation, thereby accelerating organizational *learning*.

Knowledge is the "axial principle" on which the information age is built (Tellis et al., 2000). And, increasingly, human capital is *the* critical building block of national power. An ability to invent, innovate, and diffuse innovation requires investment in human capital. EfIs offer organizational capabilities to invent, innovate, and diffuse innovation, thereby encouraging technological and *social ingenuity* as foundations upon which our national power can be generated and sustained in the future.

20.4.7   Analytical Architectures

Vision and analytics should not be seen as mutually exclusive paradigms. Rather, Strategic Analytics should link organizational vision to operational results by defining and monitoring metrics tied to enterprise objectives, and aligning incentives with those objectives.

Strategic planning and management frameworks are also essential to ensure strategies achieve intended operational results. Organizations must define and monitor metrics tied to strategic enterprise objectives that properly align behavioral incentives with these objectives. In organizations with strong cultures, especially our military services, these performance incentives must be designed to attain desired institutional outcomes. The value of an objective hierarchy is multi-fold and serves several purposes by collectively aligning strategy, processes, and metrics (Figure 20.3). And, although often neglected, such frameworks are instrumental to enable *learning* within organizations.

Strategies must then be developed to attain these desired goals and objectives. They provide "ways" to relate "means" available to desired "ends," and may consist of major programs and new policies, initiatives, procedures, or concepts. More specifically, strategies associate implementation costs with results attained in terms of objectives being pursued. They also illuminate the need for adaptation by providing mechanisms to sense the need for reacting to, as well as creating, change when necessary. Without the benefit of these analytical architectures, extra resources are more likely to be squandered than to have any visible effect.

Appropriate performance measures identify, capture, and quantify the value that has been achieved by adopting and implementing particular strategies. Effective metrics help to communicate organizational strategy, thereby improving organizational cohesion and congruency in workforce tasks. They also reinforce the pursuit of a common vision for the entire organization (Figure 20.4). In addition to defining performance, delineating accountability, monitoring progress toward strategic objectives, and providing means for management control, they also establish feedback mechanisms necessary to change a course of action when needed.

Furthermore, metrics that rely on *average* values provide little insight into what is actually driving performance, especially in large-scale, complex socio-technical systems subject to uncertainty and variability (Hax and Wilde, 2001). Systems with interdependent components are characterized by information lags, feedback delays, and nonlinearities, where small changes can amplify with large effects. Averages mask variability in performance, yet those areas most afflicted by high variability, volatility, and uncertainty clearly point to directions for improved performance.

Linking Strategy to Measurable Results

FIGURE 20.3: Enterprise strategy pyramid.

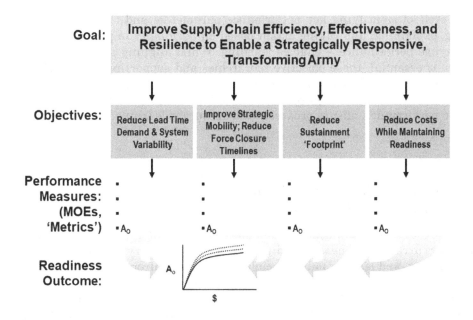

FIGURE 20.4: Objective hierarchy.

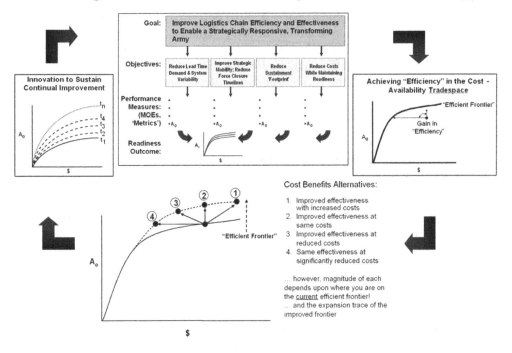

FIGURE 20.5: Innovation in a learning organization.

This cycle of identifying key performance parameters and then detecting, understanding, explaining, and taking action based on *variability* in performance – rather than *average* performance – enables learning. While aligning strategy to successful execution is essential, these objective hierarchies and supporting metrics should also be designed with the purpose of organizational learning in mind (Figure 20.5). What is needed is not merely a measurement system but a management system to motivate improved performance, ensure progress, and encourage learning.

20.5 Strategic Analytics: Applications to Present Challenges

So, how could these new concepts and methods for Strategic Analytics be applied to some of our current challenges? The final decade of my military career focused on building, developing, and leading multi-disciplinary teams confronting major challenges in large, complex commands. Central to each of these endeavors was the extensive application of Operations Research, data analytics, and management innovation for improved performance. Although the fundamental natures of these various transformational endeavors were vastly different, they all required an ability to organize, manage, and lead highly talented multi-disciplinary teams.

Among these responsibilities and accomplishments was developing and implementing the strategic framework for Army analyses during the first Quadrennial Defense Review (QDR) in 1996. In contrast, at Army recruiting command (USAREC) our immediate challenge was characterized by incipient mission failure, not only imminent but potentially catastrophic as well. A third enterprise challenge described below is the endeavor to transform our global military logistics management system which the Government Accountability Office has categorized as "high-risk" for the past three decades due to persisting inefficiency, ineffectiveness, and inadequate strategic planning. References for each of the three overviews introduced here provide detailed case studies, including conventional OR methods and creative new concepts; comprehensive briefings are also available from the author upon request.

20.5.1 Defense Resource Planning: Synchronizing Current and Future Capabilities

Today, as the US (perhaps) transitions into another post-war era, we are challenged to maintain a "balanced" force, over time and across our military capabilities, in an era of dramatic fiscal challenges, domestic turmoil, international tensions, and defense resource constraints. We have undertaken several QDRs, but the one that may be most relevant to review is the very first. Though it occurred over two decades ago, the conditions existing during the post-Cold War drawdown parallel those we must grapple with now: smaller forces, aging equipment, budget uncertainty, evolving technology, an expanding economy, a changing threat environment, and declining readiness.

By 1996 the US Army was in the middle of the post-Cold War drawdown, struggling to maintain "balance" across force structure, readiness, and modernization while ensuring that the force did not inadvertently become "hollow" or, worse yet, "broken." This condition was a historical consequence of the deleterious effects of the US Army's inevitable "boom and bust" cycle caused by precipitous force level reductions during immediate post-war periods. Regrettably, this tragic phenomenon has been a persistent, yet increasingly dangerous, pattern in American history.

Also, during the mid-1990s, much of the conventional wisdom was that national defense was entering a "revolution in military affairs" (Boot, 2005). This suggested, in the midst of the post-Cold War drawdown and declining defense budgets (the "peace dividend"), that modernization and investment accounts should be protected, even expanded, at the expense of force structure, training, and readiness during a "strategic pause." Nonetheless, while a major premise for transformation was that speed, agility, and precision could substitute for mass, the unfolding reality during the emerging new world order was that eradicating root causes of contemporary (unconventional) aggression required a completely different skill set.

Congress had also recently passed legislation mandating a series of Quadrennial Defense Reviews (QDR), the first to occur beginning in the Fall 1996. The perception then was that although Congressional intent may well have been for a comprehensive strategic defense review, within DoD the process would likely default to another huge budget battle among the Services. Indeed, the conventional wisdom then was that the Army would likely lose two more active divisions – previously reduced from 18 to 10 – when the dust settled. Finally, a lack of analytical expertise for strategic resource planning was recognized by several recent RAND studies for the Army, and increasingly acknowledged by the Army's senior leadership at the time (Parlier 1998).

To address inadequacies in land warfare modeling and especially strategic resource planning, a new Resource Plans and Analysis Division (RPAD) was created within the Office of the Chief of Staff of the Army. We developed the Army's strategic architecture for the QDR using a unique futures-ends-ways-means paradigm graphically portrayed in Figure 20.6. This analytical framework comprehensively addressed possible

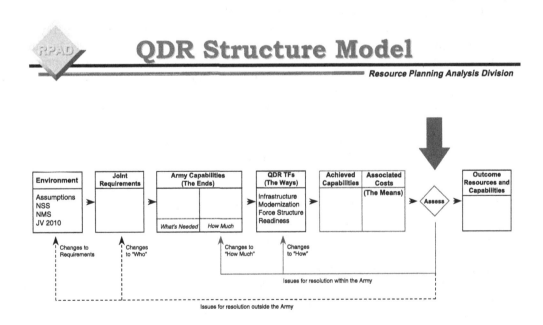

FIGURE 20.6: QDR analytical strategy.

future geopolitical environments, national security goals and objectives, force generation options, and likely resources available including emerging challenges to the All-Volunteer Force (AVF).

Conditions then were characterized by declining budgets and bureaucratic emphasis upon technocentric solutions without actually understanding or incorporating critical cause–effect relationships into models, simulations, and wargames. Consequently, RPAD undertook a comprehensive historical, theoretical, and empirical evaluation of past operations. While intuitively recognizing that training, skill, and advanced technology were important factors of success in conflict, the nature of their relationship – their respective contributions and especially how they interacted – was not well understood.

However, there were several trends then that were making a clear understanding of this interrelationship increasingly important. One of the observations that instigated this effort was the enormous gap between official model outcomes and the actual evidence of recent history. For example, medical planning factors were derived from attrition-based, theater-level campaign model casualty projections. However, when compared with actual empirical evidence from recent experiences of modern warfare, including Persian Gulf War results, these projections were overpredicting aggregate casualties by orders of magnitude, thereby creating unnecessary and unaffordable requirements for medical force structure and supply support.

The fundamental intent of our endeavor was to identify and verify patterns and relationships that would provide significant predictive power for the future. An experimental design approach was developed and pursued, grounded in relevant theory, to capture the statistically significant factors influencing future demand using all of the information provided by the empirical evidence of recent experience.

We also struggled with the critical need to maintain "balance," knowing how to define, measure, design, model, and actually achieve it – something that had not been done well throughout military history, as historical Army post-war drawdown curves illuminated. "Balance" was a multi-dimensional concept pursued from two perspectives: across time by "balancing" current and future readiness – more precisely, current force readiness, mid-term modernization, and long-term research and development investment – and by synchronizing force structure, modernization, manpower, and infrastructure within a cross-section of time (Figure 20.7).

The US Army emerged from the first QDR retaining a full-spectrum, ten division force structure. RPAD provided compelling analytical arguments to counter historical post-war drawdown inertia, developed an Army strategic resource planning capability, and applied this new concept during the first QDR. These new methods for Strategic Analytics, crude though they were then, helped to chart and resource a viable course for our leadership in what they called "an era of strategic uncertainty" (RPAD, 1997).

RPAD initiatives resulted in estimated savings for the Army of $2–$4 billion per year within a $60 billion annual budget at that time. The extensive, comprehensive analyses into the nature of modern warfare yielded major improvements (Biddle, 2004). For example, better understanding of the structure and empirical patterns of conflict resulted in improved casualty forecast accuracy and an ability to design more responsive, effective medical requirements including force structure investments, medical materiel, and supply support (Kuhn, 2010).

The World Trade Center and Pentagon attacks of 9/11 occurred just three years later, commencing a new era, the Global War on Terror. These seminal analyses, which questioned traditional views and standard practices at the time, have since been

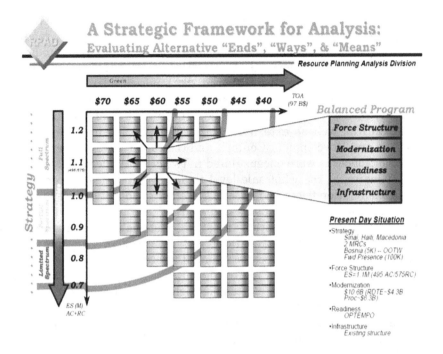

FIGURE 20.7: Strategic "balance": Defense resource planning

validated and corroborated by recent experiences in both Afghanistan and Iraq. They challenged the OR community to an introspective assessment which led to a formal *Review of Army Analysis* (Shedlowski, Shaffer & Fratzel, 2001) and to unifying themes for international venues in the MOR and defense analysis communities. This research endeavor has since been expanded to other classes of supply, including the sustainment enterprise. As the next section describes, the potential for further improvement in relating resource investments to current readiness and future capabilities is enormous.

20.5.2 Sustainment Enterprise: Transforming Materiel Supply Chains

Although budgets were beginning to stabilize in the aftermath of the post-Cold War drawdown, rapidly growing backorders for repair parts and unfunded requirements for spares were causing increasing concern although their impact on readiness was not at all clear. Especially worrisome, aggregate aviation fleet-wide readiness was beginning to decline yet expensive weapons systems across the Army were declared non-operational due to lack of relatively cheap parts. Furthermore, the promise for improved performance attributed to large investments in "IT solutions" in the form of enterprise resource planning (ERP) systems installed in recent years was not being realized. Finally, the GAO's "hi-risk list" for federal agencies had cited the Army for poor supply management since 1990 due to lack of strategic plans, poor inventory management, and fragmented supply chain operations across the materiel enterprise.

Within a year after the attacks of September 11th, 2001, to address these adverse trends the commanding generals for Army Materiel Command (AMC) and the US Army

Aviation and Missile Command (AMCOM) initiated an ambitious effort to improve logistics operations in order to better relate resource investment levels to readiness and future capabilities. In late summer 2002, shortly after I was reassigned to be AMCOM Deputy Commander for Transformation, we attacked this persisting, complex, and seemingly intractable challenge. We surveyed the landscape for logistics and inventory management expertise, tasked organized research teams, then created and implemented a comprehensive study plan. The project involved several organizations within the Army and DoD as well as external sources of expertise in academia, Federally Funded Research and Development Centers, and the commercial sector.

Initially, major concerns were encapsulated in two critical questions: (1) "How much should we be spending at the wholesale level to achieve and sustain readiness objectives for supported aviation fleets and missile systems?," and (2) "Are we buying/repairing the right combination of repair parts and spares?" A comprehensive application of advanced analytics, as well as operations research, systems engineering and analysis, and management innovation more generally, was undertaken. This initial discovery phase led to several initiatives focused on demand forecasting, inventory management policy, and retrograde operations (Figure 20.8). Early results based upon analytical demonstrations, experiments, and field tests clearly revealed that readiness could be improved with significantly reduced costs if specific "catalysts for innovation" were adopted. These catalysts were key to adopting a system-wide enterprise perspective referred to in academia and the commercial sector as "integrated supply chain management." Although the term "supply chain" was new to the Army's lexicon, the project soon came to be known as *Transforming US Army Supply Chains* (TASC).

Predictive analytic methods were pioneered. Decision support systems in conjunction with emerging ERPs, sensor-based technologies, condition-based maintenance (CBM) diagnostics and prognostics, and supply chain simulation technologies to support defense planning scenarios and pre-positioned stock requirements were all conceptualized, developed, tested, and evaluated. Costs savings for each of these catalysts were *each* estimated to be in the order of many multiples of $100M.

For example, we developed Mission Based Forecasting (MBF) with our major hypothesis asserting: "If empirically derived Class IX usage patterns, profiles, and/ or trends can be associated with various operational mission types, then operational planning, demand forecasting, and budget requirements can be significantly improved to support a capabilities-based force." We identified spare part consumption patterns and explanatory readiness factors for Army aviation fleets that either dominate, or differ significantly across operational missions and geographic locations and how they varied from peacetime training. The results from empirical studies and field tests proved remarkable, demonstrating for the first time the ability to measure, understand, and accurately predict operational mission demand with an order of magnitude improvement in forecast accuracy across major weapon systems (Parlier 2016).

Another consequential TASC initiative was developing and testing a transformation roadmap referred to as Dynamic Strategic Logistics Planning (DSLP). Within a Strategic Analytics (ends-ways-means) paradigm, DSLP provides the "ways" (concepts, policies, and plans) to link "means" (resources) with desired "ends" (objectives) in order to effectively guide supply chain transformation endeavors for the US Army and DoD. By developing this analytical architecture to sustain continual improvement, DSLP generates an efficient, increasingly effective, yet resilient global military supply chain network.

DSLP comprises four major modeling methods to sustain continual improvement: multi-stage optimization; dynamic strategic planning; risk management; and

Aligning Supply to Readiness Driven Demand

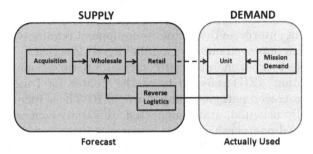

FIGURE 20.8: TASC project: Sustainment enterprise.

program development. In conjunction with testing, experimentation, and simulation, these complementary methods illuminate viable plans for implementation. Taking input from both the empirical evidence of ongoing real-world operations and new contributions derived from experimental results and operational testing, DSLP then guides enterprise transformation toward strategic goals for effectiveness, efficiency, and *resilience*.

From a global enterprise perspective, resilient concepts emphasize "building-in" flexibility by first, creating pre-positioned mission-tailored support packages designed using Readiness Based Sparing in conjunction with MBF. These tailored mission support packages can then accommodate contingency operations in locations where existing host nation sustainment is not immediately or readily available. Second, to accommodate sustained, rather than temporary, higher demand for extended operations, resilient design principles suggest creating additional capacity or relocating existing capacity closer to the demand source. Hence, the network responds quickly to initially accommodate short-term needs with built-in slack, and then adapts, if and when necessary, changing its configuration by relocating repair capacity closer to the source of demand to sustain longer term requirements.

DSLP also incorporates a sustainment early warning system (SEWS), conceptually presented at Figure 20.9. SEWS provides an anticipatory ability to recognize,

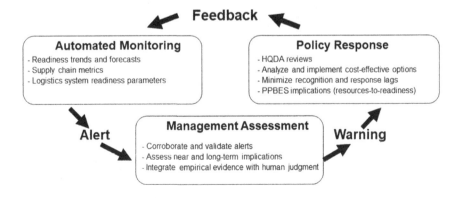

FIGURE 20.9: TASC project: Early warning system.

understand, and then pre-empt future system degradation through proactive, preventive management actions guided by supporting Sustainment Readiness Levels (SRLs) (NRC BAST, 2014). Collectively, these SRLs can yield a more effective, resilient, and efficient sustainment enterprise that achieves equipment readiness goals, is adaptive to change, and provides improved materiel availability at less cost.

The TASC project experimented with various organizational constructs for an "engine for innovation" (EfI) referred to as the Center for Innovation in Logistics Systems (CILS), portrayed notionally at Figure 20.10. These included two partnering options: a regionally oriented, state supported university-centric model; and a contract-centric AMCOM partnership with a major aerospace and defense company supporting the Aviation and Missile Research, Development and Engineering Center. A third CILS option was an organic design that spanned all of the Army's program executive offices (PEOs) and leveraged existing analytical capacity within Army Materiel Command.

The TASC project quantified major benefits and advantages for the Army. All of these results have been formally presented at recent DoD, national, and international venues and published in an award-winning book (Parlier, 2011). To the extent these TASC catalysts are fully adopted, refined, and implemented, significant savings generated from within the sustainment enterprise can then be reinvested to increase force levels, modernize capabilities, and improve readiness during a period of growing international unrest, rising competitors, and increasing regional friction. The potential magnitude of improvement is truly dramatic with tens of billions of dollars in annual savings likely. More importantly, it becomes possible to accurately forecast readiness by credibly relating investment levels to current and future capabilities (Levine, 2007). The project continues to promote international defense dialog by encouraging logistics interoperability from NATO to South Africa and the Philippines (IDA, 2010).

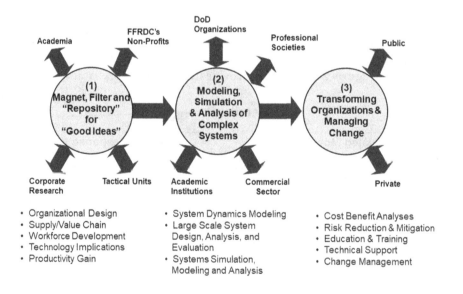

FIGURE 20.10: TASC project: Engine for innovation.

20.5.3   Human Capital Enterprise: Recruiting America's Army

One of the attributes of Strategic Analytics is alignment of methods and models with the "ends-ways-means" strategy paradigm. *Descriptive analytics* are used to systematically diagnose structural disorders, perform root-cause analysis, and identify enabling remedies – "means." Integration challenges are addressed using *prescriptive analytics* to attain policy objectives for desired "end" states. Design and evaluation then incorporate *predictive analytics* to develop "analytical architectures" ("ways") to guide the change management effort toward desired ends. One recurring observation from applying Strategic Analytics to several enterprise challenges is that confusion between "ends" (what is to be achieved) and "ways" (how it is to be achieved) can be uncovered and resolved.

For example, during our last major military recruiting crises in the late 1990s the All-Volunteer Force (AVF) was indeed at great risk, although this is not commonly remembered today. Our forecasts then for future economic conditions, demographic trends, youth propensity, and recruiting resources available made it clear we could not meet near-term recruiting requirements needed to achieve Army manpower and readiness goals. Projections were truly dire: the phrase used to describe the situation then was "imminent catastrophic failure" (Hauk and Parlier, 2000). The Army could no longer sustain the AVF in its current form.

Our most pressing task was to gain a thorough understanding of the fundamental nature of the challenge. In addition to the other armed services, the Army also competes against two other critical institutions of America – industry (the commercial sector) and post-secondary education (colleges and universities) for market share of what we referred to as the "prime market": 18–21 year old youth, medically, morally (e.g., no felony convictions), and academically qualified (high school diploma). However, it was becoming increasingly apparent that this traditional approach of "competing in the market" was not only resulting in greater failure for the Army, but by 1999 for *all* the other Services as well.

Furthermore, this destructive competition among institutions failed to prudently invest in our future – America's youth. No matter which option high school graduates pursued – joining the workforce, college, or the military – "failure" was inevitably occurring (Figure 20.11). Youth were not prepared to join increasingly hi-tech industries with only a high school diploma. College drop-out rates were a national disgrace. And enlisted first term attrition for the Army was nearly 40%. With this understanding of the underlying causes for recruiting failure, we then designed and implemented a human capital acquisition and development strategy to completely transform this previously unrecognized dynamic. We established and implemented new incentives, programs, and policies, many of which required OSD and/or Congressional approval, to rapidly convert this destructive, competitive dynamic into partnerships and alliances where our national institutions truly *complemented* one another, rather than competed against each other.

Innovative solutions that directly addressed the fundamental nature of the challenge were found and quickly implemented. For example, working with industry we created the Partnership for Youth Success (PaYS) program. PaYS partner companies, initially only Fortune 100 companies, were aligned with our military occupational categories. Each company jointly agreed with the Army and prospective recruits to provide guaranteed employment opportunities *after* completion of an honorable term of service. The individual soldier benefits; the company benefits by receiving mature, credentialed,

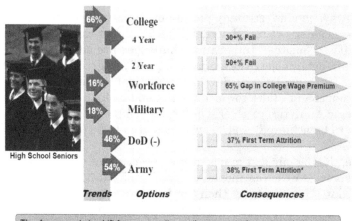

Current Market Dynamics

The Army needs to shift from competing options to complementary choices

FIGURE 20.11: Challenges to the All-Volunteer Force.

and experienced leaders; and the Army benefits by partnering with industry in a mutually beneficial way that genuinely invests in our future.

Similarly, complementary programs were developed to work collaboratively with colleges and universities, including the College First program where high school seniors could join the Army, yet postpone initial entry until *after* completing a college degree *with* financial assistance. Other education initiatives were developed and tested to capitalize on information technology and the explosion in online learning programs through our new E-Army U option to work toward completing college *while* serving on active duty. Short-term enlistment options with college loan repayment programs were developed which had significant appeal to college students and graduates. And through the new GED-Plus program, we could assist, qualify, then enlist a greater proportion of Hispanic youth.

Massive re-engineering was required and undertaken. We established the Recruiting Reinvention Lab for innovative concept design and testing, and led our strategic outreach effort through our Recruiting Research Consortium consisting of internal Army research assets, academia, FFRDCs, and national assets including the National Research Council. We developed, refined, and utilized a comprehensive recruiting research model, identified gaps in our knowledge, allocated both research expertise and funding, periodically reviewed results, and adapted both our research focus and organizational structure as we learned. We also consulted eminent experts across a wide range of disciplines: futurists, distinguished military sociologists, economists who decades earlier had advocated the creation of the AVF, and innovation theorists focused on learning organizations.

Our new Recruiting Research Consortium, Strategic Concepts Center, and Strategic Management Division created a shared vision for Army recruiting with a supporting set of goals and objectives, enabled by 126 initiatives bundled into 23 cohesive strategies with five priorities for the command. Requisite resources and authorities were acquired, and management information systems were established. We then systematically tracked progress during command-wide reviews to refine, accelerate, or cancel various efforts as needed to meet our objectives. We also developed and refined the Enlisted Early Warning System (EEWS), an econometric forecasting model that enabled us to better understand evolving market conditions, then proactively adjust recruiting resources to cost-effectively meet future requirements.

Although we were ultimately able to re-engineer and thus salvage the AVF, in the beginning there were no guarantees those new concepts, initiatives, and programs would succeed. But they were also designed to preclude catastrophic failure and allow a graceful transition to an alternative had we not been successful. What also became crystal clear then, and should be re-emphasized now given current and foreseeable trends, was recognizing that while the internal organizational objective is to "man the Army," the Army's larger purpose is to "serve the Nation." The AVF was then, and is now, a "way" to achieve those "ends." The AVF should *not* be viewed as an end unto itself.

Once again in 2018, for the fifth time in five decades, the Army substantially failed its enlisted recruiting mission. Army recruiting may again be on the precipice of "imminent catastrophic failure," imposing a strategic constraint on the use of American power. Of course, our national military manpower system should align with the requirements of defense strategy and our foreign policy objectives. However, the current AVF is but one of several military manpower systems ("ways") that can be considered to reconcile means with ends.

By broadening our perspective to also include domestic challenges to national security, we could improve social cohesion and better develop human capital to provide for the common defense and *ensure domestic tranquility* for ourselves *and our posterity*. Since the end of World War II, the United States has actually implemented or seriously considered at least five different military recruitment systems. While the worst condition may be one in which recruitment policies change constantly, thus inviting manipulation and undermining any moral basis for military obligation, we must nevertheless recognize the AVF, in its current form, has again become unsustainable.

"Rational choice theory" provided the foundation for the Gates Commission to recommend transitioning to an AVF in 1973. This economic focus views military service as an occupation requiring that government pay the market rate for military personnel. But this market rate fluctuates, subject to economic conditions, creating a regrettable countercyclical pattern where the health of the AVF is inversely proportional to the health of the US economy. This has been a persistent impediment to sustaining the AVF, and has periodically created a worrisome strategy-manpower mismatch.

This economic orientation has also diminished sociological and cultural perspectives which significantly impact propensity to serve among American youth, including human passions such as self-sacrifice which reinforce moral foundations for service (Eighmey, 2006). By unmasking these intrinsic, intangible perspectives, which emphasize institutional rather than occupational goals, a feasible alternative can possibly be found which addresses both economics and culture. "[We] need to fashion a collective identity that will resonate with and hold all Americans together. A first step could be some form of a national service" (Chua, 2018), including options for military, community, domestic, and foreign service programs.

Although strategy and manpower are both dependent variables, they are clearly not dependent on the same independent variables. Defense strategy is derived from requirements to protect national sovereignty and perceptions of foreign threats to national interests (Hooker, 2015). In contrast, military recruitment policy is constrained by demographics, economics, and political culture. Nonetheless, these domestic constraints have determined manpower policy in virtual isolation from functional imperatives that influence national security strategy (Kagan, 2006). For example, the US originally adopted the current AVF in 1973 due to domestic conditions during the 1960s that led to fulfilling a presidential campaign promise, rather than manpower levels derived from defense strategy (Foster, Sabrosky, & Taylor 1987). Soon after, during the recruiting crisis in the late 1970s, a "hollow force" developed and later, during both the Persian Gulf War and the Global War on Terror, a "back-door draft" was invoked. Indeed, in the late 1990s, Army recruiting failure became a binding strategic constraint on the use of American power (Knowles et al., 2002).

The US military today is both idealized and ignored. It remains the most impressive and trusted public institution in America. Yet the decision to maintain a small AVF throughout our longest wars has now removed any sense of obligation and shared danger from 99% of American families. America's military now represents less than one-half of 1% of our Nation (Desch, 2017) and is also fast becoming a hereditary profession. Children of veterans join the military at much higher rates than those whose parents did not serve, further isolating the military with an esprit de corps increasingly removed from the citizenry itself (Kaplan 2017). The American military should be part of America, yet is increasingly apart from it (Mazur 2010).

Although assimilation begins in local communities and schools, our American national creed could be strengthened by some form of universal national service, either military or public.

While no American has a "right" to military service, which veterans regard as an honor and privilege, every citizen does have an obligation to serve the Nation. In stark contrast to the current culture of identity politics, victimhood, and grievance, this would re-emphasize that citizenship also demands commitment and sacrifice by encouraging young Americans to work together with others from different social classes, regions, races, and ethnicities across our Nation. It would encourage mutual bonds both within and across generations, while rapidly integrating newcomers and assimilating them into our national culture.

If alternatives to the current AVF are to be considered, we should recall that universal service has neither been common nor alien to the American historical experience. For much of early American history, our republican tradition linked citizenship with public obligation expressed through military service. Today, however, given widespread concern over the lack of civic engagement (Bass, 2013), the challenge for American society is the reconstruction of both culture and citizenship through service (Krebs, 2006). Ideally, whatever military manpower system is selected should constitute both a cultural as well as economic institution in our society. By linking American concepts of civic virtue to national purpose, cultural cohesion within generations can be improved and bonds of mutual loyalty across generations can be strengthened (Hallowell, 1954).

It is now an open question whether the AVF can again be sufficiently re-engineered to meet current constraints and endure for perhaps another decade, or whether an alternative system will better serve the Nation. Importantly, it is *not* the existing recruitment system we should venerate, especially if it has been prone to failure and is now failing again. Rather it is the manpower itself, American sons and daughters, that we must invest in to pursue the common defense and to ensure domestic tranquility, for they are our Nation's posterity.

We must recognize the strategic dimensions to military manpower policy. We should also reconsider the arguments and perspectives of some of our most esteemed sociologists and political scientists. As Scottish military historian, Sir Hew Strachan, observed, in our discussions on civil-military relations, "we have tended to assume that the danger is a military coup d'état, when the real danger for western democracies today is a failure to develop coherent strategy" (Strachan, 2013). Strategic Analytics can be used to illuminate a better "way" ahead for Army recruiting – perhaps uniquely American (Parlier, 2019).

20.6 Future Directions

In their recent Final Report, the National Defense Strategy Commissioners warned:

> *Making informed decisions about strategic, operational, and force development issues requires a foundation of state-of-the-art analytical capabilities. This [current] deficit in analytical capability, expertise, and processes is intolerable in an organization responsible for such complex, expensive, and important tasks, and it must be remedied ... Repairing DoD's analytical capability is essential to meeting the challenges of the National Defense Strategy.*

> (CNDS, 2018)

We can – we *must* – do better. In this dramatic era, we *must* harness and apply the full power of analysis – OR, Strategic Analytics, management innovation, *and social ingenuity* – across our supporting institutional enterprise endeavors. By describing these defense resource planning, sustainment, and manpower enterprise applications, our extended MOR community will hopefully be encouraged to consider, adapt, and apply Strategic Analytics to our many other daunting national security challenges as well (Locher 2009).

Military organizations, especially successful ones, are renowned for their strong cultures. Yet the long history of military innovation reveals those cultures can also become impediments to necessary organizational adaptation when failure, sometimes catastrophic, looms imminent or is actually even evident. Organizational change has always provoked resistance, which should naturally be expected. However, we are reminded by Pulitzer Prize-winning author Barbara Tuchman, among our most distinguished historians, pursuing flawed and failed policies *knowing* that plausible alternatives and better options are available is truly "the march of folly" (Tuchman, 1984).

Nonetheless, the pace of technological change is not always compatible with organizational ability to absorb that change. To overcome both bureaucratic inertia and paralysis induced by disruptive chaos, military cultures must have sources of innovation they can embrace. Strategic Analytics challenges underlying logic of existing practices and demonstrates better, *credible* ways ahead that can accommodate graceful transitions rather than catastrophic disruptions or slow-motion failures.

Operations Research, with its distinctively rigorous problem-solving paradigm, emphasizes identifying, formulating, and understanding the fundamental nature of any challenge. A powerful byproduct of this approach is OR's ability to differentiate between issues that can only be managed and problems that can genuinely be solved. We must also correct erroneous perceptions that OR merely represents the most recent mathematical technique or periodic management fad. Rather the *practice* of OR incorporates existing and emerging methods that prove relevant and useful for improved decisions in operational applications and complex enterprise systems.

Of course, the practical challenge of "connecting the dots" among key bureaucratic elements is as essential as it is frustrating. These include senior policy officials responsible for regulatory guidance, program directors who control funding, test and evaluation agencies that rigorously assess plausible alternatives, and of course, operators who "own" the problem but are constrained by insufficient authority and inadequate resources to pursue better options. Strategic Analytics can help connect the dots and synchronize efforts to effect continuous improvement in increasingly competitive, stressful, and resource-challenged environments.

A final observation is the remarkable similarities between the newly emerging "collaborative enterprise" (Heckscher, 2007) and the original purpose and organizational forms that created OR during the early years of World War II. The idea for implementing a "system of *teams*" with expertise across a wide range of scientific and military disciplines, using empirical evidence from ongoing military operations in conjunction with creative scientific models for rapid learning, defined and differentiated OR. Working closely with, trusted by, and responsively advising high-level commanders and government leaders, all while operating under extraordinary pressures, was the hallmark of OR at its inception. We can learn from – we *must* fully capitalize on – the promise of our own heritage.

20.6.1   Final Thoughts

Strategic Analytics can be the critical keystone for our time, becoming stronger, more valuable in use and widespread in implementation as geopolitical challenges and domestic pressures of our modern era increase (Figure 20.12). OR can provide the "glue" to coordinate, orchestrate, and pull our defense enterprise organizations

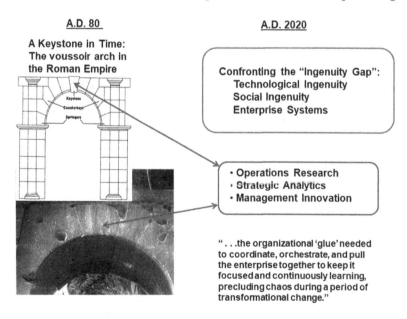

FIGURE 20.12: Operations research and the Roman Arch: The keystone analogy.

together to keep them focused, continuously improving and learning while under increasingly greater pressure, precluding chaos and decline during a period of disruptive transformation. We must develop and integrate relevant analytical capacity as a core competency for military operations, defense strategy, and international security policy.

Operations Research can provide a crucial source of American power. Strategic Analytics can help us reconcile ends with means, illuminating a path toward national renaissance. Although we need a surge of strategic imagination to confront conventional wisdom in the face of major national security challenges, our Nation's problems are not insoluble. While some may be different in kind, they are certainly not unknown in degree during our history (Herman, 2013; Kennedy, 2013; Budiansky, 2013; Dimbleby, 2016). Ultimately, despite our cultural proclivity and reliance on technology for solutions, we must remind ourselves that American power and influence ultimately take their meaning from the values and purposes they serve. Over the centuries, history cautions that our most cherished accomplishments have been "laboriously achieved but only precariously defended" (Kimball, 2012). Indeed, the veneer of civilization is very thin.

In less than 20 years Operations Research will celebrate the centennial of a unique profession. Will a retrospective then – a look-back by our professional posterity – regard this period we are entering now as our "darkest hour," our "finest hour," or perhaps both, as Churchill characterized the Battle of Britain early in World War II when our profession was invented (Churchill, 1949)?

"Come my friends,
'tis not too late to seek a newer world ...
One equal temper of heroic hearts
... strong in will
To strive, to seek, to find, and not to yield."

Tennyson, 1842

Acknowledgments

Earlier versions of this chapter were presented and refined during tutorial seminars at recent International Conferences on Operations Research and Enterprise Systems (www.icores.org). The author remains grateful for comments, suggestions, and contributions by many colleagues representing a variety of professional disciplines from nearly 50 nations. Abridged extracts from two previous publications are included herein: The State of Military Operations Research, from the Institute for Operations Research and the Management Sciences (INFORMS) February 2014 issue of *OR/MS Today* (Parlier, 2014); and Operations Research and the United States Army: A 75th Anniversary Perspective, published by the Association of the US Army (AUSA) as *Land Warfare Paper #105* (Parlier, 2015). Permission to reproduce them is gratefully acknowledged to *OR/MS Today* editor Peter Horner and AUSA's Institute of Land Warfare program director Sandra Daugherty. The author expresses gratitude to two anonymous reviewers for their constructive suggestions that improved the manuscript, and to the *Handbook* editors for this opportunity.

References

Albright, M. (2014, July 27). Former Secretary of State Madeleine Albright on "Face the Nation" interview with Bob Schieffer: "To put it mildly, the world is a mess".

Auslin, M. (2017). Asia's other great game. *The National Interest*. 152, 10–24.

Ayson, R. and Bell, D. (2014). Can a Sino– Japanese war be controlled? *Survival: Global Politics and Strategy*. 56(6), 135–165.

Bass, M. (2013). *The politics and civics of national service*. Washington, DC: Brookings Institution Press.

Biddle, S. D. (2004). *Military victory: Explaining victory and defeat in modern battle*. Princeton, NJ: Princeton University Press.

Blackwill, R. D. and Harris, J. (2016). *War by other means: Geoeconomics and statecraft*. Cambridge: Belknap Press.

Boot, M. (2005). The struggle to transform the military. *Foreign Affairs*. 84(2), 106.

Brands, H. and Edel, C. (2019). *The lessons of tragedy: Statecraft and world order*. New Haven, CT: Yale University Press.

Budiansky, S. (2013). *Blacketts' war: The men who defeated the Nazi U-boats and brought science to the art of warfare*. New York: Alfred A. Knopf.

Carney, T. P. (2019). *Alienated America: Why some places thrive while others collapse*. New York: HarperCollins.

Chua, A. (2018). Tribal world: Group identity is all. *Foreign Affairs*. 97(4), 25–33.

Churchill, W. (1949). *The Second World War: Their finest hour* (Volume II). London, UK: Cassel and Company.

Commission on National Defense Strategy (CNDS). (2018). *Providing for the common defense: The assessment and recommendations of the National Defense Strategy Commission*, United States Institute of Peace, p. x.

DeMuth, C. (2014, July). Our democratic debt. *National Review*.

de Neufville, R. (1990). *Applied systems analysis: Engineering planning and technology management*. New York: McGraw-Hill Publishing.

de Neufville, R. and Scholtes, S. (2011). *Flexibility in engineering design*. Cambridge: MIT Press.

Desch, M. C. (2017, Sep/Oct). The soldier and the state. *The American Conservative*, p. 49.

deWeck, O., Roos, D. and Magee, C. (2011). *Engineering systems: Meeting human needs in a complex technological world*. Cambridge: MIT Press;

Dimbleby, J. (2016). *The battle of the Atlantic: How the allies won the war*. New York: Oxford University Press.

Eighmey, J. (2006). Why do youth enlist? Identification of underlying themes. *Armed Forces & Society*. 32(2), 307–328.

Ferguson, N. (2013). *The great degeneration: How institutions decay and economies die*. New York: Penguin Press.

Fisher, D. (2005). *A summer bright and terrible: Winston Churchill, Lord Dowding, radar, and the impossible triumph of the Battle of Britain*. Washington, DC: Shoemaker and Hoard.

Foster, G., Sabrosky, A. and Taylor. W. (1987). *The strategic dimension of military manpower*. Cambridge: Ballinger.

Gershenfeld, N. and Vasseur, J. P. (2014). As objects go online: The promise (and pitfalls) of the internet of things. *Foreign Affairs*. 93(2), 60–67.

Giachetti, R. (2010). *Design of enterprise systems: Theory, architecture, and methods*. Boca Raton: CRC Press.

Government Accountability Office (GAO). (2018, Oct 9). *Weapon systems cybersecurity: DOD just beginning to grapple with scale of vulnerabilities*. GAO-19-128. Washington, DC: GAO. Also, GAO High-Risk Series (2018, Jul 25). *Urgent actions are needed to address cybersecurity challenges facing the nation*. GAO-18–645T.

Hallowell, J. (1954). *The moral foundation of democracy*. Carmel, IN: Liberty Fund.

Hatch, O. (2018, May 19–20). Identity politics threaten the American experiment. *Wall Street Journal*, p. A13. Former president pro tempore of the Senate, Utah Senator Orrin Hatch, observed with dismay that identity politics "threatens the American experiment" becoming "a cancer in our political culture" causing "groupthink, polarization, and gridlock." He laments a reversion to "tribalism" where "our civility will cease, our national community will crumble, and the US will become a divided country of ideological ghettos." One of the longest serving US Senators, Hatch represented Utah for 41 years. He was awarded the Presidential Medal of Freedom on Nov 16, 2018.

Hauk, K. and Parlier, G. (2000). Recruiting: Crises and cures. *Military Review.* May–Jun, 73–80.

Hax, A. and Wilde, A. (2001). Chapter 10: "Managing by averages leads to below average performance: The need for granular metrics" in *The delta project: Discovering new sources of prosperity in a networked economy.* New York: Palgrave.

Hazony, Y. (2018). *The virtue of nationalism.* New York: Basic Books.

Heckscher, C. (2007). *The collaborative enterprise: Managing speed and complexity in knowledge-based businesses.* New Haven, CT: Yale University Press.

Her Majesty's Stationery Office. (1963). *The origins and development of operational research in the Royal Air Force,* Air Ministry Publication 3368, London. Republished by the Military Operations Research Society (MORS) Heritage Series as *Operational Research in the RAF.*

Herman, A. (2013). *Freedom's forge: How American business produced victory in World War II.* New York: Random House.

Homer-Dixon, T. (2000). *The ingenuity gap: How can we solve the problems of the future?* New York: Alfred A. Knopf.

Homer-Dixon, T. (2006). *The upside of down: Catastrophe, creativity, and the renewal of civilization.* Washington, DC: Island Press.

Hooker, R. D. (2015). American land power and the two-war construct. AUSA Land Warfare Paper #106. Arlington, TX: AUSA.

Huntington, S. P. (1996). *The clash of civilizations and the remaking of world order.* New York: Simon and Schuster.

Huntington, S. P. (2005). *Who are we? The challenges to America's national identity.* New York: Simon and Schuster.

Institute for Defense Analysis (IDA). (2010). Defense resource management studies: Introduction to capability and acquisition planning processes. Document D-4021. Alexandria, VA: IDA. Also see: *Transforming a complex, global enterprise: OR and management innovation for the US Army's logistics system.* Retrieved at https://player.vimeo.com/video/64403069?title=0&portrait=0

Kagan, F. W. (2006). The US military's manpower crisis. *Foreign Affairs.* 85(4), 97–110.

Kaplan, R. D. (2014). *Asia's cauldron: The South China Sea and the end of a stable Pacific.* New York: Random House.

Kaplan, R. D. (2017). The rise of Darwinian nationalism. *The National Interest.* 151, 25–32.

Kazianis, H. J. (2017). Containing North Korea. *The National Interest.* 152, 5–9.

Kennedy, P. (2013). *Engineers of victory: The problem solvers who turned the tide in the Second World War.* New York: Random House.

Kimball, R. (2012). *The fortunes of permanence: Culture and anarchy in an age of amnesia.* South Bend, IN: St. Augustine's Press.

Knowles, J., et al. (2002). Reinventing army recruiting. *Interfaces.* 32(1), 78–92.

Kotlikoff, L. J. and Burns, S. (2012). *The clash of generations: Saving ourselves, our kids, and our economy.* Cambridge: MIT Press.

Krebs, R. B. (2006). *Fighting for rights: Military service and the politics of citizenship.* Ithaca, NY: Cornell University Press.

Kuhn, G. (2010). When will force planning finally escape its "mass logistics" past? *Phalanx.* 43(1), 35–39.

Levine, D. (2007). *Enhancing the readiness of army helicopters.* Report Number P-4245. Alexandria, VA: IDA.

Lieven, A. (2018). The forever war marches on. *The National Interest.* 157, 20–27.

Locher, J. R. (2009). Forging a new shield: The project on national security reform. *The American Interest.* 4(3), 15–26.

Mazur, D. H. (2010). *A more perfect military: How the constitution can make our military stronger.* New York: Oxford University Press.

McCraw, T. K. (2007). *Prophet of innovation: Joseph Schumpeter and creative destruction.* Cambridge: Belknap Press.

McKay, B. (2018, Sep 14). Heroin, opioids fuel surge in deaths. *Wall Street Journal,* p. A5.

Mearsheimer, J. (2018). *The great delusion: Liberal dreams and international realities.* New Haven, CT: Yale University Press.

MIT Engineering Systems Division. (2008). *Strategic Report.* Cambridge: MIT Press.

National Research Council (NRC). (2014). Board on Army Science and Technology (BAST) Special Committee: Force multiplying technologies for logistics support to military operations. Washington, DC: National Academies Press.

Parlier, G. H. (1998, May 21). *Resourcing the United States Army in an era of strategic uncertainty.* Presentation at INFORMS Military Applications Society Symposium. Huntsville, AL.

Parlier, G. H. (2011). *Transforming U.S. Army supply chains: Strategies for management innovation.* New York: Business Expert Press.

Parlier, G. H. (2014). The state of military operations research. *OR/MS Today.* 41(1), 18–24.

Parlier, G. H. (2015). Operations research and the United States Army: A 75th anniversary perspective. AUSA Land Warfare Paper #105W. Arlington, VA: AUSA.

Parlier, G. H. (2016). Mission based forecasting: Demand planning for military operations. *Foresight: The International Journal of Applied Forecasting.* 43, 32–37.

Parlier, G. H. (2019, Jan 31). *Operations research for the US Army recruiting enterprise: Past, present, and future challenges.* Final project report to the Under Secretary of the Army.

Perrow, C. (1984). *Normal accidents: Living with high-risk technologies.* New York: Basic Books.

Resource Planning and Analysis Division (RPAD). (1997). *Dynamic strategic resource planning: Toward properly resourcing the Army in an uncertain environment.* Office of the Chief of Staff, Army: Program, Analysis, and Evaluation Directorate. Washington, DC: Pentagon.

Shedlowski, D., Shaffer, D. and Fratzel, M. (2001). *Revolution in analytical affairs – XXI.* Concepts Analysis Agency Report CAA-R-01-23, DTIC# ADA399181.

Slayton, R. (2013). *Arguments that count: Physics, computing, and missile defense 1949–2012.* Cambridge: MIT Press.

Stent, A. (2019). *Putin's world: Russia against the west and with the rest.* New York: Grand Central Publishing.

Stephens, B. (2014). *America in retreat: The new isolationism and the coming global disorder.* New York: Sentinel.

Strachan, H. (2013). *The direction of war: Contemporary strategy in historical perspective.* New York: Columbia University Press.

Sullivan, J. (2019). More, less, or different: Where US foreign policy should – and shouldn't – go. *Foreign Affairs.* 98(1), 168–175.

Tellis, Bially, Layne, and McPherson. (2000). Chapter 5: Measuring national resources. *Measuring national power in the postindustrial age* (MR 1110-A). Santa Monica, CA: RAND.

Tennyson, Alfred Lord. (1842). Lines 56, 67–69, from his epic poem *Ulysses.*

Tuchman, B. (1984). *The march of folly: From Troy to Vietnam.* New York: Random House.

Vaill, P. B. (1982). The purposing of high performing systems. *Administrative Leadership: New Perspectives on Theory and Practice.* Hoboken, NJ: Jossey-Bass.

Whitaker, R. (2010). *Anatomy of an epidemic: Magic bullets, psychiatric drugs, and the astonishing rise of mental illness in America.* New York: Broadway Books.

White, B. (2014). *America's fiscal constitution: Its triumph and collapse.* New York: PublicAffairs.

Index

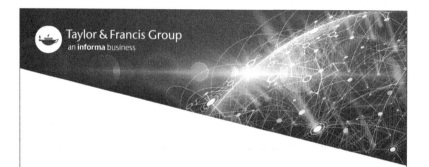

Printed in the United States
by Baker & Taylor Publisher Services